HIGHWAY DESIGN REFERENCE GUIDE

The McGraw-Hill
Engineering Reference Guide Series

This series makes available to professionals and students a wide variety of engineering information and data available in McGraw-Hill's library of highly acclaimed books and publications. The books in the series are drawn directly from this vast resource of titles. Each one is either a condensation of a single title or a collection of sections culled from several titles. The Project Editors responsible for the books in the series are highly respected professionals in the engineering areas covered. Each Editor selected only the most relevant and current information available in the McGraw-Hill library, adding further details and commentary where necessary.

Church • EXCAVATION PLANNING REFERENCE GUIDE

Gaylord and Gaylord • CONCRETE STRUCTURES REFERENCE GUIDE

Hicks • BUILDING SYSTEMS REFERENCE GUIDE

Hicks • CIVIL ENGINEERING CALCULATIONS REFERENCE GUIDE

Hicks • MACHINE DESIGN CALCULATIONS REFERENCE GUIDE

Hicks • PLUMBING DESIGN AND INSTALLATION REFERENCE GUIDE

Hicks • POWER GENERATION CALCULATIONS REFERENCE GUIDE

Hicks • POWER PLANT EVALUATION AND DESIGN REFERENCE GUIDE

Higgins • PRACTICAL CONSTRUCTION EQUIPMENT MAINTENANCE REFERENCE GUIDE

Johnson & Jasik • ANTENNA APPLICATIONS REFERENCE GUIDE

Markus and Weston • CLASSIC CIRCUITS REFERENCE GUIDE

Merritt • CIVIL ENGINEERING REFERENCE GUIDE

Perry • BUILDING SYSTEMS REFERENCE GUIDE

Rosaler and Rice • INDUSTRIAL MAINTENANCE REFERENCE GUIDE

Rosaler and Rice • PLANT EQUIPMENT REFERENCE GUIDE

Ross • HIGHWAY DESIGN REFERENCE GUIDE

Rothbart • MECHANICAL ENGINEERING ESSENTIALS REFERENCE GUIDE

Woodson • HUMAN FACTORS REFERENCE GUIDE FOR ELECTRONICS AND COMPUTER PROFESSIONALS

Woodson • HUMAN FACTORS REFERENCE GUIDE FOR PROCESS PLANTS

HIGHWAY DESIGN REFERENCE GUIDE

KENNETH B. WOODS   Editor-in-Chief
Head, School of Civil Engineering
Purdue University

STEVEN S. ROSS   Project Editor
Columbia University

McGRAW-HILL BOOK COMPANY
New York   St. Louis   San Francisco   Auckland
Bogotá   Hamburg   London   Madrid   Mexico
Milan   Montreal   New Delhi   Panama City
Paris   São Paulo   Singapore
Sydney   Tokyo   Toronto

Library of Congress Cataloging in Publication Data

Highway design reference guide/Steven S. Ross, project editor

Kenneth B. Woods, editor-in-chief
Rev. ed. of: Highway engineering handbook. 1st ed. 1960
Includes index.
1. Highway engineering – Handbooks, manuals, etc. I. Ross, Steven S. II. Woods, Kenneth B.
(Kenneth Brady), b. 1905. III. Highway engineering handbook. IV. Series.
TE145.H64   1988   625.7   dc1988-12903
ISBN 0-07-053924-3

1234567890   DOC/DOC   8954321098

ISBN 0-07-053924-3

Printed and bound by R.R. Donnelley and Sons

For more information about other McGraw-Hill materials,
call 1-800-2-MCGRAW in the United States. In other
countries, call your nearest McGraw-Hill office.

CONTENTS

Section 11. Soil Stabilization

Section 12. Highway Maintenance

Section 13. Landscaping

Index follows Section 13

CONTRIBUTORS

ROBERT F. BAKER, *Professor of Civil Engineering, The Ohio State University, Columbus, Ohio.* (SECTION 3)

Dr. DONALD S. BERRY, *Professor of Civil Engineering, Northwestern University, Evanston, Illinois.* (SECTION 2)

LEO H. CORNING, *Chief Consulting Structural Engineer, Portland Cement Association, Chicago, Illinois.* (SECTION 8)

Dr. D.T. DAVIDSON, *Professor of Civil Engineering, Iowa Engineering Experiment Station, Iowa State University of Science and Technology, Ames, Iowa.* (SECTION 11)

HARMER E. DAVIS, *Director, Institute of Transportation and Traffic Engineering, University of California, Berkeley, California.* (SECTION 1)

H. L. FLODIN, *Retired, Portland Cement Association, Chicago, Illinois.* (SECTION 8)

WILLIAM H. GOETZ, *Professor of Highway Engineering and Research Engineer, Joint Highway Research Project, School of Civil Engineering, Purdue University, Lafayette, Indiana.* (SECTION 9)

Dr. HAMILTON GRAY, *Chairman, Department of Civil Engineering, The Ohio State University, Columbus, Ohio.* (SECTION 3)

L. E. GREGG, *Chief Engineer, L. E. Gregg and Associates, Lexington, Kentucky.* (SECTION 6)

MAURICE GRUSHKY, *Associate, Tippetts-Abbett-McCarthy-Stratton, Engineers and Architects, New York.* (SECTION 1)

Dr. R. L. HANDY, *Assistant Professor of Civil Engineering, Iowa Engineering Experiment Station, Iowa State University of Science and Technology, Ames, Iowa.* (SECTION 11)

Dr. MORELAND HERRIN, *Associate Professor of Civil Engineering, University of Illinois, Urbana, Illinois.* (SECTION 11)

A. W. JOHNSON, *Engineer of Soils and Foundations, Highway Research Board, National Research Council, Washington, D.C.* (SECTION 11)

KENNETH A. LINELL, *Chief, Arctic Construction and Frost Effects Laboratory, and Foundation and Materials Branch, Headquarters, U. S. Army Engineer Division, New England, Waltham, Massachusetts.* (SECTION 5)

W. J. McCOY, *Director of Research, Lehigh Portland Cement Company, Research Laboratory, Coplay, Pennsylvania.* (SECTION 8)

Dr. JOHN F. McLAUGHLIN, *Associate Professor of Civil Engineering and Research Engineer, Joint Highway Research Project, School of Civil Engineering, Purdue University, Lafayette, Indiana.* (SECTION 7)

Dr. RICHARD C. MIELENZ, *Director of Research, Master Builders Company, Cleveland, Ohio.* (SECTION 7)

NATHAN G. ROCKWOOD, *Editor Emeritus, Rock Products, Naperville, Illinois.* (SECTION 7)

BERNARD ROSENBLUM, *Chief, Specifications, Estimates, and Contracts Section, Tippetts-Abbett-McCarthy-Stratton, Engineers and Architects, New York.* (SECTION 1)

Dr. JOHN W. SHUPE, *Associate Professor, Applied Mechanics Department, Kansas State College, Manhattan, Kansas.* (SECTION 10)

NELSON M. WELLS, *Director, Landscape Bureau, New York State Department of Public Works.* (SECTION 13)

EUGENE M. WEST, *Associate, L. E. Gregg and Associates, Lexington, Kentucky.* (SECTION 4)

REX M. WHITTON, *Chief Engineer, Missouri State Highway Commission, Jefferson City, Missouri.* (SECTION 12)

Dr. LEONARD E. WOOD, *Associate Professor of Civil Engineering, School of Civil Engineering, Purdue University, Lafayette, Indiana.* (SECTION 9)

KENNETH B. WOODS, *Head, School of Civil Engineering and Director, Joint Highway Research Project, Purdue University, Lafayette, Indiana.* (SECTION 7)

RICHARD M. ZETTEL, *Research Economist, Institute of Transportation and Traffic Engineering, University of California, Berkeley, California.* (SECTION 1)

PREFACE

The *Highway Engineering Handbook*, all 1400 pages of it, was last published in 1960. Since then, of course, drivers have seen great improvements in the design of roads they use. This is not, however, because these "new" designs are newly envisioned. While there have been some changes in detail, there are astonishingly few entirely new approaches to road design and road maintenance practice. The big changes have come in two areas.

First, there has been an expansion of what is considered "economical" in highway design, to accommodate ever-increasing traffic flows and ever-increasing lags between highway conception and completion. This design trend has produced a greater willingness to increase up-front capital expenditures to allow lower future maintenance costs and to reduce the chance of accidents. Thus, shoulders and lanes are wider, road cuts deeper (to reduce up- and down-grades), overpasses more common, and traffic circles on new roads almost nonexistent.

Second, there has been an increase in financial support and political oversight by states and, especially, by the federal government. This has led to a greater standardization in design and maintenance practice with regard to such matters as drainage, signage, geometry of exits and entrances, acceptable bearing loads, surfacing, even bridge inspection.

Thus, surprisingly little of the original work is outdated technically. In choosing what to include in this *Reference Guide*, therefore, I focused on material that is of particular interest to personnel who have been largely ignored as national attention swung to construction rather than maintenance, and to Interstate rather than local and regional roads. These are personnel who now find they must design and maintain roads with somewhat less federal funding than has been the norm.

In this *Reference Guide*, such personnel will find practical advice on everything from patching old concrete-surfaced streets to reducing the slipperiness of superhighway pavements; on efficient drainage and on subsurface conditions that can be corrected to avoid frost heave; on grades of asphalt and on soil treatments to reduce dust and increase durability on unpaved country roads.

In short, this is a practical guide, with advice that doesn't start by telling readers to to call Washington for a large grant before proceeding. This guide also preserves the language of the original – language that can be easily grasped by engineers, technologists, technicians, and administrators alike.

As for administration, this guide is based on the 1960 edition. By then, the Interstate highway system was well under way. There's more to transportation than roads and highways these days, however. Mass transportation has made a comeback. This is reflected by the old AASHO – the American Association of State Highway Officials, cited so often in the pages of this book – which has become the AASHTO, the American Association of State Highway and Transportation Officials.

On the other hand, the Highway Research Board, also frequently mentioned, of the National Research Council (the contract-performing arm of the National Academy of Sciences) has been eclipsed by the federal Department of Transportation itself.

In a few places, material had to be updated to reflect new advances. In all cases, these additions are minor. In "Landscaping" (Sec. 13), for instance, we now note that seeding can be done by blowing a slurry of wet seeds onto moistened soil and covering with a protective mulch. Grooved pavement, a fascination in 1960, is described more realistically in Sec. 10, on curing slippery surfaces.

Except for the most basic tests, we have referenced rather than reprinted. This is because tests change in detail with time and from state to state.

For ease in ordering reprints of standards, test methods, and research papers referenced in the text, the current addresses and phone numbers of relevent organizations are listed here:

American Association of Motor Vehicle Administrators, 1828 I St., NW, Suite 500, Washington, DC 20036. 202-296-1955.

American Association of State Highway and Transportation Officials, 341 National Press Building, Washington, DC 20004. 202-624-5800.

American Society for Testing and Materials, 1916 Race St., Philadelphia, PA 19103. 215-299-5585. All the standards and test methods cited in the *Reference Guide* are now in Section 4 of the annual ASTM compilations. Section 4's Volume 4.01 covers cement, lime, and gypsum; Volume 4.02 covers aggregates; 4.03, paving materials and surface characteristics; 4.04, bituminous materials; 4.08, soil and rock.

U. S. Government Printing Office, Washington, DC 20402. (Check your phone book for local USGPO book stores.)

My thanks to Susan Schwartz, who copy-edited this revision with great care and intelligence.

STEVEN S. ROSS
Columbia University

HIGHWAY DESIGN REFERENCE GUIDE

Section 1

HIGHWAY ADMINISTRATION AND FINANCE

HARMER E. DAVIS, *Director, Institute of Transportation and Traffic Engineering, University of California, Berkeley, Calif.* (Highway Administration).

RICHARD M. ZETTEL, *Research Economist, Institute of Transportation and Traffic Engineering, University of California, Berkeley, Calif.* (Highway Finance).

I. THE NATURE OF HIGHWAY TRANSPORT

Roads, simply as rights-of-way, do many essential things. They provide, for example, simple nontransportation functions such as easements of light and air. Even without motor vehicles, roads permit travel and transport, intercourse and communication. They provide access without which the institution of private property could not exist.

Except for limited, however important, purposes, society is not so much interested in highways as it is in highway transportation—the end product combining vehicles and roads, to which the efforts of the highway engineer are expected to contribute. This, then, is perhaps the fundamental purpose of highway administration: to contribute to efficient highway transportation, so that it may not only serve where other transport is not available, but that it may also complement and compete with other forms of transport where its economy and fitness can be demonstrated. However, the pursuit of efficiency in highway transportation should be kept consonant with other social and economic values.

Highway transportation must find its proper place in the over-all transportation complex of the nation. We need not here belabor the importance of the total transportation effort to the economy of a country. Almost taken for granted are facts of resource development, geographical specialization, division of labor, mass production, and other economic achievements contributing to productivity, which would not be possible without efficient transport.

A. Social and Economic Implications

Highway transportation by motor vehicle has given remarkable freedom and flexibility of movement to social and economic activity. Gone are the monopoly characteristics of transportation that gave rise to much of current policy of regulation and control of the transport business. It is fair to say that common, contract, and private carriage by motor vehicles has so altered the economics of transport that fundamental changes in transport policy must inevitably come about.

Not all highway transportation is competitive; much of it is complementary. Motor vehicles originate and terminate a large share of movements for other transport

1-1

agencies. But there is no gainsaying the fact that highway transportation has made significant inroads through direct competition with other carriers. Along with air transport for longer movements, highway transportation must be held partly responsible for current economic difficulties of the railroads. In freight carriage, not only the tonnage and ton-miles moved by motor vehicle, but the fact that highway carriers take some of the "cream," the goods commanding higher rates, has serious implications in the over-all transportation picture.

In the absence of compelling social and economic considerations to the contrary, it would appear that investment and traffic should be allocated among carriers in accordance with the relative economy and fitness of each. For then the economic resources required to provide transportation service of the quality desired would be held to a minimum. If this is to be the guiding principle, there is a presumption that government policy should be held neutral in such matters as taxation and expenditure.

In the passenger field most trips by motor vehicle are for short distances. But this very fact has given rise to a serious dilemma for our larger cities and metropolitan areas. Short but frequent trips by auto have cut seriously into the traffic of mass transit agencies, so much so that it is the consensus of many that mass transit must be publicly subsidized by one means or another if it is to survive at all in competition with the private car.

Obviously, the freedom of movement of the motor vehicle has made possible the widespread, low-density suburban development surrounding our cities and, in turn, has been responsible for business growth outside of core areas. Whether we are heading away from the concentration of a Manhattan in the direction of the "sprawl" of a Los Angeles, time alone may tell. Whether or not such a movement is desirable is a question that involves broad social and economic implications.

It is not our intention here to suggest, if indeed we could, the kinds of policy decisions that ought to be made in respect to alternative forms of transportation, either in the intercity or the urban fields. It may be emphasized, however, that the decisions that are made will have much to do with cutting out the size of the job that the highway engineer is called upon to perform.

Whatever the future may hold, the American people are now spending vast sums for highway transportation. The annual cost is conservatively estimated to be in the magnitude of $500 billion, the equivalent of one-eighth of the total value of all goods and services produced in the United States (gross national product). Very little of this expenditure is for trivial movement—much of it is for the direct satisfaction of human wants. Much of it, personal as well as goods movement, is used in the production and distribution of goods and services. Freight movement is for the purpose of adding what the economist calls place utility to goods. The demand for passenger movements not made in business activities is a derived demand—derived from the demand one has for goods and services, not the least of which may be social activity, including recreation, at the journey's end.

B. The Role of Highways

The end product of any transportation system is to provide service for which there is a demand. Three major and characteristic components of a transportation system combine to affect the cost and quality of the ultimate service provided: the fixed plant, the rolling stock, and the operational scheme of the system as a whole. In a self-contained (and idealized) system, prudent management would be expected to seek to optimize the net return from the revenues for service by maintaining an appropriate balance of plant, rolling stock, and level of operation.

Highway transportation service in the United States, however, is not provided through a single management. The components of the "system" are developed and made available in separate ways. Some agency of government provides the roads. Both government and private agencies provide terminal facilities. The rolling stock is acquired and operated by millions of separate operators, the majority of them being private individuals and businesses. The operators have an appreciable range of choice with respect to operating characteristics and operating costs of vehicles. The operational scheme is quite flexible with respect to routes and schedules, but is conditioned by operating rules (traffic laws, size and weight limits, etc.) imposed and enforced by agencies of government.

In this kind of situation, the vehicle operator, or highway user, may be said to be purchasing highway services, comprising the use of the plant (the physical highway system) and the direct operational features that go with the plant. It is estimated that the annual cost of the highway plant and its operational features is only of the order of 10 per cent of the total annual cost of highway transportation service. But it happens that the practices relating to the development, maintenance, and operation of the highway system, and the policies relating to the charges made for the highway service, have a substantial and significant influence on the nature and quality of the ultimate transportation.

Vehicle operating costs and total transportation costs are influenced by the characteristics of the fixed plant and the operating rules. Operating conditions (travel time, accident exposure, convenience, amenities), and hence actions of users and consumers, are affected by the nature of the plant and how it operates.

Highway service is, for the most part, provided through the agencies of government. Additionally, over-all regulatory conditions, some of which are not directly related to the operation of the highway system per se, are imposed through the actions of government. The activities, procedures, and devices through which the highway services are provided and the establishment and imposition of general policy controls may be said to constitute highway administration, in a broad sense. Thus the way in which the highway function is administered is a matter of considerable consequence to all concerned with highway transportation.

II. THE OVER-ALL PATTERN OF ADMINISTRATION OF HIGHWAY AFFAIRS

Since the provision of highway plant and the control of its operation is predominantly a function of government, the discussion herein will be devoted largely to the role of governmental agencies in highway affairs. It should be observed, however, that the general public, the highway users, and other interested parties can influence the conduct of highway affairs at various points. Further, it may be noted that highway administration, in the sense of the conduct of that whole range of affairs resulting in the provision of highway services, involves actions by the three branches of government (legislative, executive, and judicial) at the several jurisdictional levels of government (Federal, state, and local). Thus the administration of highway affairs is a complex thing; much more is involved than is involved in the management of a transportation system by a single organization.

Three broad aspects of administration are the development and expression of policy, the establishment of objectives and plans to carry out established policy, and the production of plant, goods, and services in accordance with the objectives and plans. These may be characterized as the directorship function, the general-management function, and the operational function, respectively. The exercise of these functions is usually not sharply and uniquely confined to separate agents of the administrative complex; rather it is possible to discern levels or orders of policy, the development of which may be largely centered in, but not confined to, a given level of administration.

A. Policy Making and Main Policy Guidance

The mode of expressing policy may be by law (constitutional law, statutory law, and judicial interpretation), by formal declarations of policy intent (resolutions by legislative bodies, commissions, boards), by administrative directives and regulations, and by adoption of standards of practice (usually relating to design or operational procedure). Because the levels and methods of highway financing have an impact on highway transportation costs, and hence on the demand for services, the pattern of finance is a potent policy instrument.

Another aspect of general policy, which is of interest in passing, is a distinction between what may be considered formulated policy and accumulated policy. In formulated policy, the intent is actively and positively to set up guide lines in an attempt to reach a recognized goal or to control some courses of action in the interest of recognized objectives. Accumulated policy which may be discerned in retrospect is the result of a series of actions in the past, often compromises, and sometimes inaction (in effect, a decision to do nothing)—an accumulation of precedent which may be said to express policy, or at least a historical policy attitude.

In general, the main features of policy in highway affairs are laid down by legislative action (in their respective realms by Congress, by state legislatures, and by boards of supervisors, commissioners, or councilmen). But the actions taken at one level of government may also become policy conditions for administration at another level. For example, both the congressional acts and the judicial interpretations and administrative extensions thereof may provide guide lines for, as well as limitations upon, actions in the states. Conversely, the simple failure of one level of government to act may induce another to take actions which establish a new direction of policy. Further, interpretations by the courts of previously unsettled points of common law (as opposed to statute law) may result in policy; examples may be found in the field of access control.

Within the scope of authority delegated by a legislative body, a board, commission, or individual may have policy-making functions; for example, a state highway commission may have been given certain policy-making powers (as in budgeting funds or in locating highways), which in turn govern the management of the highway agency.

And within the scope of broader policy established by law and by the declarations of a policy-making commission, the general management and the line supervisors in an operating organization (say, a highway agency) may adopt various standards by which to control the operations and their product (highway facilities).

III. CONTRACTS AND SPECIFICATIONS*

A. Scope of Section

Highway construction contracts, including specifications and other contract documents, are discussed in this section. The discussions, limited to practices in the United States, pertain primarily to the various contract documents and include definitions and descriptions of the more important contents. Also presented is a brief résumé of some basic elements of contract law and incidental legal aspects which may be encountered in the implementation of the contract documents.

B. Construction Contracts—General

1. General. Construction projects, in common with other business transactions, are brought about by a contract between the Contracting Agency and the Contractor. Both parties mutually agree that the Contractor will construct the project, meeting specified requirements, and the Contracting Agency will pay the specified consideration to the Contractor.

Construction contracts are generally classified in accordance with the (1) basis of payment, (2) procedure for selecting contractor, and (3) type of service by contractor.

a. **Basis of Payment.** Contracts are further classified as to (1) lump sum, (2) unit price, and (3) cost-plus-fee.

(1) LUMP-SUM CONTRACT. The Contractor is paid a fixed amount for performing all work required by the contract documents. This type of contract is suitable where the extent and other requirements of the work are known during the bidding period and a reasonably accurate estimate can be prepared by the bidder. The drawings and specifications should be comprehensive and should completely detail the requirements for the work. Incomplete drawings and specifications, in a lump-sum contract, often lead the bidder to include a substantial amount for contingencies in his proposal and to

---

* The remainder of Sec. 1 was prepared by Maurice Grushky and Bernard Rosenblum of Tippetts-Abbett-McCarthy-Stratton, New York.

make expensive changes after execution of the contract. The lump-sum contract is sometimes supplemented with additive and deductive unit prices for a portion of the work, the quantity of which is subject to variation.

(2) UNIT-PRICE CONTRACT. The Contractor is paid a total amount based on the contract unit prices and the actual quantity for each item of work. This type of contract is suitable when the exact extent of various items of construction is not known during the bidding period. Variations in actual quantity of the work from the estimated quantity can, within reasonable limits, be processed without formal change order. The drawings and specifications should completely indicate the nature and details of the work. The inclusion by the bidder of a substantial amount for contingencies, for an indefinite extent of the work, is generally avoided by the use of a unit-price contract. The unit-price contract often contains items of work which are paid for on a lump-sum basis. These particular items are generally well defined in so far as the scope of work is concerned and therefore need not be of the unit-price variety.

(3) COST-PLUS-FEE CONTRACT. The Contractor is paid his actual cost of construction plus a fee as compensation for his work. Cost-plus-fee contracts are generally negotiated as discussed in paragraph IIIB3b. Cost-plus-fee contracts are of various types based on the method of determining the fee.

(a) COST-PLUS-FIXED-FEE CONTRACT. This provides for payment to the Contractor of the cost of the work plus a fixed amount as fee. The fee is based on the nature and extent and the estimated cost of the work, among other factors. Since the fee remains fixed unless the conditions on which the fee was based are changed during the contract, it is apparent that the scope of the work should be well defined prior to agreement by the parties on the estimated cost and resultant fee.

(b) COST-PLUS-FEE CONTRACT WITH BONUS-PENALTY CLAUSE. This provides for payment to the Contractor of the cost of the work plus an amount as fee to be modified by a specified bonus-penalty arrangement. The bonus-penalty provision is often written into the contract to furnish an added incentive to the Contractor to keep the cost of construction down and to complete the work ahead of schedule. The bonus-penalty arrangement may provide that if the actual cost of the work varies from the estimated cost, if the actual time of completion varies from the specified time for completion, or both, the amount of the Contractor's fee will be increased or decreased in accordance with a specified formula. Usually a minimum limit is provided below which the Contractor's fee will not be reduced. Sometimes the bonus arrangement is specified without the corresponding penalty provision. Often a fixed amount alone is assessed against the Contractor as liquidated damages for each day of delay in completion until the work is completed.

(c) OTHER TYPES OF COST-PLUS-FEE CONTRACTS. These include (1) cost-plus-percentage-of-cost contract, (2) cost-plus-fee based on sliding scale of fees, and (3) cost-plus-fee with a guaranteed ceiling price for the actual cost of construction.

b. Procedure for Selecting Contractor. Contracts may be competitive-bid or negotiated. When competitive-bid, the Contractor is selected by the Contracting Agency after examination of his bid proposed competitively against other contractors. Public works contracts are generally open-competitive-bid wherein bids will be received from all contractors possessing at least minimum qualifications and the contract is awarded to the lowest responsible bidder. In private construction, competition is usually restricted to a selected list of bidders and no legal requirements restrain the Contracting Agency from awarding the contract to other than the lowest bidder, as in public construction. A negotiated contract is awarded to a selected Contractor after a study of the qualifications of one or more contractors. Under special conditions, such as a national emergency, special legislative measures are usually adopted permitting public contracting agencies to waive open competitive bidding, where necessary, and negotiate contracts in the interest of expediency. Competitive-bid contracts are usually of the lump-sum or unit-price type and seldom of the cost-plus type. Negotiated contracts may be either of the three types but are most often of the cost-plus type.

c. Type of Service by Contractor. In addition to the construction service, the

Contractor is sometimes required to provide planning, design, and management services. Thus contracts may be classified as management contracts, architect-engineer-management contracts, and combined engineering and construction, or "turn-key" contracts. In the management contract, the Contractor provides specified managerial services such as hiring of labor, letting of subcontracts, and purchase of materials. In the architect-engineer-management contract, the Contractor provides specified planning, design, drawing, and specification services in addition to the managerial services. In the "turn-key" or "package" contract, all planning, design, drawing, specification, and construction services are included in one contract.

C. Construction Contracts—Highways

Highway construction contracts are generally let on a unit-price basis. In very special cases, where the amount of work is small and the quantities of work are more or less fixed, the contract may be of the lump-sum type. Under special or emergency conditions, the contract may be of the cost-plus-fee type.

Most of the highways in the United States have been and are constructed under contract with the various state highway or public works departments. Generally, highway contracts and specifications are the products of the state highway or public works departments. The Federal government, through the Bureau of Public Roads and others, has contributed a minor though significant share of contracts and specifications for highway construction. The contract and specification practices presented in this section are generally based on practices of state highway and public works departments.

IV. CONTRACT DOCUMENTS—GENERAL

A. General Statement

A complete set of contract documents for a highway construction project usually includes, in one form or another, the following elements:
1. Advertisement or Notice to Contractors
2. Information for Bidders
3. Proposal or Bid
4. Agreement
5. General Requirements
6. Technical Requirements
7. Drawings
8. Bonds
9. Changes

These elements, which are described in Arts. V through VIII, have the following general meanings. The *Advertisement* or Notice to Contractors is a concise notice to contractors, placed in a suitable publication or posted at an appropriate location, announcing the project and giving bidders sufficient information to enable them to determine whether they are sufficiently interested in the project to secure a set of the bidding documents. The *Information for Bidders* generally furnishes the bidder with more detailed and additional information than furnished in the Advertisement, concerning receipt of bids, preparation of the proposal, award of contract, etc., which will enable the bidder to properly prepare and submit his bid. The *Proposal* or *Bid* is the offer of the bidder to perform the work. The *Agreement* is the executed portion of the contract wherein it is mutually agreed that the Contractor will do the work in accordance with the contract documents which are referenced in the Agreement. The *General Requirements* include all nontechnical requirements not covered in the Advertisement, Information for Bidders, and Agreement. The *Technical Requirements* include all technical requirements for materials, workmanship, and the like, and usually incorporate the *Drawings* by reference. The General Requirements often include one or more types of *Bonds.* After execution of the contract, *Changes* are made by change order, supplementary agreement, or other practice of the Contracting Agency. During the bidding period changes are made by addenda to the bidding documents.

B. State Standard Specifications, Special Provisions

In this country, where most of the highways are constructed under state specifications, the use of State Standard Specifications plus Special Provisions is almost universal among the states (see paragraph IX.B, "Outlines of Contract Documents"). Each state has its own set of Standard Specifications, which include provisions pertaining to all of the contract documents and apply to each project in so far as they are made applicable by the Special Provisions. The Special Provisions, issued for each project, indicate which of the Standard Specifications are applicable to the particular project and give all modifications of and supplementary provisions to the Standard Specifications. The drawings, Standard Specifications, and Special Provisions are intended to be complementary and provide for a complete work. A requirement in one is as binding as though occurring in all. However, in case of discrepancy between the Special Provisions and drawings and Standard Specifications, the Special Provisions generally govern.

C. Contract Forms, Guide Specifications, Standard Specifications

Because of the repetitive nature of a substantial portion of contract and specifications work generally, and particularly for highways, the agencies preparing such documents have taken various measures, described below, to facilitate the preparation and processing of these documents. Although these measures save time and effort, great care is required in their use in order to avoid discrepancies developing from inapplicable provisions in the final contract documents.

1. Contract Forms. Standard contract forms are used to a large extent, although the extent to which such forms are used by any one agency varies widely. These include forms for the Proposal, Contract, Performance Bond, Payment Bond, Bid Bond, Contractor's Financial Statement, Experience Questionnaire, Plan and Equipment Questionnaire, Invitation for Bids, Instructions to Bidders, and Request for Permission to Sublet.

2. Guide Specifications. In preparing specifications for highway and other construction, the U.S. Army, Corps of Engineers, and the U.S. Navy, Bureau of Yards and Docks, use guide specifications. These guide, or type, specifications are prepared by the department for its own use. When a specification is needed for a particular project, the guide specification is adapted to the particular project and reproduced for use in the contract.

3. Standard Specifications. Standard Specifications for construction are prepared by various contracting agencies, generally in permanently bound book form. These Standard Specifications are supplemented by special provisions for each particular project (paragraph IV.B, "State Standard Specifications, Special Provisions").

The term Standard Specifications is also applied to standard reference specifications both for materials and construction prepared by private and governmental organizations and often incorporated by reference into the job specification. Such Standard Specifications are prepared by the following organizations among others:

1. American Society for Testing Materials
2. American Association of State Highway Officials
3. U.S. Government
4. Asphalt Institute
5. American Concrete Institute

V. RECEIPT OF BIDS—AWARD OF CONTRACT

A. Advertisement

The Advertisement (paragraph IV.A, "General Statement"), or Notice to Contractors, is generally required by the law, under normal conditions, for work by public agencies. By means of the Advertisement, published sufficiently in advance of the date of the opening of the bids, the public Contracting Agency performs its responsibility in notifying prospective bidders of the work, allowing them sufficient time to prepare bids, and in inviting adequate competition. The Advertisement usually contains the information listed below. Other items, varying with the practice of the Contracting Agency, are often included.

1. Name and address of the Contracting Agency
2. Time and place of receipt and opening of bids
3. Concise description of the work, including its nature and extent with some approximate quantities of work
4. Instructions relative to obtaining drawings, specifications, and other bidding documents
5. Type of bid, such as lump-sum or unit-price
6. Nature and amount of bid guarantee required
7. Conditions relative to acceptance of bids such as the reservation of the right, by the Contracting Agency, to reject any and all bids

B. Information for Bidders

Information for Bidders (paragraph IV.A, "General Statement"), containing information relative to the preparation and submission of bids, is presented identically to all bidders for any one project. However, practice in method of presentation varies among different contracting agencies. State agencies often have most of the information as part of the Standard Specifications under headings such as Proposal Requirements and Conditions, and Award and Execution of Contract. The remainder of the information is often part of the Special Provisions adapting the Standard Specifications to the particular project. U.S. Government agencies have standard forms of Invitation for Bids and Instructions to Bidders which contain bidding information. In addition to these U.S. Government forms, the Bureau of Public Roads provides some bidding instructions with Standard Specifications, as do the state agencies.

The contents of the Information for Bidders of highway construction contract bidding documents are discussed below.

1. Receipt and Opening of Bids. Obtaining and examining the bidding documents, as covered in the Advertisement, is sometimes repeated in the Information for Bidders, depending on the practice of the Contracting Agency.

2. Examination of Contract Documents and Site of Work. Bidders are required to carefully examine the bidding documents and the site of the work and to inform themselves of all conditions which may affect the cost and prosecution of the work. It is often specified that submission of a bid will be considered prima facie evidence that the bidder has made and is satisfied with such examination.

3. Proposal Requirements. The contents of the Proposal Form are explained, and the requirements governing the preparation of the Proposal are outlined.

a. To help assure that bidders are bidding on an equal basis, it is usually mandatory that the Proposal be submitted on the prescribed form. The extent to which the Proposal Form must be filled in is given. Special instructions are sometimes necessary. For example, if the Proposal Form is divided into parts, it should be stated whether it is mandatory for all parts to be filled out or whether the bidder may propose on less than all parts. If the Proposal Form allots space for price on concrete pavement and bituminous pavement, it should be stated whether a bid may be submitted on only one of the alternates.

b. The estimated quantities in the Proposal Form are considered approximate and are usually stated to be solely for the purpose of comparing bids. The extent to which the unit prices apply in case of variation between the estimated quantities and the actual quantities of work is covered in the contract (paragraph VII.G.4, "Compensation for Changes").

c. Prices are submitted in ink and often in words in addition to numerals. In case of conflict between the written unit price and unit price in figures or between the unit price and amount, the written unit price invariably governs.

d. Instructions on how the bidder is to sign the Proposal Form should be specific.

e. Proposals are required to be submitted in sealed envelopes properly marked. Proposals received after the stipulated time are generally returned unopened.

f. A bidder is usually permitted to withdraw his bid provided his request for withdrawal is received in writing at the place designated in the Proposal before the time set for opening of bids. Some contracting agencies permit a bidder, who has submitted proposals on more than one project of any one letting, to withdraw any or all of his additional proposals after the person who opens and reads the bids has announced that such bidder has submitted the lowest proposal on a project. After the date set for opening of bids, a bidder is often permitted to withdraw his bid only after lapse of a specific period. However, if a bidder is low on more than one project, the total amount of which exceeds the bidder's rating with the Contracting Agency, the bidder is usually permitted to withdraw bids on such project or projects as will bring the remaining total to within the limit of the rating.

4. Public Bid Openings. The practice of public agencies is generally to open bids publicly so that bidders or their representatives can be present.

5. Proposal Guaranty. A guaranty that the bidder, if his proposal is accepted, will enter into a contract and furnish the required bonds is invariably requested with the bid by public agencies for open competitive bidding. Such guarantees usually are about 5 per cent and sometimes are as low as 2 per cent and as high as one-third of the total amount of the proposal and are usually in the form of certified checks. Bid bonds, currency, or bank guarantees are sometimes required. All Proposal Guarantees are eventually returned to the bidders unless the apparent successful bidder refuses to execute the contract, in which case his guarantee is kept by the Contracting Agency. Except for the guarantees of one or more (usually about three) of the lowest responsible bidders which are kept until execution of the contract, all of the guarantees are returned soon after the amounts of the bids are determined.

6. Rejection of Proposals. Proposals are often rejected and bidders are often disqualified for various reasons. Proposals may generally be rejected if they contain irregularities such as conditional bids, unbalanced bid prices, alteration of forms, incomplete bids and erasures, among other reasons. Proposals generally are rejected if (1) they are not accompanied by the required Proposal Guaranty; (2) they are submitted by an individual or firm who has submitted more than one Proposal for the same project; (3) more than one bidder can reasonably be shown to be interested in such Proposals; (4) there is collusion among bidders; (5) the Proposal is submitted by an unqualified bidder, and for other reasons of like nature.

7. Qualifications of Bidders. Bidders are required to furnish evidence indicating that they are qualified to perform the work bid upon. Bidders on public works projects are generally required to complete prescribed forms giving among other items such data as experience record, financial statement, and information on availability of equipment and key personnel for the work.

Some contracting agencies require bidders to prequalify. In such cases, qualifications must be submitted by the bidders within a certain time prior to receipt of bids. If the bidder is qualified, a certificate of qualification is issued by the public Contracting Agency fixing the aggregate amount of work which the Contractor will be permitted to have at any one time.

Where special conditions warrant and open competitive bidding is waived, prequalification of bidders appreciably in advance of the project may be used to establish a selected list of bidders.

8. Material Information. The successful bidder is sometimes required to furnish a statement describing the origin, composition, and manufacture of certain materials proposed for use in the permanent construction, together with samples of such materials for purposes of testing by the Contracting Agency.

9. Award of Contract. *a.* **Analysis of Bids.** A general statement of procedure in handling bids prior to selection of the successful bidder is often given by public contracting agencies. After the proposals are read, all arithmetic extensions and totals are verified and corrected, if necessary. If bids were taken on more than one basis, the merits of the low bid on each basis are evaluated. It is invariably specified that until final award of contract, the Contracting Agency has the right to reject any or all bids and to waive technicalities as its interests may necessitate.

b. **Award.** It should be the responsibility of public contracting agencies to state in general the criteria for selection of the successful bidder. As noted in paragraph

III.B.1.*b*, "Procedure for Selecting Contractor," public contracting agencies usually award the contract to the lowest responsible bidder whose proposal conforms to the prescribed requirements. Public contracting agencies choose the low bidder on the basis of the estimated quantities and the prices bid. The award is generally made by written notification to the successful bidder as soon as practicable after the opening of bids. It is good practice for the Information for Bidders to specify a reasonable time limit, say, 30 or 45 days after opening of proposals, within which award of contract will be made, if at all. The time within which the successful bidder will be required to execute the contract, after the papers are presented to him, should also be given. Ten days is an average figure for such purpose.

c. **Surety Bonds.** In connection with highway construction contracts, the following types of surety bonds are often required: (1) bid bond, (2) contract bond, (3) performance bond, and (4) payment bond. The bid bond has been discussed above (paragraph V.B.5, "Proposal Guaranty"). In a *contract bond*, generally required at the time of execution of the contract, the surety guarantees to the Contracting Agency performance of all obligations assumed by the Contractor in his contract. Such obligations generally include performance of the contract and payment of all proper labor and material bills. Instead of one contract bond, the practice of some contracting agencies such as the Federal government and some state agencies require a separate *performance bond* and a separate *payment bond*. The payment bond is sometimes called a labor and material bond. The requirements for the amounts of the bonds vary with different contracting agencies. Most common in highway construction contract practice in the United States is to require the contract bond in the full original amount of the contract, or when separate payment and performance bonds are used, each at 50 per cent of the original contract amount. However, a performance bond with payment bond each in the full amount of the contract is also known to be used. The Federal government to a large extent requires a 100 per cent performance bond and a lesser payment bond, depending on the amount of the contract.

d. **Execution of Contract.** The apparent successful bidder is generally required to execute the contract and furnish the required bonds to the Contracting Agency within a specified time. In the event of failure on the part of the selected bidder to execute the contract and furnish the bonds within the required time, the Proposal Guaranty becomes the property of the Contracting Agency. The Contracting Agency may either award the contract to the next bidder or reject all bids and advertise for new proposals.

C. Proposal

1. A prescribed Proposal Form is provided by the Contracting Agency generally for each bidder to fill in his prices for the work. Good practice dictates the form of proposal to assure bids on the same basis from all bidders, to help make apparent any qualifications or other deviations from the basis of the bid as understood by the Contracting Agency, and to facilitate the analysis and comparison of bids.

2. In the Proposal Form the bidder agrees to do the work in accordance with the contract documents. For highway work, proposals are invariably required on a unit-price basis. The form contains various items of work such as excavation and concrete, with an estimated quantity for each item of work. Spaces are provided for the bidders to fill in the unit price and amount for each item, based on the estimated quantity.

3. In addition to the requirements for prices, Proposal Forms often include additional statements to which the Contracting Agency requires the bidder's signature. These additional statements sometimes are repetitive of requirements elsewhere in the contract documents and are not functionally necessary in the Proposal Form if they are so covered elsewhere in the bidding documents. They include (*a*) agreement to execute the contract and furnish the required surety bond within a given time, (*b*) agreement to commence work within a given time, (*c*) agreement to prosecute the work at a rate required for completion within the specified contract time, and (*d*) statement that the required form of bid guarantee accompanies the bid; and other statements.

4. It is proper for the Proposal Form to require the bidder to list the addenda to the bidding documents which he received and considered in the preparation of the bid.

By this means the Contracting Agency is informed upon examination of bids whether all bidders have incorporated the required changes to the bidding documents in the basis of the bid.

VI. AGREEMENT

The Agreement is that portion of the contract documents wherein both parties to the contract mutually agree that the Contractor will perform the work in accordance with the contract documents for the agreed price.   Since the Agreement is the most fundamental of the contract documents, it should contain a collective reference to each of the contract documents, such as to the Proposal, Special Provisions, Drawings, and Specifications.   The prices for the work may be given therein or referenced to the accepted Proposal or by other suitable means.

In highway construction contract practice, the Agreement, sometimes labeled Contract, usually appears as a separate document within the contract documents and contains the spaces for execution by the parties.   However, the basic agreement provisions are often supplemented with many other general requirements, depending on the practice of the Contracting Agency.

The agreed contract price is understood to cover sufficient profit to the Contractor to warrant his relieving the Contracting Agency of the responsibilities in connection with management, purchase of materials, employment of labor, and construction of the work.   The Contractor's obligations are discussed in Arts. VII and VIII.

VII. GENERAL REQUIREMENTS OF THE CONTRACT

A. Definition of Terms

State highway specifications generally include an extensive list of definitions.   This practice is normally not followed in contracts for other types of construction by other agencies.   In addition to the usual definitions of Contract, Contractor, State Department, and Engineer, the state highway document definitions cover such basic terms as Materials, Equipment, Culverts, Pavement, Subgrade, and the like.   Whether an extensive list of definitions is necessary to the proper functioning of the contract is questionable since other construction contracts function properly without the extensive list of definitions.

B. Scope of Work

1. General Description.   In addition to the scope of work indicated on the drawings, the contract documents should include a description of the work from which a bidder and contractor can conveniently receive a general picture of the work required.

2. Intent of Drawings and Specifications.   The Contractor is usually informed that the intent of the Drawings and Specifications is to prescribe a complete work which the Contractor undertakes to do and that what is called for by either the Drawings or the Specifications is as binding as if called for by both.   The provision that the Drawings and Specifications are intended to prescribe a complete work generally requires the Contractor to perform work omitted from the Drawings and Specifications when the omission is in the form of details and the over-all intent of the documents is clear.   The fact that a requirement shown in either document is binding prevents an omission from one document alone from being a cause of disagreement and also minimizes repetition between the Drawings and Specifications.   Any misunderstanding arising as to the meaning or intent of the Drawings and Specifications should be resolved by the representative of the Contracting Agency under his authority to interpret the Drawings and Specifications (see paragraph VII.C.1, "Engineers and Inspectors").

3. Changes.   Prior to receipt of bids, changes are generally made by addenda to the bidding documents.

The contract should always prescribe to the Contracting Agency the right to make changes to the executed contract. Changes are often made by Extra Work Order, Change Order, or Supplemental Agreement.

Usually, changes within the general scope of the contract are made by Extra Work Order or Change Order and must be performed by the Contractor under his original contract obligations. Changes outside the scope of the original contract are often made by supplemental agreement between the parties to the contract.

The contract provides means for making adjustment in the contract price and time of completion for changes made within the general contract scope, where such adjustments are warranted. Such means are discussed in paragraph VII.G.4, "Compensation for Changes," for price adjustment, and paragraph VII.D.8, "Time Extensions," for time of completion adjustment.

On projects where the funds available for construction are limited, the amount available and other information pertinent to the limitation should be made known to bidders. The effect of the limitation of funds on the extent of the work as shown on the Drawings and Specifications, and on the contract prices, should be specified.

C. Control of Materials and Workmanship

1. Engineers and Inspectors. The authorized representative of the Contracting Agency often is referred to in the contract documents as the engineer. Contract documents prepared by some agencies often specifically delineate the differences in authority between the Head of the Department, e.g., Commissioner or Chief Engineer, Resident Engineer at the site of the work, and his Assistants or Inspectors. The Engineer is responsible to the Contracting Agency for making certain the work is performed by the Contractor in accordance with the requirements of the contract documents. The Engineer generally has the authority to decide questions that arise in connection with quality and acceptability of materials and workmanship, rate of progress, interpretation of drawings and specifications, acceptable fulfillment of the contract, and compensation to be received by the Contractor, among other authority. The Engineer, although representing the Contracting Agency, by mutual agreement between the parties to the contract is given broad powers. The specific duties of the Assistant to the Resident Engineer, or of the Inspectors, are often prescribed. These include inspection of the work and directing attention of the Contractor to and rejection of materials and workmanship which do not conform to the requirements of the contract documents. Inspectors are generally not authorized to modify any requirements of the drawings and specifications or to approve or accept any portion of the work.

In contemplating the broad authority given to the Engineer, it may be well to note that the state highway department construction contracts generally do not provide a standard procedure such as arbitration by which the Contractor may obtain relief from the Engineer's decision should the Contractor so desire. It would thus appear that in such contracts, outside of recourse to the courts of law, the Contractor's only relief from an Engineer's ruling is to appeal to the Engineer's superior within the Department and ultimately to the Head of the Department. This is also true of the Federal government, which provides for decision to settle all disputes by the Contracting Officer and for appeal by the Contractor to the Head of the Department.

2. Drawings. The drawings which govern the construction can be grouped into two classes, i.e., those furnished by the Contracting Agency and those furnished by the Contractor for approval of the Contracting Agency. Generally, the Contracting Agency furnishes complete design drawings showing details of the required permanent construction. Such drawings, covering roadway and structure, plan, alignment, profile, cross section, and construction details, would be included in the documents which formed the basis for the Contractor's bid. After execution of the contract, the Contractor is required to furnish such drawings as are necessary for the fabrication and erection of the work, which cover shop details, stress sheets, camber and erection diagrams, reinforcing bar schedules, falsework, centering, formwork, masonry layout diagrams, and the like. Although the Contractor's fabrication and erection drawings are approved by the Engineer prior to use, the contract documents generally provide that such approval does not relieve the Contractor of all responsibility for the adequacy of such drawings. The practice of some agencies is to specify that the approval of the Engineer of the contractor's drawings relates to strength and general arrangement and that such approval will not relieve the Contractor from responsibility for such factors as omissions, errors in dimensions, shop fits, field connections, and quantity of materials.

The work is required to adhere strictly to the drawings, within the limits of any construction tolerance which may be specified. However, the Engineer is usually given permission to authorize in writing deviations from the drawings which are not otherwise provided for in the contract documents.

3. Quality of Materials. Supplementing the technical-material specifications, general requirements should cover the quality of materials, control of the source of materials, samples, inspection and tests, and defective materials.

A general requirement for first-class quality of materials often assures good quality of materials, specific requirements for which have been omitted from the contract documents. Material which fails to meet the requirements of the contract documents is not permitted to be used. Promptly after execution of the contract, the Contractor is often required to furnish a complete statement of the origin, composition, and manufacture of all materials to be used in the construction.

The sources of supply of the primary materials selected by the Contractor for the work are generally required to be approved by the Engineer before delivery is started. In general, sources of materials for pits and quarries may be those selected by the Contractor and approved by the Engineer, designated by the Contracting Agency, or listed by the Contracting Agency for use by the Contractor.

For designated sources, the Contracting Agency generally assumes responsibility for the suitability and sufficiency thereof based on previous investigations and tests. For selected and listed sources, approvals are given by the Contracting Agency based on control tests during the contract.

Representative samples of materials required to be tested are specified to be submitted by the Contractor. Only materials tested and found to conform to the requirements of the specifications are permitted to be used in the work. The acceptance of a sample, however, does not mean acceptance of future materials from the same source. Nonconforming materials may be rejected at any time.

In highway work, public agencies usually perform their own testing of materials. Sampling, testing, and laboratory methods are generally required to be in accordance with the latest standards of such societies as the American Association of State Highway Officials and the American Society for Testing Materials.

In lieu of making tests, the Engineer usually has authority, at his discretion, to accept manufacturer's certified analyses for commercial products.

Materials are required to be stored in locations approved by the Engineer, off the right-of-way, if feasible, in a manner to ensure the preservation of the quality and fitness for the work and proper inspection and control.

Inspection and defective materials are discussed below.

4. Character of Workmen and Equipment. In addition to the technical specifications for workmanship and equipment for the various parts of the work, the contract documents should contain general requirements regarding the character of workmen and equipment. Proper and sufficient labor and equipment for prosecuting the work in the manner and time specified are generally required. All workmen must have sufficient skill and experience to properly perform the work assigned to them. At the request of the Engineer, the Contractor is required to remove any member of the Contractor's or Subcontractor's force guilty of misconduct.

All machinery and equipment should be required to be of such type and capacity as is necessary to produce a satisfactory quality of work. The measure of the capacity is the actual performance on the work.

5. Construction Lines and Grades. The Contractor is often required to lay out lines and grades from the original base line and bench mark established by the Engineer. However, in most highway practice, the Engineer provides the basic line and grade and also furnishes the Contractor with all necessary information relating to lines, slopes, and grades. It is not uncommon for the Engineer to lay out and set the original construction stakes establishing profile grade, slope and line stakes for road work, and center lines and bench marks for bridges and other structures. The Contractor is responsible for maintaining the given points and lines and providing any additional measurements necessary to construct the work. These lines and grades constitute the field dimension control by which the Contractor executes the work.

6. Inspection. The Engineer and his Inspectors are given free access to all parts of the work and materials at all time, including fabrication and erection. When conditions warrant, inspection of materials is made at the source, as well as at the site of the work. The work is inspected as it progresses, and the Contractor is required to provide every reasonable facility necessary for the representatives of the Contracting Agency to perform their inspection. Failure to reject defective work or materials at any time should in no way prevent its rejection when it is discovered. If work has been covered, the Contractor is required to uncover for inspection, if so ordered by the Engineer. However, adequate allowance is made to the Contractor for the additional cost of such work should the uncovered work conform to the requirements of the contract documents. Where an agency other than the Contracting Agency has an interest in the work, such as when the Federal government provides funds for a state project, and the other agency will inspect the work, the contract documents should so state, so that due allowance for such additional inspection can be made by the bidders and Contractor.

7. Defective Materials and Workmanship. Materials or work which do not conform to the requirements of the contract documents and which are rejected must be removed and replaced by the Contractor at his expense. Rejected material, the defects of which have been subsequently corrected, should not be used until approval has been given by the Engineer. If the Contractor fails to remove and replace the defective work, the Contracting Agency may withhold payment to the Contractor or have the work performed at the Contractor's expense.

8. Use of Materials Found on the Work. Highway construction contracts generally specify that the Contractor will be permitted to use, in the proposed construction, the materials found in excavation, if such materials meet the requirements of the specifications or are approved by the Engineer. Such excavation is paid for at the applicable contract unit price. However, the Contractor is required to provide sufficient suitable material to complete the portion of the work for which such material was originally intended.

D. Control of Prosecution and Progress

1. General. In addition to means for control of material and workmanship, the contract must obviously provide means to control the prosecution and progress of the work. In general, this is done by specifying times for commencement and completion of the work, providing requirements for progress and capacity which the Contractor must meet within those times, and providing means with which the Contracting Agency can enforce these requirements.

2. Time of Commencement and Completion. The time of the contract must be figured from a definite starting date. The date of receipt by Contractor of notice to proceed is a suitable and often-used starting time. Others, such as date of the contract, are also used. The Contractor is usually allowed a reasonable period of time to commence and is therefore required to start work within a specified number of days from the aforementioned starting date, e.g., 10 days. In most cases the required completion time is specified somewhere in the contract documents during the bidding stage, and all bidders bid on such basis. Where special conditions warrant and the Contracting Agency is interested in having bidders quote completion times for consideration in award of contract, bidders are required to fill in their best completion times in their Proposals. The time unit in which completion time is usually specified is discussed below in paragraph VII.D.7, "Determination of Contract Time."

3. Capacity. The contract should provide that the Contractor should conduct the work in such manner and with sufficient materials, equipment, and labor as are necessary in the opinion of the Engineer to ensure completion of the work in accordance with the requirements of the contract documents. Such requirements would include time, among the technical and other requirements for the work.

4. Progress Schedule. To determine the progress of the work throughout the entire contract, a progress schedule is prepared soon after award of contract and agreed upon by the contracting parties. The progress schedule should be consistent with the specified time or times of completion for the various parts of the work and show the times of commencement and completion of each of the salient features of the work. The Contractor is generally required to submit for the approval of the Engineer a progress schedule (see paragraph X.A, "Progress Schedule"). The Engineer is thus provided with ready means of determining early enough for proper action by the Contracting Agency whether the work is being prosecuted in a manner which will permit completion on time.

5. Limitations of Operations. A fair contract should, wherever possible, be specific about all limitations which will be placed on the method of operation by the Contractor. He should be informed as to what he will have to do with regard to maintaining and protecting existing traffic, dealing with other contractors working on the project, accommodating a special sequence of work, and other such features which will place significant restriction upon the Contractor's operations. Although the sequence of the work is generally specified to be subject to approval of the Engineer, the Contractor should be given latitude unless special conditions warrant a specific sequence.

6. Subcontracting. Contracts generally provide that no portion of the contract may be sublet or assigned without prior approval of the Contracting Agency. By this means the Contracting Agency can take such means as are necessary to determine whether the subcontractor is capable of performing as required. In order to assure an appreciable amount of work being performed by the Contractor, whose qualifications have presumably been determined by preaward scrutiny, many contracts provide that not less than a specified percentage of certain items of work shall be performed by the Contractor with his own organization. However, even if approval with reference to subcontractors is given by the Contracting Agency, the Contractor is not relieved of any of his responsibility under his prime contract.

7. Determination of Contract Time. The time which elapses after commencement of the contract is used to determine whether the Contractor is in default of his contract and the amount of damages required to be paid by the Contractor if in default. In general, the contract time is the duration in time units specified above in paragraph VII.D.2, "Time of Commencement and Completion," plus such extensions as are authorized by the contract. Three time units in use are the calendar day, the working day, and the contract day.

The most common time unit is the calendar day, in which all days including Saturdays, Sundays, and legal holidays are computed except such days of delay for which the contract permits an extension of time. Another form of the calendar-day method is the specification of a definite completion date.

When the working day is used, Saturdays, Sundays, and legal holidays are not included in the computation of time for completion, unless the Contractor utilizes Saturday as a working day. Working days are generally also not charged for delays due to causes beyond the control of the Contractor and without the fault or negligence of the Contractor, provided such delays prevent the Contractor from prosecuting major operations under the contract.

Where the contract day is used, a predetermined number of contract days in each month are allotted in the contract; for example, 12 days in January, 18 days in May, 20 days in September.

The use of the contract day and the working day tends to make the specified completion time more nearly approach the actual number of days that the Contractor is expected to work. Whether the calendar-, working-, or contract-day method is selected is a matter of preference with the Contracting Agency. In each of the three methods, a theoretical time of completion can be established at the start of the job. The actual contract time available to the Contractor would be determined from the theoretical time by the manner in which provisions of the contract granting extensions of time for delay are written and interpreted. Thus we can see that more important than the method (type of day) used is the number of days allotted and the contract provisions for extensions of time which are discussed below in paragraph VII.D.8, "Time Extensions."

8. Time Extensions. These extensions are provided for due causes, such as changes in work which require more time than the work as existing prior to the changes, delays beyond the control of the Contractor as referred to in paragraph VII.D.10, "Damages for Delay," and periods of suspension of work for the convenience of the Contracting Agency. Suspensions may apply to part of the work or all of the work, in which cases partial or total time extensions are made, respectively. Generally, the Contractor may apply for extension of time at any time prior to completion of the contract unless a definite time for such application is stated in the contract.

9. Suspension of Work. The Engineer usually has authority to suspend work temporarily by written order. The suspension may be for the convenience of the Contracting Agency or as a means of dealing with a Contractor who is not performing properly. Although the power of suspending the work is used only in rare instances, it sometimes is the only practicable measure to be taken. Such authority may be exercised when the Contracting Agency considers current conditions unfavorable to the prosecution of the work or when the Contractor fails to correct conditions which are unsafe for the workmen or the general public. Suspension of work by the Contracting Agency could lead to claims for damages by the Contractor against the Contracting Agency.

Where the suspension is due solely to the convenience of the Contracting Agency, allowances are made in the contract for such elapsed time. When suspension of work is ordered, the Contractor should be required to place the work in proper condition for the period of suspension.

10. Damages for Delay. If the Contractor does not complete the work within the contract time, including authorized time extensions thereof, the Contractor is required to pay to the Contracting Agency an amount to cover the extra cost incurred by the Agency due to such noncompletion. Such extra cost would include such items as expenses due to employment of engineers, inspectors, and other employees and such other costs as may be entailed by the Agency due to the delay in construction. Since it is difficult to accurately determine actual damages, both parties to the contract agree to a predetermined amount for each calendar day of delay until the work is completed. Since contract penalties are difficult to enforce under the law, such amounts are generally specified to be considered as the *liquidated damages* which the Contracting Agency would suffer as a result of the delay, and not as a penalty. For prior agreed amounts, Standard Specifications often have liquidated-damage tables listing the daily charges for various original contract amounts. Sometimes the amount is not fixed by prior agreement, and the damages to be paid by the Contractor are specified to be the demonstrable costs incurred by the Contracting Agency.

The injured party must justify this amount in case of a lawsuit even though it was agreed upon in the original contract. Unjustifiable liquidated damages imply a penalty, and the court reserves to itself the right to assess penalties.

Liquidated damages are generally not figured for such time where the Contractor was not able to work because of conditions beyond his control, such as acts of the Contracting Agency, Acts of God, war, flood, epidemic, and the like. With respect to bad weather, some contracts specifically exclude normal bad weather from the definition of causes beyond the control of the Contractor. Where such distinction is not made, the determination of whether bad weather was severe enough to justify an extension of time under the terms of the contract would be made by the representative of the Contracting Agency when the Contractor submits a claim for time extensions.

11. Termination for Default. The Contracting Agency generally reserves the right to terminate or abrogate the contract for proper cause by fault of the Contractor. Proper causes would be those which violate the very substance of the contract, such as failure to commence work on time, abandonment of work, unreasonable delay in the progress of the work, and assignment of the contract without approval as required. Provision is generally made for the Contracting Agency to notify the Contractor and his surety in writing of the default before action by the Contracting Agency is taken. If the Contractor does not take measures within a specified time, say, 10 days, the Contracting Agency has full authority to act.

In such cases, contracts usually provide that the Contracting Agency may use the

Contractor's materials and equipment at the site and complete the work by another agreement or otherwise. The Contractor and his surety are held responsible for the costs incurred by the Contracting Agency in completing the work.

12. Materials Furnished by the Contracting Agency. In unusual cases, the Contracting Agency may furnish to the Contractor certain selected materials or equipment, or both, for construction purposes or for incorporation into the permanent work. Such conditions would pertain where the Contracting Agency decides to place early orders for certain materials (before selecting a contractor) the delivery of which would otherwise delay the prosecution of the work. In such cases, the Contractor is required to assume responsibility for protection of the materials at the designated point of transfer of the materials to him.

E. Responsibilities of Contractor

Within the scope of the primary responsibility of the Contractor, namely, that of performing the work in accordance with the contract documents, numerous specific responsibilities, including the following, are placed on him by contracts.

1. Cooperation. The Contractor is required to cooperate with others in the performance of his work. These include other contractors, Public Utilities, the general public, and the forces of the Contracting Agency. The respective rights of the various interests are determined by the Engineer to secure harmonious prosecution and completion of the project.

The Contractor is required to conduct his operations with a minimum of interference with those of others and to always have a competent representative, as his agent, on the work to receive instructions from the Engineer and his assistants.

Where Public Utilities are involved, it is often the Contractor's responsibility to notify them in sufficient time when work is to be performed or inspected by the Utility's forces.

In general, the Contractor should be informed as to what the division of responsibility is between him and others, so that he knows what to expect while bidding and will not have to make claims, during the work, against the Contracting Agency.

2. Laws. Although the requirement for conformance with the law is inherent in every contract, the contract generally specifically provides for the Contractor to keep himself informed of, observe, and comply with the laws of Federal, state, local, and other bodies having jurisdiction in connection with the work. Laws which generally concern Contractors are those regarding hours of labor, alien labor, labor discrimination, prevailing wages, use of domestic materials, and fair labor standards, among numerous others.

3. Wage Rates. Wages are not permitted to be less than the prevailing wages as defined by law. A list of minimum wage rates is usually incorporated with the contract documents.

4. Permits and Licenses. The Contractor is generally required to procure all temporary permits and licenses and give all notices incident to the prosecution of the work. The Contracting Agency generally is responsible for providing any permanent right such as furnishing the right-of-way for construction.

5. Patents. The Contracting Agency is usually indemnified against all claims for infringement of a patent, trade-mark, or copyright which may arise because of the operations under the contract.

6. Sanitary Provisions. The Contractor is required to provide suitable sanitary accommodations for his employees and to take such measures as are necessary to avoid unhealthful conditions. The rules and regulations of Health authorities having jurisdiction in connection with the work must be complied with.

7. Public Convenience and Safety. Work must be performed so that the convenience of the general public, including residents along and adjacent to the highway and those traveling through the highway area, and the protection of persons and property are given prime importance. Observing traffic requirements as discussed in the following paragraph, keeping fire hydrants accessible, avoiding obstructions from material storage, etc., and similar duties are responsibilities of the Contractor.

8. Traffic. *a.* **Maintenance and Protection of Traffic.** Where the highway is constructed through an area with existing roads that are traveled regularly, the Contractor's responsibility of maintaining and protecting traffic through the work area and immediate vicinity becomes a prime consideration. The extent of the traffic, the general plan of operation, and specific requirements for work at individual facilities should be described, and the Contractor should be required to submit a detailed description of his intended plan of operation for approval of the Engineer. In addition to constructing and maintaining the basic plan for maintenance of traffic including detours, half-width pavements, and the like, the duties among others that may be given the Contractor are as follows:

Provide and maintain the road under construction in such condition that the public can travel the road in safety, when the road is to be kept open to traffic.

Not to close any road to the public without the Engineer's authorization.

Provide for local traffic to private property within the closed portion of the work.

Provide necessary bypasses around structures or temporary bridges over the structures to be rebuilt or extended.

Provide flagmen and watchmen to regulate traffic at hazardous places on the project.

Limit load capacity over existing bridges and roads used by the public, so as not to exceed the capacity of the road and its structures.

b. **Opening Section of Highway to Traffic.** To fulfill its responsibility to the traveling public, the Contracting Agency often reserves the right to open any completed section of highway to traffic. In such case, the contract generally provides that the Contractor maintain the highway, including repairs, on the section open to traffic, pending completion and final acceptance of the highway by the Contracting Agency. It is generally specified that necessary repairs due to defective materials and work and due to natural causes, on the opened section of highway, shall be performed at the Contractor's expense. With respect to necessary repairs due to ordinary wear and tear of traffic, it is believed to be more equitable for the Contractor to receive additional compensation therefor; however, such practice is not universal.

9. Barricades and Signs. The Contractor is generally responsible for providing and maintaining all barricades, danger signals, and signs, providing sufficient watchmen, and taking such precautionary measures as are necessary for the protection of the work and workmen and safety of the public. Sometimes certain specific signs such as those at the ends of the project may be furnished by the Contracting Agency. Among the signs generally required to be furnished by the Contractor are warning and detour signs at all closures, intersections, and along the detour routes, directing the traffic around the closed portion or portions of the highway. Barricades and obstructions must be illuminated at night. As a precaution against failure of lights, barricades should be equipped with suitable reflecting material.

10. Explosives. Requirements for the use of explosives generally emphasize the exercising of utmost care so as not to endanger life and property. The use of explosives should conform to modern standard practice and should meet the requirements of authorities having jurisdiction over the area where the explosives are used. The number and size of charges should be subject to the approval of the Engineer.

11. Restoration of Surfaces Opened in Permanent Work. If the contract documents do not indicate all of the locations in the highway where the permanent work constructed by the Contractor may be opened up by others, e.g., by authorities of a municipality through which the highway passes, the contract documents should call attention to the likelihood of such occurrence. The Contractor is required to allow authorized agencies (having permits) to make openings in the highway. If the Contractor is ordered to make repairs to such openings, additional payment therefor to the contractor should be made.

12. Protection of Property. The Contractor is responsible for the prevention of damage to existing public and private property along and adjacent to the highway due to his operations under the contract. Among the items often required to be protected by the Contractor are overhead and underground structures, land monuments, trees survey monuments, and property marks. Of these items, those that are required for future reference should not be removed until the Contractor has obtained proper direction for this disposal from the Engineer. Survey monuments within the limits of construction that must be removed are generally required to be reset with proper notice being given by the Contractor to the agency involved. All existing property not required to be removed that is damaged by the Contractor must be restored at his expense. Where the work is adjacent to state or national forests the Contractor is required to abide by all sanitary, fire, and other regulations of the proper authorities.

13. Responsibility for Damage Claims. Highway construction contracts generally place responsibility for protecting the Contracting Agency against claims and for carrying sufficient liability insurance by the Contractor. Although the former practice seems to be universal, the latter requirement sometimes is not specified. The Contractor is thus required to indemnify and save harmless the Contracting Agency and its representative from claims brought by third parties arising out of the work under the contract.

The types of insurance required include Workmen's Compensation Insurance, to the extent required by law, and liability insurance. Some types of liability insurance are Public Liability and Property Damage Liability, Protective Public Liability and Property Damage Liability, Railroad's Protective Liability and Property Damage Liability, and Completed Operations Liability Insurance.

14. Responsibility for Work. The Contractor is responsible for the work until final acceptance by the Contracting Agency. The Contractor is generally required to make good all damage to the work occurring prior to final acceptance, from any cause except for certain specified causes beyond the control of the Contractor. Among such causes are often included Acts of God, of the public enemy, and of governmental authority.

15. Termination of Contractor's Responsibility. In general, the Contractor's contract obligations end when all of the work is deemed to be satisfactorily completed during final inspection and acceptance by the appropriate representative of the Contracting Agency. However, the Contractor would still be bound under the terms of the contract bond. Also, under provisions discussed in paragraph VII.F.3, "No Waiver of Legal Rights," the Contracting Agency maintains the right of action in case any provision of the contract was not carried out.

16. Final Cleaning Up. Upon completion of the work and before final acceptance and final payment is made, the Contractor is required to clean up the roadway, restore all property damage during the work, and leave the site of the work, including each structure, in a neat and presentable condition.

17. Subsurface Information. In addition to surface physical conditions, the contract documents generally make available to the bidders the subsurface information, including borings which had been obtained as a result of previous investigations by the Contracting Agency. Thus the drawings indicate, among other subsurface data, boring logs which characterize the materials below ground surface at each boring location. The subsurface-exploration drawings and the actual boring samples are made available to the bidders. The contract documents of state and Federal contracting agencies generally provide that the Contracting Agency will not be responsible for any interpretations and conclusions drawn from the subsurface information by the Contractor or that the borings show conditions at boring points only and do not necessarily show nature of material to be encountered in the work. However, Federal government contracts do provide for an adjustment in contract price if subsurface physical conditions at the site differ materially from those indicated in the contract.

F. Miscellaneous General Requirements

1. Right-of-way. The contract should be specific as to which party furnishes the right-of-way for the work. Usually it is furnished by the Contracting Agency without cost to the Contractor.

2. Personal Liability of Public Officials. Representatives of the governmental contracting agencies are generally considered to be only representatives in performing their functions under the contract and shall bear no personal liability for their actions so conducted.

3. No Waiver of Legal Rights. Contracts usually incorporate provisions allowing the Contracting Agency to take future action against the Contractor if it is found, even after final acceptance, that payment had been incorrectly made or that the contract had been violated. To this end it is stated that:

a. The various actions of the Contracting Agency or its representatives, such as those in paying for and accepting the work of the Contractor, shall not operate as a waiver of any provision of the contract, or any power reserved therein to the Contracting Agency, or of any right to damages provided in the contract.

b. The Contracting Agency shall not be precluded from showing that any payment had been incorrectly made nor from recovering from the Contractor, or his sureties, or both, damages sustained by reason of the Contractor's failure to comply with the terms of the contract.

4. Federal Participation. Where a portion of a state project is paid for out of Federal funds, Federal laws require that the work be subject to inspection by the appropriate government agency. When this condition obtains, the contract usually states that the work will be subject to inspection by the Federal agency. However, in such cases, the Federal government is not a party to the contract.

5. Escalation Clauses. Although the use of escalation clauses in highway construction contracts is not the rule, where special conditions warrant, a contract may contain provisions for adjustment of the contract price should any one of a number of factors which formed the basis of the accepted bid change. Among the factors for which escalation clauses may be used are common-carrier rates, wage rates, and cost of materials. If an escalation clause is used, the exact formula for adjustment of contract price must be prescribed.

G. Measurement and Payment

1. General. In order that the cost of the work may be properly estimated by the bidder, a proper procedure established for making payments to the Contractor, and a basis established for determining what is beyond the scope of the work on which the bid was based, the contract should be very clear on the subject of measurement and payment. Since highway contracts are mostly performed on a unit-price basis, the method of measurement for payment of the work done under each contract item must be defined. The scope of payment included under each contract item must be delineated so that it is clear as to what the price for that item includes, in case of future change of contract. The method of making partial-progress payments as work is completed throughout the contract and the method of compensation for extra work and change to the contract must be established.

2. Measurement. The specific method of measurement of the work under each contract item is usually provided under the section of the contract that prescribes the technical requirements for that item. However, state highway and public works department contracts usually contain general provisions regarding measurement that are intended to supplement the specific measurement paragraphs in the technical requirements. The following information is often contained in the general provisions regarding measurement: who will perform the measurement, e.g., the Engineer; the system of standard on which measurement will be based, e.g., United States Standard; the general method for computing earthwork, e.g., average-end-area method; definition of "ton," e.g., 2,000 lb; method of measuring gallonage of bituminous material, e.g., measurement at 60°F with formula for conversion to other temperatures; requirements for vehicles to be used for volume "truck" measurement; when linear measurements are intended to be surface, horizontal, or theoretical neat-line measurements.

3. Scope of Payment. With the measurement paragraph for each item in the technical specifications, the contract provides a corresponding payment paragraph that defines what is included for payment under the item of work measured. To supplement these individual measurement paragraphs, state highway and public works department contracts usually contain general provisions regarding payment. Such provisions often include a general statement of the scope of the obligations of the Contractor, for the fulfillment of which he receives and accepts the compensation provided in the contract. Such obligations generally include furnishing all materials, labor, and equipment, performing the entire work embraced under the contract, absorbing all expenses, and taking all risks connected with the contract unless otherwise specifically provided in the contract documents.

4. Compensation for Changes. In paragraph VII.B.3, "Changes," it was noted that changes to the contract are made by Extra Work Order, Change Order, Supplemental Agreement, or other means depending on the practice of the Contracting Agency. Whatever the method used, it is generally required that no extra work or any other change will be paid for unless authorized in writing by the proper representative of the Contracting Agency.

In general, the intent of the contract is to give the Contractor additional compensation for changes increasing the cost of the work and to reduce the contract price for changes decreasing the cost of the work.

With respect to the existing contract unit prices which are applicable to the work involved in a change, highway construction contracts usually provide a limit of change from the original work within which the original contract prices apply and beyond which the contract prices are required to be adjusted. It is often provided that an adjustment of contract price(s) will be made due to any change resulting in (1) an increase or decrease of more than 20 or 25 per cent in the original contract amount, in the quantity of a major (sometimes any) item, or in the length of the project, or (2) a substantial change in the character of the work that materially increases or decreases the cost of its performance.

If the parties to the contract cannot agree on a price or prices for the work covered by the Extra Work Order or other document covering the change, contracts usually provide that the Engineer may order such work done and the Contractor shall do such work on a force-account basis.

For work on a force-account basis, the Contractor is required to furnish periodic detailed statements to the Engineer of the cost of the force-account work, including receipted material bills and transportation charges. The statement should be in prescribed form so that Engineer and Contractor can readily check the data therein and come to an agreement as to its correctness.

Force-account procedures generally provide for the Contractor to receive a form of percentage of the costs to cover the Contractor's overhead and profit. The Contractor's compensation is often computed as follows: the actual amount of wages for direct labor, plus the amount paid for employee contract costs, insurance, taxes, plus a percentage of the actual cost of the wages and the other costs; plus the actual cost of material including transportation charges plus a percentage thereof; plus a reasonable rental price for equipment with no percentage; no allowance being made for general superintendance, small tools, and manual equipment. The Contractor's and Engineer's representative should compare force-account records at the end of each day's work.

5. Partial Payments. Partial-progress payments are generally made periodically based on the Engineer's estimate of the value of the completed work done by the Contractor during the period. The Engineer is often given authority to include in the partial estimate the value of materials delivered to the site. Sometimes the estimate may be made by the Contractor and approved by the Engineer. Usually, the period of payment is the month, and payment is made in the amount of 90 per cent of the estimate with 10 per cent retained. It is often specified that the Contracting Agency may make partial payments semimonthly if the amount of work so warrants; also the Contracting Agency may release a portion of the retained percentage prior to final acceptance if most of the work is completed, but at least a minimum, say, 2 per cent, is retained until final acceptance.

The payment of a partial periodic estimate does not mean that the Contracting Agency accepts the work. The Contractor bears complete responsibility for the work as described in paragraph VII.E.14, "Responsibility for Work," and all periodic estimates are considered approximate and subject to revision until the final estimate.

6. Final Examination, Acceptance, and Payment. When the entire work is completed and final cleaning up is performed, the work is given a final inspection by the appropriate representative of the Contracting Agency. If the work is found satisfactory, a certificate of acceptance is generally issued by the Contracting Agency and final payment, including all retained percentages as described in paragraph VII.G.5, "Partial Payments," is made to the Contractor, after the Contracting Agency satisfies itself that there are no outstanding liens or claims against the contract. The Contractor is still under obligation, however, as explained in paragraph VII.E.15, "Termination of Contractor's Responsibility."

VIII. TECHNICAL REQUIREMENTS OF THE CONTRACT

A. General

The technical requirements of the contract are generally indicated on the drawings and in the technical specifications. The drawings show what is to be done, including the extent thereof. The specifications define the requirements for scope of work, materials, construction procedures, and methods of measurement and scope of payment for each of the contract items. Our discussion here is limited primarily to a brief general description of the technical specifications for highway and bridge construction contracts, since the scope of technical specifications is as broad as the scope of the technical information in this handbook. Such treatment is intended to supplement the technical qualitative and quantitative requirements of highway and bridge construction indicated throughout the handbook.

B. Presentation

Since the great majority of highways are constructed under the specifications of State Highway and Public Works Departments, the method of presentation of highway specifications, in the United States, is generally that used by the state departments. A typical grouping of the standard technical specifications into five parts is shown under Division II—Construction Details in the outline of a typical set of State Standard Specifications in paragraph X.C, "Outline of State Standard Specifications." Examples of the kind of specifications covered under each part are as follows:

Part 1—Earthwork: Clearing and Grubbing, Roadway and Borrow Excavation, Excavation and Backfill for Structures, Overhaul, Embankment, and the like.

Part 2—Bases: Aggregate Base, Bituminous and Portland Cement Concrete Bases, and the like.

Part 3—Surfacing and Pavements: Aggregate Surfacing, Bituminous Prime, Tack and Seal Coats, Bituminous-surface Treatments, Bituminous Road Mix and Plant-mix Surfacing, Bituminous Concrete and Portland Cement Concrete Pavements, and the like.

Part 4—Structures: Piling, Concrete Structures, Stone Masonry, Steel and Timber Structures, Waterproofing, Pipes, Culverts, and the like.

Part 5—Incidental Construction: Riprap, Guardrails, Fencing, Monuments, Topsoil, Seeding, and the like.

In the outline presented in paragraph X.C, "Outline of State Standard Specifications," there is no separate division for materials of construction. The requirements for materials are specified throughout Division II—Construction Details, under the item with which the particular materials are concerned. The standard specifications of some states provide a separate Division—"Materials of Construction" in which all of the materials are specified and incorporated by reference as applicable under each item of the technical specifications under the Division—"Construction Details."

The remainder of the technical specifications required to supplement those in the above standard specifications would be incorporated in the special provisions (paragraph X.B.1, "State as Contracting Agency").

C. Type of Specifications

In general, the technical specifications control the work of the Contractor in the following manner.

1. Bid-item Scope. The scope of work under each bid item is defined. The work which the Contractor is required to do under each item of the unit-price contract is described. For example, clearing specifications generally require the removal and disposal of trees, logs, brush, debris, and similar obstructions from the entire right-of-way. Removal of structures is generally included under a separate item of the specifications, although occasionally when the quantity of structure removal is small, it is included with clearing. Grubbing specifications generally require the removal and disposal of stumps, roots, and other materials objectionable in foundation or subgrade from within the limits of the cut and fill slopes and other areas that may be designated, such as borrow areas. Removals should be made to at least a specified depth. Where embankments are above a minimum height, the requirements for grubbing are generally relaxed.

2. Material Requirements. The quality and other characteristics of materials necessary for the desired end facility are specified. For example, portland cement is generally specified by reference to standard specifications such as American Association of State Highway Officials Standard Specifications for Portland Cement Designation M 85. These standard specifications cover the chemical and physical requirements of five different types of cement in addition to sampling, testing, and other necessary requirements. Air entrainment, when required by the design, is specified by providing maximum and minimum limits of entrained air in the concrete mixture. Two methods of obtaining the required air content are generally permitted, namely (1) the use of normal portland cement plus an air-entraining admixture introduced at the mixer, and (2) the use of air-entraining portland cement (AASHO Designation M 134). If laboratory tests show the air content of the concrete with the air-entrained portland cement not to be within the specified range, adjustments are generally permitted to be made by the addition of normal portland cement or an air-entraining admixture as necessary to bring the percentage of air entrainment within range.

3. Construction Procedure. Specifications for workmanship and construction methods are given to produce a finished structure, from the specified materials, which will perform the required service for the desired duration. For example, embankment construction specifications generally include requirements for the preparation of the surface on which the embankment is to be placed, placement, including equipment and compaction of material, and preparation of the embankment surfaces, such as subgrade of pavement, to receive other material. Among other means of attaining the required compaction, embankment specifications usually specify the required density of the completed embankment.

4. Measurement and Payment. Methods and units by which the quantity of work under unit-price items are measured for payment are given. In addition, what is included in compensation for each item is described. For example, for a bid item such as bituminous prime coat, the unit of measurement may be the gallonage of bituminous material actually used in the accepted work measured at 60°F or converting gallonage measured at other temperatures to gallonage at 60°F by a specified method. Payment may be specified to constitute compensation for furnishing and placing the material, including all labor, equipment, tools, and incidentals necessary to complete the work prescribed under the item.

IX. LEGAL ASPECTS

A. General

In addition to requirements of the contract documents, there are rights and duties imposed by law which bind the parties to the agreement. Here are presented some principles of civil law applicable to highway construction, namely, the laws of contracts, agency, torts, property, and labor and Federal-Aid Highway Acts.

B. Contracts

A contract is an agreement made between two or more parties enforceable at law. It must contemplate a legal obligation and be based on an offer and acceptance, supported by valid consideration, to accomplish a lawful purpose by competent parties. The parties may be individuals or business organizations such as partnerships, corporations, etc. Individual firms may contract, or there may be a joint venture type of contract in which several firms combine their assets, plants, and personnel to undertake a project. The contract may be express or implied. In the former, the agreement by each party is expressly stated. In the implied contract, the actions or conduct of the parties indicate a mutuality of understanding and agreement with regard to certain subject matter.

Certain agreements are required under the Statute of Frauds to be in writing, "signed by the party to be charged," before they can be enforced at law. The Statute of Frauds does not require that a contract for the construction of a highway, bridge, or structure be in writing. Such contracts would be legal and binding even if oral. However, good business practice dictates that they be in writing. The following types of contracts, among others, must be evidenced in writing: (1) contracts for the sale of land or any interest in land or a lease of more than one year; (2) contracts which by their terms are not to be performed within the space of one year from the making thereof; or (3) contracts for the sale of goods in excess of a certain value (varying from $25 to $500 in the different states) unless the buyer accepts part or makes part payment.

Contracts should clearly express the intentions of the parties. These intentions are determined by a consideration of the entire contract and not by an isolated portion. However, where a contract is ambiguous or indefinite and the parties have given it a practical construction by their conduct, the courts will treat the contract as having that meaning. Technical terms will be given their technical meaning. The validity and operation of a contract is generally governed by the law of the state where the contract was made, unless the contract provides otherwise or by its terms is to be performed in another state.

As a general rule, only parties to a contract acquire any rights under it. However, beneficiary contracts and assignments of contracts are exceptions to this general rule. The beneficiary has only a derivative right and acquires no better rights against the promisor than the promisee had. A contract may be assigned provided it is not personal in its nature and is one that can be performed by an agent just as well.

1. Discharge of Contracts. *a.* **Specific Performance.** A construction contract completed in accordance with drawings and specifications and paid for at the contract price constitutes specific performance by both parties and terminates the contract.

b. **Agreement.** The parties may agree to waive all their rights and obligations under the contract at a specific time. It may be based on a substituted agreement; on payment in lieu of performance; or on accord and satisfaction wherein one party agrees to accept a substitution in the obligations of the other.

c. **Breach.** This results when there is nonperformance, without justification, of the obligations required by the contract; when one party prevents or hinders performance or repudiates contractual duties. In construction contracts, the injured party may sue for damages resulting from the breach or the reasonable value of performance. When the performance of a contract is guaranteed by a bonding company, the surety becomes liable for any damages or for the performance of the contract.

d. **Impossibility of Performance.** The general rule is that the mere fact that a party cannot perform his promise because it has become impossible for him to do so will not excuse him. The courts, however, have allowed many exceptions, including the following: (1) where performance becomes impossible by operation of law, (2) where the contract relates to a specific subject matter and that subject matter ceases to exist before the time for performance arrives, (3) illness or death of one of the parties to a contract for personal services, or (4) if the contemplated means of performance are destroyed or cease to exist.

C. Agency and Independent Contractor

1. Agency. This is a legal relation whereby one party, the agent, is authorized to represent the other party, the principal, in business dealings with third persons. In construction contracts, the Engineer is generally the agent of the Contracting Agency. As an agent he must exercise care, skill, and diligence in performing his specified duties. The Engineer must show reasonable skill, but perfection is not demanded.

2. Independent Contractor. In construction contracts, the contractor is generally considered to be an Independent Contractor and not an agent. The choice of means and methods of accomplishing the work is generally left by the Contracting Agency to the Contractor.

3. Engineer-Contractor Relations. Generally the Engineer is responsible for the design and specifications, and the Contractor for the execution of construction. It is the Engineer's duty to control the work, and the contract usually provides him with the necessary authority. No contractural relationship exists between the Engineer and the Contractor. The responsibilities of the Contractor and the duties of the Engineer are generally described in this section, and within this framework they should cooperate to provide a satisfactory project.

4. Liability. The general rule is that the principal is liable for all acts of his agent done in the scope and course of the employment. However, in an Independent Contractor relationship, the principal is not liable for the acts and torts of the Independent Contractor, except in certain cases. Whether one or the other relationship exists, the question is one of right to control. Sometimes, however, the person for whom the work is being done is liable, even though the person doing the work is an Independent Contractor. Liability occurs (1) when the person for whom the work is being done interferes and directs the act causing damage or orders the work done in such a manner that harm results; (2) when the thing contracted to be done is unlawful and creates a nuisance; or (3) where a statute places a duty on the Contracting Agency and the injury is the result of an act which the statute made it the duty of the Agency to refrain from doing or to do, for in such case the Agency delegating its duty must see that it is properly performed.

In general, a Subcontractor acting as an Independent Contractor and not as an agent is liable to the Contractor with respect to torts just as the Contractor is liable to the Contracting Agency.

D. Torts

A tort is a breach of duty established or imposed by law, independent of contract or agreement, proximately resulting in damage for which an action for damages may be maintained. A tort is distinguished from a crime in that a tort is an invasion of the private rights of a person, whereas a crime is an offense against the state. The same act may be both a tort and a crime, as in criminal negligence. The torts of special interest in the construction of highways are presented in the following discussion.

1. Trespass. This tort pertains to the unlawful disturbance of one's property. In trespass to lands, the tort is generally an unlawful entry onto the land of another. Intent is no factor in this tort; thus the inundation of premises by a defective sewer and the accidental throwing of stones and debris on the lands of another in blasting are trespasses. In some cases, the law permits an entry, as in the case of an impassable highway; one may enter upon adjoining land. Where private property is taken under the process of eminent domain, entry on the land becomes lawful.

2. Nuisance. This is a term applied to wrongs arising from an unreasonable or unwarrantable, and therefore unlawful, use by a person of his property and resulting in actual or threatened material discomfort or hurt to another. The following have been judged as nuisances: spite fences, pollution of the air or running water; diversion of a flowing stream or damming it up through lengthy periods; changing the natural condition of one's land so that water drains from it onto that of another person where none did before the change; the unlawful obstruction of navigable waters, highway, sidewalk, or right-of-way, or the maintenance of a structure which is likely to give way and cause damage. The storage of explosive fuels, or any dangerous things, in a place where their escape or discharge will do serious injury or menace health may be judged a nuisance.

3. Negligence. The failure to observe the necessary care, precaution, and vigilance to protect the interests and rights of another person constitutes the tort of negligence. For example, the failure of a contractor constructing a highway or trench to provide lighting and to properly protect the obstruction by a barrier whereby injury results is a tort.

4. Violation of Riparian Rights. The title to land under tidal waters below the high-water mark is generally in the state. The owners of the adjoining upland own only to the high-water mark. In some states, however, the shore between high- and low-water marks belongs to the owner of the upland. The beds of small nonnavigable streams belong to the riparian owners to the middle, or thread, of the stream.

The riparian owner has the right of access to the water and the reasonable use of water flowing past his land. An unreasonable interference with this right is classed as the tort of nuisance.

5. Violation of Right of Lateral Support. An owner of land has the right to have his land supported in its natural condition by the land of an adjoining owner. An excavation on one parcel resulting in a cave-in on the other is a violation giving an action for damages. If the caving is caused by the weight of a building or other improvement on the adjoining land, there is no violation, the burden imposed on each parcel being limited to the support of the adjoining parcel in its natural condition, without the pressure of buildings.

Ordinances have been generally enacted imposing conditions for protecting adjoining property in the case of excavations below a certain depth. In the City of New York, excavators must apply to the adjoining owner for permission to enter and shore up or brace his building so as to prevent injury thereto from excavation extending 10 ft or more below the curb. Where permission is granted, he must protect the building. However, if permission is denied, he may excavate without incurring liability, provided he uses due care.

6. Violation of Right of Subjacent Support. The right of subjacent support exists where the property of one owner is located above that of another. The owner of the lower stratum has the duty to support the upper stratum. For example, the lower stratum of lands containing mineral deposits may be granted to one party, the grantor retaining the surface. The owner of the mine must not excavate in such a manner as to cause a settling of the surface. Subsidence caused by acts of the lower owner makes him liable.

7. Infringement of Patents, Trade-marks, and Copyrights. The following infringements are torts: (1) the unauthorized making, using, or selling for practical use, or for profit, of an invention covered by a valid claim during the life of a patent; (2) unauthorized use of colorable imitation of a trade-mark, already appropriated by another, on goods of a similar class; and (3) unauthorized copying, more or less servile, of a copyrighted work.

8. Discharge of Torts. Torts may be discharged by acts of the parties or by operation of law. The following are the common methods of settlement: (1) Satisfaction of a judgment. (2) Agreement or release. (3) Death of either or both parties, which discharges only some torts. Generally, an action to redress a property tort may be maintained against an executor or administrator. (4) Bankruptcy of the tort-feasor discharges some torts. His liability for willful or malicious injuries to person or property continue. (5) Operation of a Statute of Limitations which discharges the tort automatically if no action is started to recover damages within a specified period. The period varies in different states.

E. Property

The exclusive right of possession and use is ownership; and property includes all those things and rights which are the objects of ownership. Property is classified as real or personal. Real property includes lands and other interests concerning lands. Land includes all those things attached to the land permanently; buildings, trees, fences, bridges, mines and minerals; also standing water and percolations beneath the surface. Running waters are not included, since the owner of the land has only the right to the use of such streams.   Personal property includes such things as are of a personal or removable nature, or chattels.

1. Deeds.   A deed is evidence in writing wherein the grantor conveys real property to the grantee.   The several types of deeds include full covenant and warranty, bargain and sale, quit claim, executor's and referee's.   Among other things, the deed must contain a description of the property, which may be described in a number of ways; there is no prescribed form.   It is the surveyor's duty to establish the original boundaries of the property and not to correct the original survey.

"Absolute fee" or "fee simple" conveys to the grantee absolute ownership of real property without qualifications or restrictions.

2. Easements.   An easement is the right to make use of the land of another for definite and limited purposes; it is distinct from ownership of the land.   A right-of-way which one party has over the land of his neighbor for the purpose of reaching his own land is an easement.   Lots are sometimes sold with the covenant that the purchaser will not build within a given distance from the street.   This creates an easement in favor of the seller.

F. Eminent Domain

This is the right of the state to condemn private property for public use upon the giving of just and adequate compensation.   The details of such proceedings are covered by state statutes.

G. Highways

1. Nature.   A highway is a right-of-way existing in favor of the public.   The right extends broadly to every reasonable public use to which the highway may be put in the public service which does not unreasonably interfere with rights of the owners of the land over which the highway exists.   "Highway" is sometimes synonymous with public way, but in its more limited sense, it refers to country roads under the control of local authorities; it does not include city streets.   Streets refer to avenues and thoroughfares of cities and villages and include roadway gutters and sidewalks.

2. Establishment.   Highways may be acquired where: (1) The public uses private land as a public way for 20 years or more with the knowledge of the owner.   In such case, a right of way by prescription is created.   (2) An owner throws open or appropriates his land for use as a public way, and the public authorities accept the same as such.   In such case, a public way by common law dedication is created (state statutes provide expressly for the manner of statutory dedication).   (3) Land is acquired for use as a public way by condemnation or direct purchase.

3. Rights of Abutters.   An owner of property abutting on a public way is presumed to own to the middle of the way.   Irrespective of fee ownership, an abutting owner has an easement for access, light, and air.   He is entitled to compensation for damage to his easement, unless such damage results from normal street use.

4. Maintenance.   The control of public ways rests primarily with the state, but the state generally charges municipalities with the duty of caring for them.   The municipality must exercise reasonable care to see that the public ways are in a safe condition.

5. Taxpayers' Actions.   To protect the interests of taxpayers and to facilitate the prosecution of officers for illegal acts, or to prevent waste or injury to public property, states have enacted legislation permitting persons to bring a taxpayer's action against public officers.

H. Right-of-way

A tract of land acquired for the operation of a highway or public utility in which each parcel of property is dependent upon the other property for its use or operation as distinguished from a holding for purposes in connection with the utility but not strictly necessary for its operation is the right-of-way.   Such land required for a public utility is usually obtained by purchase or lease.   A right-of-way acquired by a municipality or a public utility by the right of eminent domain is only an easement for the purposes for which it is secured.

If the public has acquired title to the land used for highway in fee, it may use the land for any public purpose which does not interfere unreasonably with the right of passage or with the easements of light, air, and access in the highway which exist in favor of abutting owners.

If a highway is impassible, any person using the road has a license, given him by law, because of the necessity of the case, to enter on the land of the abutting owner to the extent necessary to get past the obstruction. In exercising this right, he must enter only so far as is strictly necessary and do as little damage as possible.

I. Labor Laws in Construction Work

The labor laws of particular interest in construction and highway contracts are noted below:

Employer's Liability Acts place additional responsibilities for injuries on the employer.

Workmen's Compensation Acts vary somewhat from state to state, aim to place the cost of industrial accidents on the employer, regardless of negligence.

Social Security Act (1935) provides old-age insurance for the benefit of workers.

National Labor Relations Act (1935) is designated to protect workmen against unfair labor practices.

Fair Labor Standards Act (1938), amended 1949, is designed to eliminate conditions detrimental to the maintenance of the minimum standard of living necessary for health, efficiency, and general well-being of workers.

Labor-Management Relations Act (Taft-Hartley Act) (1947) is aimed primarily to provide additional methods for the mediation of labor disputes and to equalize the legal responsibilities of labor organizations and employers.

Special Labor Laws Applicable to U.S. Government Contracts:

Davis-Bacon Act, March 3, 1931, amended August 30, 1935, makes minimum wage rates a condition of a contract.

Copeland Act, June 13, 1934, outlaws the kickback of wages.

Walsh-Healey Act, June 30, 1936, and subsequent amendments implement Fair Labor Standards Act of 1938 and provide for minimum wages, overtime compensation, etc.

Buy American Act, Mar. 3, 1933, amended Oct. 29, 1949, requires the use of items produced in the United States.

J. Federal-aid Highway Act of 1956

This legislation was signed into law on June 29, 1956. The Act authorized Federal-aid funds to states for rebuilding of the strategic national system of interstate and defense highways within 13 years. Because of its primary importance to the national defense, the system is named the National System of Interstate and Defense Highways and referred to as the *interstate system.* The Act also authorizes a stepped-up Federal-aid program for primary, secondary, and urban systems. In 1957, there were some 755,300 miles of roads made eligible for Federal grants. Thirty years later, the system is still incomplete.

By the Federal-aid Road Act, approved July 11, 1916, the Federal government began to assist states in the construction of roads. At first, the funds could be spent on any roads used to carry mail. But in 1921, use was restricted to the *primary system*, the general network of main highways. Federal funds were granted to states on the basis of their population, area, and mileage of rural mail routes.

The Federal-aid Highway Act of 1944 provided for Federal aid for the *secondary highway system* and *urban roads.* The former system consisted of farm-to-market rural roads designated by the states and coordinated by the Bureau of Public Roads. The latter system provided for extension of Federal-aid roads within urban areas. This Act also provided for a *national system* of *interstate highways,* a supernetwork of the most important 40,000 miles of the *primary system* (subsequently increased to 41,000 miles).

The Act of 1956 provided a new basis for appropriating funds for the construction of the interstate system. For the first three years, apportionment is based two-thirds on population, one-sixth on area, and one-sixth on the mileage of rural post roads. For the final 10 years, funds are to be apportioned to states by the estimated cost of completing the interstate system in each state. Funds for Federal-aid highways other than the interstate system will be divided on the old basis, one-third each on population, area, and post-road mileage.

X. ILLUSTRATIONS

A. Progress Schedule

Figure 1-1 indicates a typical progress schedule which may be established for a highway construction project.

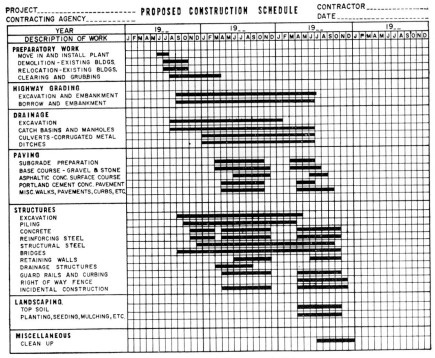

Fig. 1-1. Typical progress schedule.

B. Outlines of Contract Documents

Outlines of contract documents for highway construction by various contracting agencies are given below. Since actual practices in preparing and processing contract documents vary widely in detail but are essentially the same in principle, the outlines are simplified to illustrate only general principles in contract-document format. The outlines of basic documents would also be supplemented by miscellaneous documents such as bonds and other contents incident to the practice of the Contracting Agency. The documents are intended to be those in the contract at time of execution. Some additional documents such as special instructions to bidders and proposal forms that are used during the bidding stage may or may not form part of the executed contract documents.

The outline in paragraph X.B.1 below, "State as Contracting Agency," manifests the use of Standard Specifications with Special Provisions prepared to adapt the Standard Specifications to the particular job as discussed in Art. IV, "Contract Documents." Paragraph X.B.2, "Federal Government as Contracting Agency," outlines the contract documents for the Corps of Engineers and the Bureau of Public Roads. The Corps of Engineers does not use Standard Specifications; specifications are completely prepared and adopted for each project. The Bureau of Public Roads uses Standard Specifications with Special Provisions according to the predominating state practice.

1. State as Contracting Agency. The following is an outline listing of the basic contract documents as may be prepared by a state:

    a. Contract or Agreement
    b. Schedule of Prices
    c. Special Provisions
    d. Addenda to Special Provisions
    e. Standard Specifications
        General Requirements and Covenants
        Construction Details
    f. Drawings

2. Federal Government as Contracting Agency.   *a.* **U.S. Army, Corps of Engineers.** The following is an outline listing of the basic contract documents as may be prepared by the U.S. Army, Corps of Engineers:

    (1) Contract—Standard Form 23
    (2) General Provisions—Standard Form 23A
    (3) Schedule of Prices
    (4) Specifications
        Statement of Work
        General Conditions
        Special Conditions
        Technical Provisions
    (5) Addenda to Bidding Documents
    (6) Drawings

    b. **Department of Commerce, Bureau of Public Roads.** The following is an outline listing of the basic contract documents as may be prepared by the Bureau of Public Roads of the Department of Commerce:

    (1) Contract—Standard Form 23
    (2) General Provisions—Standard Form 23A
    (3) Schedule of Prices
    (4) Special Provisions
    (5) Addenda to Special Provisions
    (6) Standard Specifications
    (7) Drawings

C. Outline of State Standard Specifications

The following is an outline of the contents of Standard Specifications which is typical of that used by many state highway and public works departments:

    a. Division I—General Requirements
    b. Division II—Construction Details
        (1) Part 1—Earthwork
        (2) Part 2—Bases
        (3) Part 3—Surfacing and Pavements
        (4) Part 4—Structures
        (5) Part 5—Incidental Construction

BIBLIOGRAPHY

Abbett, R. W.: *Engineering Contracts and Specifications*, 3d ed., John Wiley & Sons, Inc., New York, 1954.

Federal-aid Highway Act of 1956, Public Law No. 627, 84th Cong.

Highway and bridge standards, contracts, and specifications used in various state and Federal government projects.

Mead, D. W., H. W. Mead, and J. R. Akerman: *Contracts, Specifications, and Engineering Relations*, 2d ed., McGraw-Hill Book Company, Inc., New York, 1956.

Spencer, W. H., and C. W. Gilliam: *A Casebook of Law and Business*, 2d ed., McGraw-Hill Book Company, Inc., New York, 1953.

Tucker, J. I.: *Contracts in Engineering*, 4th ed., McGraw-Hill Book Company, Inc., New York, 1947.

Section 2

TRAFFIC ENGINEERING

Dr. DONALD S. BERRY, *Professor of Civil Engineering, Northwestern University, Evanston, Ill.*

I. SCOPE OF TRAFFIC ENGINEERING

A. Definition

Traffic engineering is defined by the Institute of Traffic Engineers as "that phase of engineering which deals with the planning and geometric design of streets, highways, and abutting lands, and with traffic operations thereon, as their use is related to the safe, convenient and economic transportation of persons and goods."

B. The Traffic Problem

Construction of new street and highway facilities has not kept pace with the rapid growth in traffic volumes. The result has been increased congestion and delay, especially in metropolitan areas. In addition, the accident rates on many sections of street and highway are higher than on sections which meet modern design standards and/or are properly protected with adequate traffic-control devices.

C. Traffic-engineering Approaches

Traffic engineering attacks the problems of traffic accidents and congestion from two approaches: (1) the constructive approach and (2) the restrictive approach. The constructive approach includes the planning and geometric design of new street, highway, transit, and parking facilities to meet estimated future desires for transportation and termination. The restrictive approach implies the obtaining of maximum efficiency from existing streets and highways through application of traffic regulations and traffic-control devices. These controls, by their very nature, place restrictions on the driver's freedom. Traffic studies and analyses provide the basis for both the constructive and the restrictive approaches.

D. Coverage

This chapter covers traffic characteristics and studies, traffic regulations and control devices, and traffic-engineering administration.

II. TRAFFIC CHARACTERISTICS AND STUDIES

Traffic data provide the basis for both the constructive and restrictive approaches to traffic problems. This section reviews present data on factors affecting traffic flow and the characteristics of traffic variations and outlines techniques for traffic study and analysis.

A. Human Limitations

Human, vehicular, and roadway factors are elements in every traffic problem (Fig. 2-1). Although the engineer is concerned primarily with the roadway elements, he must take into account the human and vehicular elements in planning, in geometric design, and in traffic regulation and control.

Drivers and pedestrians are individuals. Each individual has certain physical limitations, motivations, and personal desires. The range in ability and performance is great. Traffic-engineering design is often based on the performance of the 85th-percentile driver (on such items as vision, reaction time, and passing performance), rather than on the performance of the average driver.

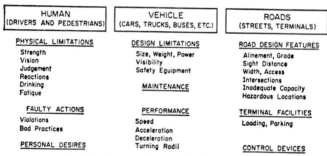

ELEMENTS
IN EVERY TRAFFIC PROBLEM

HUMAN (DRIVERS AND PEDESTRIANS)	VEHICLE (CARS, TRUCKS, BUSES, ETC.)	ROADS (STREETS, TERMINALS)
PHYSICAL LIMITATIONS Strength Vision Judgement Reactions Drinking Fatigue	DESIGN LIMITATIONS Size, Weight, Power Visibility Safety Equipment MAINTENANCE	ROAD DESIGN FEATURES Alinement, Grade Sight Distance Width, Access Intersections Inadequate Capacity Hazardous Locations
FAULTY ACTIONS Violations Bad Practices	PERFORMANCE Speed Acceleration Deceleration Turning Radii	TERMINAL FACILITIES Loading, Parking
PERSONAL DESIRES		CONTROL DEVICES

Fig. 2-1. Elements in every traffic problem.

The success of many traffic-engineering improvements and controls depends on the extent to which the remedies change the behavior of drivers and pedestrians. Examples are installation of signs to control speed or lane placement, use of signals to allocate right-of-way, and construction of left-turn lanes to remove left-turning vehicles from the through-traffic stream. Such improvements and controls are much more likely to be successful if they are based on traffic study and appear to be reasonable to the average driver.

1. Physical Factors. Physical factors of the more permanent types which affect driver ability include vision, strength, hearing, and motor coordination. Minor deficiencies in these factors can normally be compensated for, and thus do not appear in accident summaries as contributing to many traffic accidents.

Fig. 2-2. Effects of variation in driver vision on legibility distances for different sizes of capital letters on overhead traffic signs. [*Courtesy of Highway Research Board* (3).]

a. **Vision.** Good vision is important in perception and identification and in speeding up the process of judgment and reaction in traffic situations. Visual acuity may be less important than other visual abilities, according to Brody (1),* but some basic minimum is needed such as 20/40 in one eye or 20/100 in the poorer eye. Peripheral vision relates to the field of vision and is 150 to 180° for the average driver. Clear

---

* Numbers in parentheses refer to corresponding items in the bibliography at the end of this section.

seeing is in the narrow cone of 5 of 10°. There are few cases of "tunnel vision." Color perception is not an important factor contributing to accidents. Depth perception is quite important because of its relation to ability to estimate speeds and distances (2). Glare at night affects some drivers more than others, with glare-recovery time a critical factor. Visual fatigue and attention are more important than most other vision factors (1). One illustration of the effect of variations in vision on driver performance is shown in Fig. 2-2, which presents data on legibility distances for different heights of capital letters on overhead traffic signs. Frequency distributions are shown for observations of legibility distances using three different letter heights. Also shown is information on median and 85th-percentile legibility distances. In this example (3) words with capital letters 13 in. high could be read by 85 per cent of the drivers at 830 ft, while only 50 per cent could read the words from a distance of 950 ft.

b. **Motor Coordination.** Motor coordination affects the time required for drivers to react to an external stimulus such as a sign message. Simple reaction time (when the individual is anticipating a stimulus and presses a key in response) may average 0.15 sec (4). Brake reaction time (with a single stimulus) is longer, because of time needed to shift the foot, and may average 0.35 sec (4).

Total driver response time (including time for perception, judgment, and reaction) increases with the number of stimuli, the range of choices, and the complexities of the judgments to be made. Laboratory tests using multiple stimuli and requiring selection of one of three responses indicate that the average response time exceeds 1.0 sec (4). In addition, a driver needs 0.6 to 1.0 sec to turn his eyes to the left, fixate his gaze long enough to see, and then look to the right. Thus traffic situations which require a driver to check hazards simultaneously on two approaches to an intersection (as at a four-way intersection) may need an additional second of driver response time.

Driver-response-time values used for design differ from values used for emergency-stopping computations or for educational purposes. The AASHO design policy (5) uses 2.5 sec for sight-distance design and 2.0 sec for intersection design. The perception-reaction time recommended by several national organizations for stopping distance calculations is 1.5 sec. The safe-approach-speed chart for computing safe approach speed at intersections (Fig. 2-4) is based on a 1.0-sec perception-reaction time. (Note the summary update material at the end of this chapter.)

c. **Temporary Impairments.** Temporary impairments are much more important as factors contributing to accidents than are minor defects of a more permanent nature. Fatigue, illness, and the effects of alcohol are in this classification. Fatigue is an increasingly important factor on long sections of expressway, where motorists may maintain high speeds for a long period of time. Alcohol was reported as a factor in about 30 per cent of the fatal traffic accidents in 1956 (6) and 50 per cent in 1986.

2. Knowledge and Skill. Knowledge of driving rules and driving experience are important attributes of safe drivers. Drivers who have taken high-school driver-training courses have lower accident rates than similar drivers without such training. However, the attitude of trained drivers may have been improved, as well as their knowledge and skill. Judgment is related to skill, experience, and maturity.

Table 2-1. Comparison of Accidents per Driver in Two Periods of Time, State of Connecticut

Driver group	Accidents per driver, 1931–1933	Accidents per driver, 1934–1936
Group 1.............	0	0.101
Group 2.............	1	0.199
Group 3.............	2	0.300
Group 4.............	3	0.484
Group 5.............	4	0.700

Source: Ref. 9.

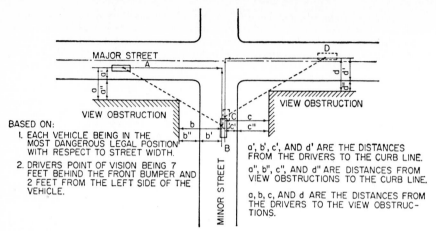

FIG. 2-3. Layout for determining safe approach speeds on minor streets, for different approach speeds on major streets, and different positions of obstruction to view.

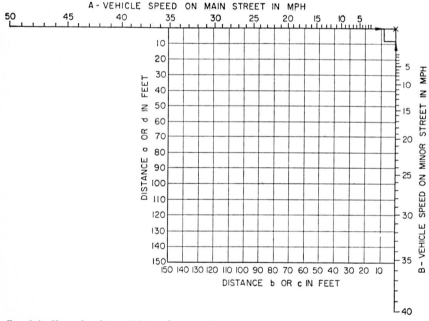

FIG. 2-4. Chart for determining safe approach speeds. First, locate the view obstruction on chart. Then select a speed value for main street. Draw straight line from this speed value through point of view obstruction. The point where this line intersects the speed line for vehicle *B* is the safe approach speed for vehicle *B*. Chart based on AAA method.

Assumptions:

1. Vehicles in most dangerous legal position.
2. Reaction time is 1 sec.
3. Deceleration rate is 16 ft per sec per sec.
4. Driver's eye is 7 ft back of bumper.
5. Vehicle can stop 8 ft from point where paths cross.

3. Attitude.   Many faulty actions of the driver result from his attitudes and emotions (7).   Controlled studies of groups of accident repeaters revealed that these repeaters had significantly poorer attitudes than comparable groups of accident-free drivers (8).   Serious repeaters also tended to be more upset by frustrations and annoyances than comparable accident-free drivers (8).

4. Accident Repeaters.   On a group basis, some groups of drivers have a greater number of accidents for equal exposure than others.   A Connecticut study showed that the high-accident group for one 3-year period of time also tended to be in the high-accident group during the following 3-year period of time (9).   However, the elimination of those drivers who experienced more than one accident in both periods would reduce by only a very small number the total of accidents which were reported for the 6 years (9).

5. Driver Characteristics.   Data are available also on performance of median- and 85th-percentile drivers on such factors as legibility distances for highway signs (3), distances required in passing at different speeds (10), headways between following vehicles on freeways, sizes of time gaps in traffic on a through street which drivers consider adequate in entering past a stop sign, lateral placement, speeds, travel times, and reaction time.

6. Pedestrian Characteristics.   Pedestrians can be expected to cross streets at 3.5 to 4.5 ft per sec walking speed.

B. Volume Counting

1. Methods.   Manual counts are generally used to obtain turning movements at intersections and for classification counts.   Hand counters facilitate recording.

Portable recording counters are operated by a pneumatic road tube and a diaphragm switch, with the electrically operated counter printing totals on a paper tape each hour or at 15-min intervals.   Nonrecording mechanical counters operate in a similar manner, except that the count appears on a dial and must be read at desired intervals. Permanently located counters are used to record at hourly intervals the traffic volume past a point on a continuous basis (365 days a year).   The actuating element usually is a photoelectric cell, or a treadle placed in the roadway.   These mechanical counters are self-recording and reduce the cost of counting.   However, care must be taken in their operation and maintenance in order to obtain reliable results.

2. Scheduling Counts.   Selection of times and locations for taking volume counts depends on the purpose of the counts, the method used, and the sampling procedure.

The Bureau of Public Roads has described a procedure (16) for sampling traffic volumes on rural highways which utilizes continuous-count stations and "control-count" stations for establishing seasonal and daily traffic-volume patterns.   Counts then are made for 24 or 48 hr at numerous coverage stations.   Expansion factors which were developed for nearby control-count stations are used to convert the coverage-count results to ADT values representative of the entire year.   Classification counts are also made at control stations.

Cordon counts and screen-line counts are counts so located as to permit counting of all vehicles (and persons) crossing a line.   A screen-line count usually intercepts all streets extending in one general direction.   The cordon count consists of a series of screen lines entirely surrounding an area such as a central business district.   Counting all stations on one line simultaneously permits an analysis of total movements into and out of the area and a determination of the accumulation of vehicles and persons in the area.   Figure 2-5 shows a typical graph of accumulation of persons and vehicles in a central area.

Many volume counts are taken in urban areas in connection with studies of problem locations, such as at intersections where traffic signals are under consideration.   In addition to these counts, a systematic schedule of volume counts should be developed in each large urban area to aid in planning and to provide better correlation between isolated counts.   The National Committee on Urban Transportation has recommended a count program which includes permanent count stations, control stations, a "moving-vehicle" count for coverage, and a series of afternoon peak-hour counts at intersections of major streets for use in capacity-volume analysis (17).

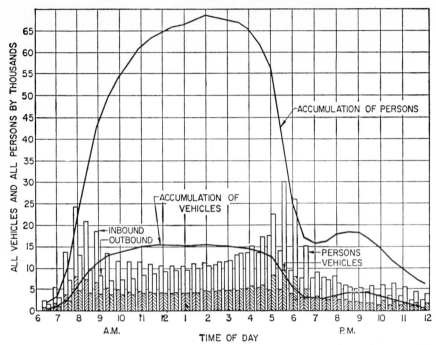

FIG. 2-5. Typical graph of hourly variations in cordon-count accumulation for a central business district.

C. Speed

1. Definitions

SPOT SPEED: The instantaneous speed of a vehicle as it passes a point on a street or highway.

OVER-ALL TRAVEL SPEED: The average speed of a vehicle over a given section of street or highway, determined by dividing total distance traversed by the total travel time.

RUNNING SPEED: Similar to over-all travel speed, except that the distance traversed is divided by the running time (excluding stopped time).

Spot-speed data are used in studying high-accident locations, in determining speed limits for speed zones, in timing traffic-signal systems, in determining proper types and locations of traffic-control devices, and for many other similar purposes.

Over-all speeds, travel time, and delay are important as relative measures of the efficiency of a roadway or intersection. Travel-time data are used also in economic studies, in estimating potential traffic on proposed facilities, in identifying and correcting causes for delay and congestion, and in evaluating effectiveness of changes in physical conditions or traffic controls.

2. Speed Characteristics. Average speed on open rural highways has been increasing since 1946, as shown by results of Bureau of Public Roads compilations of spot-speed observations (Table 2-2). The percentage of vehicles exceeding 60 mph has also increased, with 15 per cent exceeding 60 mph in 1958 as compared with 7 per cent in 1946. Average speeds in Central and Western regions in 1958 were 5.0 mph higher than the average for Eastern regions. In Eastern regions the average speed for trucks was 3.2 mph lower than for passenger cars; the differential in the Central and Western regions was 6.9 mph. Average speeds for buses generally exceeded the averages for passenger cars in most states (18).

Table 2-2. Trends in Vehicle Speeds on Main Rural Highways

Region and type of vehicle	Average speed, mph			Percentage over 60 mph		
	1946	1952	1958	1946	1952	1958
All vehicles, by region:						
Eastern region............	42.5	45.7	48.2	2	4	5
Central and Western......	46.7	51.1	53.2	10	16	21
All states................	45.1	49.5	51.5	7	12	15
All states, by type:						
Passenger vehicles.........	46.1	50.8	52.6	...	15	18
Trucks....................	40.0	44.8	47.1	...	2	3
Buses....................	47.9	51.9	53.2	...	18	20

Source: Bureau of Public Roads (18).

The variation in speeds of vehicles passing a point can be shown graphically on a cumulative-speed distribution curve. Figure 2-6 shows such a curve, with values shown for the median (50th-percentile) and the 85th- and 15th-percentile speeds. For spot-speed studies on uncongested highways, the distribution curve is practically normal (19).

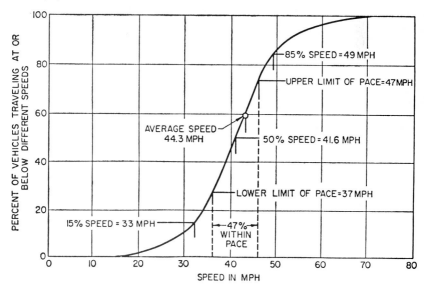

Fig. 2-6. Typical cumulative distribution curve for observations of spot speeds of vehicles.

The dispersion, or spread, in speeds may be determined as the percentage of vehicles in the *pace* (the 10-mph speed range containing the highest percentage of vehicles) or by computing the standard deviation. The standard deviation may be estimated for normal distributions from the 85th- and 15th-percentile speed values, (19) using the formula

$$\text{Standard deviation } \sigma = \frac{P_{85} - P_{15}}{2.07}$$

Pneumatic-tube detectors and electronic timers are also used in spot-speed studies, with timing sections as short as 11 ft in length. This equipment permits timing a greater percentage of the vehicles. Detectors may also be placed so as to detect traffic separately for each lane of travel.

Radar-type speed meters are now widely used for speed checking. These meters, by use of reflecting-wave techniques, permit measurement of speeds without use of roadway detectors. A graphic recorder can be used for recording automatically the speed of each vehicle and groups of vehicles.

The Bureau of Public Roads has developed equipment for recording speeds and lateral placement of every vehicle in each of several lanes of traffic simultaneously, with results recorded on punched tape so that the data can be placed on punch cards automatically. Time-lapse photos, utilizing single-frame exposures at time intervals, can also provide accurate data for determining speeds on freeways. When using the photographic method, considerable time is required for the reduction of the data.

In sampling spot speeds, care is needed in obtaining a representative sample of adequate size. If the standard deviation of the spot speeds of all the vehicles can be estimated, it is possible to determine approximately from Table 2-3 the minimum numbers of vehicles which should be checked so that an average speed can be determined which will be within the true average by $\pm E$, in miles per hour, with a confidence level of 95 times in 100 (19).

Table 2-3. Approximate Relation between Number of Spot-speed Observations and Ranges of Allowable Error for Different Standard Deviations of the Population and for a Confidence Level of 95 Per cent

E = range in allowable error, mph	Number of vehicles to be checked, standard deviation, mph			
	$\sigma = 4$	$\sigma = 6$	$\sigma = 8$	$\sigma = 10$
± 0.5	240	560	980	1540
± 1.0	60	140	245	385
± 2.0	20	35	60	95

Source: Ref. 19.

Thus if an accuracy in the average speed value of ± 1 mph is desired, and the standard deviation of the speeds of all vehicles is estimated at 6 mph, then a sample of 140 observations could be considered adequate.

D. Travel-time Studies

Travel times of vehicles in a traffic stream are usually obtained by utilizing test vehicles which "float" with traffic. An observer in the test vehicle records the travel times and the causes and durations of all stopped-time delays. Equipment is also available for automatic recording of elapsed time and distances, with codes for indicating causes for delays (22) (Fig. 2-7).

Travel times may also be obtained by the license-number-matching method, in which the license number of each vehicle passing two stations is recorded, along with the exact times the vehicle passed each station. Data for all vehicles passing each of the two stations are recorded, and later the license numbers are matched and a travel time is obtained for each vehicle.

Speed and delay data usually are illustrated in bar charts. Figure 2-8 shows that stops for traffic signals constitute a major portion of urban delay. Time contours may also be prepared to illustrate travel times in rush or nonrush hours for routes radiating from one point in a city (34).

E. Traffic-stream Flow

Vehicles in traffic streams affect one another and are also affected by such boundary conditions as width of lane and shoulder, nearness to obstructions, and the merging and crossing of traffic lanes. A knowledge of longitudinal distribution, lateral placement, gaps in the traffic streams, and merging and weaving characteristics is helpful in understanding traffic flow.

1. Longitudinal Distribution. Headways between vehicles in a traffic stream (measured from head to head of successive vehicles) are important in capacity analyses

FIG. 2-7. Meter used in test car to record times, distances, and codes for indicating causes of delay. (*Courtesy of University of California.*)

TOTAL TRAVEL TIME

31%	9%	60%
INTERSECTION DELAY	MID-BLOCK DELAY	RUNNING TIME

SIGNAL DELAY	60%
LEFT TURNS	12%
STOP SIGNS	8%
RIGHT TURNS	4%
PEDESTRIANS	8%
OTHER	8%

INTERSECTION DELAY

FIG. 2-8. Typical distribution of travel times and intersection delays in urban areas.

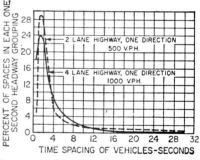

FIG. 2-9. Distribution of headways between vehicles on typical two-lane and four-lane highways. [*Adapted from Figs. 9 and 10 of Highway Capacity Manual (23).*]

and in evaluating intersection performance. Vehicles tend to be spaced at random in a traffic stream, with the distribution of headways conforming generally to a Poisson distribution for low traffic volumes. At traffic volumes near capacity, there is an increase in the percentage of vehicles spaced at 1 to 2 sec.

Figure 2-9 shows the distribution of actual headways for a traffic volume of 500 vehicles per hour (vph) in one direction on a two-lane highway and a volume of 1,000 vph in one direction on a multilane highway (23). The two distributions differ, even though the lane volume is the same in the two cases. The two-lane road is operating at or above practical capacity. There is less freedom in passing on the two-lane highway.

2. Intersecting Flows. When a driver desires to enter a traffic stream on a merging maneuver, he will enter when a gap in the traffic seems adequate to him. In merging from a ramp where there is no stop sign, the average driver will accept a 3-sec gap in the traffic stream. If the merge is controlled by a stop sign, the acceptable gap size may be as long as 6 sec for the average driver (24). The average driver desiring to enter an intersection from a side road controlled by a stop sign will also need a 6-sec gap in the main-street traffic before he will attempt to enter or cross.

A chart showing the probability of waiting not more than 40 sec for gaps of different sizes for different traffic volumes under rural highway conditions in California (25) is

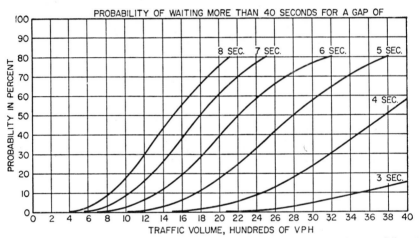

Fig. 2-10. Probability of waiting more than 40 sec for a gap of 3, 4, 5, 6, 7, or 8 sec in different traffic volumes on a four-lane rural highway in California. [*Courtesy of Highway Research Board* (25).]

shown in Fig. 2-10. When the traffic volume on the four-lane highway was 1,600 vph, a driver approaching on the side road and desiring a gap of 6 sec or more would need to wait at least 40 sec about 20 per cent of the time.

F. Capacity

"Possible capacity" is defined as the maximum number of vehicles that can pass a point in one hour under prevailing roadway and traffic conditions. "Practical capacity" is the maximum number that can pass the point without unreasonable delay or restriction to the average driver's freedom to pass other vehicles.

1. Uninterrupted-flow Condition. On roadways where the practical capacity is not limited by intersections, the practical capacity is related to the over-all speed of operation which is acceptable to the average driver. Possible capacity is attained when all vehicles are traveling at the same speed. Shown in Table 2-4 are roadway capacities for uninterrupted flow, for ideal traffic and roadway conditions (23).

The procedure for computing practical capacity for the uninterrupted-flow condition is as follows:

1. Select an operating speed which is acceptable for the class of highway, the terrain, and the drivers.

2. Determine the appropriate capacity for ideal conditions from Table 2-4 below or from Fig. 2-11.
3. Determine reduction factors for conditions which reduce capacity below ideal values (widths, commercial vehicles, alignment, grade, sight distance).
4. Multiply these factors by ideal capacity value (from step 2).

Table 2-4. Capacities for Uninterrupted Flow under Ideal Road and Traffic Conditions

Operating speed, mph	Two-lane highway, both lanes, vph	Four-lane highway, per lane, vph
50–55....................	600	
45–50....................	900	1,000
35–40....................	1,500	1,500
Possible capacity.........	2,000	2,000

Source: *Highway Capacity Manual*, 1950.

Restricted passing sight distance (less than 1,500 ft) is taken into account on two-lane highways by using Fig. 2-11. Reduction factors for restricted widths are found from Table 2-5.

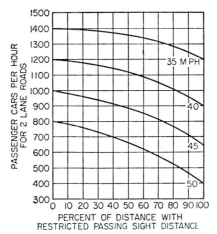

Fig. 2-11. Practical capacity under ideal conditions for two-lane roads as affected by sight distance and operating speed. (*Courtesy of California Division of Highways.*)

Reduction factors for commercial vehicles may be selected from Table 2-6 for the appropriate terrain. When grades are steep and of considerable length, the correction factors may be computed from the formula

$$F_c = \frac{100}{100 + P(E - 1)}$$

where F_c = correction factor for commercial vehicles and grades
E = passenger-car equivalent per truck for the grade conditions
P = percentage of dual-tired vehicles
The value of E may be selected from Table 2-7 or from more complete tables given in the *Highway Capacity Manual*.

Table 2-5. Combined Effect of Lane Width and Lateral Clearances on Highway Capacities

Clearance from pavement edge to obstruction, ft	Capacity in percentage of that for two 12-ft lanes with no restrictive lateral clearances							
	Obstruction on one side				Obstruction on both sides			
	12-ft lanes	11-ft lanes	10-ft lanes	9-ft lanes	12-ft lanes	11-ft lanes	10-ft lanes	9-ft lanes
Possible capacity of two-lane highway, %								
6	100	88	81	76	100	88	81	76
4	97	85	79	74	94	83	76	71
2	93	81	75	70	85	75	69	65
0	88	77	71	67	76	67	62	58
Practical capacity of two-lane highway, %								
6	100	86	77	70	100	86	77	70
4	96	83	74	68	92	79	71	65
2	91	78	70	64	81	70	63	57
0	85	73	66	60	70	60	54	49
Possible and practical capacity of multilane highway, %								
6	100	97	91	81	100	97	91	81
4	99	96	90	80	98	95	89	79
2	97	94	88	79	94	91	86	76
0	90	87	82	73	81	79	74	66

SOURCE: *Highway Capacity Manual*, 1950.

Table 2-6. Effect of Commercial Vehicles on Practical Capacities for Uninterrupted Flow

Percentage of commercial vehicles	Capacity in percentage of passenger-car capacity on level terrain			
	Multilane highways		Two-lane highways	
	Level terrain	Rolling terrain	Level terrain	Rolling terrain
0	100	100	100	100
10	91	77	89	71
20	83	63	77	54

SOURCE: *Highway Capacity Manual*, 1950.

Practical capacity, in terms of average daily traffic (ADT), is much higher for four-lane than for two-lane highways and is highest for urban conditions. Table 2-8 compares practical capacity for three sets of conditions (all assuming uninterrupted flow, 5 per cent of trucks, and a directional distribution as given below).

2. Intersection Capacity. Capacity at signalized intersections may be computed by using curves given in the *Highway Capacity Manual* (23) and applying corrections for commercial vehicles, turns, bus stops, and signal timing. Computed values corre-

Table 2-7. Value of Passenger-car Equivalent for Trucks on Grades on
Two-lane Highways

Percentage and length of grade		Percentage of highway with restricted passing sight distance less than 1,500 ft			
Per cent grade	Length, miles	0	30	50	70
3	1	4.6	6.3		
	2	5.0	6.7		
	5	5.1	7.0		
5	0.4	4.9	...	9.6	
	1	6.0	...	11.0	
	2	6.5	...	11.9	
	5	6.7	...	12.1	
7	0.4	5.5	...	....	14.6
	1	7.4	...	....	17.4
	2	8.0	...	....	18.4
	5	8.3	...	....	19.1

Source: *Highway Capacity Manual,* 1950.

Table 2-8. Practical Capacity by Type of Highway

Type highway	Directional distribution, %	Peak hour, % of ADT	Operating speed, mph	Capacity, vehicles per day
2-lane, modern design..........	.....	12	50	5,700
4-lane, divided...............	67–33	12	50(rural)	21,700
4-lane, divided...............	60–40	10	40(urban)	41,000

spond to average conditions and may deviate somewhat from actual values because of local conditions.   The procedure for two-way streets is as follows:

1. *Select chart capacity.* Use Fig. 2-12 in selecting a capacity value for the street width, type of area, and parking condition which exists at the intersection being studied.
2. *Determine adjustment factors.*
   a. *Commercial vehicles.* Subtract 1 per cent for each 1 per cent by which commercial vehicles *exceed* 10 per cent of the total number of vehicles, or add 1 per cent for each 1 per cent that commercial vehicles are *less* than 10 per cent of the total.
   b. *Right turns.* Subtract ½ per cent for each 1 per cent by which traffic turning right exceeds 10 per cent of the total traffic, or add ½ per cent for each 1 per cent that traffic turning right is *less* than 10 per cent of the total (maximum reduction for right turns not to exceed 10 per cent).
   c. *Left turns.* Subtract 1 per cent for each 1 per cent by which traffic turning left exceeds 10 per cent of the total traffic, or add 1 per cent for each 1 per cent that traffic turning left is less than 10 per cent of the total (maximum deduction for left turns not to exceed 20 per cent).
   d. *Bus stops.*
      (1) *Parking permitted, no bus stop.* Deduct ¼ per cent for each 1 per cent that right and left turns combined are of total traffic entering from that approach (max. 6 per cent).
      (2) *Parking permitted, near-side bus stop.* Add ¼ per cent of each percentage that right and left turns combined are of total traffic (max. 6 per cent).

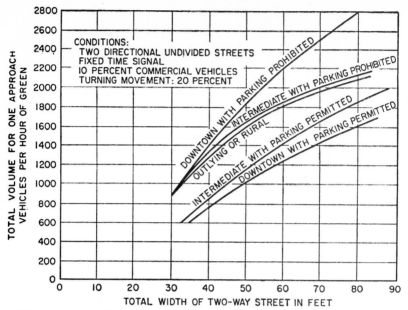

FIG. 2-12. Average capacities of approaches to signalized intersections on two-way streets. [*Source: Highway Capacity Manual* (23).]

(3) *Parking permitted, far-side bus stop.* No correction.

(4) *Parking prohibited, no bus stop.* Add 5 per cent.

(5) *Parking prohibited, near-side bus stop.* Deduct 10 per cent.

(6) *Parking prohibited, far-side bus stop.* Deduct 3 per cent in downtown areas and 15 per cent in intermediate areas.

e. Signal timing. Correct for the percentage that the green indication is of the total signal cycle. Multiply by the percentage figure.

f. Practical capacity. Practical capacities are about 10 per cent lower than values shown on the chart, so deduct 10 per cent by multiplying by 0.90.

g. Possible capacity. If it is desired to compute possible capacity, use a +10 per cent factor, with adjustment factors *a, b, c, d,* and *e*.

3. *Apply adjustment factors.* Multiply the chart capacity value by the adjustment factors to obtain the practical (*or* possible) capacity for traffic entering that approach of the intersection, in vehicles per hour.

The capacity of signalized intersections may be estimated for design purposes from charts prepared by the Bureau of Public Roads (26).

III. TRAFFIC REGULATIONS

Traffic regulations provide the controls which are required for efficient use of streets and highways. Some controls have area-wide application (such as driving on the right), whereas others are applied at specific locations (such as one-way control or parking prohibitions). Practically all regulations are based on state laws and local ordinances.

A. Traffic Laws and Ordinances

The states have the main responsibility for providing the legal basis for traffic control through the state motor-vehicle code of each state. These codes generally provide the following:

1. General rules on driving and walking applicable to all public highways (speed limits, right-of-way, making of turns, meaning of signals, etc.)
2. Rules on accident reporting, vehicle equipment, licensing of drivers, financial responsibility, vehicle registration
3. Enabling legislation to permit local authorities to adopt regulations applicable to specific locations (speed zones, one-way streets, parking control, turn prohibitions)

The Uniform Vehicle Code (29) has been prepared as a guide for states in adopting uniform traffic laws. The Model Traffic Ordinance (30) is the pattern for guidance of cities in preparing traffic ordinances. Universal adoption by states of uniform general rules for driving and walking would reduce chances for misinterpretation of law, with its resulting confusion, delays, and hazards.

The Model Traffic Ordinance supplements the state-law provisions by providing the following:

1. Regulations applying throughout the city, such as pedestrian control, bicycle control, parking control, and emergency vehicle control.
2. Schedules of special regulations applying to specific locations (through streets, one-way streets, parking prohibitions and time limits, speed zones, stop signs, yield control, and traffic signals).
3. Administrative provisions establishing the position of traffic engineer, the police traffic division, the traffic-violations bureau, and delegating authority to the traffic engineer.

B. Intersection Controls

1. Need for Intersection Controls. Control devices are needed at intersections to allocate right-of-way, cut speed of movement, prohibit undesirable turning movements, allocate lane use, and channel pedestrian and vehicular flow. Conflicts may be separated by time or by allocation of space. Proper control can accomplish the following:

a. Prevent Accidents. Almost half of all urban traffic accidents occur at intersections. By proper control of speed, placement, and/or timing of movements, the potential conflicts of crossing or converging movements can be minimized.

b. Utilize Intersection Capacity. By providing orderly movements of streams with separation by time, signal controls can increase the capacity. Other controls (turn prohibitions, parking control at approaches, etc.) can also increase capacity.

c. Protect Arterial Streets. Stop signs and signals can provide for continuous movement along arterial streets with safety. This encourages through traffic to use arterial streets and thus provides more safety for local streets.

d. Protect Pedestrians.

2. Types of Control. Conflicting movements at intersections may be controlled by one or more of the following methods:

a. Right-of-way Rule Only. Satisfactory at minor intersections where sight distance is adequate.

b. Yield Sign. Helpful in allocating right-of-way at light-volume intersections where a stop is not usually necessary.

c. Two-way Stop. Useful on minor approaches to an arterial street, or at isolated intersections where view is obstructed and accidents have occurred.

d. Four-way Stop. May be helpful from a safety standpoint at locations not warranting a signal, when volumes from opposing directions are about equal.

e. Traffic-control Signal. Can aid in providing orderly movement of traffic at busy intersections by alternately assigning right-of-way.

f. Police-officer Control. An expensive method of allocating right-of-way, but needed in some instances where pedestrian flow is heavy.

g. Coordinated-signal System. Use of signals at several intersections along major streets, for segregating traffic into platoons, for orderly, continuous movement along the street.

h. Turn and Lane-use Control.

Judicious use of the above methods is related to the street system. A system of through streets should be designated to channel major flows from one section of the city to another. Intersection controls should be related to this through-street system.

Signal installations should be planned with a view toward future coordinated control on the arterial street system. The control method used should provide a minimum of delay, consistent with safety and economy.

3. Control of Right-of-way. Most intersection controls are designed to control right-of-way conflicts (signals, stop, yield, and right-of-way rule). The choice of methods depends on engineering studies and application of warrants.

Traffic signals are used primarily at high-volume intersections where there are insufficient gaps in opposing streams to permit other vehicles or pedestrians to cross without undue waiting. Gaps of 6 sec are usually needed for the average driver. Warrants for use of traffic signals, in terms of volumes, speeds, accidents, etc., are given in the *Manual on Uniform Traffic Control Devices* (31) and are summarized in paragraph IVD of this section.

Stop signs on two approaches assign preferential right-of-way to the other approaches and thus expedite flow and cut accidents when used properly. Most stop signs are used along arterial streets. They may also be used at isolated intersections where high speed, restricted view, and/or serious-accident record indicate the need for control by a stop sign.

Warrants for use of two-way stop signs at an isolated intersection include the following:

1. The view across a corner is so restricted that the safe approach speed (Fig. 2-3), from the minor leg, is 8 mph or less.
2. Some right-angle collisions have occurred.
3. Less restrictive remedies (such as removing a hedge) are not feasible or effective.

Four-way stop control is somewhat controversial because it requires all vehicles to stop. However, such control may be warranted at locations satisfying *all* the following:

1. Volume of traffic on the intersecting roads should be about equal.
2. Traffic volume should be less than that requiring a signal (but probably at least half of the volume required for a fixed-time signal under the warrants of the *Manual on Uniform Traffic Control Devices*).
3. Intersection should be hazardous as determined by accident experience and/or near accidents.

Yield right-of-way control is relatively new, and warrants for its use are not given in the *Manual*. However, studies have indicated some possible warrants, as follows:

1. The safe approach speed on the street carrying the minor volume is greater than 8 mph and is less than the normal speed of traffic.
2. There is a demonstrated hazardous condition existing at the intersection such as an accident record with at least two right-angle collisions in a 12-month period.
3. The traffic volumes on the approaches to be signed are less than on the street carrying the major flow.
4. A traffic signal or a stop sign is not warranted.

4. Control of Turns. The Model Traffic Ordinance authorizes the traffic engineer to control turns. Turn prohibitions should be based on careful study and used only for the hours when they are needed. Indiscriminate prohibition of turns may increase travel distance, shift the problem to another intersection, and/or encourage hazardous violations. All locations where turns are prohibited should be posted with appropriate signs or signals.

Left-turn prohibitions are commonly used at heavily traveled signalized intersections of two-way streets where capacity is a problem. They also are used to prohibit a left turn into a one-way street which enters from the left.

Prohibiting left turns during the rush hours (or at all times when necessary) increases the capacity of a two-way street and also reduces delay and hazard. The procedure for determining whether to prohibit left turns is as follows:

1. Obtain needed data on street pattern, flow pattern, traffic volume, turning movements by hour, intersection capacity, signal timing, accident pattern, speeds and delays, possible alternate routes for left-turning vehicles, and possible effects on other intersections.
2. Prepare alternate plans for dealing with the problem, and evaluate each from standpoint of delay, hazard, cost, and benefits to motorists and pedestrians.

   a. Compare existing plan with:

     (1) Left-turn prohibition and rerouting of left-turn traffic.

     (2) Widening or redesigning intersections.

     (3) Retiming the signal.

   b. Do not prohibit turns if it can be avoided.

Right-turn prohibitions are normally used at the following types of locations:

1. To prohibit turns into a one-way street which enters from the right.

2. At those downtown signalized intersections during times when pedestrian volume is very heavy and vehicles waiting to turn right would block a lane most of the time.

Turns in one direction from more than one lane may be permitted when necessary. Such turns are warranted when the volume of turning traffic at the peak hour requires more than one lane and additional lanes are available. This commonly occurs at intersections of one-way streets, or two-way streets over 64 ft wide.

Methods for posting the multilane turn control are not yet standardized. Arrow and word markings on the pavement include "Left turn only" in one lane, and "Left or thru" in the other lane. Overhead signs with arrows and messages such as "Left only—this lane" and "Left or thru—this lane" are desirable and may be necessary in order to legalize the controls.

Right turns on red signals after a stop are permitted at all locations where not posted otherwise, in California and some other Western states. The success with this control is due primarily to the willingness of drivers to yield to pedestrians lawfully in the crosswalk.

With some forms of lane-use controls, a lane is reserved for the exclusive use of a certain movement (such as a left-turn lane). These controls are operable when using:

1. A left-turn median lane which has no outlet across the intersection.

2. A left-turn or right-turn lane where the turning movement is heavy, and it is desired to avoid allowing two lanes to cross to the opposite side of the intersection.

3. Lanes reserved for exclusive use of buses (as on Washington Boulevard in Chicago).

Figure 2-13 shows typical lane-use controls as recommended for use at widened intersections in Appleton, Wis. (32).

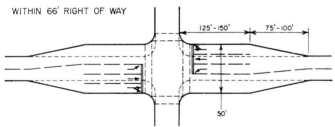

Fɪɢ. 2-13. Typical lane-use allocations recommended for widened intersections in Appleton, Wis. (32).

C. One-way Streets

One-way streets are used widely in congested areas of cities to facilitate traffic flow. They are used also in pairs to provide added capacity for traffic moving to and from central business areas. Advantages include the following:

1. Provides multilane efficiency, which reduces running time.

2. Reduces conflicts and delaying action of left turns.

3. Provides added intersection capacity, especially when a pair of one-way streets is established and traffic is diverted to a little-used street parallel to an existing major street. Also, one-way streets are often needed to provide sufficient capacity to receive the flow from a freeway ramp in a downtown area.

4. Facilitates timing of signal systems for progressive movement without stopping.
5. May utilize more efficiently streets with an odd number of lanes.

Disadvantages of one-way streets include the following:

1. Travel distance may be increased 10 to 15 per cent on the average.
2. One-way streets may produce difficulties in transit routing and movement of fire equipment.
3. Motorists may have difficulty finding destinations.
4. Some merchants may be affected adversely.

Engineering studies to determine the feasibility of establishing a pair of streets as one way may include the following:

1. Traffic volumes, by direction, by hour, with turning movements obtained at critical intersections.
2. Travel-time studies, parking studies, accident analyses, signal-timing studies, and capacity analyses.
3. Physical-condition survey to determine whether each street can carry the needed volume and whether end connections at the terminals of the one-way streets can be designed to provide adequate capacity.
4. Study of added travel distance, possible effects on roadside business, problems of transit routing, movement of emergency vehicles, and pedestrian crossings.

One-way streets are often established in pairs when studies reveal that congestion relief is needed and it can be provided best by one-way streets. Other uses of the one-way principle include:

1. Very narrow streets where two-way movement is not feasible.
2. Rotary movement around a public square.
3. In connection with multilane highways (roadways on divided highways and frontage roads and ramps and to provide the downtown distribution system for traffic from high-capacity ramps of a freeway).

Other adaptations of the one-way principle may be useful where pairs of one-way streets cannot be established yet there is need for added capacity. These include the following:

1. The reversible one-way street, in which the street is one-way in one direction for one part of the day and one-way in the opposing direction at another period of the day. Thirteenth Street in Washington, D.C., is one way inbound in the morning, two-way in midday, and one-way outbound in the evening. Illuminated neon arrows convey the message.
2. Off-center lane movement, in which the number of lanes allocated to each direction is changed for A.M., P.M., and midday periods to accommodate rush-hour volumes.

D. Off-center Lane Control

Off-center lane control on a two-way street is used primarily when there is an unbalanced distribution of traffic by direction (such as 70 to 80 per cent of the traffic moving in one direction on a two-way street). Usually, the number of lanes allotted to each direction is changed as the traffic volume varies. For example, four lanes of a six-lane street may be used for inbound traffic in the morning rush hours. In the midday, three lanes are used in each direction. In the evening rush hours, four lanes are used for outbound traffic.

Off-center lane-movement regulations may be conveyed to the motorist by lane direction signals, movable pedestals or cones placed in the roadway, or by overhead signs. In Detroit, two types of signs are erected over the two center lanes of a six-lane street, with messages (for inbound traffic), "Use both center lanes 7–9 A.M." and "Keep off both center lanes 4:30–6:30 P.M."

Off-center lane movement generally is used on streets which have at least five lanes and where no other method is feasible for providing the needed additional capacity. Capacity studies are needed also to determine whether the minor-flow movement can be accommodated by the reduced number of lanes provided in that direction.

E. Speed Control

1. Elements in Speed Regulation. Speed control is accomplished by a combination of elements, including:
1. Adequate speed laws (patterned after provisions in the Uniform Vehicle Code)
2. Speed zoning
3. Proper posting of speed limits
4. Enforcement of speed regulations
5. Public education and driver training
6. Highway design, with appropriate design speeds

2. Legal Authority. General speed limits are established by the state legislatures for highways in business, residence, and other districts. These speed limits are not always suited to physical and traffic conditions existing on many arterial streets and highways. The Uniform Vehicle Code contains suggested provisions to authorize state and local authorities to establish special zoned speed limits after engineering and traffic investigations. Establishment of such special speed limits is called speed zoning. Most states have such provisions in their state laws to delegate this speed-zoning authority to the state highway department and to local authorities.

3. Types of Speed Zones. Types of speed zones are as follows:
1. *Transition zones.* On highways approaching built-up areas, where transition speed zones aid in changing from rural to urban speed limits (example: 60 to 50 to 40 mph).
2. *Urban arterial streets.* All arterial streets in urban areas should be studied and zoned with speed limits suited to the conditions.
3. *Rural highways.* Dangerous sections where lower limits are needed and also controlled access highways where the state-wide speed limit may be too low.
4. *Curves.* Where advisory speeds can be posted.
5. *Temporary zoning.* For construction or repair zones, school zones, and for unusual conditions (fog, ice, etc.). Such zones may be posted by temporary signs or by signs which are illuminated or otherwise displayed only when the regulation is effective.
6. *Minimum speed limits.* Sometimes posted on heavily traveled roadways to aid in maintaining high capacity.

4. Procedure in Speed Zoning. The engineering and traffic investigations for the purpose of determining the appropriate numerical speed limit include the following:
1. Inventory of physical conditions (alignment, grade, sight distance, roadside development, driveways, etc.).
2. Critical-speed calculations (curves, hillcrests, other sight-distance limitations).
   a. Curves:

$$V^2 = 15R(e + f)$$

where V = critical speed, mph
   R = radius of curve, ft
   e = superelevation, ft per ft
   f = side-friction factor (.10 to .25)
   b. Nonpassing sight distance:

$$D = \frac{V^2}{30f} + RV$$

where R = perception + reaction time
   V = speed, mph
   f = coefficient of friction (.40 to .50)
   D = nonpassing sight distance, ft
3. Spot-speed studies at two or more locations to get the average and the 85th-percentile speeds. The zoned speed limit usually lies between the average speed and the 85th percentile.

4. Study of accidents (types, rates, location).
5. Trial runs over the section.   Use a ball-bank indicator when making trial runs on curves.   A reading of 10 per cent corresponds to acceptable speed for rural conditions.
6. Consultation with enforcement officers.
7. Consideration of all the above factors in selecting the appropriate numerical limits.
8. Establishment of the speed zone by resolution or order of body having authority.
9. Posting of the speed zone with appropriate signs ("Speed zone ahead," "Speed limit 35," "End speed zone") (31).

F. Curb-parking Control

Curb-parking controls are used for two general purposes: (1) to provide additional street space for movement of vehicles, or (2) to give priority to short-time users by time-limit parking.

1. Prohibiting Parking.   In allocating use of curb space, movement should have first priority.   Whenever the curb space is needed for providing additional street capacity, consideration should be given to prohibiting parking, standing, and/or stopping of vehicles.   Prohibition of parking can almost double the practical capacity of a 50-ft two-way street in a downtown area.

Studies to determine the feasibility of prohibiting parking along a street include volume-capacity studies, travel-time studies, and studies of parking occupancy and turnover.   If there is a large demand for short-time parking which cannot be taken care of at other nearby locations, consideration should be given to providing alternate routes for traffic movement instead of prohibiting parking.   In many cases it will be necessary to compromise between conflicting demands for movement and for parking by prohibiting parking and standing during the rush hours only (such as 7 to 9 A.M. and 4 to 6 P.M.).

Parking should also be prohibited where necessary to provide space for bus stops, taxi stands, loading zones, and at locations near crosswalks and fire plugs.

When the curb space is urgently needed for movement, as on a bridge, the sign may read "No stopping."   When occasional stopping to pick up or discharge a passenger can be allowed, the sign can read "No standing."   If some loading and unloading can be tolerated, the sign may read "No parking."   When the regulation is effective for certain hours only, the hours should be specified.

2. Angle Parking.   Parking at the curb at a 45° angle reduces the street width for movement by 10 ft, as compared with parallel parking.   There are few business streets where angle parking can be tolerated because of the reduction in capacity.   Angle parking also is more hazardous than parallel parking.   Careful study should be given to volume-capacity relations before reaching a decision to permit angle parking on a section of a street.

3. Loading Zones.   Designation of commercial and passenger loading zones is a traffic-engineering function.   Such zones should have priority over use of the curb for time-limit parking, but the curb space should not be used as a loading zone during the hours that the curb lane is needed for traffic movement.

Loading zones for commercial vehicles are designated when there are no alley or off-street entrances available.   In Wichita, Kans., zones are established only when there are at least 15 deliveries per day along the section of street to be served by the loading zone.   Off-street loading facilities should be required for new business establishments as part of zoning ordinances.

Bus stops may be established on the near side of intersections except where far-side locations are advisable because of a turn in the bus route, or to facilitate transfers to another bus line, or where there are heavy right-turning movements.   Near-side stops on streets with parking usually add to the capacity of signalized intersections.   On streets where parking is prohibited, far-side stops would have less adverse effect on street capacity than near-side stops.   Lengths of the bus stop should be adequate to

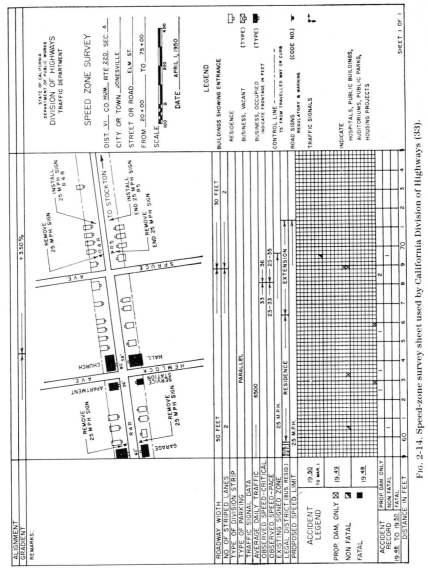

Fig. 2-14. Speed-zone survey sheet used by California Division of Highways (33).

permit the bus to pull to the curb for loading. Bus-stop lengths of 80 ft (near-side) or 60 ft (far-side) will accommodate a single bus of 40 to 50 passengers.

4. Time-limit Parking. Time-limit parking is used in areas where the demand for short-time parking exceeds the supply. The supply of curb parking space should be considered in relation to the supply of off-street parking space.

Studies of curb-space occupancy and duration should be made before determining whether to change a time limit, extend the time-limit zone, or install parking meters. These studies can reveal whether there is a real shortage of space for short-time parkers or whether the problem is one of existing short-time space being used illegally by long-time parkers. Occupancy studies are made by cruising the area and recording periodically the number of "open" spaces per block or portion of block. Turnover studies are made by recording the last four digits of the license numbers of each vehicle in each stall every 15 min and later determining the duration of stay of each vehicle (34).

Results of turnover studies can be summarized in tables similar to Table 2-9, which shows the length of time parkers used various types of parking space in Wheaton, Ill., in 1957. Metered curb spaces were controlled with 12-min, 1-hr, and 2-hr meters. Of the parkers in metered curb spaces, 22 per cent stayed for more than 2 hours, with 6 per cent staying over 8 hr.

Table 2-9. Length of Time Parkers Used Various Types of Space

Length of time parked	Percentage of parkers at these facilities remaining for the indicated periods		
	Metered curb	Free curb	Off-street
0–15 min	17	1	3
15–30 min	9	2	3
30–60 min	33	3	7
1–2 hr	19	3	5
2–4 hr	9	7	3
4–8 hr	7	11	23
Over 8 hr	6	74	56

SOURCE: Parking Survey in Wheaton, Ill., by George Barton and Associates, 1957.

Parking meters aid in enforcement of parking time limits by automatically indicating a violation as soon as the time limit has expired. Meters also provide revenue which can be very helpful in financing off-street parking. It may still be necessary to enforce time limits by marking of tires in order to discourage "meter feeding."

IV. TRAFFIC-CONTROL DEVICES

A. Standards

Traffic-control devices are placed by official agencies to serve the following functions:
1. To display regulations (the regulatory type, such as the stop sign and traffic signals)
2. To warn of hazard (such as the curve sign)
3. To guide or inform traffic (route markers, lane lines)
4. To control or segregate traffic streams (islands)

Traffic-control devices should be designed and placed in such a manner as to compel attention, convey a simple, clear meaning at a glance, allow ample time for the required response, and command respect. In order to accomplish these objectives, it is essential that the different official agencies follow uniform standards in design and application of traffic-control devices. Uniformity is needed in design of the device (size, shape, color), in its position (height, lateral and longitudinal placement), and in its application (conditions under which it is used).

The *Manual of Uniform Traffic Control Devices* contains standards for design, placement, and application of traffic-control devices for the United States. This manual was first issued in 1928.

UPDATE

The 1-sec perception time lag noted on page 2-3 has been refined considerably in recent years. Perception time varies with trip length, driver age and experience, weather, and type of stimulus. Design of an intersection in an area frequently bedeviled by fog, off an Interstate highway, in a part of the country with a substantial number of over-65 drivers, for instance, might be based on a 2-sec perception time lag.

One advance is the new "triple" taillight that has been mandatory on new autos starting with the 1987 model year. It improves perception time by a quarter-second, on the average.

REFERENCES

1. Brody, Leon: "The Role of Vision in Motor Vehicle Operation," *International Record of Medicine and General Practice Clinics*, June, 1954.
2. Forbes, T. W.: "Driver Characteristics and Highway Operations," *Proceedings, Institute of Traffic Engineers*, 1953, pp. 68–73.
3. Forbes, T. W., K. Moskowitz, and G. Morgan: "A Comparison of Lower Case and Capital Letters for Highway Signs," *Proceedings, Highway Research Board*, 1950.
4. Forbes, T. W., and M. S. Katz: *Summary of Engineering Research Data and Principles Related to Highway Design*, American Institute for Research, Pittsburgh, Apr. 30, 1957.
5. American Association of State Highway Officials: *A Policy on Geometric Design for Rural Highways*, Washington, D.C., 1954.
6. *Accident Facts*, National Safety Council, 1957.
7. Eno Foundation: *The Motor Vehicle Driver: His Nature and Improvement*, 1949.
8. Eno Foundation: *Personal Characteristics of Traffic Accident Repeaters*, 1948.
9. Forbes, T. W.: "The Normal Automobile Driver as a Traffic Problem," *Journal of General Psychology*, vol. 20, pp. 471–474, 1939.
10. Matson, T. M., and T. W. Forbes: "A Study of Overtaking and Passing Requirements as Determined from a Moving Vehicle," *Proceedings, Highway Research Board*, 1938.
11. Stonex, K. A.: "Lessons Learned by the Proving Ground Engineer in Highway Design and Traffic Control," *Proceedings, Institute of Traffic Engineers*, 1955.
12. Moyer, R. A.: "Roughness and Skid Resistance Measurements of Pavements in California," *Highway Research Board Bulletin* 37, 1951. Also "Skid Resistance Measurements," *ITTE Reprint* 10, University of California, 1953.
13. Normann, O. K.: "Braking Distances of Vehicles from High Speed," *Public Roads*, June, 1953.
14. Petring, F. W.: "Stopping Ability of Motor Vehicles Selected from the General Traffic," *Public Roads*, June, 1957.
15. Dimmick, T. B.: "Traffic and Travel Trends, 1956," *Public Roads*, December, 1957.
16. U.S. Bureau of Public Roads: *Traffic Counting, Classification and Weighing in Rural Areas*, 1951.
17. National Committee on Urban Transportation: *Procedure Manual—Measuring Traffic Volume*, Public Administration Service, 1958.
18. U.S. Bureau of Public Roads: *Traffic Speed Trends*, March, 1959, and earlier editions.
19. Berry, D. S., and D. M. Belmont: "Distribution of Vehicle Speeds and Travel Times," *Proceedings, Symposium on Mathematical Statistics and Probability*, University of California, 1951.
20. American Association of State Highway Officials: *Road User Benefit Analyses For Highway Improvements*, 1952.
21. Taragin, A.: "Driver Performance on Horizontal Curves," *Proceedings, Highway Research Board*, 1954.
22. May, A. D., and E. T. Kaneko: "Comparative Study of the Highway Research Board Statistical Instrument and a Modified Speed and Delay Recorder," *Proceedings, Highway Research Board*, 1958.
23. U.S. Bureau of Public Roads: *Highway Capacity Manual*, U.S. Government Printing Office, 1950.
24. Wynn, F. H., S. M. Gourlay, and R. I. Strickland: *Studies of Weaving and Merging Traffic*, Yale Bureau of Highway Traffic, 1948.
25. Moskowitz, K.: "Waiting for a Gap in a Traffic Stream," *Proceedings, Highway Research Board*, 1954.
26. Leisch, J. E.: "Design Capacity Charts for Signalized Street and Highway Intersections," *Public Roads*, February, 1951.
27. U.S. Bureau of Public Roads: *Motor Vehicle Traffic Conditions in the U.S., Part V.* House Document 462, 1937.
28. Prisk, C. W.: "How Access Control Affects Accident Experience," *Traffic Engineering*, March, 1958.
29. *Uniform Vehicle Code*, U.S. Government Printing Office, 1956.
30. *Model Traffic Ordinance*, U.S. Government Printing Office, 1956.

31. National Joint Committee on Uniform Traffic Control Devices: *Manual on Uniform Traffic Control Devices*, U.S. Government Printing Office, 1948, rev. 1954.
32. *A Primary Street Plan for Appleton, Wisconsin*, George Barton and Associates, Evanston, Ill., 1955.
33. California Division of Highways: *Planning Manual of Instructions, Part 8, Traffic*, loose leaf, rev. 1958.
34. *Manual of Traffic Engineering Studies*, Association of Casualty and Surety Companies, New York, 1953.
35. American Association of State Highway Officials: *Signing and Pavement Marking for the Interstate Highway System*, 1958.
36. Webster, F. V.: *Delays at Traffic Signals: Fixed-time Signals*, RN/2374, Great Britain, Road Research Laboratory, February, 1955.
37. Berry, D. S.: "Field Measurement of Delay at Signalized Intersections," *Proceedings, Highway Research Board*, 1956.
38. Illuminating Engineering Society: *American Standard Practice for Street and Highway Lighting*, approved Feb. 27, 1953, American Standards Association.
39. *Highway Transportation for the Indianapolis Metropolitan Area*, George Barton and Associates, Evanston, Ill., August, 1957.

Section 3

DESIGN OF FOUNDATIONS, EMBANKMENTS, AND CUT SLOPES

ROBERT F. BAKER, *Professor of Civil Engineering, The Ohio State University, Columbus, Ohio* (Landslides and Design of Slopes).

DR. HAMILTON GRAY, *Professor of Civil Engineering and Chairman, Department of Civil Engineering, The Ohio State University, Columbus, Ohio* (Embankment Foundations).

I. INTRODUCTION

Proper consideration of the engineering properties of soils is required to provide the most economical design of highway facilities. The development and acceptance of the theories of soil mechanics have permitted improved analyses of the influence of the soil on the structure. The empirical relationships and arbitrary building codes are quite often overly conservative, sometimes inadequate, and usually not economic, and gross assumptions that were required thirty years ago are no longer necessary. The most troublesome shortcoming of applying solely experience to this type of analysis is the probability of *overdesign*. In fact, unless failures have occurred, overdesign must have been inherent in the solutions. In the highway field, *underdesign* is not infrequent, and the result is excessive maintenance. In addition, soil failures can lead to physical dangers. Newspapers repeatedly refer to accidents caused by landslides. Rockfalls are a constant threat in some areas (2).* Settlement of highway embankments and bridge approaches produces surface irregularities which undoubtedly have resulted in highway accidents.

The services of a competent soils engineer may not always be available to a practicing engineer. The following section is intended as a guide, and with the relatively simple principles which are provided, the mechanics of soils can be utilized. However, just as it is impossible to provide a simple analysis for a major structure, so it is unreasonable to expect a simple technique for solving complex soil problems. For major designs, every effort should be made to obtain the services of an engineer competent in soils. Certainly such an engineer will provide the best balance between overdesign and underdesign.

II. FUNDAMENTAL CONSIDERATIONS

Engineering soil studies can be divided into two major groups—one involving *stability* and the other concerned with *settlement* (Fig. 3-1). Stability problems deal

---

* Numbers in parentheses refer to corresponding items in the bibliography at the end of this section.

3-1

with the ability of soil to resist shearing forces as indicated in Fig. 3-1a, while settlement analyses consider the amount of densification the soil foundations will undergo as shown in Fig. 3-1b. While Fig. 3-1 is for an embankment, the same principles apply to any engineering soils problem.

Two other types of slope problems that will be encountered by the highway engineer can be classed as *gravitational* (tension) and *erosion*. The latter is a hydraulic and sedimentation phenomenon and is discussed in Secs. 5 and 13. Gravitational failures or downward displacements due to the force of gravity refer to those types of earth and rock movement which develop as a result of tension cracks. Examples would be the failures that are associated with fissured clays or exfoliated bedrock. Gravitational movement in soils is not a common type of failure, but is frequently encountered in rock slopes. Chemical and mechanical weathering, as well as the force of gravity, can contribute to the development of tension forces.

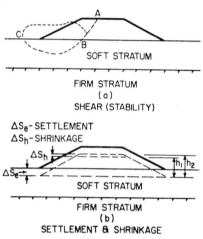

FIRM STRATUM
(a)
SHEAR (STABILITY)

ΔS_e - SETTLEMENT
ΔS_h - SHRINKAGE

FIRM STRATUM
(b)
SETTLEMENT & SHRINKAGE

Fig. 3-1. Stability and settlement.

Shear failures and excessive settlement are the two principal sources of difficulty related to the engineering behavior of soils. Such a statement should serve to simplify the understanding of the soil-mechanics approach. However, many factors tend to confuse the problems. With shear failures, it is difficult to estimate for the arc *ABC* of Fig. 3-1a (1) the stresses that exist, (2) the shearing resistance of the soil, and (3) the location of the surface upon which shear will occur. The determination of these factors causes most of the difficulty in estimating soil-shear behavior.

Natural soil deposits upon which a load is placed will tend to become more compact (densified). In addition, within the soil mass, which is an integral part of the structure, some added densification or shrinkage can be expected. The latter problem, which is indicated by the differences in heights h and h_2 in Fig. 3-1b, is much the simpler of the two to control. The condition of the compacted soil can be specified, and excessive densification eliminated. In fact, under good construction practice, little or no settlement within the compacted mass is to be anticipated. Within the natural soil beneath a structure, accurate settlement estimates are difficult to make. The principal sources of uncertainty arise from difficulties in determining (1) the extent and uniformity of the in-place soil designated as "soft-strata" in Figure 3-1b, (2) stress conditions that exist, and (3) the reaction of saturated soils to the stresses applied.

A. Field Investigations

The conducting of field studies is essential to most soil problems. The specific requirements for a given type of problem are covered briefly in subsequent pages, while the following general comments will also be applicable:

1. Ground-surface measurements are constantly required for soil studies and consist primarily of instrument surveys, observations of ground-water conditions, evaluation of vegetative cover, and qualitative-stability considerations. The results will be of value in determining (a) the stress conditions that will exist (i.e., depths to various layers), (b) inferences as to shearing forces and resistance (i.e., past and present loading), and (c) quantitative data on design quantities (i.e., cubic yards of excavation).

2. Subsurface studies represent the essential first step of a soils investigation. The results are necessary for any evaluation of stability or settlement. The depth to which the studies should be carried is a function of the magnitude of the applied loads, uniformity of the soils in the area, and the severity of the problem as reflected by cost or safety aspects. For settlement considerations, the depth will be greater than generally required for stability problems. Soft-clay materials are quite susceptible to settlement even though at considerable depth beneath the structure. In general, subsurface studies are needed (a) for a description of the character and dimensions of the layers, (b) to identify and to locate weak or compressible strata, (c) to obtain samples for tests or to test *in situ* the engineering properties, and (d) to develop moisture and seepage data.

3. Field testing is becoming increasingly more important because of the difficulties in duplicating field conditions in the laboratory. The added advantages of simplicity and of rapidity in obtaining test results are also attractive. The most frequently used field tests are the standard penetration resistance, vane-shear apparatus, soil-loading tests, and field permeability. These tests are conducted in order to obtain the specific engineering properties that are required for the solution of the problem at hand.

The extent of the field work will be governed by such economic factors as the size of the project and the hazards that would be created by a failure. Another controlling factor is the variability of the soil profile; exacting computations are of no value when nonuniformity precludes the use of suitable soil-mechanics theories. The use of preliminary, exploratory studies will be valuable in determining the degree of detail that will be justified.

It is important that the field studies be designed to obtain an area concept of the conditions rather than a single line or profile. The latter will be of considerable value in the analysis, but are of more significance if an areal variation is considered, particularly if the soil is not uniform.

The role of geology and other sciences also should be considered. Guidance in the planning and the conducting of field studies can be infinitely better if sound geologic principles are applied. The type and extent of subsurface strata, the implications of geologic history, and the inferences of relative strength or stability of the material are typical of the values to be derived.

B. Laboratory Studies

For a specific problem related to stability or settlement, laboratory testing is needed to evaluate the soil properties that are related to stresses, shearing resistance, and densification. Classification tests, in-place unit weights, and in-place shearing resistance of the soil are of use in determining the uniformity of the soil strata. Moisture-density tests will be of value in establishing the ultimate strength properties of the soil as a construction material and for controlling the conditions under which the material is to be compacted. Other laboratory tests determine (1) the shearing resistance, (2) the compressibility, and (3) the permeability.

The importance of obtaining undisturbed samples for the determination of strength, permeability, and consolidation properties is to be emphasized. It is generally recognized that it is impractical to obtain absolutely "undisturbed" samples. However, every care must be exercised in order to ensure as little disturbance as possible.

C. Slope- and Foundation-stability Analyses

The study of the resistance of a soil mass to shear requires that (1) the unit shearing resistance of the soil be known, (2) the surface upon which shear occurs be established, and (3) the stresses in the soil mass be computed. The shearing resistance of the soil can be estimated from field tests (vane-shear or standard-penetration tests) or by laboratory techniques (direct shear, triaxial shear, unconfined compression, or vane shear). The tests are conducted on undisturbed samples of the soil, from the general area in which a knowledge of shear resistance is required.

The expression for the unit shearing resistance of soils is as follows:

$$s = p \tan \phi + c \tag{3-1}$$

where s = unit shearing resistance, psf
$\quad p$ = normal stress, psf
$\quad \phi$ = angle of internal friction of soil
$\quad c$ = cohesion of soil, psf

The frictional component of the shear is comparable with that of any friction condition; i.e., with greater normal pressure, the frictional resistance increases. Cohesion is resistance imparted by the fine particles (clay size) and is reasonably constant for a given soil at a specified unit weight and moisture content. Since the coarse particles exhibit no cohesion, the following equation represents the shearing resistance for granular soils:

$$s = p \tan \phi \qquad (3\text{-}2)$$

Also, since cohesive materials have no appreciable friction, the unit shearing resistance of clays is expressed by

$$s = c \qquad (3\text{-}3)$$

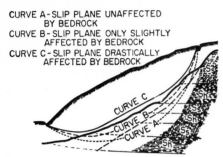

CURVE A - SLIP PLANE UNAFFECTED BY BEDROCK
CURVE B - SLIP PLANE ONLY SLIGHTLY AFFECTED BY BEDROCK
CURVE C - SLIP PLANE DRASTICALLY AFFECTED BY BEDROCK

FIG. 3-2. Effect of firm stratum on the location of the slip surface. *(Ref. 1, fig. 8, courtesy of Highway Research Board.)*

The last equation is based upon the assumption of $\phi = 0°$ and is frequently used where homogeneous clays are encountered.

For homogeneous soils, shear failures tend to develop along curved surfaces, generally assumed to be circular in shape on a two-dimensional basis as shown by Curve C in Fig. 3-2. Therefore, when a stability problem is being studied, one can assume that shear will occur along a circular path, unless there are conditions present that indicate the probability of a different surface. While the latter statement appears ambiguous, the factors tending to change the shape of the shear surface are quite logical. Soils, like all materials, tend to fail along the path of least resistance. In homogeneous materials, the path can be approximated by a circle. On the other hand, an extremely weak stratum will provide an easier path, as shown in Fig. 3-3, and divert the surface from the circle it would normally take. The same is true if a relatively firm layer is encountered. For example, firm bedrock within the area of influence of the "normal" circle will offer more resistance to shear than another path that remains within the weaker (soil) material, as indicated by curve B in Fig. 3-2. Curve A represents an estimate of a composite surface made up of circle arcs and straight lines.

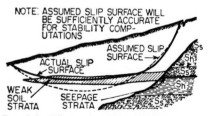

NOTE: ASSUMED SLIP SURFACE WILL BE SUFFICIENTLY ACCURATE FOR STABILITY COMPUTATIONS

FIG. 3-3. Effect of weak stratum on the location of the slip surface. *(Ref. 1, fig. 9, courtesy of Highway Research Board.)*

Locating accurately the slip surface of a landslide is quite difficult, but some indicators are available in the field. Rotation at the toe for slides will normally provide some idea of its location in the bottom 10 to 50 ft (horizontally) of the slide. At the top of the slide, the location is also evident. Within the middle of the sliding mass, subsurface exploration will normally give some indication of a change in material. If not, it is well to remember that a circular failure will be a reasonable estimate, and if this circle is drawn tangent to the slip plane at the top and toe, a reasonable approximation is available.

The stress conditions for the complex slip surfaces, such as curve C in Fig. 3-3, are such that there is no possible combination for a "block" or single-unit movement. At some point in the system a tension failure or a zone of compression must develop in order to account for the differential downslope movement. Several failures of this type are illustrated in Fig. 3-4. Rigorous and rational analyses of these types of landslides are not warranted on the basis of current knowledge of applying the elastic theory to soils. In most cases, a reasonable estimate can be achieved by approximating the shear surface with a circle and by assuming that the block moves as a unit. It has been shown that such assumptions do not materially affect the size of the structure required, at least for the relatively small to medium landslides involving less than 1 million cu yd.

Stability analyses of slopes involve estimating the relative size of the two forces acting—the driving force of the mass and the shearing resistance of the natural material. Referring to Fig. 3-4, just prior to failure, a balanced (equilibrium) condition exists, expressed as follows:

Driving force = shearing resistance

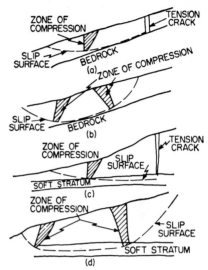

Fig. 3-4. Complex slip surfaces.

The simplest form of slope shear, represented in Fig. 3-5, involves straight slip surfaces. Such types occur most frequently in tilted bedrock, usually at the interface between two strata. Straight shear surfaces can occur on soil slopes but require (1) close proximity between ground surface and an underlying, resistant layer such as bedrock, and (2) the removal of the transported material at the toe of the movement (Fig. 3-5). If the resistant layer is not close to the ground surface, a circular shear surface will probably develop; and if the material at the toe is not removed, rotation and a circular slip surface will develop at the lower extremities of the slide.

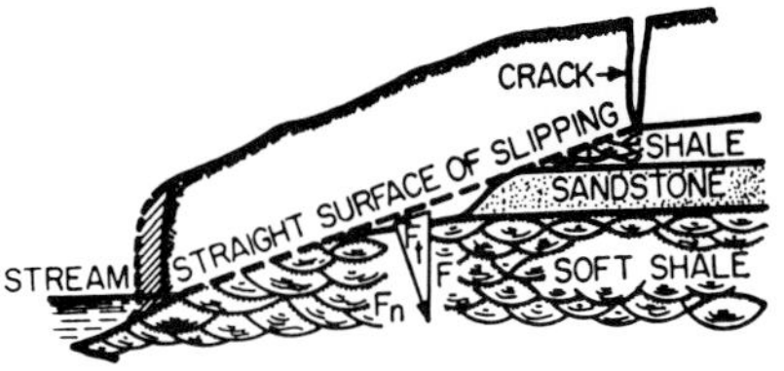

The stresses existing in the straight-slip-surface condition of Fig. 3-5 involve shearing forces that consist of the component of the weight parallel to the slip force, as follows:

$$T = W \sin \beta \qquad (3\text{-}4)$$

and the normal force that contributes to the shearing resistance is

$$N = W \cos \beta \qquad (3\text{-}5)$$

Fig. 3-5. Mechanics of landslide with straight slip surface. *(Ref. 4, fig. 3, p. 674, courtesy of American Society of Civil Engineers.)*

where W = total weight of sliding mass
β = angle between slip surface and horizontal
The ratio of shearing resistance to shearing force is the degree of stability and is commonly referred to as the safety factor G_s:

$$G_s = \frac{W \cos \beta \tan \phi + cl}{W \sin \beta} \qquad (3\text{-}6)$$

For noncohesive soils and sheared bedrock, $c = 0$; therefore:

$$G_s = \cot \beta \tan \phi \qquad (3\text{-}7)$$

Equation (3-7) states in effect that for noncohesive material, the stability is independent of the weight of the sliding mass.   To state it in another manner, there is a limit to the slope for granular materials or for sheared bedrock.   The angle ϕ is normally in the range of 26 to 35°.   Assuming an intermediate ϕ value of 30° and a safety factor of 1.25:

$$\cot \beta = \frac{1.25}{\tan 30°}$$

$$\beta = 24°48' \tag{3-7a}$$

Therefore, for general conditions of $c = 0$, shear surfaces sloped in excess of 25° (approximately 2:1, horizontal:vertical) with the horizontal are not likely to provide sufficient shearing resistance within themselves.   More liberal values for ϕ would lead to the conventional "angle of repose" of 11/2:1(33°41').

The preceding estimates neglect the very important factor of seepage forces.   From Fig. 3-5, the effect of these forces is seen to be related to the hydrostatic pressures at a given point.   The basic-shearing-resistance equation [Eq. (3-1)] is as follows, for the seepage condition:

$$s = (p - \mu_i) \tan \phi + c \tag{3-8}$$

where μ_i = excess hydrostatic pressure, psf
The excess hydrostatic pressure can be estimated by the development of a flow net. However, a conservative estimate can be obtained by establishing the ground-water level (water table) and assuming that it remains at a constant level with no flow. Thus, assuming that the unit weight of water is equal to 62.4 lb per cu ft,

$$\mu_i = 62.4 h_h \, \Delta 1 \tag{3-9}$$

where h_h = average vertical distance from shear surface to ground-water line, ft
Thus, Eq. (3-5) becomes

$$N = (W - \mu_i) \cos \beta \tag{3-5a}$$

Since μ_i can have any value, depending upon the magnitude of the seepage forces, Eq. (3-7a) is not valid where hydrostatic pressures are suspected.   For such cases, Eq. (3-6) becomes

$$G_s = \frac{(W - \mu) \cos \beta \tan \phi + cl}{W \sin \beta} \tag{3-6a}$$

where μ = total hydrostatic pressure, lb

Therefore
$$G_s = \frac{W - \mu}{W} \cot \beta \tan \phi \tag{3-7a}$$

For cohesive material, and $\phi = 0$, Eq. (3-6) becomes

$$G_s = \frac{cl}{W \sin \beta} \tag{3-10}$$

Assuming a relatively uniform thickness of soil, $\sin \beta$ is equal to $h/1$, where h is the average vertical height (Fig. 3-5) and l is the length of the slip surface:

$$G_s = \frac{cl^2}{hw} \tag{3-10a}$$

and
$$G_s = \frac{cl^2}{htb\gamma_s} \tag{3-10b}$$

where t = average thickness of soil layer, ft
  b = horizontal length of slide, ft
  γ_s = unit weight of soil mass, lb per cu ft
Equation (3-6), the general expression, can be simplified in a similar manner to

$$G_s = \frac{b}{h}\tan\phi + \frac{cl^2}{hW} \qquad (3\text{-}6a)$$

and
$$G_s = \frac{bW\tan\phi + cl^2}{hW}$$
$$= \frac{p_s t b^2 \tan\phi + cl^2}{hW} \qquad (3\text{-}6)$$

Where a uniform thickness of soil cannot be assumed, the preceding simplifications are not possible and the basic equation (3-6) must be used.

For the analysis of circular slip surfaces, the "method of slices" can be used, as illustrated in Fig. 3-6. In this case, the shearing forces can be assumed to result solely from a component of the weight of the mass above the shear surface (area ABC). The total weight W is estimated by $hb\gamma_s$, where γ_s is the unit weight of the soil and a unit slice width is assumed. The component that acts parallel to the surface of shear is designated T, or the tangential forces. The normal force N contributes to the shearing resistance of the soil. Therefore, for each increment in the soil mass, assuming a unit of width perpendicular to the cross section,

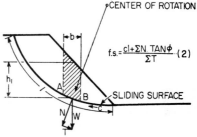

FIG. 3-6. Mechanics of landslides with circular slip surface. (*Ref. 8, fig. 116, p. 192, courtesy of Highway Research Board.*)

$$\text{Shearing resistance} = \Delta N_i \tan\phi + c\,\Delta l \qquad (3\text{-}11)$$
$$\text{Shearing force} \quad= \Delta T_i$$

where ΔT_i = increment of tangential force, lb
  ΔN_i = increment of normal force, lb
  Δl = increment of shear-surface length, ft
For the entire length of the shear surface, the sum of the individual increments is obtained as indicated by Fig. 3-7 and the following:

$$\Sigma_a^n T = \Sigma_a^n N \tan\phi + cl \qquad (3\text{-}12)$$

The basic stability equation becomes

$$G_s = \frac{\Sigma_a^n N \tan\phi + cl}{\Sigma_a^n T} \qquad (3\text{-}13)$$

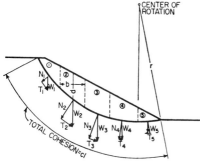

FIG. 3-7. Solution by "method of slices." (*Ref. 8, fig. 117, p. 192, courtesy of Highway Research Board.*)

A method for computing the average unit shearing resistance required of the soil consists of taking moments about the center of rotation O, as shown in Fig. 3-8.

$$s = \frac{W_1 d_1 - W_2 d_2}{rl} \qquad (3\text{-}14)$$

where W_1 = weight of driving mass W_1, lb
  W_2 = weight of resisting mass W_2, lb

d_1 = distance from center of gravity of mass W_1 to center of rotation, ft
d_2 = distance from center of gravity of mass W_2 to center of rotation, ft
r = radius of circle of rotation, ft
l = length of shear surface, ft

The total shearing force T is equal to

$$T = sl = \frac{W_1 d_1 - W_2 d_2}{r} \tag{3-15}$$

The value of the shearing resistance sl thus obtained carries the combined effect of friction and cohesion and is of little value where estimates of the frictional part of the shearing resistance is required. The problem of the frictional component is of particular concern where seepage or groundwater movement is involved. In such cases, the equation for the degree of stability involves the excessive hydrostatic pressure μ and

$$G_s = \frac{\Sigma(N - \mu)\tan\phi + cl}{\Sigma T} \tag{3-16}$$

Values for μ can be determined by Eq. (3-9), which involves the assumption that the hydrostatic head h is equal to the vertical distance between the shear surface and the water table (Fig. 3-8).

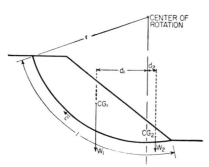

FIG. 3-8. Shearing resistance by moment equation.

CG_1, CG_2 = centers of gravity, respectively, of driving and resisting masses
W_1 = weight of driving mass
W_2 = weight of resisting mass
d_1, d_2 = lever arms, respectively, of W_1, W_2
c = cohesion per unit of length and width of slice
l = length of slip surface

(Ref. 8, fig. 121, p. 201, courtesy of Highway Research Board.)

III. EMBANKMENT FOUNDATIONS

Some geologic deposits are so weak and compressible that they will cause undesirable behavior of any new embankment placed on them, regardless of how small it may be. On the other hand, embankments can be built to such heights that they will cause undesirable behavior in an underground that consists of relatively compact materials. Consequently, it is fitting to speak of "soft-ground behavior" rather than of "soft ground" since unsatisfactory behavior may develop on ground that is not normally considered "soft." By way of definition, Fig. 3-1 illustrates the most common types of undesirable behavior produced by construction on ground that is incompetent to properly support a superimposed embankment.

A. Settlement

Settlements result from a subsoil which is *compressible* under the weight of the embankment. Pressures caused by the weight of an embankment may produce volume changes in the subsoil, the original ground surface then subsides, and, of necessity, the superimposed embankment moves downward after it. If the settlement occurs rapidly, it may not be noticed, and the only evidence may be the increased yardage required to bring the top of the fill to desired grade. If the settlement occurs slowly, it may become evident only after completion of the fill. This latter type of behavior, while often not particularly serious, may result in unpleasant or even dangerous differences in grade when a roadway passes from cut to fill areas since these changes in grade can be sufficiently abrupt to constitute a hazard for high-speed traffic.

At the top of the embankment, the effect of "shrinkage" may be indistinguishable from that of subsidence of the subsoil. Shrinkage occurs wholly within the fill and can be controlled by properly compacting the fill.

B. Shear Strength

If a supporting soil stratum is deficient in *shear strength*, a portion of the embankment, or even a section of the entire embankment, may displace the underlying material in a manner illustrated by the lower portion of Fig. 3-1. Such action usually occurs rapidly, and the results are spectacular, much to the chagrin of engineers associated with the project. This type of soft-ground behavior is more costly to remedy but is of less frequent occurrence than is settlement.

The two types of undesirable behavior are dependent upon entirely different soil properties, and means for ameliorating the tendency toward such behavior differ with the type of behavior anticipated. Soft-ground behavior is observed over soils of widely varying consistencies ranging upward from swamp or marsh materials with extremely high compressibility and very low strength, which will settle under fills as low as 1 or 2 ft in height and which may rupture under fills less than 4 or 5 ft in height. Some of these marsh deposits are themselves still subsiding under their own weight so that the surface tends to move downward year after year without requiring the influence of a superimposed embankment. On the other hand, if we consider an embankment 40 or 50 ft high, the shearing strength of supporting soil required to prevent rupture may amount to something of the order of 1 ton per sq ft yet such a material, if of cohesive nature, is normally deemed a highly competent stratum. The great majority of embankment problems arise in connection with cohesive or organic subsoils. It is fortunate that for such soils it is possible to measure the mechanical properties, either *in situ* or by testing suitable samples.

C. Subsoil Explorations

A rational approach to the behavior of the subsoil under a surface load requires, first, the acquisition of certain data, which can be summarized as follows:
1. Stratigraphy
   a. Thickness, continuity, extent of distinct subsurface layers
   b. Descriptions, based on competent examination of adequate samples, preferably substantiated by the results of routine identification or classification tests
   c. Consistency, preferably based on boring data such as penetration resistance or field vane-shear resistance
2. Mechanical properties of each stratum
   a. Indicative: water content, porosity, gradation, plasticity, organic content
   b. Specific: compressibility, permeability, shear
3. Stresses, each stratum, existing and predicted
   a. Intergranular, compressive
   b. Hydraulic
   c. Shear

1. Wash Borings. Since questionable subsurface materials may extend to great depths, it is necessary ordinarily to resort to wash borings in order to secure the necessary stratigraphic information. These borings, if carefully made, can serve to reveal the general nature and extent of subsurface strata. Sometimes the underground conditions are so favorable that no further investigations are required, it being evident to an experienced person that the ground is competent to support the proposed loads. This is true when the resistance to drilling is relatively great.

2. Thin-walled Sampling. In many instances, however, there is doubt as to the competency of the underground, and it therefore becomes necessary to determine explicitly its mechanical properties. Cohesive soils which are often responsible for soft-ground behavior frequently occur with a very "sensitive" type of grain structure.

When such soils are in any way distorted they undergo weakening, and this becomes particularly apparent if such soils are vigorously manipulated in the fingers. Sensitive clays which may occur naturally at a "medium" or even "stiff" consistency may be reduced in this manner to a viscous fluid state. In order to secure realistic values of mechanical properties when dealing with such soils, every effort must be made to avoid disturbance of the structure and consequent weakening of the material. To obtain reasonably reliable samples for testing, it is common practice to utilize thin-walled sampling tubes having a diameter of several inches. These thin tubes when forced into the ground displace but little material, and hence cause relatively small distortion of the penetrated soil. In addition, if some means, such as a piston, is provided to prevent any displaced material from entering the tube, the soil disturbance is further minimized. Figure 3-9 shows schematically the arrangement of

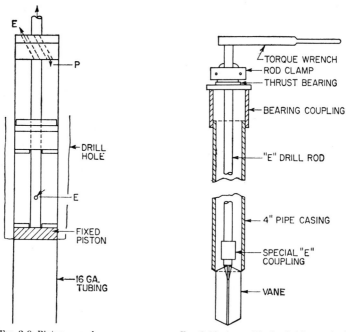

FIG. 3-9. Piston sampler.   FIG. 3-10. Assembly for field vane testing.

a "piston"-type sampler which is lowered to the bottom of a drill hole so that the piston comes in contact with undisturbed underlying soil. The piston is held firmly in place while the thin-wall sampling tube is forced into the underlying soil by means of hydraulic pressure. After the sampling tube has penetrated to its full length the piston serves to create a vacuum at the top of the sample and thus prevents a loss of material when the sampling device is removed from the drill hole. Experience has shown that such sampling devices often serve to reduce sample disturbance materially, and consequently to improve the reliability of test results. In some areas where highly sensitive soils predominate, the use of such samplers is an outright necessity.

Samples taken in this manner can be used in laboratory tests for determining the compressive and strength characteristics of cohesive soils.

3. Vane Tests. Strength characteristics can be measured also in the natural deposit by relatively simple means, and results so obtained are less likely to be affected by sample disturbance than values obtained from samples. Figure 3-10 illustrates schematically the arrangement of a four-bladed vane which can be inserted into the undisturbed soil below the bottom of a drill hole and rotated by means of a torque wrench in such a manner as to determine the shearing resistance of the soil in which it is embedded. Such testing is relatively economical, and the results are believed in many cases to be considerably more reliable than those obtained by testing ordinary soil samples. Figure 3-11 illustrates how data obtained directly from the field drilling operations, together with the results of routine index tests made in the laboratory upon adequate samples, may be compiled and summarized in graphical form to present a reliable picture of the stratification encountered in any one drill hole.

D. Stratigraphic Profiles

From the results of the borings it is possible to plot profiles showing the thickness and location of the various typical materials underlying a construction site. Such profiles permit ready determination of those areas where conditions are likely to be most severe. Stratigraphic profiles are interpolated between borings and to that extent are somewhat speculative. One must realize that the usual drilling operation is not highly precise and that even at boring locations the actual elevations of the boundaries between various strata may not be determined with an error less than several inches. Obviously, the more closely the borings are spaced and the more uniform the stratification appears to be, the more reliably may interpolation be made. Even with closely spaced borings, small pockets of material frequently may be missed entirely. It is, therefore, impossible to state in advance with what frequency or to what depths borings should be drilled. Rather, it should be expected that after preliminary investigations have revealed the general character of the stratification, additional borings will be required in order to fill in gaps and supplement the data initially secured. Knowledge of geologic formations can be helpful in planning the boring program and in extending its scope.

E. Mechanical Properties

Mechanical properties of soil as determined by laboratory tests will be no more reliable than the samples upon which they are based; that is, if material is distorted and thus weakened in extracting it from the ground, the tests made upon it will not reliably reflect the properties which it possesses in its natural bed. Consequently, the best available methods of obtaining samples should be employed, and the engineer should constantly be aware of the possibility that the test results are not truly representative of soil *in situ*. An experienced soils engineer can usually determine from the character of the results whether soil samples have been subject to serious disturbance. However, the exact degree of disturbance and its effect on soil properties cannot be expressed in numerical terms.

F. Stress Determinations

The determination of existing stresses in the underground is usually relatively straightforward, requiring only such basic data as the thickness of the various layers and the corresponding void ratios or densities of these materials. The use of Fig. 3-12 in this connection will conserve much time. On the other hand, the prediction of future stress intensities is generally accomplished by having recourse to the theory of elasticity, it being assumed that the surface-load intensities are proportional to the depth of fill at any point. This assumption is not strictly correct, nor is the application of elastic theory to the underground fully rational. Nevertheless, at the present time there exists no more satisfactory substitute, and it so happens that the propagation of stresses through cohesive, highly compressible soils is more nearly in agreement with elastic theory than is the case for cohesionless granular soils which are normally much less compressible than the cohesive varieties.

By way of illustrating a logical sequence of steps in the analysis of embankment behavior, reference is first made to Fig. 3-13, which represents a typical stratigraphic profile. The classification of the various soil layers in this figure is based upon the examination of many properly taken samples and upon Atterberg limits. The average water content of each of the layers is shown. This can be converted to void ratio for use with Fig. 3-14 or to unit weights, whence it is possible to plot the existing intergranular stress as shown in Fig. 3-10, which likewise illustrates the

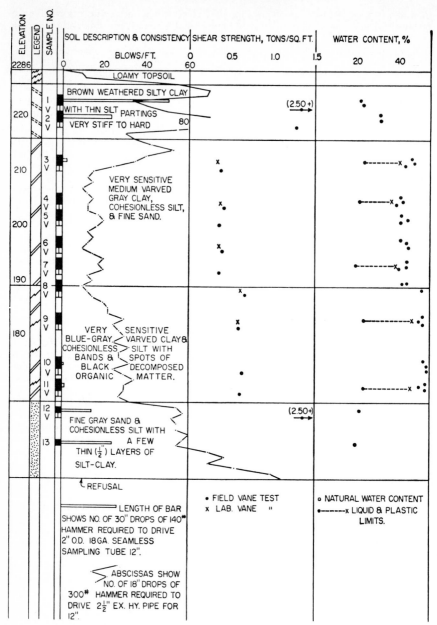

FIG. 3-11. Soil consistency, moisture, and stratification versus elevation.

cross section of the proposed fill.   By making a larger drawing it is possible to show in the profile more complete data, such as appeared on Fig. 3-11.   Thus minor changes in the properties of any one stratum can be represented.

By means of stress charts such as those given in Figs. 3-15 to 3-17, it is possible to estimate the magnitude of compressive stresses and also of shearing stresses which the proposed embankment is likely to induce in the subsoil. By properly combining the diagrams of loading such as shown in Figs. 3-15 and 3-16, it is possible to establish the stresses for many different combinations of loads. Some of the possibilities are illustrated in Fig. 3-18, which shows the procedure used. For the case of the embankment cross section shown in Fig. 3-14, the stress increment produced by the embankment load at any depth beneath its center line is equal to the horizontal distance between the two curves A and B and can be obtained by the use of

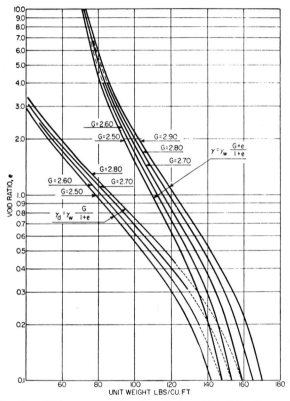

Fig. 3-12. Unit soil weights versus void ratio. Unit weight, lb/cu ft versus unit weight of soil, as a function of void ratio e and specific gravity G:

γ_w = unit weight of water
    = 62½ lb/cu ft
γ = unit weight of completely saturated soil
γ_d = unit weight of dry soil
    Note. To obtain buoyant weight of solids in each cubic foot of soil γ_b, subtract 62½ from saturated unit weight, $\gamma_b = \gamma - \gamma_w$.

Fig. 3-16. Since this difference is added to curve A in order to yield curve B, the latter curve represents the normal compressive stress which is presumed to develop at the completion of the embankment construction. In a similar manner the magnitude of the computed maximum shearing stress at any depth can be plotted and, when compared with the shearing resistance, indicates whether the soil is likely to be overstressed at any point by the proposed load. Strictly speaking, this procedure neglects any shearing stresses that may exist in the soil prior to the application of the embankment load, but in the case of those soils which are generally found to be

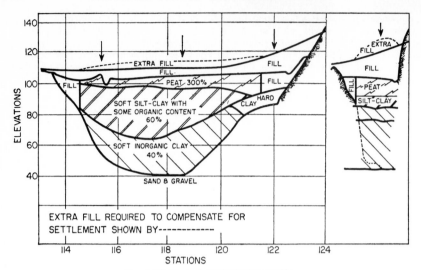

FIG. 3-13. Typical stratigraphic profile.

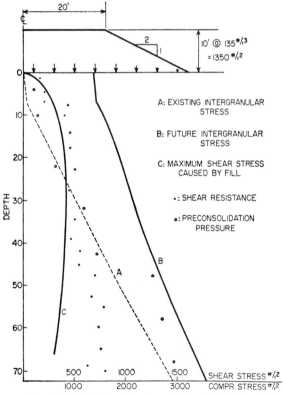

FIG. 3-14. Variation of various stresses with depth.

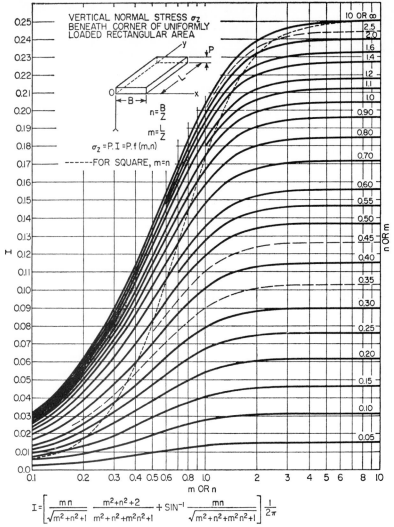

$$I = \left[\frac{mn}{\sqrt{m^2+n^2+1}} \; \frac{m^2+n^2+2}{m^2+n^2+m^2n^2+1} + SIN^{-1} \frac{mn}{\sqrt{m^2+n^2+m^2n^2+1}} \right] \frac{1}{2\pi}$$

Fig. 3-15. Normal stress chart for uniformly loaded rectangular area.

weakest in their resistance to shear, the shearing stress prior to application of the embankment load is normally of very small magnitude.

G. Analysis of Probable Behavior of Subsoil

When a comparison of the future and existing intergranular stresses has been prepared it is possible to make a very rough preliminary estimate of possible future compression on the basis of indicator tests alone. For example, if only the natural water content of the soil is known, one can arbitrarily and conveniently *assume* that the deposits are normally consolidated and, therefore, since the pressure and void ratio are both known, estimate the nature of an appropriate pressure–void-ratio diagram. Figure 3-19*a–c* shows a number of typical pressure-void-ratio diagrams in which it will be seen that the slope increases in a general way as the initial void ratio at an arbitrary pressure of, say, ⅒ ton per sq ft increases. By way of example, observe that one of the substrata (inorganic clay, Fig. 3-13), at which the existing intergranular pressure is 1.25 tons per sq ft, has a natural water content of 40 per cent. This corresponds approximately to a void ratio of 1.1. A point corresponding to 1.25 tons per sq ft and $e = 1.10$ lies close to curve P on Fig. 3-19a–c, and therefore an assumed

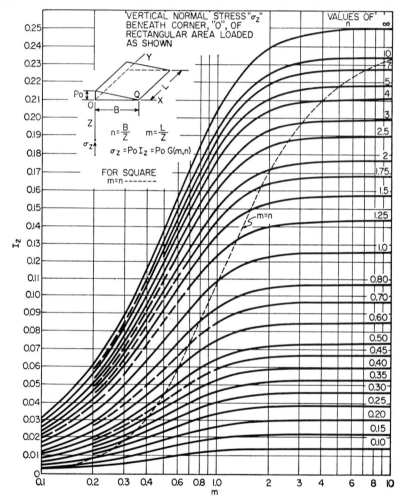

FIG. 3-16. Normal stress chart for rectangular loaded area.

pressure–void-ratio diagram passing through this point will have a slope closely approximating that of curve P. If the future load is estimated to amount to approximately 1.75 tons per sq ft at the same point, the consequent change in void ratio as read from curve P will amount to approximately $\Delta e = 0.10$, and this can be used for estimating the settlement of this particular portion of that soil layer. Actually, of course, the soil may have been overconsolidated beyond the current overburden stress of 1.25 tons per sq ft, and in this case the increase in the load would not result in as great settlement and possibly the settlement would be of inconsequential magnitude. Hence the crude approximation based upon the foregoing assumptions is usually conservative. Only rarely is a deposit of inorganic soil encountered which is still consolidating under its own weight, but organic (peat) deposits frequently are still "growing," and hence have not attained equilibrium.

1. Prediction of Settlement. To establish a more reliable prediction of settlement it is necessary to actually determine pressure–void-ratio diagrams for each stratum in question and to ascertain from these diagrams the probable preconsolidation loads. Undisturbed samples of large diameter are ordinarily required for this purpose. The cost of obtaining and taking such samples is vastly greater than that of securing samples from which reliable water-content determinations can be made. Here the

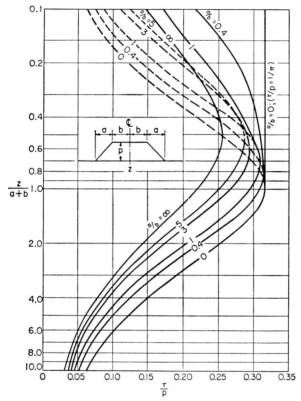

Fig. 3-17. Maximum shearing stress at any depth beneath embankment load. (Dotted curves give stress along centerline only.)

more reliably a sample represents the stratum from which it comes, the more accurately can the subsidence be predicted. The change in void ratio produced by increased load is determined from the pressure–void-ratio diagram of the particular stratum in question, and this change extended over the thickness of the stratum will provide a measure of the total settlement thereof.

The compression of a stratum, or of a portion of a stratum, with a thickness $= H_1$ is $\Delta s = H_1 \Delta e/(1 + e_1)$, where Δe is the change in void ratio which will result from an increase in load, e_1 is the initial void ratio just prior to the load increase, and Δs the change in thickness, is expressed in the same units as H.

The settlement of the ground surface will be equal to the sum of the compressions experienced by each stratum or portion thereof.

Settlement analysis is extended to each compressible stratum in the subsoil, and if one or more strata are very thick, say, in excess of 15 ft, one customarily determines the void-ratio changes which will be produced at two or more elevations in each such stratum and combines the corresponding settlements to determine the future downward motion of the ground surface.

Because of unavoidable sample disturbance, it often happens that the *estimated* preconsolidation load is somewhat less than the *true value* of this load. Furthermore, the stresses computed on the basis of elastic theory are often likely to be somewhat more intense than those which actually exist in the real subsoil. The combination of these two effects usually leads to a prediction of a settlement magnitude somewhat greater than that which actually develops. However, the amount of the discrepancy depends in large measure upon the integrity of the samples, and, therefore, it is not feasible to indicate by what percentage the predicted settlement is likely to exceed the actual settlement which will be observed.

Because of the usual low permeability and high compressibility of cohesive soils, the phenomenon of consolidation is frequently of considerable significance in connection with the subsidence of embankments or other structures placed upon such soils.

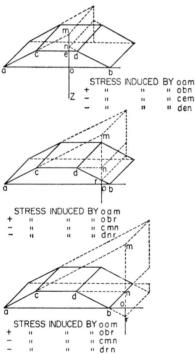

STRESS INDUCED BY oam
+   "   "   " obn
−   "   "   " cem
−   "   "   " den

STRESS INDUCED BY oam
+   "   "   " obr
−   "   "   " cmn
−   "   "   " dnr

STRESS INDUCED BY oam
+   "   "   " obr
−   "   "   " cmn
−   "   "   " drn

FIG. 3-18. Establishment of stresses for different combinations of load.

This means that the subsidence cannot occur as rapidly as the loads are applied, but rather is a gradual process which may continue for many months at a decreasing rate. Analysis of the consolidation characteristics of soil strata is based on studies of the results of consolidation tests, and predictions of settlement rates are based upon knowledge of subsurface stratification. Wherever beds of compressible soil are of substantial thickness, the lag in the progress of settlement must be expected to continue for many months or years, whereas normal construction is completed in a matter of weeks. Consequently, it may happen that only a small portion of the total settlement actually has occurred by the time the embankment is finished. In order to counteract the effects of gradual subsidence on the profile of an embankment, it is necessary to anticipate and predict the amount of gradual subsidence that will occur after completion of the fill. Guided by such a prediction, it is possible to build the fill to an excess height, so that while temporarily the elevation of the surface will be above the design grade, ultimately the subsidence will bring the surface into reasonable conformity with the design grade. As long as the magnitude of settlement varies gradually along the fill, this mode of solution is satisfactory; however, when one passes abruptly from cut to a fill placed over highly compressible soil, the rate of change in grade per station induced by settlement may be rather extreme and consequently unacceptable from the standpoint of safety.

One mode of meeting this problem is to complete the fill to the design grade and surface it temporarily with a thin bituminous pavement. As soon as subsidence has occurred to such an extent that the grade changes are not tolerable, additional fill is placed on the existing surface and again a thin temporary pavement laid on top. This process may be repeated several times in an effort to maintain a grade which acceptable and at the same time accommodate the increasing subsidence. Such a

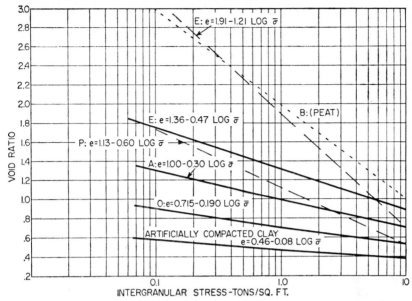

FIG. 3-19a. Typical pressure-void-ratio diagram for clays and silty clays (cohesive soils)

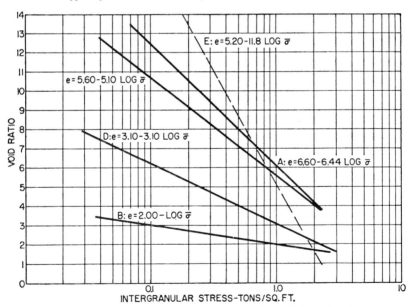

FIG. 3-19b. Typical pressure-void-ratio diagrams for peats (organic soils).

procedure may prove to be the most economical, although to persons unfamiliar with the problem it may appear as a "makeshift."

Some advantage may be gained by placing fills on compressible materials as early in the construction program as possible so that the weight of the fills may have a longer period in which to act prior to the opening of the project. This is manifestly better than delaying the construction of fills until midway through or even toward the end of the project.

Since the rate of subsidence depends in part upon the magnitude of the surface load, it is feasible in many cases to construct the fill to a height substantially greater than the design grade. The additional material, usually referred to as "surcharge," may in some cases amount to 50 or even 100 per cent of the ultimate fill height. The settlement produced by the fill and its surcharge in a given period of time will be considerably greater than the settlement which would have been produced by the fill acting alone, and consequently when the surcharge is removed just prior to finishing the surface of the embankment, a greater percentage of the total settlement has already occurred than would otherwise be the case. Again it is evident that the sooner the fill and its surcharge can be applied to the compressible soil, the more completely will the subsidence have been accomplished prior to completion of the

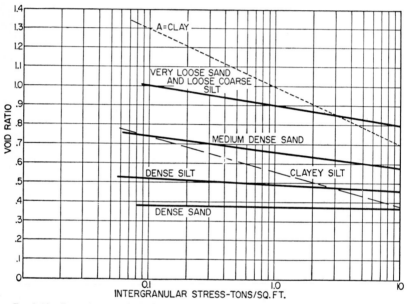

Fig. 3-19c. Typical pressure-void-ratio diagrams for sands and silts (cohesionless soils).

project. Under the most ideal circumstances, the amount of surcharge and available time would permit the amount of settlement attained by the end of the construction period to equal the total amount of settlement that would develop under the weight of the fill alone. Hence, when the surcharge is removed, settlement ceases and the surface of the fill will retain its elevation permanently.

The subsidence of a fill can also be accelerated by the introduction of additional drainage into the compressing soil. Inasmuch as the rate of consolidation is strongly influenced by the opportunities available for drainage, introduction of vertical sand drains will often enable the consolidation to take place at a greatly increased rate.

More commonly vertical sand drains are employed in an effort to increase shearing resistance of the subsoil, and the acceleration of subsidence is merely incidental to this aim. Consequently, although such drains can be highly effective in accelerating consolidation, a detailed discussion will be postponed for consideration in connection with strength of the subsoil.

2. Rupture of Subsoil. If it is found in accordance with Figs. 3-14 and 3-17 that the shearing resistance is everywhere in excess of the maximum shearing stress which the fill induces in the subsoil, one may neglect further consideration of the danger of rupture of the subsoil. However, when the shearing stresses exceed soil strength in a given stratum or even in a portion of a stratum, it becomes advisable to determine the over-all resistance of the subsoil to the shearing tendency produced by the embankment load. This determination is made by means of what is commonly referred to as a "circular-slide" analysis in which an assumed surface of rupture is studied with regard to the comparison between rupture tendency and available shearing resistance. The analysis is similar to that which is applied to slopes in cohesive homogeneous materials. Figure 3-20 illustrates the potential elements involved in such an analysis.

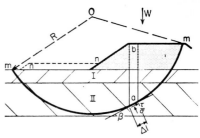

In this figure W is the vector which represents the weight of the shaded portion of the embankment extending a distance of 1 ft parallel to the embankment axis. This portion of the embankment is bounded by part of an assumed surface of shearing failure. In this particular instance a potential surface of failure in the form of a

FIG. 3-20. Circular-slide analysis.

circular cylinder has been assumed, but it is sometimes feasible to consider a potential shearing surface consisting of a cylinder of some other shape and sometimes even of a combination of such cylinders.

M_0 denotes the moment of this weight W about the center 0 of the assumed shearing surface. The tendency of M_0 to produce rotation about the center is resisted only by the shearing reaction along the surface of failure, and the intensity of this reaction can nowhere exceed the maximum available shearing strength or resistance of the soil. The manner in which this shearing resistance can be expressed will depend upon the nature of the material through which the assumed shearing surface passes. For certain soils the shearing resistance will depend heavily upon the effective normal pressure on the potential sliding surfaces, while in other cases the shearing resistance will be much less influenced by, or even be independent of, this pressure. In such cases the analysis is greatly simplified. For the general case it is customary to express the shearing resistance in the following form:

$$\tau = c + \bar{\sigma} \tan \phi$$

The various terms in this expression can be interpreted in numerous ways. For example, although in all cases τ represents the shearing resistance and $\bar{\sigma}$ may represent the effective or intergranular normal pressure on the surface of failure, there is no unanimity of opinion with respect to the definition of C, the cohesion, and ϕ, the angle of internal friction. If a graph is plotted of shearing resistance as a function of effective normal pressure and if this is assumed to be approximately a linear relationship, the magnitudes of ϕ and of C may merely represent the properties of the graph. Moreover, somewhat different values of C and ϕ would suffice to relate shearing resistance and *total* normal stress. In such cases C would equal the shearing resistance when the normal stress is zero and tangent ϕ would represent the slope of the diagram. It is not necessary, then, that C and ϕ denote intrinsic properties of the material. Thus it is known that for cohesive soils the value of C as defined may be expressed as a function of void ratio, and consequently, inasmuch as void ratio is in turn a function of intergranular pressure, the diagram illustrating the variation of shear resistance with normal pressure has a slope which is compounded of two parts, namely (1) internal friction and (2) increasing cohesion. A thorough appreciation of the significance of the term "cohesion" and the appropriateness of utilizing values of cohesion in determining shear resistance is essential to successful utilization of this property of the soil.

Here we shall consider only that the shearing resistance is made up of one part, cohesion, which does not vary directly with the intergranular normal pressure and another part, internal friction, which develops in proportion to the intergranular stress. We shall, in other words, use the foregoing expression for the shearing resistance without attempting to define the precise significance of each term.

The equation of equilibrium may be written in a variety of ways, all of which signify essentially the following:

$$M_0 \lesseqgtr R\Sigma(c + \bar{\sigma} \tan \phi) \Delta l = R\Sigma\tau \Delta l$$

where, in general, the shearing resistance, $\tau = c + \bar{\sigma} \tan \phi$, is a function of position along the sliding surface. The relationship between "driving" and resisting moments can also be expressed in terms of a safety factor:

$$F = \frac{R\Sigma\tau \Delta l}{M_0}$$

For a stratum of plastic, cohesive soil it is often permissible to assume $\tau = c$ where c represents the shearing resistance of the material under the prevailing conditions of pressure and void ratio. Because of the consolidation phenomenon, application of the embankment load can alter neither intergranular stress $\bar{\sigma}$ nor the void ratio e immediately. $\bar{\sigma}$ will gradually increase and e will gradually diminish under the new load and both these tendencies will contribute to an increase in shearing resistance, but the increase ordinarily will not be available to support the fill when it is first completed. The assumption $\tau = c$ greatly simplified the analysis.

Whenever the shearing resistance of a stratum is considered to be a function of $\bar{\sigma}$, the magnitude of this quantity along the sliding surface must be estimated. Figure 3-21 illustrates the means of accomplishing this. If p represents the vertical pressure of the column ab of overlying material (Fig. 3-20) and σ_w the hydrostatic, or more generally the fluid, stress at the same point, $\bar{\sigma}_1$ is given by $p - \sigma_w$ and is the intergranular vertical normal stress at the point in question. The corresponding intergranular component of the lateral pressure $\bar{\sigma}_2$ depends both upon the soil type and the stress history of the deposit. Normally it will not be much *less* than $\frac{1}{2}\bar{\sigma}_1$ for any material possessing internal friction, $\phi > 0$.

Equilibrium of the triangular prism under the action of $\bar{\sigma}_1$ and $\bar{\sigma}_2$ is obtained if, on the diagonal surface defined by α,

$$\frac{\bar{\sigma}}{\bar{\sigma}_1} = \sin^2 \alpha + \frac{\bar{\sigma}_2}{\bar{\sigma}_1} \cos^2 \alpha$$

$$\frac{\tau}{\bar{\sigma}_1} = \left(1 - \frac{\bar{\sigma}_2}{\bar{\sigma}_1}\right) \sin \alpha \cos \alpha$$

A graph such as shown in Fig. 3-22 aids in the determination of $\bar{\sigma}$ for various values of α. However, it is rather common to simplify matters by making the unrealistic but conservative assumption that $\bar{\sigma}_2 = 0$.

The analysis consists of:

(1) Tracing a possible failure arc.

(2) Determining the driving moment M_0 about the center of this arc.

(3) Determining the resistance along the arc and thence the resisting moment.

(a) If the shear resistance is taken as $\tau = c + \bar{\sigma} \tan \phi$, the horizontal distance between extremities mm of the surface is divided into approximately 10 equal parts to define columns like ab in Fig. 3-20.

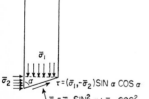

Fig. 3-21. Shearing resistance along a sliding surface.

The shearing resistance on the base of each of these columns is then determined:

$$S = (c + \bar{\sigma} \tan \phi) \Delta l$$

where Δl is the length of the base of the column and $\bar{\sigma}$ is determined by the aid of Fig. 3-22. If $s = c$, the process involving $\bar{\sigma}$ is omitted.

(4) Determining the safety factor with respect to shear along the assumed arc:

$$F = \frac{R\Sigma(c + \bar{\sigma} \tan \phi) \, \Delta l}{M_0}$$

(5) Repeating the foregoing steps for other possible assumed failure arcs. The positions of the assumed arcs can be varied methodically, and arcs of different radii should be investigated. By this means, the smallest F for each arc center can be determined, and when a number of centers have been investigated, the location of the one most prone to failure is closely defined. A safety factor of 1.25 or more is usually considered adequate.

Since many cohesive soils suffer a great loss in resistance when severely strained, it is necessary to recognize that the full resistance of all strata through which a potential failure arc passes may not develop simultaneously. For example, suppose in Fig. 3-20 that stratum I is a very stiff (almost brittle) material which fails when the shear strain is rather small, whereas stratum II is a soft plastic clay which deforms greatly before failing. The small shearing strain required to stress stratum I to its capacity is insufficient to develop more than a small part of the resistance of stratum II, while strains adequate to evoke the maximum resistance in stratum II would have already produced failure in stratum I, with consequent loss of resistance therein. Hence if the maximum resistances available in stratum I alone or in stratum II alone are inadequate to oppose M_0, it is necessary to combine the maximum resistance of stratum II with the *remolded strength* of stratum I.

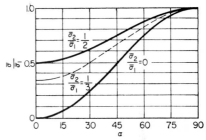

Fig. 3-22. Graph for determination of $\bar{\sigma}$.

If inadequate resistance is available, the tendency to slide can be reduced by means of a "toe load" or "counterweight," such as nn in Fig. 3-20, since this opposes a counterclockwise tendency to the clockwise moment M_0. However, it should be recognized that the weight of this toe load will tend to increase the subsidence of the main body of the embankment.

Since the application of the embankment load will tend to cause settlement of a normally consolidated stratum, the increase in $\bar{\sigma}$ and decrease in e in this stratum will enhance its shearing resistance. The rate at which this will occur depends upon the consolidation characteristics of the stratum as well as the relationship between $\bar{\sigma}$ and τ and e. Where a shear analysis shows that the resistance is inadequate, one may partially complete the embankment and allow the consequent consolidation to strengthen the soil prior to attempting to complete the construction. In this way the soil can be strengthened during construction. However, if the soil consolidates very slowly, the completion of the work will be unduly delayed. Therefore the planning of this sort of work must be done with great care and in the light of very complete knowledge of soil conditions.

a. **Field Control.** If it appears feasible to construct in stages so as to avoid overstressing the subsoil in shear, means should be provided for assessing the progress of consolidation in the field. These means consist of settlement plates placed on the original ground surface, of pore pressure gauges embedded in the consolidating stratum, and of alignment stakes accurately set parallel to the alignment. Estimates of the increase in strength which will result from consolidation under various loads will permit the preparation of a tentative schedule of construction, but these estimates should be confirmed by observations of the progress of settlement, dissipation of excess pore-water pressure, and shear tests made after some consolidation has taken place. These latter data will indicate whether consolidation is occurring as rapidly as expected and whether its progress has the anticipated effect on shear resistance. The alignment stakes are useful when construction is being prosecuted at the maximum permissible rate, since soil slips are likely to be preceded by slight lateral motions of the earth outside the toes of the embankment. Such slight motions can readily be detected with a transit, the construction operations stopped, or counterweights placed, to reduce the shearing tendency and permit an increase in strength prior to further increase in shear stresses. The analysis and control of such construction must be in the hands of an experienced soils engineer, since it is essential that the data be interpreted with understanding if the work is to be prosecuted with success.

b. **Accelerated Drainage.** Where the estimated time required to complete construction in the foregoing manner exceeds that deemed reasonable, recourse may be had to methods of accelerating the consolidation and its beneficial effects on strength. Artificial drainage can permit a great reduction in consolidation time and in some instances offers a feasible solution to the problem of "soft ground" because both an increase in shear resistance and the final subsidence will be attained relatively quickly.

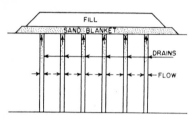

The most practical method of accelerating drainage of a consolidating stratum is through the use of vertical "sand drains." Figure 3-23 illustrates the nature of such installations. The drains, being filled with sand, which is thousands of times more permeable than the surrounding soil, readily accept water expelled from the natural soil voids by the pressure induced by the embankment load. This water flows readily upward in the drains to escape via the sand blanket. Such drains increase in effectiveness as their diameters are increased or spacing decreased. Fur-

FIG. 3-23. Illustration of sand-drain installation.

thermore, they tend to be most effective in deposits of clay varved or stratified with silt or fine sand. Spacing of the drains is usually between 10 and 20 ft, and they should completely penetrate the compressible strata.

Because of the numerous variables involved, a detailed analysis of drain and soil behavior is required to establish a rational construction schedule. Drains may be installed by driving pipe with a closed valve at its lower end, using pile-driving equipment. When it has penetrated to the required depth, it is filled with coarse sand and the valve at the bottom opened. The upper end is closed, and compressed air applied therein to raise the pipe out of the ground and simultaneously expel the sand out of the bottom of the pipe. In such a procedure the earth is displaced during the driving of the pipe and horizontal seams of highly permeable sand or silt may be broken up or impermeable clay smeared over the inner surface of the hole. Either event will decrease the permeability in the immediate vicinity of the drain, where for geometric reasons resistance to flow is greatest. To avoid this detrimental effect, the hole may be drilled with an auger so that soil is *removed* rather than *displaced*. This more nearly preserves the original soil structure and enhances the effectiveness of the drains, but is likely to prove more costly.

c. **Alternative Methods.** When subsoil is particularly unstable or the loads particularly heavy, the magnitude of subsidence and danger of shear failure may be so great as to warrant (1) relocation, (2) bridging, (3) complete or partial removal and replacement of the offensive material. The cost of any of these solutions may be very great, but where there exist requirements for early completion of a project with no subsequent detrimental behavior such costs may be acceptable.

Where the thickness of unstable material is not greater than 25 ft, removal thereof by dredging, displacement, or blasting may be feasible, and occasionally greater thicknesses are so treated.

In dredging, the tendency of undredged material to slide into the dredged area increases with increasing height and steepness of the slopes. Furthermore, slopes may fail after standing for a few hours, and hence it is advantageous to place fill as soon as possible after dredging occurs. The sliding tendency increases as the unit weight of undredged slope material increases. Consequently, it is always helpful, and often essential, to maintain the water level in the dredged area at a high elevation since this renders the surrounding soil buoyant. Particularly in the case of organic soils with unusually large void ratios, this device will be very potent in diminishing the sliding tendency.

For example, suppose cohesive inorganic material with a void ratio of 3.0 is being excavated. The unit weight, saturated, is approximately 90 lb per cu ft, but the buoyant weight of the mineral is only $90 - 62\frac{1}{2} = 27\frac{1}{2}$ lb per cu ft, and hence the sliding tendency of the buoyant soil is less than one-third that of the nonbuoyant material. The stability of slopes in dredged material is analyzed in essentially the same manner as other cut slopes.

In removal by blasting, the most reliable procedure is to excavate a trench through the unstable soil wide enough to accommodate the new fill. Occasionally, when the thickness of unstable soil is great, fill is first placed on the natural surface and explosive charges are then detonated in the underlying soil. The confining effect of the fill tends to cause the unstable subsoil to be displaced laterally so that the fill can replace it by settling downward. Possible trapping of unstable compressible soil beneath the fill is one disadvantage of such a procedure. Nonetheless, the method has been widely used. Handbooks published by explosives manufacturers usually contain data and recommendations relative to blasting techniques.

Any fill material which is to be placed in or through water must be coarse-grained in order to achieve equilibrium quickly. Fine-grained soils, particularly if cohesive, may, if so used, be little better than the material they replaced. Any material placed through water is unlikely to quickly acquire density approaching that to which it could be compacted by customary methods above water. Hence, minor shrinkage is likely to occur for some time after completion of an underwater fill.

Studies should be made of soil properties in order to determine the behavior of excavations in unstable deposits and the action of fill material placed in these excavations.

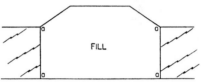

FIG. 3-24. Illustration of excavation and backfilling in unstable deposits.

For example, it may be possible in some cases to excavate a rectangular prism as shown in Fig. 3-24.

This would require that the unstable soil have sufficient strength to resist sliding into the excavation and that it also be able to oppose sufficient reaction to the fill along lines aa to maintain the shape of the fill. This means that the passive resistance of the soil must exceed the active pressure of the fill.

Where several strata of questionable soils exist, careful appraisal of mechanical properties is required to indicate how much material must be removed and discarded and what materials can be relied upon to properly support the desired embankment.

While it is true that those soils which are most prone to give unsatisfactory performance are organic or cohesive in character, it occasionally happens that granular soils in an extremely loose state are encountered. Such soils are sensitive to the effect of vibrations which tend to break down the soil structure, and this, if the soil happens to be below the water level, results in a temporary liquefaction of the material, during which it loses substantially all of its shearing resistance. Consequently, whenever extremely loose granular materials underlie embankment areas, the possibility of their becoming liquefied should be investigated and means adopted to compact them prior to placement of embankment material or at least prior to completion of the fills. Such compaction can be achieved through the medium of explosive charges, vibratory devices, and occasionally through the use of water jets. However, by far the most effective procedure is through the medium of relatively strenuous vibrations. Because of the difficulties inherent in sampling cohesionless soils without disturbing the grain structure, the effectiveness of efforts to compact such soils is most conveniently measured in terms of increased resistance to penetration by a sampling spoon or other device. Consequently, comparison is made of drilling resistances before and after efforts have been made to compact the loose material.

IV. DESIGN OF EMBANKMENTS

Embankments can be classified into those constructed on (1) flat terrain, or (2) sidehill locations. The former will be discussed in this section, while embankment problems on slopes will be considered under landslides.

The problems associated with embankments involve (1) settlement, (2) stability, and (3) erosion. Settlement considerations that relate to the foundations are discussed in the preceding paragraphs. Erosion problems and the attendant drainage and maintenance questions are discussed in Secs. 4, 12, and 13. Therefore, this section will be restricted to densification or shrinkage within the embankment (Fig. 3-1) and the stability of slopes for earth fills in level terrain.

A. Materials and Tests

Soils samples, typical of the material to be used in an embankment, are brought into the laboratory for study. Routine classification and the moisture-density tests are normally conducted. The purpose of the classification tests is to establish (1) the adequacy of the soil and (2) preliminary estimate of the relative quality of the material for embankment purposes. The results of the moisture-density test are used for the field control of earthwork (Sec. 6) and for establishing the density at which laboratory strength tests should be conducted.

In some state highway departments, materials are accepted or rejected on the basis of the classification tests or in other arbitrary manner. From considerations of settlement and stability of the embankment proper, the only types of soil that should generally be excluded are those that contain a substantial amount of organic material. The decision as to the acceptability of a soil rests principally upon economics. For the poorer soils, flatter slopes may be required for stability, and therefore such a material could be rejected because of the availability of a less costly and better soil.

For most areas, it can be assumed that an adequate embankment can be made from any nonorganic material. Analyses as to the proper slope may be required for the more questionable types. For comparative purposes, the lower the classification number and index value in the Highway Research Board classification system, the better is the material for embankments. Experience with local materials, particularly if organized within a system such as the pedalogic soil system, will form a good basis for preliminary considerations.

Bedrock materials that are to be used in an embankment are rarely tested in the laboratory. Materials that weather rapidly upon exposure to air and moisture are troublesome. There is no reliable test procedure for predicting the rate or extent of the disintegration of such materials if partial protection from air and water is provided, such as in a fill. However, it is certain that the rate is greatly reduced over that for complete exposure. Concern over rapid disintegration of bedrock materials in fills is rarely justified to the point of excluding such material from use in an embankment. Claystone (variously referred to as indurated clay, precompressed clay, massive shale, etc.) or similar materials that can be excavated with power equipment and compacted in a manner similar to soils will rarely be a problem.

B. Field Explorations

Field studies of embankment foundations are discussed in detail in Art. III of this section. The field investigation of materials from which the embankment is to be constructed requires first a systematic sampling of the slope excavations on the project. Frequently, data for soil or rock profiles are obtained. Soil areas are utilized for the basis of the field evaluation as well as for an approximation of the engineering characteristics.

Where the embankment cannot be made from the material excavated for slopes, borrow sites must be examined. The most important single factor is the availability and accessibility of the area (see Sec. 6). In order of importance, the following are the factors to be considered: (1) haul distance, (2) accessibility for earth-moving equipment, (3) cost in place, and (4) type and condition of the material. The first three are critical economic factors. The fourth may control in many instances but not frequently, since embankments can be satisfactorily constructed from almost any earth material. In cases where the first three are equal, the best construction material should be utilized. The final answer as to the borrow site is available only after analyses have been made of settlement and slope stability. For embankments in excess of 50 ft in height, the quality or strength properties of the earth borrow become more critical.

Rock borrow will rarely be competitive with soil borrow even though a considerable savings in haul distance can be obtained. On the average, soil can be hauled ½ to 1 mile at less cost than the excavation of hard rock.

The condition of the soil can be critical with regard to moisture content. Earth materials that are too wet may be difficult to handle with power equipment. In addition, the cost of reducing the moisture content to the optimum may be excessive. On the other hand, very dry material may be quite costly to handle in hot, dry weather because of water evaporation as the optimum moisture content is developed.

On occasion, wet clay material can be utilized in "sandwich construction," wherein the wet layers are alternated with a granular, permeable material (Fig. 3-25). The clay is spread in 12- to

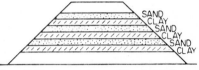

FIG. 3-25. Layered, or "sandwich," construction. *Note:* Wet clay deposited in 1-ft layers; no compaction; clean, permeable sandy layers permit rapid consolidation of clay.

15-in. layers, and no direct effort is expended in compaction. The presence of the sand above and below the clay permits its rapid drainage and densification.

C. Design Analysis

Analyses are required for (1) settlement, (2) slope stability, and (3) economics of materials usage. Foundation problems have been discussed, and so the following considerations are restricted to the behavior of the embankment itself.

Soil and rock material used in an embankment is similar to all construction materials in that the strength properties can be varied to achieve the proper balance between the cost and the stress and durability requirements. Overindulgence in earthwork compaction is as wasteful as the utilization of high-strength concrete or steel where it is not needed. The density or unit weight of a soil reflects its strength and its settlement characteristics. The absolute density is not so critical as the amount of void space (nonsoil volume) remaining in the material after compaction. The degree of compaction is normally expressed as a percentage of maximum density for cohesive soils or as relative density for granular material. The economics of material usage, therefore, includes consideration of the amount of compaction required. Currently, two procedures for cohesive soils have been accepted, the *standard* and the *modified* moisture-density tests. Within each of these two methods of test, several variations in density requirements are possible through expressing the compaction desired as a percentage of the maximum density.

The other element, the economics of compaction of cohesive soils, is involved principally with embankments in excess of 50 ft in height. For these higher fills, different materials can produce significantly different results as to slope requirements for stability. In Fig. 3-26, the effect of the shear properties on slope design is indicated as permitting steeper slopes for the better clay soils. For granular materials, a slope angle of 1½:1 should be adequate for any height. However, because of the nonhomogeneity of materials, a flatter slope (2:1) is recommended for the higher hills.

For estimates of the proper slope angle for highway embankments, a simplified procedure can be used if seepage is not a factor. Assuming that soft clays will rarely be encountered in a completed highway embankment because of current compaction requirements and that, therefore, the compacted material will have both frictional and cohesive resistance to shear, Fig. 3-27 provides a *stability factor* N_s that is defined as follows:

$$N_s = \frac{\gamma H_c}{c} \tag{3-17}$$

where γ = unit weight of soil
H_c = maximum or critical slope height
c = unit cohesion of soil

The values for N_s are dependent upon the slope angle β and the internal friction of the soil ϕ. Knowing the value of ϕ for the soil, the stability factor can be determined for a number of different slope angles. Given the stability factor N_s, and knowing the cohesion of the soil c, it is possible to compute the critical height H_c. Figure 3-28 gives the values for the reciprocal of the stability factor, known as the *stability number*, which permits analyses of higher slopes.

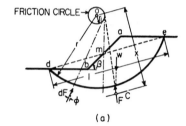

In Fig. 3-29 the influence of the shearing resistance of the soil is shown. For a slope height of 50 ft, Fig. 3-29a shows the soil characteristics required for various slope angles. For example, for a $1\frac{1}{2}:1$ slope, the value for ϕ must be at least 5° if the cohesion is 800 lb per sq ft, or ϕ must be greater than 10° if c is equal to 500 lb per sq ft. On the other hand, for a $3:1$ slope, ϕ need only exceed 4° for c values of only 400 lb per sq ft.

A further comparison of the effect on slope height of the shear properties is given

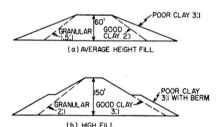

Fig. 3-26. Effect of type of material on slope design (average values).

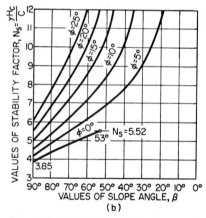

Fig. 3-27. Relation between slope angle and the stability factor for soils with cohesion and friction. (*Ref. 30, fig. 83a and b, p. 190, courtesy of John Wiley & Sons, Inc.*)

in Fig. 3-29. As the height is doubled, it will be noted that the unit cohesion required of the soil for stability is also doubled, assuming the same values for ϕ as well as the slope angle.

From the preceding and from experience, the following conservative guides to the design of stable slopes are recommended. For embankments less than 50 ft in height, a slope of $1\frac{1}{2}:1$ should be specified if the soil is compacted to 90 per cent of standard maximum density (ASTM D 698-42T, AASHO T 99-49). The same density requirements will also be adequate from settlement considerations unless a high-type pavement is to be placed upon the completed fill less than 6 months follow-

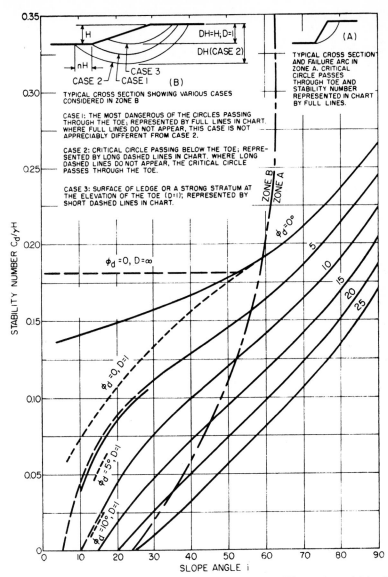

FIG. 3-28. Chart of stability numbers. *(Ref. 27a, fig. 16.26, p. 459, courtesy of John Wiley & Sons, Inc.)*

ing the completion of the lower half of the fill. In the latter instance, 95 per cent of *modified* maximum density should be specified. Very few soils that will be encountered would not be satisfactory for fills of this height, and these soils will generally be excluded on the basis of organic content or moisture.

Embankments of cohesive soils, greater in height than 50 ft, require more care if the best economics are to be achieved. Slopes of 2:1 for heights up to 75 ft and 3:1 for heights up to 100 ft should be adequate, although the fills for 75 to 100 ft should require 95 per cent modified maximum density. Experience with a material or a detailed stability analysis could result in steeper slopes and a more economical design. It is improbable that a more conservative (flatter) slope would be required unless a foundation problem is involved. From a settlement viewpoint, compaction requirements of 95 per cent standard maximum density should be adequate for the group of

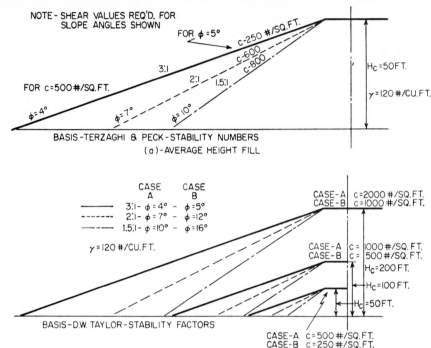

FIG. 3-29. Typical shear values required for stability.

fills 50 to 75 ft in height, unless a high-type pavement is to be placed on the completed fill less than 6 months after the construction of the lower two-thirds of the embankment.

For granular materials or rock fills a slope of $1\frac{1}{2}:1$ should be adequate for heights up to 75 ft. For 75 to 100 ft, a slope of $1\frac{3}{4}:1$ is recommended. Embankments in excess of 100 ft should utilize at least $2:1$ slopes, but will normally require a detailed analysis, because of foundation considerations.

A special type of stability problem that is caused by the presence of a weak layer in the foundation is pictured in Fig. 3-30. An approximate approach to such failures has been suggested by Terzaghi and Peck.

A trial-and-error procedure is used to determine the position of points b and c that produce the lowest safety factor, as follows:

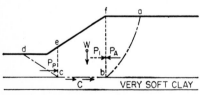

FIG. 3-30. Failure of slope underlain by weak layer. (*Ref. 30, fig. 84, p. 191, courtesy of John Wiley & Sons, Inc.*)

$$G_s = \frac{P_p + cl}{P_A} \qquad (3\text{-}18)$$

where P_p = passive earth pressure, lb
 P_A = active earth pressure, lb
The passive earth pressure is conservatively assumed to be acting horizontally. The problem is solved in the following manner:

1. A first estimate of the position of the line ec is made, and P_p is computed.
2. A preliminary estimate of the position of the line bf is made, and P_A is determined.
3. The total shearing resistance of the material in the weak strata is determined under the assumption that $\phi = 0$.
4. New positions of ec and bf are assumed until the lowest value of G_s is assured.

To conduct a quantitative study of an embankment, the stability analyses discussed under Fundamental Considerations should be used. The shearing resistance of the embankment material can be obtained from laboratory shear tests with the sample remolded to the density expected in the constructed embankment. The first shear tests should be conducted at 90 per cent of standard maximum density unless settlement is of special concern. Samples need not be saturated or the moisture content increased beyond the optimum moisture content unless the fill is likely to impound water for an extended period of time.

Quantitative estimates of settlement under various degrees of soil compaction are not too reliable. The most critical factor with reference to settlement is related to the timing of the placement of the pavement and to the type of pavement. If a high-type pavement is involved and if the pavement is to be placed upon the fill immediately following construction (less than 6 months), requirements for compaction should be 95 per cent of modified maximum density.

V. DESIGN OF SLOPES IN BEDROCK CUTS

The excavation of bedrock can lead to either *slides* or *falls*. The latter are far more common and consist principally of the phenomenon more conventionally referred to as *weathering*. Slides along curved slip surface are not common and, because of the inherent strength of even the weakest bedrock (particularly as compared with soils), will rarely occur in cuts less than 100 ft in height. Slides in bedrock will most likely develop where the strata are tilted (Fig. 3-31), and the excavation exposes an interface along which shear failures can develop. Quantitative theory of rock mechanics is not well developed, nor is the chemistry of rock weathering. With very little in the literature with reference to slope design in bedrock, current engineering practice requires judgment and experience with the material in the locale of the project.

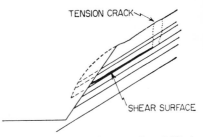

FIG. 3-31. Shear along undercut-tilted-bedrock strata.

A. Theories of Design

The design of slopes in rock cuts embodies the normal engineering principles that (1) construction costs must be held to a minimum; (2) future maintenance costs for pavement protection and for traffic safety must be minimized; and (3) the solution is based upon the characteristics of the materials involved. The point of balance between high construction costs and future low maintenance expenditures varies with the agency concerned. Practice has shown, however, that unit construction costs are one-fourth to one-third as great as are maintenance costs. Furthermore, additional economic value can be attributed to excavating the rock during construction because of traffic safety.

There are several factors that confuse the problem of rock-slope design. The slopes for soils can normally be specified and, with proper vegetative cover, will not erode or weather. Such a condition is not generally true for bedrock. Except for very hard material (and for certain others which undergo a type of "casehardening" after exposure to air and moisture) a finite amount of the exposed slope-surface material will disintegrate and erode or fall to the ditchline. Blasting will produce additional areas of weakness unless very careful control is exercised. Certain geologic-structure characteristics such as fractures and exfoliation can be classed as weak spots or discontinuities in the bedrock and will result in material moving down the slope. Finally, bedrock strata will frequently disintegrate at different rates within a single cut, so that if a material that weathers rapidly underlays a more resistant material, the upper layer will be undermined (Fig. 3-32). Such action is termed *differential weathering*.

(a) UNIFORM SLOPE   (b) VARIABLE ANGLE

There are four theories currently employed for rock-slope design: (1) *uniform slope*, (2) *variable angle*, (3) *permanent berm*, and (4) *temporary berm* (Fig. 3-33). The most advantageous theory for a spe-

(c) PERMANENT BERM   (d) TEMPORARY BERM

FIG. 3-32. Differential weathering.

FIG. 3-33. Comparison of four bedrock cut-slope design theories.

cific slope is a function of the characteristics of the bedrock and the general policy of the constructing agency relative to economic and safety considerations.

The first procedure shown in Fig. 3-33a is recommended when no significant quantity of weathering products is anticipated and a relatively uniform cross section is involved. The variable slope angle approach is illustrated in Fig. 3-33b and can be utilized where differential weathering of the bedrock is anticipated and where fairly precise knowledge is available for the proper slope of each stratum of bedrock. Such designs should not be used if weak spots or discontinuities are present in some of the strata.

Berms, or benches, utilized in the third and fourth theories, offer the only certain solution to keeping debris out of the ditch lines and off the pavement. Where fractures and exfoliation exist, berms are essential. The permanent bench procedure shown in Fig. 3-33c provides a slope for the rock such that little further weathering is expected. The temporary berm of Fig. 3-33d oversteepens the excavated slopes so that immediate weathering will take place, thus providing a cover for the lower strata, with an attendant elimination of further weathering.

B. Materials and Tests

There are no reliable laboratory or field tests for measuring the weathering characteristics of rock. Visual classification and comparisons with outcrops in the area will provide the best information. Experience and judgment will be required in the estimating of the weathering properties.

Where sloping bedrock is encountered, slides along bedding planes must be considered. Here again, experience will be the best guide. Shear tests could be utilized to determine the shearing resistance at the bedding plane. However, such procedures are not beyond the research stage.

C. Field Investigations

The objective of the field studies is to provide information concerning the weathering properties of the bedrock (falls), the probabilities of slides, and the relative quantity of bedrock and soil. Thus detailed data are needed on the various strata as follows: thickness, type of rock, elevation, dip, and depth of soil overburden. Evidences of weak spots and discontinuities should also be noted. The studies should not be restricted to the area of cut, but should be extended uphill to a height that is considered well above any possible influence of the excavation. Core drilling will furnish the most precise data, but because of high costs, auger drilling and geophysics should also be used. Observations of exposures provide excellent data. Highway or railroad cuts, stream beds, sidehill drainageways, and natural outcrops are frequently good sources of information as to weathering properties. Geologic maps should be utilized for structure and stratigraphic purposes.

The amount of drilling and distribution of the holes are open to question. For highway work, with the great number of cut sections spread over several square miles of area, it will commonly be impractical to core-drill as frequently as would be needed for precise design purposes. Furthermore, limitations on the ability to predict weathering properties place a practical limit on the subsurface investigation. At least one drill per cut should be obtained for the design of major facilities unless (1) good exposures are available in the immediate area and (2) very small cut sections are involved.

D. Design Analysis

The design of slopes in rock cuts requires analyses of two types of failures, *falls* and *slides*. Consideration of the slide problems is treated in a subsequent subsection on landslides. Therefore, the following discussion applies solely to falls which develop through tension-type failures followed by free fall under the force of gravity.

From the field studies and the examination of rock samples, two factors must be established: (1) the resistance of the material to weathering and (2) probability that discontinuities in the rock strata will lead to rock fall. The best approach to determine quantitative answers for each of these factors is a field examination of bedrock outcrops. In fact, there is no other reliable technique, particularly with reference to the establishment of discontinuities. If the field study indicates that there is a reasonable possibility that such irregularities exist, or are likely to exist following the blasting required for construction, it should be assumed that rock falls will develop. The resistance of a rock to weathering leads to the assignment of the angle of slope to be constructed. Table 3-1 contains a number of such values that are recommended for very general use, and then only after extensive efforts have been made to establish values based upon reliable experience in the area.

Table 3-1. Average Slope Values for Bedrock Excavations

Slope
(*horizontal: vertical*)

I. Igneous
  Granite, trap, basalt, and lava.......................... ¼ : 1–½ : 1
II. Sedimentary
  Massive sandstone and limestone....................... ¼ : 1–½ : 1
  Interbedded sandstones, shales, and limestones.......... ½ : 1–¾ : 1
  Massive clay stone and silt stone...................... ¾ : 1–1 : 1
III. Metamorphic
  Gneiss, schist, and marble............................ ¼ : 1–½ : 1
  Slate... ½ : 1–¾ : 1

Consideration can be given to the type of slope to be used after the factors of weathering and discontinuities have been analyzed. The first question is which of the four types of slopes are to be used, i.e., uniform, variable-angle, permanent-berm, and temporary-berm. If discontinuities are suspected, one of the berm solutions should be selected in order to intercept the falling rock fragments that are to be expected. If no discontinuities are likely, the selection of one of the four types is a function of (1) the resistance of the rock to weathering, (2) the number of types and elevation of formations, (3) the vertical height of the cut, and (4) the economic policies of the constructing agency.

The use of a uniform slope, as shown in Fig. 3-34a, is recommended when the vertical height of cut is less than 20 to 30 ft, resistant bedrock is involved, there are no discontinuities, and uniform weathering is anticipated. For cuts greater than 20 to 30 ft, a uniform slope will rarely prevent rock debris from collecting in the ditchline or from bounding onto the pavement. However, for rock strata with very uniform weathering characteristics and no discontinuities and for which good stratigraphic information is available, uniform slope as great as 30 to 50 ft can be used. For heights greater than 50 ft, uniform slopes should rarely be used.

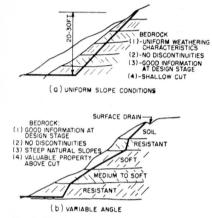

(a) UNIFORM SLOPE CONDITIONS

(b) VARIABLE ANGLE

Fig. 3-34. Principles of uniform-slope and variable-slope design in bedrock cuts.

Variable-angle slopes, indicated in Fig. 3-34b, are a modification of uniform slope in that each stratum is sloped to its proper angle. However, since no intercepting berms are involved, the approach is not recommended if discontinuities are present, because rock fragments bounding down the slope will collect in the ditchline or be a hazard on the pavement. In order to properly apply the variable-angle theory, precise data are needed on the weathering properties of the individual rock stratum. In areas of interbedded, sedimentary rocks (or similar geologic conditions), such data are rarely practical for a highway designer to obtain. Furthermore, where intergrading exists, slope angles can vary within a single stratum. Because of the complexities introduced by highway construction, variable-angle slopes will not be too frequently utilized. On a highly localized condition such as is afforded by a very high cut (greater than 100 ft) with steep natural slopes, economy may require a more elaborate investigation and the use of variable angle slopes without berms. However, even in such cases, hazards produced by rock falls due to discontinuities require that the strata be highly uniform. For variable-angle designs, the general data of Table 3-1 are not sufficiently accurate for slopes in excess of 30 to 50 ft.

The use of berms is recommended for highway conditions, particularly in stratified sedimentary deposits. The principal advantage that is associated with berm construction is the positive protection that is afforded against falling debris. In general, berms should be used where (1) discontinuities exist in the rock stratum, (2) rock is encountered that disintegrates readily under exposure to air and moisture for any reasonable angle of slope, or (3) uncertain information is available as to the type, thickness, elevation, and weathering properties of the bedrock. The final decision as to the use of berms is closely related to long-range economics. Berms will usually require greater construction costs, but savings on future maintenance expenditures and values derived from safer driving conditions will frequently offset the added capital investment.

Three variables (Fig. 3-35) control the slope design when berms are utilized: (1) berm width, (2) height between berms, and (3) angle of slope between the berms. For the permanent berm solution, the primary variable is the angle of slope to which the rock will tend to weather. Figure 3-36 illustrates a single bench design in the sedimentary-rock formation near Pittsburgh, Pa. Multiple benches could have been used, but if good data are available on the weathering properties, only one is needed, and that is for protection against rock-fragment falls. Normally, such berms will be cleaned off periodically, so they should be located at a relatively low elevation. If strata containing discontinuities are located more than 30 ft above the berm, a single bench solution is not recommended. Also, more exact data than the general type included in Table 3-1 are needed.

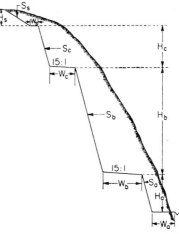

For multibench designs, the angle of slope is still the most critical consideration for the permanent-bench theory. Zones that contain discontinuities or bedrock cuts for which the weathering data are somewhat uncertain offer the best conditions for several berms. Generalized weathering data are not recommended if a clean bench is a controlling factor in the design. However, as a guide, or for comparative purposes, suggested heights and widths are included as maximum values in Table 3-2 for various slopes.

The philosophy of the temporary benches is quite different from that of the permanent type, as can be seen in Fig. 3-37. Debris on the berm is desired for the temporary-berm theory, so that part of the slope is protected from exposure to air and moisture.

FIG. 3-35. Principles of multibench design in bedrock cuts. (*Ref. 8, fig. 99, p. 167, courtesy of Highway Research Board.*)

Thus any one of the three variables of bench width, height between benches, and the slope can be varied. The controlling variable is the quantity of debris that can be accommodated on the berm, assuming that a portion of that berm will be lost because of weathering. The constructed slope, then, will be steeper than the ultimate weathered slope. The advantage to the technique is economy, in that a part of the overly steep slope will be protected and never will weather to the "natural slope." Basically, the temporary berm design can be used under any condition for which the permanent approach qualifies. In addition, where very little data are available as to

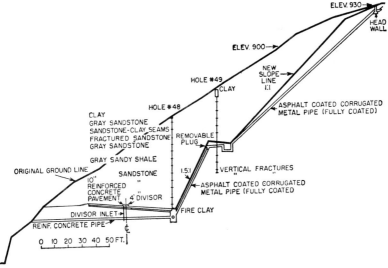

FIG. 3-36. Permanent berm design in bedrock cut. (*Ref. 8, fig. 98, p. 164, courtesy of Highway Research Board.*)

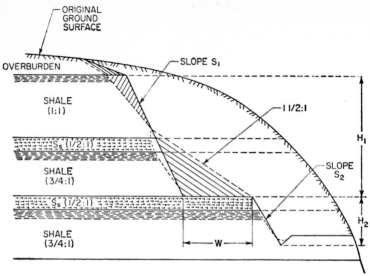

Fig. 3-37. Principles of benching rock cuts. (Ref. 3.)

discontinuities or weathering properties, the temporary approach offers more flexibility. The major disadvantage of the temporary-berm design is the appearance of the slope for the first year or two following construction. The irregular, perhaps disfigured, conditions lead to the impression of inadequacy in the design. However, the approach has been used extensively in West Virginia, Pennsylvania, and on one section of the Ohio Turnpike. Values for heights, widths, and slopes are suggested in Table 3-2 as reasonable ranges of values. More precise information should be obtained whenever practicable.

Table 3-2. Recommended Slope Variables for Multibench Design in Bedrock

Type of rock	Height between benches, ft		Width of benches, ft		Backslopes (horizontal : vertical)	
	H_a	$H_{b'}$, $H_{c'}$, etc.	W_a	$W_{b'}$, $W_{c'}$, etc.	S_a	$S_{b'}$, $S_{c'}$, etc.
1. Major cut in shale with interbedded sandstone........	5–20*	20–30	0–30	20–35	½ : 1	¼ : 1–1½ : 1
2. Major sandstone cut......	10–30*	30–40	0–20	20–30	¼ : 1	¼ : 1
3. Major cut in sandstone underlain by shale...........	10–30*	20–40	0–25	20–35	¼ : 1	¼ : 1–½ : 1
4. Moderate cuts in sandstone and shale.................	10–40*	20–40	0–20	20–30	½ : 1	¼ : 1
5. Major cuts in shale........	10–25*	20–30	0–30	20–30	1 : 1	½ : 1–2 : 1

* Use minimum if $W_a = 0$.
Source: Ref. 8, p. 167.

The design of benches has other important considerations such as the longitudinal and transverse slopes, minimum width, and location with reference to strong or weak strata. Generally, the longitudinal slope should parallel the grade line, principally for construction or aesthetic reasons, as shown in Fig. 3-38a. For hard-rock stratum, however, the longitudinal slope of Fig. 3-38b can conveniently follow the surface of the layer. The minimum bench width is a function, in part, of the type of material being excavated.   Where relatively soft to medium material is encountered, the minimum width should be 20 ft, particularly for thin, interbedded strata requiring blasting for removal.   The first bench above the road elevation should be particularly wide (25 to 35 ft) if rock-fragment falls are anticipated.   The location of the berm with reference to weak or strong layers is not particularly critical, although problems of differential weathering (and subsequent undermining) can be controlling.   Generally, the design should include a bench at the top of a weak layer that lies immediately beneath a significantly more resistant layer.

Where high design standards for the highways are being utilized, a small bench, 5 to 10 ft in width, should be constructed at roadway elevation.   The purpose of the bench is to provide a more positive interceptor for weathering products.   If the bench is not used at roadway elevation, flatter slopes and a smaller height to the first bench above the roadway should be utilized.   One of the more argumentative features of the use of berms in bedrock slopes is the direction of slope of the transverse bench.   The conventional approach has been to slope the bench away from the road in order to avoid erosion of the slope below the bench (Fig. 3-36). Since erosion of most bedrock material does not pose as serious a problem as the positive interception of the detritus or weathered products, the retention of the material on the berm is of more consequence.   For the permanent-bench procedure, the transverse slope on the bench is not particularly critical since little or no debris is anticipated on the berm.   However, if debris comes down the slope onto the bench, then longitudinal drainage of the berm will be blocked periodically.   In order to achieve stability of the debris on the bench, drainage of the surface runoff and subsurface water flow should be removed in as positive a fashion as possible.   Longitudinal drainage of this water is complicated.   In such

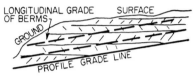

(a) WEAK BEDROCK

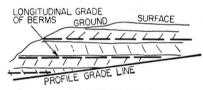

(b) HARD BEDROCK

Fig. 3-38. Longitudinal grade for bedrock berms.

cases the most certain design for removal of the water is to slope the berm toward the road.   The prompt removal of water is more critical where the detritus will consist of disintegrated shale, since the low permeability of the clay that results from the weathering of the shale will become quite unstable when saturated.   The transverse slope of the berm can be quite small (15:1).   The stability of the detritus (Fig. 3-37) is not affected by the sloping toward the road, since the shearing resistance of the material in contact with the bench is approximately equal to the shearing resistance along a similar plane within the detritus.   Therefore, it is recommended that benches be sloped toward the roadway unless (1) the material on the bench will be granular and free-draining, or (2) no debris will be accumulating on the berm, or (3) the slope below the berm is highly susceptible to surface erosion.

Certain generalities can be made with reference to the height between benches and the angle of slope between benches where a temporary bench design is involved. Basically, quantitative values for these two factors are concerned with the amount of debris that can be tolerated or is designed to be retained on the berm.   In Fig. 3-35, if flat slopes (S_b, S_c) are used in an effort to reduce the amount of the detritus to a minimum, then the height (H_b, H_c) between benches can be increased considerably. On the other hand, the presence of a massive formation with discontinuities such as exfoliation, joints, or fractures would tend to reduce the height between benches in order to eliminate possibilities of major rockfalls onto the highway.   For the temporary bench solution, the slopes (S_b, S_c) can be quite steep ($\frac{1}{4}$:1) with the anticipation that the material will weather rapidly.   In such cases the height between benches (H_b, H_c) will be held to the minimum.   The lowest slope (S_a) is particularly critical since the debris that falls from the slope should not block the roadway drainage.   A large bench (W_a) at roadway elevation is needless and greatly increases the quantity of excavation. Therefore a narrow bench (W_a), lower height (H_a), and flatter slope (S_a) are recommended.

An important factor that will control, to some extent, the design of rock slopes is the over-all vertical height of the cut. In general, for slopes less than 50 ft. in vertical height, a more conservative approach can be used without undue increase of construction costs. Even so, it is rare that more than one berm will be required. The exception would be for very weak shales or massive, soft claystone, for which two berms might be considered essential. For vertical cuts of 50 to 100 ft, the height between benches, the bench width, and the cut slopes can be reasonably conservative without increasing the cubic yards of excavation materially. For rock slopes involving vertical heights greater than 100 ft, the decision is more difficult. Generally, it will not be practical to obtain precise information on all the individual stratum. Furthermore, the hazards from rockfalls are increased. If conservative values are used for the multiple-bench design, the construction costs can be significantly increased. Since these heights will not be too frequently encountered by many highway departments, it might be well to refer such designs to a specialist for consideration.

The presence of massive layers greater than 25 ft in thickness tends to simplify the design. The prediction of the probable slope to which these materials will weather is considerably easier. Where the massive layers are hard materials, the minimum values for bench width, height between benches, and angles of slope can be utilized. Conversely, massive layers of soft material will require minimum heights between benches, wider berms, and rather flat backslopes even with the temporary berm solution. On occasion, the presence of a massive layer will control the location of a berm. Generally, it is poor practice to undercut a massive hard formation if a weak, easily weathered material underlays the resistant layer, particularly if the massive layer has discontinuities.

Use has been made of horizontal drainage for reducing weathering effects in the sedimentary rocks of the Allegheny Plateau. In stratified bedrock, differential permeability is not unusual and the removal of subsurface water can reduce erosive effects. Figure 3-36 contains such a design. Generally, the use of horizontal drainage in bedrock formations will not be economical. However, for very high vertical cuts in relatively weak bedrock or in tilted strata, the use of horizontal drainage can be effective in increasing the natural shearing resistance.

REFERENCES

1. Baker, R. F.: "Determining the Corrective Action for Highway Landslide Problems," *Highway Research Board, Bulletin* 49, 1952.
2. Baker, Robert F., and Robert Chieruzzi, "Regional Concept of Landslide Occurrence," presented 37th Annual Meeting, Highway Research Board, Washington, D.C., 1958.
3. Baker, R. F.: "The Design of the Slope of Highway Rock Excavations in West Virginia," *Proceedings, 3d Annual Symposium on Geology as Applied to Highway Engineering,* February, 1952.
4. Baker, R. F.: "An Analysis of Corrective Actions for Highway Landslides," *Transaction,* American Society of Civil Engineers, 1953.
5. Jenkins, D. S., D. J. Belcher, L. E. Gregg, and K. B. Woods: "The Origin, Distribution, and Airphoto Identification of United States Soils," *Civil Aeronautics Administration, Technical Development Report,* 52, 1946.
6. Collin, Alexandre: *Landslides in Clays,* trans. by W. R. Schriever, University of Toronto Press, Toronto, 1956.
7. Culter, T. H.: "Slipping Road-Bed Held Up by Steel Suspenders," *Engineering News-Record,* vol. 109, pp. 684-685, 1932.
8. Eckel, E. B., et al.: "Landslides and Engineering Practice," *Highway Research Board, Special Report,* 29, 1958.
9. Fenneman, Nevin M.: "Physiographic Divisions of the United States," *Annals of the Association of American Geographers,* vol. 6, 1917.
10. Fenneman, Nevin M.: "Physical Divisions of the United States," map, U.S. Geological Survey, 1949.
11. Hennes, R. G.: "Analysis and Control of Landslides," *University of Washington Engineering Experimental Station, Bulletin* 91, 1936.

12. Krynine, D. P.: "Landslides and Pile Action," *Engineering News-Record*, vol. 107, no. 122, Nov. 26, 1931.
13. Krynine, Dimitri P., and William R. Judd: *Principles of Engineering Geology and Geotechnics*, McGraw-Hill Book Company, Inc., New York, 1957.
14. Ladd, G. E.: "Methods of Controlling Landslides," *Engineering and Contracting*, vol. 67, 1928.
15. Ladd, G. E.: "Landslides and Their Relation to Highways," *Public Roads*, vol. 8, 1927, vol. 9, 1928.
16. Ladd, George E.: "Landslides, Subsidences and Rock Falls," *Bulletin, American Railway Engineering Association*, vol. 37, no. 337, July, 1935.
17. Laurence, R. A.: "Stabilization of Some Rockslides in Grainger County, Tennessee," *Economic Geology*, vol. 46, 1951.
18. Liang, Ta: "Landslides: An Aerial Photographic Study," unpublished thesis for the degree of Doctor of Philosophy, Cornell University, Ithaca, N.Y., 1952.
19. Lobeck, A. K.: *Geomorphology*, McGraw-Hill Book Company, Inc., New York, 1939.
20. Palmer, L. A., J. B. Thompson, and C. M. Yeomans: "The Control of a Landslide by Surface Drainage," *Highway Research Board, Procedures*, vol. 30, 1950.
21. Parrotte, W. T.: "Control of a Slide by Vertical Sand Drains," *Highway Research Board, Bulletin* 115, 1955.
22. Peck, R. B., and H. O. Ireland: "Investigation of Stability Problems," *American Railroad Engineers Association Bulletin* 507, 1953.
23. Price, Paul H., and K. O. Lilly: "An Investigation of Landslides Affecting Roads in West Virginia," unpublished report to the State Road Commission of West Virginia, 1940.
24. Root, A. W.: "Problem of Slipouts Studied by State Highway Engineers," *California Highways and Public Works*, March, 1938.
25. Root, A. W.: "Correction of Landslides and Slipouts," *Transactions ASCE*, vol. 120, 1955.
26. Sharpe, C. F. S.: *Landslides and Related Phenomena*, Columbia University Press, New York, 1938.
27. Smith, Rockwell: "Roadbed Stabilization," *American Railroad Engineers Association, Bulletin* 479, 1949.
27a. Taylor, D. W., *Fundamentals of Soil Mechanics*, John Wiley & Sons, New York, 1948.
28. Terzaghi, Karl: "The Mechanics of Shear Failures on Clay Slopes and the Creep of Retaining Walls," *Public Roads*, vol. 10, 1929.
29. Terzaghi, Karl: *Theoretical Soil Mechanics*, John Wiley & Sons, New York, 1943.
30. Terzaghi, Karl, and Ralph B. Peck: *Soil Mechanics in Engineering Practice*, John Wiley & Sons, Inc., New York, 1948.
31. Terzaghi, Karl: "Mechanism of Landslides," in *Application of Geology to Engineering Practice*, Geological Society of America, Berkey volume, 1950.
32. Tompkin, J. M., and S. H. Britt: "Landslides, A Selected Annotated Bibliography," *Highway Research Board Bibliography*, 10, 1951.

Section 4

DRAINAGE FOR HIGHWAYS AND AIRPORTS

EUGENE M. WEST, *Associate, L. E. Gregg and Associates, Lexington, Ky.*

I. INTRODUCTION

Since water is almost certain to be encountered over the entire range of any highway or airport project, drainage is basic in every phase of the design. Whether it be subsurface water originating below the pavement, storm water falling upon the pavement and adjacent areas, or water courses intersected by the project, a systematic drainage design must be made in order to ensure adequate and lasting performance.

There are two basic steps in the solution of surface-drainage problems: *hydrologic analysis* and *hydraulic design.* A good drainage design involves an accurate prediction of the magnitudes of peak rates of runoff for various intervals of expectancy (hydrologic analysis) as well as the design of facilities to accommodate the runoff (hydraulic design). A close correlation between these two basic parts is necessary in order to arrive at a balanced design whereby costs can be weighed against the protection provided.

Recent increases in highway construction and the development of higher types of highways have resulted in more emphasis on the drainage phases of highway design. With approximately 25 per cent of the total construction cost of highways being spent on drainage structures, there is little room for many of the antiquated concepts and rules of thumb of the past. In past years the highway engineer had very little or no hydrologic data to base hydraulic design upon and justifiably used empirical formulas and rules of thumb. However, in recent years drainage research and hydrologic data collecting have been accelerated, thus reducing many of the unknowns. For a good many locations ample hydrologic data are available, and recent studies have made possible the application of gauged data to some areas for which direct records are available. Research and further expanded data collecting presently under way undoubtedly will extend present knowledge and bring about still better methods of approaching the problems of hydrologic analysis.

II. TYPES OF DRAINAGE PROBLEMS

A. Bridge Waterway Openings

In a majority of cases the height and length of a bridge depend solely upon the amount of clear waterway opening that must be provided to accommodate the flood waters of the stream. Actually, the problem goes beyond that of merely accommodating the flood waters and requires prediction of the various magnitudes of floods for given time intervals. It would be impossible to state that some given magnitude is the maximum that will ever occur, and it is therefore impossible to design for the maximum, since it cannot be ascertained. It seems more logical to design for a predicted flood of some selected interval—a flood magnitude that could reasonably be expected to occur once within a given number of years. For example, a bridge may be designed for a 50-year flood interval; that is, for a flood which is expected (according to the laws of probability) to occur on the average of one time in 50 years.

Once this design flood frequency, or interval of expected occurrence, has been decided, the analysis to determine a magnitude is made. Whenever possible, this analysis is based upon gauged stream records. In a number of the streams in each state a series of gauges are maintained and periodic measurements are available. (This work, done principally by the U.S. Geological Survey and the U.S. Corps of Engineers, is published annually in the *Water Supply Paper*.) In addition, for several states, flood-frequency curves for gauged streams have been published by the U.S. Geological Survey.*

In areas and for streams where flood frequency and magnitude records are not available, an analysis can still be made. With data from gauged streams in the vicinity, regional flood frequencies can be worked out; with a correlation between the computed discharge for the ungauged stream and the regional flood frequency, a flood frequency curve can be computed for the stream in question. Methods for dealing with flood-frequency problems are presented at length in several sources, such as Dalrymple, "Regional Flood Frequency" (1).†

B. Highway Culverts

Any closed conduit used to conduct surface runoff from one side of a roadway to the other is referred to as a culvert. Culverts vary in size from large multiple installations used in lieu of a bridge to small circular or elliptical pipe, and their design varies in significance. Accepted practice treats conduits under the roadway as culverts, for all sizes and shapes up to 20 ft in total span. Structures of greater span are treated as bridges. Although the unit cost of culverts is much less than that of bridges, they are far more numerous, normally averaging about eight to the mile, and represent a greater cost in highways. Statistics show that about 15 cents of the highway construction dollar goes to culverts, as compared with 10 cents for bridges. Culvert design then is equally as important as that of bridges or other phases of highways and should be treated accordingly.

C. Municipal Storm Drainage

In urban and suburban areas, runoff waters are handled through a system of drainage structures referred to as storm sewers and their appurtenances. The drainage problem is increased in these areas primarily for two reasons: the impervious nature of the area creates a very high runoff; and there is little room for natural water courses. It is often necessary to collect the entire storm water into a system of pipes and transmit it over considerable distances before it can be loosed again as surface runoff. This collection and transmission further increase the problem, since all of the water must be collected with virtually no ponding, thus eliminating any natural storage; and through increased velocity the peak runoffs are reached more quickly. Also, the shorter times of peaks cause the system to be more sensitive to short-duration, high-intensity rainfall. Storm sewers, like culverts and bridges, are designed for storms of various intensity-return-period relationships, depending upon the economy and amount of ponding that can be tolerated.

The hydraulics of storm sewers and appurtenances is probably more critical than that of bridges and culverts, since the systems are more sensitive and complex. It is not uncommon to find cases where ponding is excessive on a roadway or street and yet the drainage system is not taxed to its full capacity, indicating improper inlet spacing or other improper hydraulic considerations.

* These, as well as all other USGS publications referred to herein, are available from the Superintendent of Documents, U.S. Government Printing Office.

† Numbers in parentheses refer to corresponding items in the References at the end of this section.

D. Airport Drainage

The problem of providing proper drainage facilities for airports is similar in many ways to that of highways and streets. However, because of the large and relatively flat surfaces involved, the varying soil conditions, the absence of natural water courses and possible side ditches, and the greater concentration of discharge at the terminus of the construction area, some phases of the problem are more complex.

For the average airport the over-all area to be drained is relatively large and an extensive drainage system is required. The magnitude of such a system makes it even more imperative that sound engineering principles based on all of the best available data be used to ensure the most economical design. Overdesign of facilities results in excessive money investment with no return, and underdesigning can result in conditions hazardous to the air traffic using the airport.

In order to ensure surfaces that are smooth, firm, stable, and reasonably free from flooding, it is necessary to provide a system which will do several things. It must collect and remove the surface water from the airport surfaces, allowing a minimum to enter the subsurface; intercept and remove surface water flowing toward the airport from adjacent areas; collect and remove any excessive subsurface water beneath the surface of the airport facilities and in many cases lower the ground-water table; and provide protection against erosion of the sloping areas. To design facilities which will achieve these ends requires a proper understanding of the factors involved in drainage and the use of all available hydrologic data. This text should be interpreted merely as a guide, and local Civil Aeronautics Administration, Environmental Science Service Administration, and U.S. Corps of Engineers Offices should be consulted for the information pertinent to the particular design problem.

E. Roadway Surface Drainage

Roadways are built with crown or cross slopes to provide lateral or oblique drainage flow to the sides of the pavement, and in the case of curves, lateral or oblique flow due to superelevation of the pavement and shoulders. The most common practice is to allow the water to flow from the pavement, across the shoulder, and down the side slopes to the natural drainage area or to side ditches. Where proper slope protection is provided and the flow from the pavement is distributed to give a fairly uniform sheet of water, erosion is not excessive. In places where sheet flow is not possible and concentrations of flow occur, it is necessary to provide other means of protecting the side slopes and shoulder.

Several techniques used to prevent harmful erosion are employed in conducting the storm water from the roadway. One of these is to provide a curb along the outer edge of the pavement to conduct the flow to catch basins or other collecting devices, from which it is removed and led down the slopes to a ditch or to a natural drainage area. This type of curb has its disadvantages, since the water on the pavement edge creates some hazard to traffic, especially when frozen.

F. Ditches and Cut-slope Drainage

A highway cross section normally includes one and often two ditches paralleling the roadway. Generally referred to as side ditches these serve to intercept the drainage from slopes and to conduct it to where it can be carried under the roadway or away from the highway section, depending upon the natural drainage. To a limited extent they also serve to conduct subsurface drainage from beneath the roadway to points where it can be carried away from the highway section.

A second type of ditch, generally referred to as a crown ditch, is often used for the erosion protection of cut slopes. This ditch along the top of the cut slope serves to intercept surface runoff from the slopes above and conduct it to natural water courses on milder slopes, thus preventing the erosion that would be caused by permitting the runoff to spill down the cut faces.

Extreme care must be taken in the use of crown ditches, however, to prevent a condition worse than that which an attempt is being made to cure. In cases where crown ditches are very long and slopes excessive, unless properly considered in the design, erosion can become concentrated and severe.

Side ditches should be designed not only to accommodate the flow, but also to provide waterway sufficient to prevent erosion of the roadway shoulder and undercutting of the toe of the slope. The design procedures are the same as for open-channel flow, using Manning's formula and conservative values of the roughness factor, which are included in this section under Hydraulic Design Principles, Art. IV.

G. Subsurface Drainage

To prevent excess moisture in the subgrade, which will ultimately cause a loss in stability through low resistance to wheel-load displacement, it is necessary to ensure proper subbase drainage facilities.

Moisture in the subbase falls into two categories: *free water* and *capillary moisture*, both of which the engineer must consider in a proper design.

Free water either originates under the pavement or has penetrated the subbase, through perviousness of the pavement or of the shoulders and pavement edges. Capillary moisture is the result of capillary action, by which small particles of water are drawn from free water or from wetter strata and become attached to the soil particles.

The initial step in preventing poor subsurface conditions is to control the free water, thus reducing the source of capillary moisture. This can be accomplished by intercepting surface water before it is allowed to enter the subgrade. The serious effects of capillary water can further be reduced by providing subdrains to lower the water table, by providing a more pervious subbase material, and by using subgrade soils which require relatively low amounts of moisture for binding. Volumetric changes of soils, the effects of frost action, and the like are presented in more detail in Sec. 5.

Varying types of subdrainage installations can be used, depending upon the conditions which they are intended to control or correct. Detailed recommendations for various types can be found in highway standards and in other publications listed in the References at the end of this section.

H. Channel Relocations

It often becomes necessary to relocate portions of natural water courses to make way for a roadway or an airport. Often drainage structures, both large and small, can be eliminated by relocating reaches of existing streams to keep them on one side of the roadway rather than crossing back and forth. At many structure sites the field investigations indicate that channel relocations and improvements above and below the structure may provide a better situation from the hydraulic standpoint. A decrease in initial cost of the structure installation is possible; however, maintenance and possible future cost should not be overlooked in considering the feasibility of this choice. Also, considerable caution should be taken in dealing with the changing of natural water courses. Such an undertaking requires thorough engineering analysis.

There has been a tendency to measure, from a cross section of the existing channel taken at some point, the area below some high-water mark whose validity may be questionable and to design a new channel by merely providing equal or greater cross sections. Yet this alone does not represent a design by any approximation and can in some cases produce disastrous results. The consideration of area alone is by no means sufficient, since channel capacity and flow characteristics depend upon many other factors. These factors may not be the same at all locations on the new channel, and some velocities can cause severe damages regardless of how large the cross section. Velocity, for instance, depends upon roughness, shape of the cross section, slope, and discharge, in addition to area.

In channel relocations where long reaches of the stream are eliminated, the effect of changes in storage can become a severe problem. An alteration in storage capacity could jeopardize the high-water clearance for the roadway and the structures. Providing more uniform shapes and slopes of the new channel with smoother channel linings will tend to cause higher velocities, and these may present scouring problems in the new channel and at the toe of the embankment. Generally, natural channels have become more or less stabilized through the normal course of nature, and experience indicates that new channels rarely stay as constructed. They have a tendency to undergo a change, the degree depending upon the similarity of the conditions provided to those required by nature. This change can cause an appreciable maintenance problem.

The relocation of channels requires a hydrologic and hydraulic analysis similar to that required for bridge and culvert design.

I. Erosion Protection

Generally, if the design of a drainage structure is properly made, there will be little or no source for erosion. In the special cases where the structure itself cannot be adequately designed to prevent erosive conditions, the design should provide additional protective facilities.

Whenever possible, culverts should be designed for discharge velocities within a range that will not scour the channel at the outlet. The same principle holds for channels and ditches. Whenever the hydraulic design permits, the best protection is to provide a situation that will not produce scouring velocities; however, in many cases the velocities cannot be controlled in the design and additional protective measures become necessary.

The protection of shoulders and slopes from runoff from the roadway surface should also be kept in mind in the design stage. Normally, collection and distribution of the surface water can protect these areas; but when the analysis indicates that scouring conditions are prevalent, special provisions become necessary. These may include providing a special channel lining, providing ditch checks, and installing stilling basins at culvert outlets. Additional information on erosion protection is included in Art. IV on design in this section, and specific treatments are included with the discussion of the other drainage problems in Sec. 6.

III. HYDROLOGIC ANALYSIS

By definition, "hydrology is the science that deals with the processes governing the depletion and replenishment of the water resources of the land areas of the earth" (2). Since it deals with the occurrence and movement of water upon and beneath the earth's surface, this science is of utmost concern to drainage engineers. A thorough treatment of the field is beyond the scope of this handbook, but references are given at the end of this section (2–5).

Some of the basic elements of hydrology with which the drainage engineer must deal are those concerning rainfall and runoff. Of particular concern are the relationships between peak rates of runoff and the frequency and intensities of the rainfall producing these runoff peaks.

A. Rainfall

Practically all applications of hydrology, and particularly those pertaining to the design of drainage structures, are dependent upon correlations between rainfall and ultimate surface runoff. Hydrologic "analysis" for this relationship involves as many direct measurements as feasible, estimates of conditions that are not directly measurable, and calculations of the probable occurrence of rainfall based on past records.

The three features of rainfall precipitation fundamental to hydrologic problems are:

INTENSITY: The rate at which rain falls for a given period, usually expressed as inches per hour.

DURATION: The time during which rainfall prevails at that rate, usually expressed in minutes.

FREQUENCY: The probable period of time within which combinations of intensity and duration repeat themselves, usually expressed in years. Frequency should not be interpreted to mean an even periodicity. For example, the assignment of a 50-year frequency to an event means that the event will reoccur or be exceeded on the average of once in 50 years, and if past records are repeated, the chances are 1 in 50 that it will occur in a particular year. The prediction depends upon the laws of probability, and, actually, the chances are only about 64 in 100 that the event will occur in a given 50-year period.

Through a network of automatic rainfall-recording devices the U.S. Weather Bureau and others have compiled a rather long period of record of rainfall duration and intensities. From these records a number of publications of rainfall-intensity-frequency curves are available, most of which are for specific localities. However, a nationwide coverage was published by the U.S. Department of Agriculture in 1935, "Rainfall Intensity-Frequency Data" by David L. Yarnell (6). Subsequently the U.S. Weather Bureau published rainfall-intensity-frequency relationships for the entire United States (7, 8). In some cases it is desirable to compute rainfall-intensity-frequency curves from data taken at the site or in its vicinity. To explain the probability theories involved in computing such curves is beyond this text.

B. Runoff

A portion of the precipitation occurring as rain, frozen rain, or snow is returned to the atmosphere through evaporation and transpiration, and some infiltrates into the ground as ground water. The remainder flows over the surface, beginning with a thin sheet of flowing water and progressively increasing from small tributaries to major streams. This portion is referred to as *surface runoff* (or "discharge," as it is called by the U.S. Geological Survey and other agencies that measure and record the data at various locations).

There are a number of factors which influence surface runoff. Some of the more prominent ones are topography, soil type, size and slope of the area, land use, and antecedent moisture conditions, all of which have considerable effect upon the amount of the total rainfall that infiltrates into the ground.

The *infiltration capacity* for a specific area is the maximum rate at which rain can be absorbed into the soil; it varies with the conditions of the rainfall and the antecedent moisture condition. During a storm of considerable duration, the rate of infiltration is usually large at the beginning, becomes lower, and finally remains virtually constant after a prolonged period. The intensity has an effect also in that a smaller portion of the total precipitation has an opportunity to enter the ground when rainfall is of high intensity and short duration than when intensity is lighter but with a long duration.

From the standpoint of rainfall-runoff characteristics, infiltration represents a reduction in the percentage of the total precipitation falling on a drainage area that contributes to the surface runoff which the area produces.

Surface runoff is that portion of the total precipitation remaining after the losses have been deducted. Waters that originated as surface runoff, plus the subsurface runoff entering a flow channel, constitute the total quantity that must be accommodated by a structure during a period of rainfall. However, the problem in the design of drainage structures is not so much concerned with the total amount of water to be accommodated as with the peak rate at which it must be handled. Drainage structures must be designed to carry *peak rates* of runoff (*peak discharge*). Neither the duration of a peak rate nor the total amount of runoff is of any actual concern in determining the required waterway opening.

A number of methods are used in estimating runoff, some of which have more merit than others. The engineer should understand the limitations of each method and select the one which is most suitable for his particular situation. It must always be remembered that runoff predictions for the future are subject to the laws of chance.

C. Use of Stream-flow Records

The most reliable basis for determining runoff from a given drainage area is a long-term record of actual measurements. An increasing number of streams are being gauged by the U.S. Geological Survey and other agencies. Many of these long-term records are available through the publication of *Water Supply Papers*. However, a proposed drainage structure seldom falls at the location of a gauging station. In many instances it is not even on a stream that is being gauged. Also stream gauging has been somewhat limited to the larger tributaries, and a relatively small amount of data for streams having less than 10 sq miles of drainage area is available. However, a program for gauging small watersheds has been developed, and data are becoming available.

A detailed discussion of the applications of gauged data to the design of drainage structures is too lengthy for presentation in this book. A thorough treatment of the subject has been presented by Dalrymple in the *Proceedings* of the Highway Research Board (9).

D. Observations of Existing Structures

In the absence of actual flow data, structures already in place on the stream, above and below the proposed location, may provide sufficient information for a design. This method should be used with caution, however, since the record of past performance is often of questionable reliability. It is also difficult to make any approximation of the flood frequencies involved; and factors which might have affected the stage-discharge relationship may not be apparent—such factors as excessive drifts and ice jams, changes in the watershed, and many others.

Whenever such methods are used, the engineer should carefully analyze the existing structure and channel. Size, shape, and slope of structure, entrance conditions, skew, and channel conditions above and below the structure are especially important. Methods for computing the peak flow through existing structures are dependent upon the type of structure, whether it has been operating with full or part-full flow and all conditions which control its performance. For culverts, methods of making such an analysis are published by the U.S. Geological Survey, in Circular 376 (10), and for bridges by the U.S. Geological Survey, in Circular 284 (11). Other sources are listed in the References.

E. Analysis of the Natural Channel

When no existing structures are available, the *slope-area method*, based on Manning's formula, will provide a means of estimating the discharge for a stream at any selected point. In using this method it is first necessary to determine the slope of the energy gradient (S, in Manning's formula) for a selected reach of the channel near the proposed structure site. This is done by constructing a profile of high-water marks on both sides of the channel, extending for some distance from each end of the reach. The formula is then used to compute the discharge for several cross-sectional areas, normally taken at breaks in the high-water profile.

The length and location of the reach selected should be based upon the apparent accuracy of the high-water marks and the accuracy desired for the computed discharge quantity. A length of 200 ft is generally accepted as the necessary minimum. Friction losses are computed from estimates of the channel roughness factor (Table 4-7).

The slope-area method is at best only a source of a fairly reliable approximation. If it is to be used to its maximum effectiveness several precautions are mandatory:

1. Care must be exercised in selecting the particular channel reach to be used as the prototype in the calculations. The straightest and most uniform reach in the vicinity of the site—and one that is, preferably, contracting—should be selected. An adjustment must be made to determine discharge at the actual structure site when either above or below the reach selected. This is done by multiplying the discharge obtained for the prototype reach by the two-thirds power of the ratio of the drainage area at the structure site to the drainage area at the selected reach.

2. The high-water marks should be selected for clear definition and high reliability, preferably at points where velocities are not appreciable.

3. The reach selected should have fairly uniform channel lining and banks, to facilitate the selection of a roughness factor.

The reach having been selected, the average area and wetted perimeter can be determined and the hydraulic radius R computed. From the profile of high-water marks the fall over the length of the reach can be computed and can be assumed to be the same as S for Manning's formula. The value of n is selected from tables of roughness coefficients (Table 4-7).

Substituting these variables in the formula

$$Q = \frac{1.486}{n} A R^{2/3} S^{1/2}$$

A first approximation of discharge can be made. This discharge divided by the cross-sectional area at the upper end and lower end of the reach will give the velocity at each end, and thus the velocity head $V^2/2g$ for each.

The difference in velocity head is computed:

$$\frac{\Delta V^2}{2g} = \left[\frac{Q/A \text{ (upper end)}}{2g} \right]^2 - \left[\frac{Q/A \text{ (lower end)}}{2g} \right]^2$$

In the first approximation it was assumed that the fall in feet was equal to the slope of the energy gradient. Subtracting the difference in velocity head $\Delta V^2/2g$ from the fall used in the first trial gives a new energy gradient S from which a second approximation of discharge is made.

This process can be continued using the discharge in the second approximation as an assumed discharge in the third calculation and the third for a fourth, until the assumed values are essentially the same as those calculated. An average is then made of the final trial calculations for all of the subreaches between the cross-sectional areas and accepted as the discharge for the site in question.

In addition to the formal method given, several procedures are available for making quick design approximations. In these procedures Manning's or the Chezy formula for open-channel flow and the orifice-and-sluice-gate formula for conduits are used. These are valuable only for preliminary study. Formulas, charts, and nomographs are presented under Hydraulic Design Principles, in Art. IV of this section.

F. Rainfall-runoff Methods

Frequently, when there is no gauged record and other structures are not present peak flow quantities must be estimated in some other manner. The most common methods utilize rainfall intensity-duration-frequency relationships and make adjustments for infiltration rate in determining the peak flow rate.

Several basic assumptions are necessary to use these methods. It must be assumed that peak rates of runoff coincide with peak rates of rainfall, that all portions of the watershed contribute to the peak rate of runoff, that the infiltration rate used will be constant, and that the rainfall intensity is uniform over the entire watershed; and factors such as antecedent moisture and channel storage are neglected. These basic assumptions indicate the limitations of such methods and the limited manner in which they should be used.

The initial step in this type of design is to select the rainfall intensity that corresponds to the frequency of the peak runoff storm which the structure is to accommodate. A considerable amount of rainfall data has been collected for a number of years, and several sources of intensity-frequency curves are available. Perhaps the most widely known and used of these in the past has been *U.S. Department of Agriculture Miscellaneous Publication* 204, published in 1935 (6). More recent studies by the U.S. Weather Bureau have resulted in a broader coverage of rainfall relationships and utilize a longer period of record. One of these publications is *Technical Paper* 25 (8). Various state agencies and others have published material on rainfall in specific areas (7, 12, 13) which usually represent a more dense coverage of stations and are not merely based on a few first-order statistics in each state. When the area in question is an appreciable distance from a first-order station shown in a general publication for the entire United States, it will be more practical to use frequency relationships from locally published sources or to prepare these from data at local stations. Methods of preparing curves are not included; however, a number of methods have been presented by others (2–4, 13, 14).

Before any attempt can be made to select a numerical value for rainfall intensity it is necessary to arrive at a *duration*, or length of rainfall at a given intensity, since rainfall intensity varies with the duration.

One of the basic assumptions in a rainfall-runoff method is that the design storm will be of a duration great enough to allow water to be arriving at the outlet of the watershed simultaneously from all parts of the drainage area. The minimum duration of the rainfall intensity selected then should be the time it takes for the farthermost water in the drainage area to reach the structure site, usually referred to as time of concentration. Otherwise, part of the area will not have contributed to the peak flow at the site when the intensity ceases.

The "time of concentration," then, is the time interval required for water to flow from the most remote point in the drainage area to the outlet. The time of flow in the channel can be approximated by computations of velocity, and some data are available in approximating the overland flow portion of the time. Formulas and charts for approximating times of concentration have been developed (15–19), some of which are included in this section.

The selection of the frequency for the storm on which the design is to be based is identical with the selection of peak discharge frequency discussed in Art. III. The tables included herein (Table 4-3) should be used only as a guide in this selection, with individual situations taking precedence according to the risk of economic loss and damage involved.

Two methods of evaluating some of the variables relating rainfall and runoff will be discussed in the following pages. One of these has been called a "rational" and the other a "semirational" method (20).

Based on a direct relationship between rainfall and runoff, the rational formula used is municipal and airport drainage design (14, 21, 22). Since it rests on the same assumptions as other rainfall-runoff methods, care should be taken to limit its use. Also, it neglects factors which may be of small importance in some instances and of great importance in others. It does not take into account either the retarding effect of channel storage or the variation of rainfall intensity throughout the area during the time of concentration and consequently can produce exaggerated runoff values even when exact values for rainfall and imperviousness factors have been used. Considering also that antecedent watershed conditions are ignored, the error would increase with watershed size. The method, therefore, is more reliable for smaller built areas. Solutions by this method should be confined to areas under 1,000 acres.

The method is expressed by the equation:*

$$Q = CIA$$

where Q = runoff, cfs, for total drainage
   C = a coefficient representing the ratio of runoff to rainfall
   I = intensity of rainfall, in. per hr (taken from curves similar to Fig. 4-1)
   A = drainage area, acres

The value of C is selected on the basis of the type of drainage area, and a weighed— or "averaged-out"—value is often used where the cover varies widely. Tables are included as an aid in the selection of this value (Tables 4-4 and 4-5).

G. Peak Rate of Runoff Curve

The curve shown in Fig. 4-2 is based upon studies conducted by the U.S. Department of Agriculture, Soil Conservation Service (23), and adapted to highway culvert drainage by Carl F. Izzard of the Bureau of Public Roads (24, 25). It is recommended that the use of this curve be limited to watersheds smaller than 1,000 acres and to

---

* Actually, the formula is not dimensionally correct, in equating 1 in. per hr per acre to 1 cu ft per sec. The error is within 0.8 per cent of the numerical results, however, and can be considered as correct for all practical purposes.

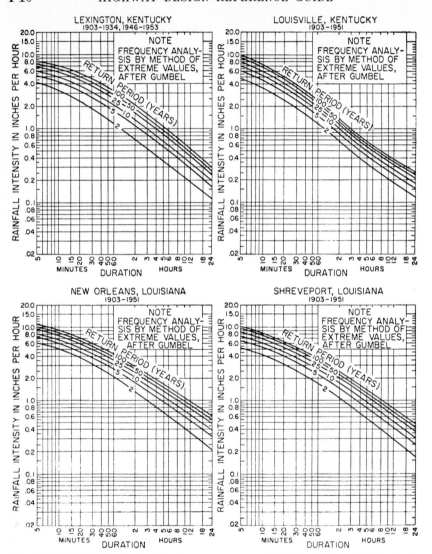

Fig. 4-1. Rainfall intensity-duration-frequency curves. (*Ref. 8. Courtesy of U.S. Weather Bureau.*) These are only 4 of 203 rainfall intensity-duration-frequency curves for stations distributed throughout the United States and Puerto Rico included in the reference cited.

farmed or wooded lands in the Eastern and Middle Western United States where the rational method is not applicable because of the wide variation of the runoff-coefficient conditions of the watershed at the time the storm occurs.

The curve gives peak rates of runoff that may be expected to be equaled or exceeded on the average of once in 25 years on mixed-cover agricultural watersheds in the humic section of the United States and in localities where the rainfall factor is 1.0.   By using the suggested factors the curve can be adjusted for other conditions of rainfall (Fig. 4-3), land use, and design frequencies, as shown in the table and example in Fig. 4-1 (24).

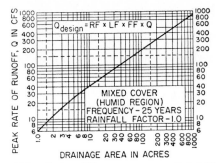

DRAINAGE AREA IN ACRES

Land-use and Slope Factors (LF)

Rainfall Factor (RF): See Fig. 4-3

Land slope	Steep, over 2%	Flat, 0.2%	Very flat, no ponds
100% cultivated (row crops)..	1.2	0.8	0.25
Mixed cover.................	1.0	0.6	0.2
Pasture.....................	0.6	0.4	0.1
Woods, deep forest litter......	0.3	0.2	0.05

Example:
100 acres near Nashville, Tenn., cultivated land sloping about 0.5%, design frequency 10 years

Solution: (see equation on graph)
$Q_{10} = 1.2 \times 0.9 \times 0.8 \times 190$
$= 170$ cfs
(Accuracy of basic data does not justify carrying more than two significant figures)

Frequency Factors (FF)

Frequency, years..........	5	10	25	50
Factor...................	0.6	0.8	1.0	1.2

Fig. 4-2. Peak rates of runoff for watersheds under 1,000 acres. (*Source: Derived in part from Potter "Surface Runoff from Small Agricultural Watersheds," Research Report No. 11-B, Highway Research Board, 1950. Land use and slope factors for flat and very flat land slopes are estimated and subject to revision when observed data become available.*)

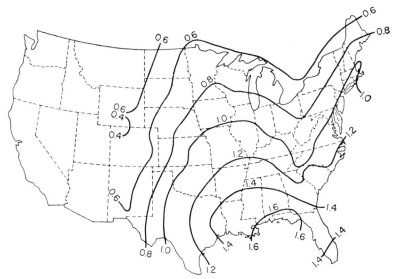

Fig. 4-3. Rainfall factors for use with Fig. 4-2 in estimating peak rates of runoff. (*Ref. 24*, Izzard. *Courtesy of ASCE.*)

Although this curve is based on stream-flow data, it should not be substituted for such data when available for a particular site. Also, its records of runoff from smaller watersheds are not extensive enough to furnish proof that the same empirical relationships apply in arid as in humid regions.

The values of some of the factors are tentative and subject to change; therefore, the results should be used with caution (25).

Similar methods for use on watersheds within areas of specific soil types have been published for the Allegheny-Cumberland Plateau and the glaciated sandstone and shale areas (26). These should be consulted for drainage problems within those areas.

H. Talbot's Formula

Empirical formulas for estimating the area of waterway opening required for culverts have been widely used in the past. These have been attractive to many because of the simplicity and because of the lack of anything better. Previously their use was justifiable because of the very limited knowledge of the hydrology and hydraulics of culvert design. However, they are gradually being replaced by more accurate and scientific methods.

Perhaps the best known of the empirical formulas is Talbot's formula:

$$a = C \sqrt[4]{A^3}$$

where a = required waterway opening
A = drainage area, acres
C = a runoff coefficient (varies from 1 to $\frac{1}{5}$ or less, depending upon the characteristics of the drainage area)

This formula, proposed near the end of the nineteenth century, approximates the area of opening required without any consideration of hydraulic principles and various design frequencies. Limited investigations have indicated that its use yields a waterway opening that is approximately adequate to accommodate the 10-year flood and a 10-ft-per-sec velocity through culverts located in the Middle Western United States; but it has little or no scientific verification and should be considered strictly as a rule-of-thumb method.

I. Time of Concentration and Inlet Time

Time of concentration refers to the length of time required for waters from all portions of a drainage area above a given point to concentrate at that point. It is thus normally equivalent to the time required for water at the farthermost portion of the area to reach the point of reference. In culvert design this point of reference is almost always understood to be the inlet of the culvert in question, which makes it identical with what is called "inlet time" (time for water from all parts of the area to begin reaching a given inlet). In storm-sewer and airport-drainage design, however, it often refers to the time required for all water from the area and from any contributing subareas to attain a point below the inlet, and thus will be equal to the inlet time plus the time of flow from the inlet, sometimes through a complex network of conduits, to the given point. Time of flow can be calculated by use of Manning's formula.

Inlet time may be derived from use of the methods which follow.

Pickering (18) suggests the use of either of the following equations for estimating time of concentration t_c for rural areas in California:

$$t_c = 7.25 \left(\frac{L^2}{S}\right)^{\frac{1}{3}}$$

where L = horizontally projected length of watershed, miles
S = average slope of watershed, ft per ft

$$t_c = \left(\frac{11.9L^3}{H}\right)^{0.385}$$

where L = horizontally projected length of watershed, miles

H = difference in elevation between site and farthermost point on watershed, ft

Time of concentration may also be determined by the use of charts, two of which are presented herein as examples (Figs. 4-4 and 4-5).

In addition to the formulas and charts, some tables are available for use in making quick estimates of time of concentration for certain general types of terrain. One of these, developed for "rolling terrain" and for watersheds with about 5-ft fall per 100 ft and a length about twice the average width, is presented in Table 4-1.

The curves of Fig. 4-6 and Table 4-2 are recommended for estimating inlet time.

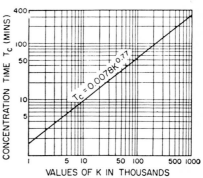

FIG. 4-4. Chart for estimating the time of concentration for small agricultural watersheds. (*Ref. 17. Rouse, Hunter: "Engineering Hydraulics," John Wiley & Sons, New York, 1950.*)

L = maximum length of travel, ft

H = difference in elevation between most remote point and outlet, ft

S = slope = H/L

$K = L/\sqrt{S} = \sqrt{L^3/H}$

J. Design Frequency

Although the final selection of the frequency to be used for a particular structure must be left to the judgment of the engineer, the following considerations should be kept in mind: (1) value of road or street, whether of high-type or low-type construction; (2) possible damage to adjacent property; (3) maintenance cost; (4) amortization cost of drainage structure during service life of roadway; (5) inconvenience to traffic, and, most important, (6) danger to human life. In all cases it is good practice to check the performance of the facility being designed for the more severe storms of lesser frequency.

Table 4-1. Time of Rainfall Concentration for Typical Agricultural Watersheds in Rolling Country

Size of watershed, acres	Time of minimum concentration, min	Size of watershed, acres	Time of minimum concentration, min
1	1.4	100	17
3	3.0	200	23
5	3.5	300	29
10	4.0	400	35
20	4.8	600	47
30	8.0	800	60
50	12.0	1,000	75

SOURCE: Armco Drainage and Metal Products, Inc., *Handbook of Drainage and Construction Products,* The Lakeside Press, Chicago, 1950, p. 200.

Design frequencies are specified by various agencies for certain types of projects. For example, the 5-year storm frequency is generally recommended for estimating runoff for airport projects. For the Federal Interstate System all culverts and bridges should be designed to accommodate peak discharge of the 50-year frequency or the greatest on record, whichever is greater, and with runoff estimates based on future land developments expected for 20 years hence. All other facilities in the interstate system

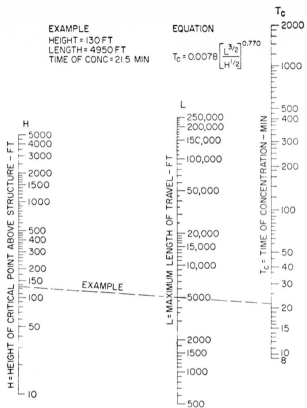

EXAMPLE
HEIGHT = 130 FT
LENGTH = 4950 FT
TIME OF CONC = 21.5 MIN

EQUATION

$$T_c = 0.0078 \left[\frac{L^{3/2}}{H^{1/2}} \right]^{0.770}$$

FIG. 4-5. Nomograph solution used by Kentucky Department of Highways. (*Prepared by R. D. Hughes.*)

should be designed to keep the traveled ways usable during floods at least as great as the 10-year frequency, except that a 50-year frequency should be used at underpasses or other depressed roadways where anticipated ponded water can be removed only through storm drains.

K. Runoff Coefficients

Tables 4-4 and 4-5, which follow, can be used in estimating a numerical value of the factor C to be used in the rational formula. When the drainage area is made up

Table 4-2. Inlet Times Used for Design in Several Major Cities

City	Inlet Time, Min
Buffalo	5–10*
Cincinnati	8–10
Cleveland	8
Detroit	15
District of Columbia	5–12
Milwaukee	5–10
New York City—Manhattan	4
St. Louis	20†

* Minimum time of concentration used in design, 15 min.
† Minimum rainfall time including inlet time.
SOURCE: Calvin V. Davis, *Handbook of Applied Hydraulics,* 2d ed., McGraw-Hill Book Company Inc., New York, 1952, p. 1037.

of several types of cover and paved surfaces, the coefficient used should be a weighed value assigned in accordance with the values of the various component areas, as taken from a reliable source such as Table 4-4 below. Slope must be considered in selecting the value within the ranges suggested.

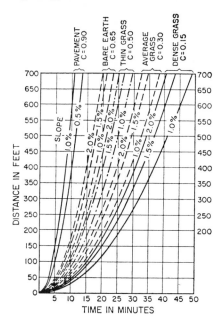

Fig. 4-6. Curves for estimating inlet time. (*From CAA, Airport Drainage, U.S. Government Printing Office, November, 1956, p. 13.*)

Table 4-3. Design Frequency for Various Types of Structures

Type of Structure	Design Return Period, Years
Bridges on important highways or where backwater may cause excessive property damage or result in loss of the bridge	50-100
Bridges on less important roads, or culverts on important roads	25
Culverts on secondary roads, storm sewers, or side ditches	5-10
Storm-water inlets, gutter flow	1-2*

* If of short duration, ponding can be tolerated.
Source: E. M. West and W. H. Sammons, *A Study of Runoff from Small Drainage Areas and the Openings in Attendant Drainage Structures*, Highway Materials Research Laboratory, Kentucky Dept. of Highways, Lexington, Ky. July, 1955, p. 40.

Table 4-4. Value of C for Different Types of Surfaces

Type of Surface	Value of C
All watertight roof surfaces	0.75-0.95
Asphalt runway pavements	0.80-0.95
Concrete runway pavements	0.70-0.90
Gravel or macadam pavements	0.35-0.70
Impervious soils (heavy)*	0.40-0.65
Impervious soils with turf*	0.30-0.55
Slightly pervious soils*	0.15-0.40
Slightly pervious soils with turf*	0.10-0.30
Moderately pervious soils*	0.05-0.20
Moderately pervious soils with turf*	0.00-0.10

* For slopes from 1 to 2 per cent.
Source: *Handbook of Drainage and Construction Products*, Armco Drainage and Metal Products, Inc., Middletown, Ohio, Table 23-5, p. 201, 1955.

Table 4-5. Runoff Coefficients Used in Various Cities

City	Type of area	Runoff coefficients
Buffalo, N.Y.	Residential	0.48–0.58
	Residential (apartments)	0.60–0.65
	Commercial	0.60–0.70
	Industrial	0.55–0.60
Cincinnati, Ohio	Suburban (large lots)	0.30
	Residential	0.35–0.40
	Residential (apartments)	0.50–0.60
	Tenements and industrial	0.70
	Commercial	0.80–0.85
Cleveland, Ohio	Residential	Varying 0.60–1.00
	Downtown areas	1.00
Detroit, Mich.	Impervious	Varying 0.50–0.95
	Pervious, clay	Varying 0.10–0.70
	Pervious, sand	Varying 0.01–0.55
District of Columbia		0.60–0.85
Louisville, Ky.	Impervious	Varying 0.60–0.95
	Pervious	Varying 0.20–0.70
Milwaukee, Wis.	Residential	0.40–0.50
	Local business	0.65
	Commercial and industrial	0.80
	Special cases	0.90
New York City—Manhattan	Central Manhattan	0.75
	More open sections	0.60
	Parks	0.25
	Streets	1.00
Rochester, N.Y.	Residential	0.25–0.40
	Commercial	0.50–0.85
	Industrial	0.60
St. Louis, Mo.	Impervious	Varying 0.50–0.95
	Pervious	Varying 0.22–0.60

SOURCE: Calvin V. Davis, *Handbook of Applied Hydraulics*, 2d ed., McGraw-Hill Book Company, Inc., New York, 1952.

Many metropolitan areas have developed a more or less standard range of local values through practice. Although Table 4-5 suggests some of these, it is advisable to consult with local city and state officials on such matters.

Table 4-6 is recommended for use in determining the value C for use in Talbot's formula. It must be remembered, however, that these are merely values considered representative in a very general way, and judgment should be exercised in their use.

Table 4-6. Deductions from Unity to Obtain a Runoff Coefficient for Agricultural Areas, for Use in Talbot's Formula

Type of Area*	Value of Deductions
Topography:	
Flat land, with average slopes of 1 to 3 ft per mile	0.30
Rolling land, with average slopes of 15 to 20 ft per mile	0.20
Hilly land, with average slopes of 150 to 250 ft per mile	0.10
Soil:	
Tight impervious clay	0.10
Medium combinations of clay and loam	0.20
Open sandy loam	0.40
Cover:	
Cultivated lands	0.10
Woodland	0.20

Example:
Given: Flat land with average slopes of 1 to 3 ft per mile, open sandy loam, and woodland.
Find: C for above given conditions.
Solution: 1. $1.0-0.3 = 0.7$
2. $1.0-0.4 = 0.6$
3. $1.0-0.2 = 0.8$

$$C = \overline{2.1} \div 3 = 0.7 \text{ (average for the area)}$$

SOURCE: M. M. Bernard, "Discussion of Run-off Rational Run-off Formulas, by R. L. Gregory and C. E. Arnold," *Transactions ASCE*, Paper 1812, p. 1038, 1932.

IV. HYDRAULIC DESIGN PRINCIPLES

For clarification in the use of the basic hydraulic principles,* design problems are divided into two categories: (1) those involving open-channel flow and (2) those involving closed conduits flowing full. (The same hydraulic design principles apply to a closed conduit flowing partially full as to an open channel.) In dealing with culverts it is necessary to make an additional classification according to the manner of operation (see following section on culvert design).

A. Flow in Open Channels

Open-channel flow is defined as flow in any conduit, closed or open, where the water surface is free. This flow may be either laminar, turbulent, steady, uniform, steady uniform, or steady nonuniform. Definitions of these terms are given at length in Woodward and Posey (28). Most flow formulas assume steady uniform flow when the discharge quantity and velocity are the same at all points along the channel.

The hydraulic principles applicable to flow in open channels are used in the hydraulic design of channels, ditches, gutters, flumes, storm sewers, and in some cases culverts.

The *depth of flow* in an open channel depends upon the discharge quantity, slope of channel, roughness of channel, and shape of the cross section. The depth of flow under conditions where these four factors remain constant is termed *normal depth*. On flat slopes the normal depth is relatively great, but as slope increases, it becomes less. At a certain value of slope, termed *critical slope*, the point of control for the quantity of discharge shifts from downstream to upstream. At this point the normal depth of flow equals the *critical depth*. The critical slope is the slope of the conduit that will maintain flow at critical depth. Critical depth is the depth of flow in a conduit at which the minimum *energy head* is required for a given discharge. Critical depth occurs when the *velocity head* is equal to half the mean depth. *Critical velocity* is the velocity of flow at conditions of critical flow.

These relationships are basic in evaluating the operation of conduits; from them location of the control and flow criteria are determined. When control is located downstream, changes in cross section or obstructions or any change in the flow character-ter will be reflected upstream as backwater. When control is upstream, changes in cross section and the like downstream will not be reflected upstream. The control is upstream when slope is greater than critical and downstream when less than critical. The velocity of flow is subcritical (tranquil) when control is downstream and super-critical (rapid) when upstream. These features are extremely important in con-sidering the effects of obstructions or changes in cross section as well as in considering erosion protection.

The relationships for various cross sections and shapes of conduits can be worked out mathematically; however, this can be greatly simplified by use of tables and charts. Numerous tables and charts of this nature, too voluminous for this text, can be found in other publications (28, 31–33).

Manning's formula for velocity of flow is used for channels or conduits with assumed steady, uniform flow. For its use the channel itself should be uniform, and the slope of the water surface constant and parallel to the slope of the bottom of the channel. To some extent the effect of irregularities in the channel or conduit can be taken into account in the selections of the roughness value; but lack of parallelism between the water surface and the general grade line of the bottom may cause direct application of the formula to give grossly inaccurate results (28). The formula is stated as

$$V = \frac{1.486}{n} R^{2/3} S^{1/2}$$

* For background material concerning the principles used and for further information, various texts on hydraulics may be consulted (14, 17, 28–31).

where V = average velocity, ft per sec
  n = Manning's roughness coefficient (Table 4-7)
  R = hydraulic radius, ft = area of flow cross section divided by wetted perimeter
  S = slope of channel, ft per ft
  Nomographs have been developed for the solution of this formula, and one of these, prepared by the Bureau of Public Roads, is presented in Fig. 4-7.

$$\text{EQUATION } V = \frac{1.486}{n} R^{2/3} S^{1/2}$$

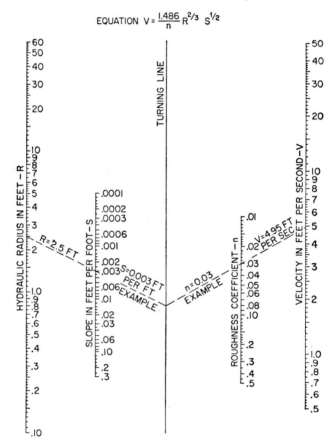

Fig. 4-7. Nomograph for solution of Manning equation.

Discharge quantities for open-channel flow may be determined by combining Manning's formula with the general flow formula $Q = AV$.

$$Q = \frac{1.486}{n} A R^{2/3} S^{1/2}$$

where Q = discharge, cu ft per sec
  A = area of flow cross section, sq ft
  Tables for solution of Manning's formula both for velocity and for discharge can be found in King's *Handbook of Hydraulics* (31) and other sources (28, 32, 33). Charts based upon Manning's formula for solving open-channel problems have been prepared by the U.S. Bureau of Public Roads (32). One of these charts and an example are presented in Fig. 4-8.

Manning's formula can also be used in estimating velocity and discharge quantity for natural channels. By taking a series of typical cross sections and superimposing them, an average cross-section area can be arrived at for use in the formula. In some instances this cross section will be geometrically similar to channels listed in the charts, and the solution can be made directly. Otherwise the hydraulic relationships can be taken from the tables for solution of Manning's formula.

When Manning's formula is used for natural channels, considerable care should be taken in estimating the value of n, the coefficient of roughness. Table 4-7 should be consulted as a guide, and special consideration should be given to the condition of the channel, type of material in the channel bed and banks, and aquatic growth in the channel and the meandering of the stream.

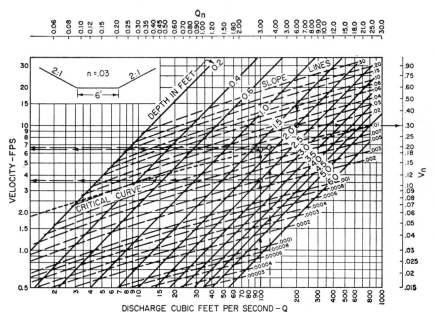

FIG. 4-8. Chart for solution of trapezoidal-channel-flow methods. Channel chart, 2:1; $b = 6$ ft.

Example A
Given: Discharge $Q = 100$ cfs
Roughness $n = 0.03$
Slope $S = 0.003$ ft per ft
Solution: Read vertically from $Q = 100$ to intersection of slope line 0.003 and note normal depth $d_n = 2.5$ ft; reading horizontally to the left the normal velocity $V_n = 3.65$ ft per sec. Continuing vertically upward from $Q = 100$, $S = 0.003$ to critical curve, read critical depth $d = 1.7$ ft, critical velocity $V_c = 6.40$ ft per sec, and critical slope $S_c = 0.015$.

Example B
Given: Discharge $Q = 120$ cfs
Roughness $n = 0.05$
Slope $S = 0.03$
Solution: $Qn = 120 \times 0.05 = 6.00$. Read down from $Qn = 6.00$ to intersection of slope line $S = 0.03$ and note $d_n = 2.0$ ft. Read horizontal to the right $Vn = 0.30$; thus $V_n = 0.30/0.05 = 6$ ft per sec.* Reading vertically from $Q = 120$ cfs to inter-

* Critical depth is independent of roughness, and therefore critical depth and velocity must be read from the Q scales.

Table 4-7. Roughness Coefficient n for Use with Manning's Formula

Manning's n

Excavated ditches and channels:
Earth, unlined, straight and uniform	0.020–0.025
Earth, unlined, irregular	0.025–0.035
Earth, with light vegetation	0.035–0.045
Earth, with fairly heavy vegetation	0.040–0.050
Earth, drag-line excavation	0.028–0.033
Rock cuts, smooth and uniform	0.030–0.035
Rock cuts, jagged and irregular	0.035–0.045

Lined channels and ditches:
Concrete, float finished	0.013–0.017
Concrete, Gunite	0.016–0.022
Riprap	0.020–0.030
Concrete sides only, gravel bottom	0.017–0.020
Riprap sides only, gravel bottom	0.023–0.033
Asphalt	0.013–0.016

Natural streams:
Clean, straight banks, fairly uniform bottom, full stage	0.027–0.033
Clean, straight banks, fairly uniform bottom, full stage, with some vegetation	0.033–0.040
Clean, meandering, with minor pools and eddies	0.035–0.050
Sluggish streams, with deep pools and winding channels	0.060–0.080
Sluggish streams, with deep pools and winding channels with dense vegetation	0.100–0.200*
Rough, rocky streams in mountainous terrain	0.050–0.080
Overflow areas adjacent to regular channel	0.030–0.200*

Structures:
Concrete pipe	0.013–0.015
Vitrified-clay pipe	0.012–0.014
Cast-iron pipe	0.013–
Corrugated-metal pipe	0.019–0.024
Concrete (monolithic)	0.013–0.017

* Wide range in value depending upon land use, with higher value for floods of sufficient stage to submerge tree branches during time of heavy foliage.

SOURCE: S. M. Woodward and C. J. Posey, *Hydraulics of Steady Flow in Open Channels,* table 101, p. 5, John Wiley & Sons, New York, 1941.

section of critical curve, note d_c = 1.8 ft and V_c = 6.7 ft per sec. Reading diagonally for d_c = 1.8 to intersection of Qn = 6.0, note that critical slope S_c = .04, or 4 per cent.

B. Flow in Closed Conduits

The principal objective in culvert design is to provide the most economical means, within specific limits of head-water elevation and velocity, of transmitting a given discharge from one side of the roadway to the other. To accomplish this, many factors must be taken into account even though some may not influence the functioning of the culvert that is finally designed and built. For a given situation, a number of alternate structures may be equally satisfactory from the hydraulic standpoint, and the choice would then be made on the basis of economic factors, such as first cost and durability, or even upon aesthetic value.

In the hydraulic analysis for culvert design the first thing to be considered is the stream channel. A culvert placed in a stream has little effect on the normal stream characteristics above the ponding area at the inlet, and below the outlet the distance of influence is short. The extent of this distance is governed by the amount of turbulence or other disturbances caused by the structure.

The quantity of water approaching a culvert in a given time is assumed to be uniform and continuous during the peak rate of runoff; and by this assumption, an equal amount of water would be leaving the culvert location by way of the downstream channel. Actually, there may be momentary retardation of the flow, lasting until the storage capacity of the ponding area is reached; but this will be followed by equilibrium between rate of supply and downstream runoff. If sufficient opening in the culvert is not provided, equilibrium will be attained through overflowing of the roadway.

Since the elevation of the water at the inlet end of a culvert (headwater elevation) is a limiting factor in the design, it is necessary to evaluate the relationship between headwater depth and discharge—a relationship which depends upon the manner in which the culvert operates.   For this evaluation culvert operation can be divided into two major classifications: (1) full flow at the outlet and (2) partial full flow at the outlet.

The normal operation of the outfall channel plays an important role in determining under which of these classifications a culvert will operate, as well as the method by which the headwater discharge relationship is analyzed.

A culvert will fall into the first general classification (full outlet flow) if laid on a grade no greater than that of the friction slope and if the outlet end is submerged by the tailwater elevation.   (Normal depth of flow in the outfall channel is greater than the height of the culvert.)   When the tailwater depth is not sufficient to submerge the structure at the outlet and the structure is laid on a slope equal to or greater than that sufficient to overcome friction, the structure will not flow full.

The headwater depth of culverts flowing full can be computed by Bernulli's equation:

$$H = \left(1 + K_e + \frac{29n^2L}{R^{4/3}}\right) \frac{V^2}{2g}$$

where H = difference in elevation between headwater elevation and elevation of tail-water surface, or difference between headwater elevation and crown at outlet where culvert is flowing full without tailwater being above crown

$V^2/2g$ = velocity head, ft

K_e = coefficient of entrance loss, varying from 0.4 for square-edged entrance to 0.1 for well-rounded entrances

n = Manning's coefficient for roughness (see Table 4-7)

L = length of culvert, ft

R = hydraulic radius, ft

Whenever the inlet of a culvert is submerged by the headwater depth, a convenient method of determining whether the culvert will flow full or part full is by finding the location of the control section.   This section represents the critical point of the system, since the principal flow characteristics are determined by it and by its location, whether at the inlet or in the barrel of the culvert.

Whenever the control is at the inlet—as is usual for standard culverts on mild to steep grades with tailwater below the crown of the culvert at the outlet—the head-water-elevation–discharge relationship is not affected by the friction in the barrel or conditions at the outlet.   The operation in this case is analogous to an orifice, and an orifice formula can be used.

Since the flow in culverts can be any of several types, the methods of classification and solutions for headwater depth prepared by the U.S. Bureau of Public Roads (32) are presented, in slightly modified form, in the material which immediately follows.

In order to simplify discussion of design procedure, the methods of operation of standard culverts have been divided into two classes, as follows:

Class I: Culverts that flow with a free-water surface through the control section and with the entrance *not* submerged ($HW \leq 1.2D$)   (Types 1 to 4, Fig. 4-9).

Class II: Culverts that flow with a submerged entrance ($HW > 1.2D$)   (Types 5 to 8, Fig. 4-9).   In this class, the method of operation is generally independent of the slope.

The four types of flow in each of these classes are described in the following paragraphs.

Class I, Type 1.   The entrance is free ($HW \leq 1.2D$), the slope at design discharge is less than critical ($S_o < S_c$), and tailwater is below critical depth at the outlet ($TW < d_c$).   This is a common condition of flow for culverts where the natural channels are on relatively flat grades and have wide, flat floodplains.   The control is critical depth at the outlet, and headwater depth is found by use of the equation

$$HW = d_c + \frac{V_c^2}{2g} + h_e + h_f - S_oL$$

where HW = headwater depth above culvert invert
$\quad\quad d_c$ = critical depth at culvert outlet
$\quad V_c{}^2/2g$ = velocity head at critical depth
$\quad\quad h_e$ = entrance loss
$\quad\quad h_f$ = friction loss in barrel
$\quad\quad S_oL$ = fall in length of culvert

A simplified solution of this equation requires the use of certain approximations in the determination of h_e and h_f; however, the results are reasonably accurate. The

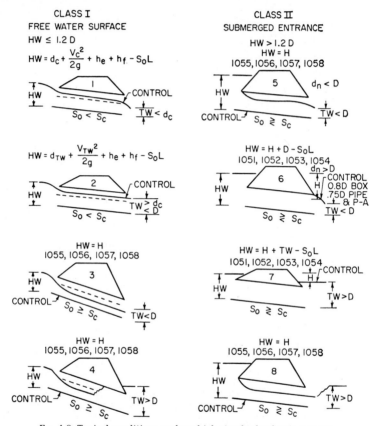

FIG. 4-9. Typical conditions under which standard culverts operate.

solution of this equation is facilitated by the use of various charts. Figures 4-10 to 4-13 will give the approximate depth of headwater for culverts of this type without requiring solution of the above equation.

It is possible for the flow to submerge the entrance with the headwater between $1.2D$ and $1.5D$ without the control moving to the entrance, and it is not illustrated in Fig. 4-9. The operation remains a Type I case, but the upstream portion of the barrel flows full.

Outlet velocity is the discharge divided by the area of flow at critical depth (critical velocity).

Class I, Type 2. The entrance is free ($HW \leq 1.2D$), the slope at design discharge is less than critical ($S_o < S_c$), and the tailwater is between critical depth at the outlet and the crown of the culvert at the outlet ($TW > d_c$, but $< D$). This is believed to be

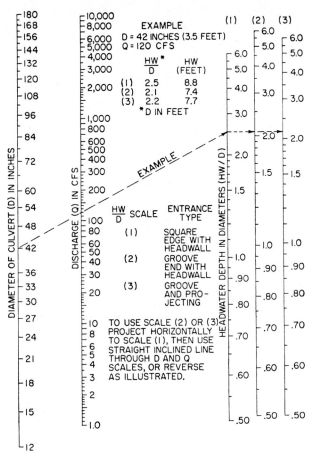

FIG. 4-10. Headwater depth for concrete-pipe-culverts with entrance control.

a case not commonly requiring solution because it will be seldom that tailwater can be estimated sufficiently accurately to make certain that culvert operation falls in this category. This type of operation is most likely to be encountered with a deep narrow outlet channel on a flat slope. The control is at the outlet, and headwater depth is found by use of equation

$$HW = TW + \frac{V_{TW}{}^2}{2g} + h_e + h_f - S_oL$$

where HW = headwater depth above culvert invert
$\quad TW$ = tailwater depth above the culvert invert at outlet
$\quad V_{TW}{}^2/2g$ = velocity head at tailwater depth
$\quad h_e$ = entrance loss
$\quad h_f$ = friction loss in culvert barrel
$\quad S_oL$ = fall in length of culvert

Like Type 1, it would be possible for the entrance to be submerged with headwater depth between $1.2D$ and $1.5D$ without changing the location of the control. Outlet velocity is the discharge divided by the area of flow in the culvert at tailwater depth.

Class I, Type 3. The entrance is free ($HW \leq 1.2D$), the slope is equal to or greater than critical ($S_o \geq S_c$), and the tailwater is below the crown of the culvert at the outlet

$(TW < D)$.   This is a common method of operation for culverts in rolling or mountainous country when the flow does not submerge the entrance.   The control is critical depth at the entrance, and headwater depth cn be determined by use of Figs. 4-10 or 4-11 for pipe culverts, 4-12 for pipe-arch culverts, and 4-13 for box culverts.   Flow in the culvert will accelerate from critical velocity near the entrance and will approach normal velocity for the barrel slope.   Outlet velocity may usually be estimated between these limits, approaching normal velocity in longer culverts.

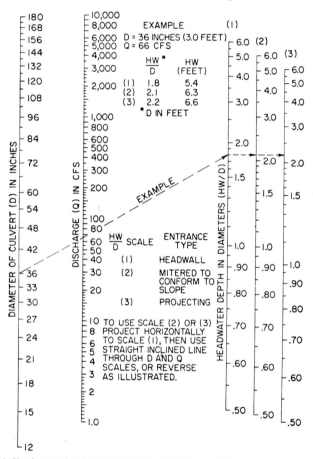

Fig. 4-11. Headwater depth for corrugated-metal pipe culverts with entrance control.

Class I, Type 4.   The entrance is free ($HW \leq 1.2D$), the culvert slope is equal to or greater than critical ($S_o \geq S_c$), and the tailwater submerges the outlet ($TW > D$). This type of operation will convert to the Class II, Type 7, full flow, if the hydraulic jump travels up the barrel to the entrance, or to Class II, Type 8, if the jump is pushed out of the barrel.   To move the jump to the entrance requires that the pressure line be above the entrance crown.   Therefore the limitation on Type 4 operation is that the tailwater elevation plus the friction loss in the full barrel must be less than the elevation of the crown at the entrance ($TW + S_f L) < (D + S_o L)$.

Type 4 is not a common method of operation, because in rolling or hilly country with natural outlet channels it is seldom that tailwater would be sufficient to submerge the outlet.

In Type 4 operation the control would be critical depth at the entrance, and head-water depth can be determined by use of Figs. 4-10 or 4-11 for pipes, 4-12 for pipe arches, and 4-13 for box culverts. Outlet velocity should be based on the full area of cross section of the culvert outlet.

Class II, Type 5. The entrance is submerged ($HW > 1.2D$), the tailwater elevation is below the crown of the culvert at the outlet ($TW < D$), and the normal depth is less than the height of the barrel ($d_n < D$). This is a common form of culvert operation.

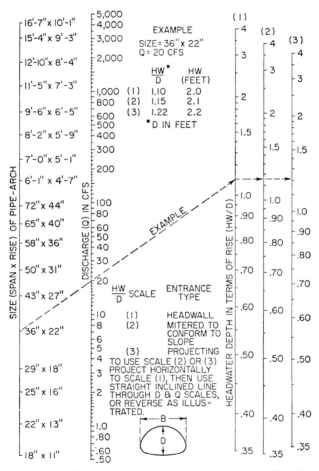

FIG. 4-12. Headwater depths for corrugated metal pipe-arch culverts with entrance control.

With normal depth less than the height of the culvert barrel, the culvert cannot flow full. The culvert entrance produces flow similar to that under a sluice gate. The control is at the entrance, and headwater depth can be determined by use of Figs. 4-10 or 4-11 for pipe culverts, 4-12 for pipe-arch culverts, and 4-13 for box culverts. The headwater depth in this case is independent of the roughness of the culvert. The depth of water downstream from the contracted flow at the entrance of Type 5 culvert will increase or decrease to approach normal depth for that particular culvert, slope, and discharge. An increase in the discharge will increase the normal depth. If this normal depth equals or exceeds the height of the culvert barrel, the culvert could fill and the method of operation would change to Type 6. Outlet velocity is the discharge divided by the area of flow at the outlet, which may be expected to be somewhere between two-thirds of the total area and the area at normal depth.

Class II, Type 6. The entrance is submerged ($HW > 1.2D$), the tailwater is below the crown at the outlet ($TW < D$), and normal depth is greater than the height of the culvert ($d_n > D$), or pipe charts indicate a pipe would flow full. At the present time,

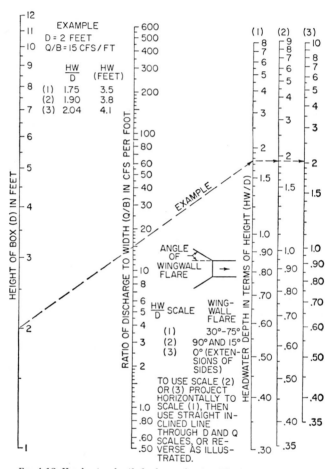

FIG. 4-13. Headwater depth for box culverts with entrance control.

we are not able to predict positively whether a standard culvert meeting the requirements of Type 6 will operate as Type 6 or Type 5. It would be quite possible for a long smooth pipe to operate as Type 5 while a rougher pipe under the same conditions may fill and operate as Type 6. The hydraulics involved are quite complex. When operating as Type 6, the control is at the outlet. Headwater will depend upon velocity head and losses upstream from the outlet and also on the elevation of the pressure line at the outlet. While the actual location of the pressure line may be anywhere above the center of the outlet, for practical design purposes it is assumed to be 0.75D above the invert for pipes and pipe arches and 0.8D for boxes. The assumption produces a conservative value for headwater and permits determination of headwater depth by use of Figs. 4-14 to 4-18. Since the method of operation of a culvert with the above characteristics is not definite, the culvert should be checked for headwater depth for both Type 5 and Type 6 operation and the higher value of headwater should be considered as possible. The outlet velocity for Type 6 is the discharge divided by the full area of cross section. To be on the safe side for design in doubtful cases, however, it is suggested that outlet velocity be computed as for Type 5.

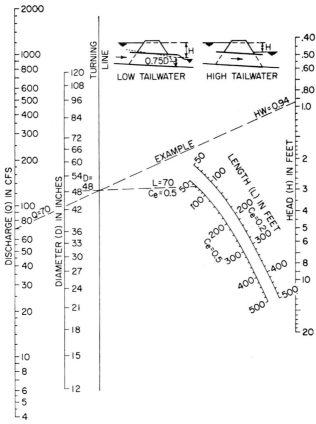

FIG. 4-14. Head for concrete pipe culverts flowing full, $n = 0.015$.

$$H = \left[\frac{2.5204\ (1 + C_e)}{D^4} + \frac{466.18n^2L}{D^{16/3}}\right]\left(\frac{Q}{10}\right)^2$$

where H = head, ft
C_e = entrance loss coefficient
D = diameter of pipe, ft
n = Manning's roughness coefficient
L = length of culvert, ft
Q = design discharge rate, cfs

Class II, Type 7. The entrance is submerged ($HW > 1.2D$), and tailwater is above the crown at the outlet ($TW > D$). Under this condition, the culvert will flow full. This method of operation is considered to be certain only if the tailwater elevation plus the friction loss in the full barrel is above the elevation of the crown of the culvert at the entrance ($TW + S_fL) > (S_oL + D)$. The control is tailwater elevation at the outlet, and headwater depth can be determined by use of Figs. 4-14 to 4-18. Outlet velocity is the discharge divided by the full cross-section area of the culvert.

Class II, Type 8.   The entrance is submerged ($HW > 1.2D$), and the tailwater is above the crown at the outlet ($TW > D$), as in Type 7, but the velocity of the flow at the culvert outlet is high enough to keep the tailwater from submerging the outlet. This type of operation probably results when, with an increasing discharge the rising tailwater reaches the elevation of the crown of the culvert at the outlet after the flow

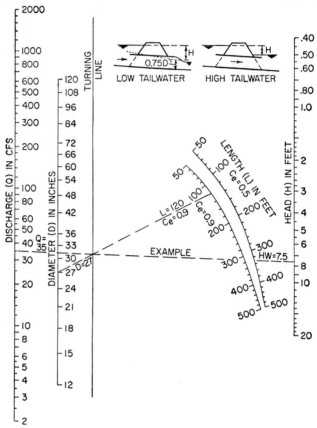

Fɪɢ. 4-15. Head for corrugated-metal pipe culverts flowing full, $n = 0.024$.

$$H = \left[\frac{2.5204(1 + C_e)}{D^4} + \frac{466.18n^2L}{D^{16/3}} \right] \left(\frac{Q}{10} \right)^2$$

where H = head, ft
 C_e = entrance loss coefficient
 D = diameter of pipe, ft
 n = Manning's roughness coefficient
 L = length of culvert, ft
 Q = design discharge rate, cfs

has submerged the entrance.   If the tailwater submerged the outlet before the entrance was submerged, the operation would probably go to Type 7.   With Type 8 operation, the control is at the entrance, and headwater depth can be determined by use of Figs. 4-10 or 4-11 for pipe culverts, 4-12 for pipe-arch culverts, and 4-13 for box culverts. Outlet velocity is the discharge divided by the cross-section area of flow, which is between two-thirds the area of the culvert and the area at normal depth for the culvert slope and roughness.

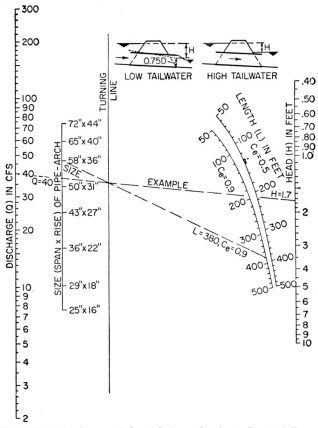

FIG. 4-16. Head for riveted corrugated-metal pipe-arch culverts flowing full, $n = 0.024$.

$$H = \left[\frac{2.460(1 + C_e)}{BD} + \frac{350.7n^2L}{(BD)^2D^{4/3}} \right] \left(\frac{Q}{10} \right)^2$$

where H = head, ft
C_e = entrance loss coefficient
B = span of pipe arch, ft
D = rise of pipe arch, ft
n = Manning's roughness coefficient
L = length of culvert, ft
Q = design discharge rate, cfs

C. Erosion and Silting

The primary objective in channel or conduit design is to find the most economical cross section that is hydraulically satisfactory. In channel work, consideration must be given to the problems of erosion and silting. Velocities must be sufficiently high to prevent silting and yet not great enough to cause harmful erosion. For earth channels, side slopes of 2:1 or flatter are desirable. For channel in rock or other hard material or for channels that are to be lined, this side-slope limitation does not apply.

Velocities of flow calculated from the flow formulas should be compared with the allowable velocities for various channel linings as shown in Table 4-8. To prevent erosion, when velocities are found to be in excess of those recommended, either the channel should be redesigned or erosion protection should be provided. Generally,

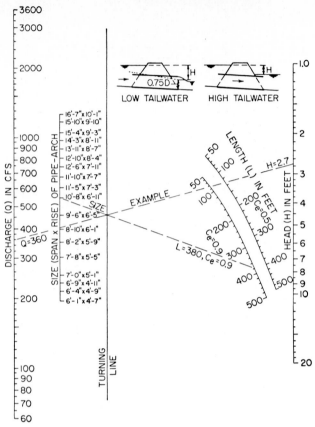

Fig. 4-17. Head for field bolted corrugated-metal pipe-arch culverts flowing full, $n - 0.025$.

$$H = \left[\frac{155.5(1 + C_e)}{A^2} + \frac{4530 n^2 L}{A^2 R^{4/3}} \right] \left(\frac{Q}{100} \right)^2$$

where H = head, ft
C_e = entrance loss coefficient
A = area of pipe-arch opening, sq ft
n = Manning's roughness coefficient
L = length of culvert, ft
R = hydraulic radius, ft
Q = design discharge rate, cfs

when sufficient velocity reduction cannot be attained through decreasing the slope, the alternative is to design for artificial channel linings.

Table 4-8. Maximum Mean Velocities for Erosion Prevention

Material	*Mean Velocity, Ft per Sec*
Sand...	Up to 2.0
Loam...	2– 3
Grass...	2– 3
Clay..	3– 5
Clay and gravel.....................................	4– 5
Good sod, coarse gravel, cobbles, soft shale...........	4– 6
Conglomerates, hard shales, soft rock..................	6– 8
Hard rock, concrete-lined.............................	10–15

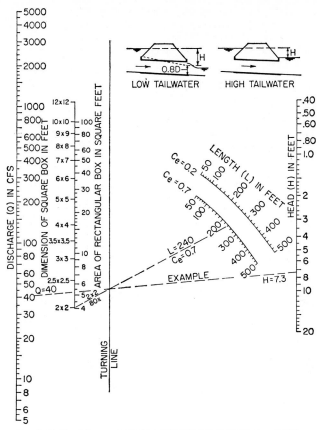

FIG. 4-18. Head for concrete box culverts flowing full, $n = 0.013$.

Equation for square box: $$H = \left[\frac{1.555(1 + C_e)}{D^4} + \frac{287.64 n^2 L}{D^{16/3}} \right] \left(\frac{Q}{10} \right)^2$$

where H = Head, ft
C_e = entrance loss coefficient
D = height, also span, of box, ft
n = Manning's roughness coefficient
L = length of culvert, ft
Q = design discharge rate, cfs

Silting prevention and debris control are essential to good culvert operation. Silting can usually be prevented by velocity control and by designs which minimize eddies and dead-water pockets. In some cases desilting check dams are constructed in the upstream portion of the channel. Various standard types of debris collectors are available, and these are usually included in the standard designs used by highway departments and other government agencies.

V. DESIGN OF DRAINAGE FACILITIES

A. Ditches

The selection of the shape of a ditch cross section depends primarily upon the limitations of spread and depth of water that can be tolerated and whether the ditch is to be paved or unpaved. When velocities are not excessive and earth or sodded ditches can be used (Table 4-8), triangular or trapezoidal shapes are somewhat standard. It is often necessary to pave the ditches where the ditch gradient is steep and where ordinary or sodded channel linings are likely to become eroded. The use of paved ditches creates a wider variety of possible shapes, and rectangular, semicircular, and other cross sections can be used.

The expected runoff can be estimated by the methods outlined previously in Art. III.

Successive solutions of Manning's formula for discharge for various trial cross sections and slopes can be used in the hydraulic design.

The hydraulic design of ditches has been greatly simplified by charts prepared by the U.S. Bureau of Public Roads (Figs. 4-19 to 4-21). From these charts the depth and velocity of flow for various trial cross sections and slopes can be read directly, and visual comparisons of sections can be made. Similar charts have been prepared for

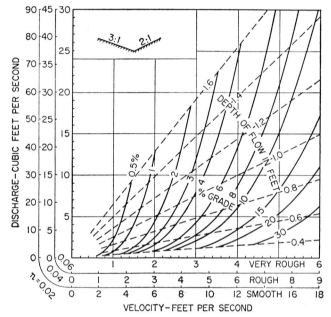

FIG. 4-19. Channel chart, 3:1, 2:1; $b = 0$.

trapezoidal shapes with different bottom widths and side slopes, as well as for rectangular and circular shapes (32).

B. Depressed Median Drains

Depressed median drains are designed in much the same manner as side ditches. Usually located along the center of the median strip, these drains collect runoff from the median itself and from the adjacent half of the pavement on each side when the pavements have their crown at the center.

The rational formula $Q = CIA$ is recommended for computing the design discharge at selected stations along the median. It is usually convenient to estimate a weighed C value, based upon the percentages of the area that are paved and made up of grass and shoulder material, and to multiply this value by the area between 100-ft stations, giving a value of CA per 100-ft station. The discharge at any selected station can then be computed by multiplying this value by the number of stations involved and by the rainfall factor for the selected station.

The time of concentration should take into account the time of flow over half a pave-

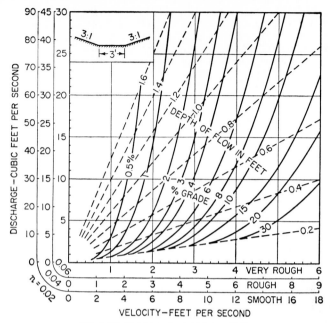

Fɪɢ. 4-20. Channel chart, 3:1; $b = 3$ ft.

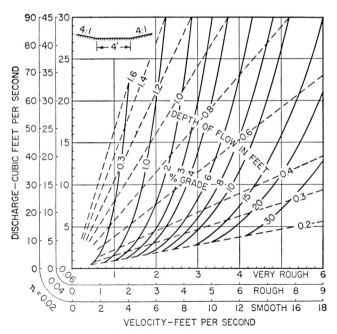

Fɪɢ. 4-21. Channel chart, 4:1; $b = 4$ ft.

ment width, over the shoulder, and along the median to the selected station.   It is also convenient to compute the time of flow in the median for one 100-ft station length. Successive times of concentration along this line can be estimated by adding this additional time increment for each station.   Rainfall intensities for the selected return period can then be read from rainfall intensity-duration-frequency curves (8).

The depth of flow, spread, and velocity can be computed from Manning's formula or read direct from charts similar to Fig. 4-7.

Inlets should be spaced at intervals sufficient to limit the spread to within the allowable amount.   When the velocity reaches the maximum permissible for unprotected soil, the drain should be sodded (see Table 4-8, for allowable velocities).   To prevent

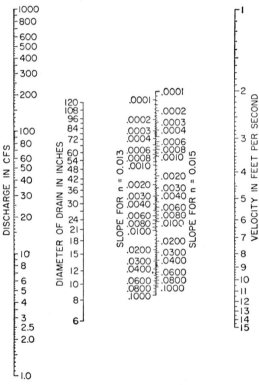

FIG. 4-22. Nomograph for computing required size of circular drain, flowing full, $n = 0.013$ or 0.015.   (*Ref. 34.   Courtesy of U.S. Army, Corps of Engineers.*)

ponding, the longitudinal slope of the drain should be sufficient to cause velocities of at least 0.5 ft per sec.

C. Channel Changes and Relocations

Changes or relocations of natural channels should be designed in accordance with the principles of open-channel flow.   Methods for determining the design discharge are outlined in Art. III.   The hydraulic design consists of estimating the design discharge for the selected return period and sizing a channel which will accommodate this discharge without undesirable overflowing, will provide adequate protection to the highway embankment, and will not have excessive scour.

Perhaps the most important of these is the consideration of velocity, a factor which in the hydraulic design can be controlled to some extent.   There are situations, however, where velocities cannot be held to within the recommended allowables for a new channel and where protective linings become necessary. It is often necessary to provide protection for highway embankments adjacent to a relocated channel, particularly when these are of new construction. This can be accomplished by having rock-fill material spilled down the embankment, installing riprap, or paving; but seeding and sodding are generally not satisfactory for embankment protection along channels.

D. Storm Sewers

Storm sewers used in conjunction with highway drainage are generally designed to carry the runoff for storms of 5- to 10-year frequency with the conduit flowing slightly less than full. The selection of the exact design frequency is subject to several considerations and is often specified by the city, county, state, or Federal agency involved. In all cases the performance of the entire storm-sewer system should be checked for storms of lesser expectancy in order to evaluate possible damage.

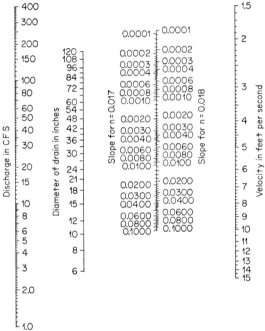

Fig. 4-23. Nomograph for computing required size of circular drain flowing full, n = 0.017 or 0.018. (*Ref. 34. Courtesy of U.S. Army, Corps of Engineers.*)

The rational formula is recommended for estimating the peak discharge. Each inlet is designed to intercept all the approaching runoff. Although the inlet seldom intercepts all this runoff, treating it this way allows for the addition of future inlets if and when they become necessary.

The hydraulic design of the conduit is based upon the principles of open-channel flow since these conduits are to be designed to flow slightly less than full. If, however, it is necessary for the conduit to be designed to flow full, the flow can still be considered to be in accordance with Manning's equation and the formula or the following nomographs can be used (Figs. 4-22 to 4-25).

A complete treatment of the project area having been prepared, a preliminary layout of the drainage system should be made. This should begin with the location of possible outfalls and their flow-line elevations. The inlets necessary to drain the project properly should be located, and their elevations determined. A series of tangent lines connecting these points will represent the preliminary system.

When a profile of the surface has been plotted for each of the lines, the location of the inlets, outlets, and manholes required can be plotted with the proposed sewer connecting them.   As a general rule, the slope should be held constant between manholes and between inlets and manholes.   Abrupt changes in slope from steep to mild should be avoided, and the over-all average slope should be sufficient to maintain velocities of 2.5 ft per sec to avoid silting.

The next step is to calculate the area contributing to the flow at each inlet.   After these areas have been determined and plotted on the plan sheet, a runoff factor for each can be selected by using tables of recommended values given in Art. III of this section as a guide.

The actual design of the storm sewer system is begun at the uppermost inlet.   It is therefore convenient to begin the designation numbers or letters for inlets and manholes at this point.   It is also convenient to set up a computation sheet similar to Fig. 4-26 to simplify the record keeping involved in the calculations.

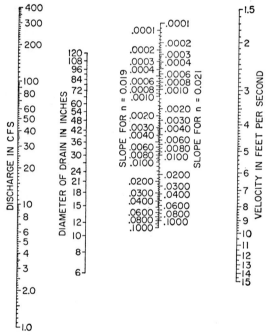

Fig. 4-24. Nomograph for computing required size of circular drain, flowing full, $n = 0.019$ or 0.021.   (*Ref. 34.   Courtesy of U.S. Army, Corps of Engineers.*)

The time of concentration for the upper end, the first inlet, will be the overland flow time plus the gutter flow time.   For subsequent manholes the time of flow in the pipe is added to the time on the inlet to get the time of concentration.   Using the time of concentration, the rainfall factor can be read from rainfall intensity curves, and the design discharge computed from the rational formula.

From Manning's formula, nomographs, or tables, the pipe size can be selected for the desired slope and type of pipe.   The mean velocity and pipe capacity can then be computed.   From the design slope selected and the length, the fall can be computed and elevations of the inverts established.   Usually the pipe size for the first section, from the inlet to the manhole, will be governed by specifications of minimum pipe sizes (varying from 12 to 18 in.).

This system of calculation is repeated for the next manhole.   The time of concentration will now be longer, however, since the time of flow in the pipe must be added.   Additional discharge will be picked up from the increment area coming in at the manhole.   The hydraulic design in selecting the pipe size and slope is, of course, the same as in the first calculation.   Using the slope and the invert elevation at the manhole, the invert elevation for the next manhole is computed, and so on, down the drainage system.

The final step is to check the hydraulic gradient of the system for a discharge greater than that produced by the design storm, when at least part of the system will be operating under pressure.   Flow under pressure will begin in the section of line where the pipe begins to flow full.   To compute the height of rise of water in the manholes or inlets, the calculations should commence at the outlet end of the lower section.   The control elevation will be the water-surface elevation at the outlet if the pipe is submerged, or at a point 0.8 of the diameter above the invert at the outlet if the pipe is unsubmerged and flowing full.   Starting at this elevation and working back, the hydraulic grade line for each section of pipe is computed.   This computation can be

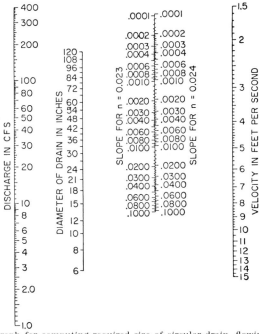

Fig. 4-25. Nomograph for computing required size of circular drain, flowing full, $n = 0.023$ or 0.024.   (*Ref. 34.   Courtesy of U.S. Army, Corps of Engineers.*)

made by solving for the friction slope and adding the minor losses according to Bernulli's theorem (see hydraulic text for references on basic pipe flow).

E. Gutters

In designing curbs and gutters, the width to which the water can spread must first be considered.   A maximum allowable is usually specified by the city, state, or Federal agency involved.   The width will depend upon the cross section of the gutter and its depth of flow.   When the depth of flow in the gutter reaches a point where the spread is too great, collection into storm drainage is made possible by placing inlets along the gutter.

The discharge capacity for various types of triangular gutters can be computed by the formula below or by the nomograph given in Fig. 4-27.

$$Q = 0.56 \frac{Z}{n} S^{1/2} Y^{8/3}$$

where Q = discharge, cu ft per sec

Z = transverse slope of bottom of channel, or horizontal distance corresponding to unit vertical distance (reciprocal of cross slope)

N = coefficient of roughness in Manning's formula (Table 4-7)

S = longitudinal slope of channel

Y = maximum depth of water, ft

F. Storm-water Inlets

There are three types of storm drain inlets in common use: a curb opening, a gutter-grating, and a combined curb-opening and gutter-grating type. The capacity of either of these depends on the depth of flow in the gutter adjacent to the inlet. The amount of storm water to be diverted can be estimated by use of the rational formula.

The capacity for the *curb-opening* type of inlet when intercepting 100 per cent of the approaching flow in the gutter can be calculated by the formula (32)*

$$Q = 0.7L(a + y)^{3/2}$$

Line No.	Location of sewer			Tribu-tary area, (acres)		Time of flow, min		Rate of rainfall I, in./hr	Runoff factor C	Discharge $Q = CIA$	Design					Profile			
	Street	From M.H. No. or sta.	To M.H. No. or sta.	Increment	Total	To upper end	In pipe				Pipe dia., in.	Slope, ft/ft	Coeff. of roughness N	Full capacity, cfs	Mean velocity, fps	Length, ft	Fall, ft	Invert elevation upper end	Invert elevation lower end
(1)	(2)	(3)	(4)	(5)	(6)	(7)	(8)	(9)	(10)	(11)	(12)	(13)	(14)	(15)	(16)	(17)	(18)	(19)	(20)
1	U.S. 60	1	2	1.80	1.80	15	1.6	4.45	0.6	4.80	18	.0060	.015	7.30	4.30	400	2.4	102.40	100.00
1	U.S. 60	2	3	2.30	4.10	16.6	1.5	4.35	0.7	12.48	24	.0040	.015	13.0	4.50	400	1.6	99.33	97.73
1	U.S. 60	3	4	1.50	5.60	18.1	1.1	4.05	0.65	14.75	24	.0070	.015	17.0	5.85	400	2.8	97.06	94.26

Cols. 1, 2, 3, and 4. Location and designation of various parts of system.
Col. 5.   Increment area to be added at each point in system.
Col. 6.   Total area = original area + increment.
Col. 7.   Total time to a point in system. For first inlet this is overland flow + gutter time. For other points in system time of flow in pipe is added (see Hydraulic Analysis, Art. III).
Col. 8.   Time of flow in this section of pipe.
Col. 9.   Rate of rainfall—read from intensity-frequency-duration curves (see Hydraulic Analysis, Art. III).
Col. 10. Weighed or average runoff coefficient.
Col. 11. Discharge $Q = CIA$.
Col. 12. Trial pipe size for slope Col. 13 and roughness Col. 14.
Col. 13. Slope in feet per foot.
Col. 14. Roughness coefficient selected for pipe material (see Table 4-7).
Col. 15. Calculated or from nomographs for solution of Manning's formula (Figs. 4-22 to 4-25).
Col. 16. Mean velocity computed or read from charts.
Col. 17. Length of pipe section between manholes or inlets.
Col. 18. Fall in section = slope × length.
Col. 19. Invert elevation upper end—selected at the beginning, computed thereafter.
Col. 20. Computed by subtracting the fall.

FIG. 4-26. Computation sheet for storm-sewer design.

* For more thorough discussion and derivations see Carl F. Izzard, "Tentative Results on Capacity of Curb Opening Inlets," Highway Research Board, *Research Report* 11-B, 1950.

where y = depth of flow in approach gutter, ft
  a = depression in curb inlet, ft
  L = length of curb-inlet opening
  Since the capacity of an inlet varies with the depth of flow, it can be increased by allowing part of the flow to go past the inlet and be intercepted at the next inlet.   For this reason curb inlets should be designed and spaced to intercept only approximately 85 to 90 per cent of the flow.
  The *gutter-grating* type of inlet, if designed with efficient openings, will intercept the water flowing over the entire width of the grating, and where the cross slope is greater and the grating is depressed and of sufficient length, some water will flow in along the edge.   For highways it is not usually practical to depress the gratings, and the flow outside the area of the grating will flow past the grating and on to the next inlet.   As

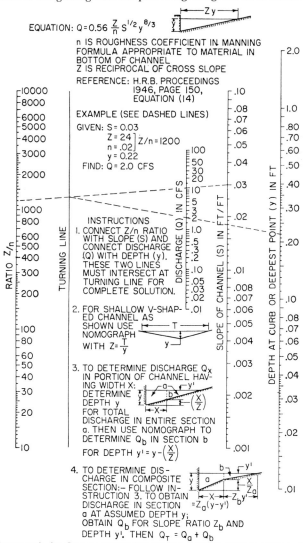

FIG. 4-27. Nomograph for flow in triangular gutters. (*From Carl F. Izzard, Surface Drainage, Highway Research Board, Research Reprint 11-B.*)

in the case of curb inlets, the grating gutter capacity will be increased if part of the water is allowed to flow past the inlet, thereby increasing the depth in the cross section flowing over the grating.

The bars in the grating should be parallel to the direction of flow. Their spacing is usually not a matter of design by the drainage engineer, since the gratings used will normally be of some standard type. The length of the grating should be checked to ensure that it is long enough for the jet of water to fall clear of the downstream end of the slat. This length can be estimated by the formula (18, 32)

$$L = \frac{V}{2} (y + d)$$

where V = mean velocity of flow in approach gutter
y = depth of flow in approach gutter
d = thickness of grate

The carry-over portion can be computed by recalculating the gutter flow, with the grating width neglected, from the cross section. In highway design it is not usually practical to attempt to intercept all the side flow, and it will therefore give additional safety against undesirable spread to space inlets closer and omit side-flow consideration.

Grate inlets are placed in low points in the grade and must be designed to carry all of the approaching discharge. The operation of the inlet for depths up to 0.4 ft is similar to a weir, and the weir formula is used (18, 32):

$$Q = CPH^{3/2}$$

where Q = discharge, cu ft per sec
C = a coefficient experimentally determined, usually regarded as 3.0
H = depth of water, ft
P = perimeter of grate opening, ignoring the bars (where grating is against curb, that side is omitted)

At depths between 0.4 and 1.4 ft, disturbances caused by vortices and turbulance make the operation indefinite, and thus both the orifice formula and the weir formula should be used. For depths above 1.4 ft the orifice formula should be used (18, 32):

$$Q = CA \sqrt{2gh}$$

where A = total area of clear opening, sq ft
H = depth of water, ft
C = an experimentally determined coefficient, usually taken to be 0.67

The capacity of a combined curb opening and grating inlet can be computed by treating each part as if it were placed separately, one below the other on a continuous grade. The curb-opening capacity can be computed from the gutter flow, and the grating opening from the carry-over. Combination inlets are desirable in sumps, since the grating can be designed to intercept the flow and the curb opening will serve as an overflow where ice and debris clog the grating.

The spacing between inlets will be governed by the capacity of the gutter, within the limits of the allowable spread of water on the pavement, and the capacity of the type of inlets used. Using standard procedures an inlet can be designed to intercept the flow at any predetermined point; however, it is usual practice to use a standard inlet and vary the runoff to be intercepted. This is done by altering the spacing to limit the contributing drainage area.

Inlet spacing based upon capacity can be computed by dividing the balanced capacity for the type of inlet for the roadway grade by the runoff per station:

$$L_s = \frac{Q_c}{CIWL/436}$$

where L_s = inlet spacing in 100-ft stations
Q_c = balanced capacity for inlets
C = runoff coefficient, rational formula
I = intensity, in. per hr
W = width of contributing area, ft
L = length of contributing area, in 100-ft stations

When inlet spacing is governed by the allowable spread on the roadway, the limiting gutter flow divided by the runoff per station will give the inlet spacing in stations, or

$$L_s = \frac{\text{limiting gutter flow}}{\text{runoff per station}}, \text{cfs}$$

On sag vertical curves three inlets are recommended, one at the low point and the other two on either side. These two should be placed at the point where the grade is about 0.2 ft above the low point. Through this arrangement water is intercepted before it begins to spread in the low point, and the amount of sediment deposited in the sump area is reduced.

G. Flumes

Flumes are designed according to the principles of open-channel flow. The usual procedure is to select a standard cross section, either trapezoidal or rectangular in shape, and after computing the design discharge by the rational formula, determine the vertical height required at the inlet and for the sides of flume.

The critical section is at the top, or inlet, of the flume. This is the section at which the slope changes from relatively flat to steep and at which supercritical velocities are reached. Above this point the required vertical height is equal to or greater than the critical depth d_c, plus the velocity head. Below the critical section, the vertical height will be less.

The first step in the design is estimating the discharge. From the discharge and the flume cross section chosen, the critical depth and critical velocity v_c are then computed. These can be read from charts similar to Fig. 4-7, if available; otherwise they may be computed from tables (31, 33). The depth of the flume at the inlet should not be less than the sum of $d_c + V_c^2/2g$. Although the depth of flow in the flume will decrease below the critical section, the usual practice in highway design is to carry this section all the way down the slope rather than vary the height of the sides.

Some type of special treatment will usually be required for protection of the channel into which a flume discharges. Since all slope flumes discharge at a supercritical velocity, in many instances unless some sort of stilling basin or channel protection is provided, undesirable scour will result. The design of stilling basins is not included in this handbook, but can be found in Highway Standards and in Refs. 18, 32, and 34.

H. Culverts

Any drainage structure with less than 20 ft of total span is generally referred to as a culvert, and the principles of design discussed under Hydraulic Design Principles, Art. IV of this section apply to the design regardless of the shape or type of structural material used.

Before any attempt is made to design a culvert for a given site, certain field investigations are necessary. The required information is usually taken along with the survey work. Data relevant to centerline location, skew of structure, and suggested possible channel changes are recorded along with the field survey. These data should include information on the existing channel sufficient to provide cross sections and profile, the elevation to which the water can rise without damage to adjacent property, sufficient description of the channel to estimate a roughness factor and the maximum allowable velocities, foundation conditions at the culvert site, notations concerning backwater from streams below the structure, and a description of the drainage area sufficient to permit selection of runoff factors for the hydrologic analysis.

For a given site, the discharge can be estimated by the methods outlined in Hydraulic Analysis. The slope of the structure can be taken from the stream profile; the length of the conduit computed from the roadway cross section; elevations of the invert at the inlet and outlet determined from the profile; the maximum allowable headwater determined from the desired freeboard below roadway grade and elevation of existing property; and the maximum allowable outlet velocity determined from Table 4-8.

An analysis is then made of the existing channel to determine the depth of flow, after which the tailwater depth is computed. This will establish for a given trial size whether the culvert outlet will discharge freely or submerged. The classification of the type of flow then can be established, and the appropriate chart chosen from Figs. 4-10 to 4-18 for determining the headwater elevation.

Various trial sizes are selected on the basis of the design discharge and the allowable velocity, and headwater depth is computed. Trial should be continued until the smallest size has been found that will satisfy the limitations of headwater elevation and outlet velocity. Sometimes it is not practical to limit the outlet velocity to the maximum allowable for the natural channel, in which case protective measures for the outlet and channel are more feasible.

In order to illustrate the procedure more fully, an example of the hydraulic design of a typical culvert is presented.

Given: Design discharge Q = 300 cfs
Length of culvert L = 100 ft
Slope of channel S_o = 0.045 ft/ft
Tailwater depth estimated TW = 2.5 ft
Allowable headwater depth HW = 9 ft

In order to design for these conditions, several materials and types suggest themselves. Not all these can be demonstrated here, but three will be discussed. The first of these will be reinforced-concrete pipe, square entrance, flush head walls. For a first trial size, enter Fig. 4-10, scale No. 1. For a 60-in. reinforced-concrete pipe, read HW/D = 2.6, HW = 13.0 ft. This will obviously give too much headwater depth. For a 72-in. reinforced-concrete pipe, read HW/D = 1.43, HW = 8.6 ft.

Try a 72-in. reinforced-concrete pipe:

$$\frac{HW}{D} = 1.43 > 1.2$$

$$TW = 2.5 < D \text{ (diameter of pipe)}$$
$$d_n \text{ (72 in.)} = 2.6 \text{ ft} < D \text{ (diameter of pipe)}$$

Note: The normal depth of flow can be computed by analytical methods; however, nomographs for flow in circular and rectangular conduits are much more convenient and allow for rapid calculations of alternate sizes. It is not possible to include all the charts available for this purpose in this text; however, they can be found in drainage manuals published by various states and the Bureau of Public Roads (18, 32, 35, 36).

Thus control is at the entrance. Operation is Class II, Type 5. Figure 4-10, scale No. 1, is the appropriate chart, and the headwater depth HW, as read from this chart, is 8.6 ft.

For a corrugated-metal pipe the same procedure is followed in determining the size and headwater depth, but Fig. 4-11 is used.

Try a box-culvert design for the same given conditions, using 45° wing walls. For the first trial size, use Fig. 4-13, scale No. 1, as a guide. If the height D is 6 ft, HW/D cannot exceed 1.5. For D = 6 ft and HW/D = 1.5, a line connecting these on the nomograph (Fig. 4-13) indicates that Q/B cannot be greater than 66.

Then, B = 300/66 = 4.55 ft. Try B = 5 ft.

Try a single 5 × 6:

$$TW \text{ depth} = 2.5 < D$$
$$d_n = 2.5 < D$$

Control is at the entrance.

Operation is Class I, Type 5. Using Fig. 4-13 for Q/B = 300/5 = 60 and D = 6 ft, read HW/D = 1.33. HW = 1.33 × 6 = 7.98 (use 8 ft). This is within the allowable HW depth.

REFERENCES

1. Dalrymple, Tate: "Regional Flood Frequency," in *Surface Drainage*, Highway Research Board, *Research Report* 11-B, 1950.
2. Wisler, C. O., and E. F. Brater: *Hydrology*, John Wiley & Sons, Inc., New York, 1949.

3. Linsley, R. K., et al.: *Applied Hydrology*, McGraw-Hill Book Company, Inc., New York, 1949.
4. Johnstone, D., and W. P. Cross: *Elements of Applied Hydrology*, The Ronald Press Company, New York, 1949.
5. Meinger, Oscar E.: *Physics of the Earth*, McGraw-Hill Book Company, Inc., New York, 1942.
6. Yarnell, David L.: "Rainfall Intensity-Frequency Data," U.S. Department of Agriculture, *Miscellaneous Publication* 204, 1935.
7. "Rainfall Intensity for Local Drainage," *U.S. Weather Bureau Technical Paper* 24, 1954.
8. "Rainfall Intensity-Duration-Frequency Curves," *U.S. Weather Bureau Technical Paper* 25, 1955.
9. Dalrymple, Tate: "Use of Stream-flow Records in Design of Bridge Waterways," *Highway Research Board, Proceedings*, 1946.
10. Carter, R. W.: "Computation of Peak Discharge at Culverts," *U.S. Geological Survey Circular* 376, 1957.
11. Kindsvater, C. E., R. W. Carter, and H. J. Tracey: "Computation of Peak Discharge at Contractions," *U.S. Geological Survey Circular* 284, 1953.
12. Sanderson, Earl E.: "The Climatic Factors of Ohio's Water Resources," *Division of Water Bulletin* 15, Columbus, Ohio, May, 1950.
13. West, E. M., and W. H. Sammons: "A Study of Runoff from Small Drainage Areas and the Openings in Attendant Drainage Structures," *Highway Materials Research Laboratory, Research Report*, Lexington, Ky., July, 1955.
14. Davis, C. V.: *Handbook of Applied Hydraulics*, 2d ed., McGraw-Hill Book Company, Inc., New York, 1952.
15. Ramser, C. E.: "Runoff from Small Agricultural Areas," *Journal of Agricultural Research*, vol. 34, no. 9, May 1, 1927.
16. Kirpich, V. P.: "Time of Concentration of Small Agricultural Watersheds," *Civil Engineering*, vol. 10, no. 6, p. 362, June, 1940.
17. Rouse, Hunter: *Engineering Hydraulics*, John Wiley & Sons, Inc., New York, 1950.
18. Pickering, H. P.: *Drainage Design in Highway Practices*, University of California, Institute of Transportation and Traffic Engineering, 1955.
19. Tilton, G. A., and R. Robinson Rowe: "Culvert Design in California," *Highway Research Board, Proceedings*, 1943.
20. Ritter, L. J., Jr., and R. J. Paquette: *Highway Engineering*, The Ronald Press Company, New York, 1951.
21. Armco Drainage and Metal Products, *Handbook of Drainage and Construction Products*, Middletown, Ohio, 1955.
22. Civil Aeronautics Administration: *Airport Drainage*, U.S. Department of Commerce, November, 1956.
23. Potter, W. D.: "Surface Runoff from Agricultural Watersheds," in *Surface Drainage*, Highway Research Board, *Research Report* 11-B, 1950.
24. Izzard, Carl F.: "Peak Discharges for Highway Drainage Design," *Transactions ASCE, Paper* 2709, 1954.
25. Hewes, L. T., and C. H. Oglesby: *Highway Engineering*, John Wiley & Sons, Inc., New York, 1954.
26. Potter, W. D.: "Use of Indices in Estimating Peak Rates of Runoff," *Public Roads Magazine*, April, 1954.
27. State of California, Dept. of Public Works, Division of Highways: *California Culvert Practice*.
28. Woodward, S. M., and C. J. Posey: *Hydraulics of Steady Flow in Open Channels*, John Wiley & Sons, Inc., New York, 1949.
29. Vennard, J. K.: *Elementary Fluid Mechanics*, 3d ed., John Wiley & Sons, Inc., New York, 1954.
30. King, H. W., C. O. Wisler, and J. G. Woodburn: *Hydraulics*, John Wiley & Sons, Inc., New York, 1952.
31. King, H. W., and E. F. Brater: *Handbook of Hydraulics*, 4th ed., McGraw-Hill Book Company, Inc., New York, 1954.
32. U.S. Bureau of Public Roads: *Highway Drainage Manual*.
33. U.S. Army, Corps of Engineers: *Hydraulic Tables*, 2d ed., War Department, 1944.
34. U.S. Army, Corps of Engineers: *Engineering Manual*, War Department, 1945.
35. Kentucky Department of Highways: *Manual of Drainage*, Frankfort, Ky., 1954.
36. Ohio Department of Highways, Bureau of Location and Design: *Drainage Guide Plan Preparation Manual*, Columbus, Ohio.
37. Bernard, M. M.: "Discussion of Runoff Rational Run-off Formulas, by R. L. Gregory and C. E. Arnold," *Transactions ASCE, Paper* 1812, p. 1038, 1932.

Section 5

FROST ACTION AND PERMAFROST

KENNETH A. LINELL, *Chief, Arctic Construction and Frost Effects Laboratory, and Foundation and Materials Branch, Headquarters, U.S. Army Engineer Division, New England, Waltham, Mass.*

I. INTRODUCTION

Section 5 on Frost Action and Permafrost covers definitions and detailed discussions on the occurrence, properties, and frost susceptibility of soils, with special emphasis on the design and construction of highways (and airfields) in cold regions.

II. DEFINITIONS

The following specialized terms are used in connection with frost and permafrost road and highway engineering:

A. Regions

ARCTIC: The northern region in which the mean temperature for the warmest month is less than 50°F and the mean annual temperature is below 32°F. In general, the Arctic coincides with the tundra region north of the limit of trees.

SUBARCTIC: The region adjacent to the Arctic in which the mean temperature for the coldest month is below 32°F and the mean temperature for the warmest month is above 50°F and where there are less than 4 months having a mean temperature above 50°F. In general, the Subarctic coincides with the circumpolar belt of dominant coniferous forest.

SEASONAL FROST AREAS: Those areas of the earth in which significant freezing occurs during the winter season but without development of permafrost.

B. Soil and Frost Terms

ANNUAL FROST ZONE: The top layer of ground subject to annual freezing and thawing. In arctic and subarctic regions where annual freezing penetrates to the permafrost table, suprapermafrost and the annual frost zone are identical. (Sometimes referred to as active zone.)

BASE, OR BASE COURSE: As used in this section of the manual, the total thickness of non-frost-susceptible materials used between the bottom of the wearing surface and the top of the subgrade.

DEGRADATION: Progressive lowering of the permafrost table, occurring over a period of years. The opposite condition is called aggradation.

FROST ACTION: A general term for freezing and thawing of moisture in materials and the resultant effects on these materials and on structures of which they are a part or with which they are in contact.

FROST HEAVE: The raising of a surface due to the formation of ice in the underlying soil.

FROST-SUSCEPTIBLE SOILS: Soils in which significant ice segregation will occur when the requisite moisture and freezing conditions are present.

NON-FROST-SUSCEPTIBLE MATERIALS: Cohesionless materials such as crushed rock, gravel, sand, slag, and cinders in which significant ice segregation does not occur.

GROUND ICE: A body of more or less soil-free ice within frozen ground.

ICE SEGREGATION: The growth of ice as distinct lenses, layers, veins, and masses in soils, commonly, but not always, oriented normal to the direction of heat loss.

ICE WEDGE: A wedge-shaped ice mass in permafrost, usually associated with fissure polygons.

PERMAFROST: Perennially frozen ground.

PERMAFROST TABLE: An irregular surface within the ground which represents the upper limit of permafrost.

RESIDUAL THAW ZONE: A layer of unfrozen ground between the permafrost and the annual frost zone. This layer does not exist where the annual frost zone extends to permafrost.

SUPRAPERMAFROST: The entire layer of ground above the permafrost table.

VARVED CLAYS: Alternate layers of clay and silt and, in some instances, fine sand. The thickness of the layers rarely exceeds ½ in., but occasionally much thicker varves are encountered. They are likely to combine the undesirable properties of both silts and soft clays. Varved clays may differ considerably in performance, and local experience with these soils should be taken into account in assigning a frost-susceptibility classification.

C. Terms of Temperature and Season

AVERAGE ANNUAL TEMPERATURE: The average of the average daily temperatures for 1 year.

MEAN ANNUAL TEMPERATURE: The average of the average annual temperatures for several years.

AVERAGE DAILY TEMPERATURE: The average of the maximum and minimum temperatures for one day or the average of several temperature readings taken at equal time intervals during one day, generally hourly.

MEAN DAILY TEMPERATURE: The average of the average daily temperatures for a given day for several years.

DEGREE-DAYS: The degree-days for any one day equal the difference between the

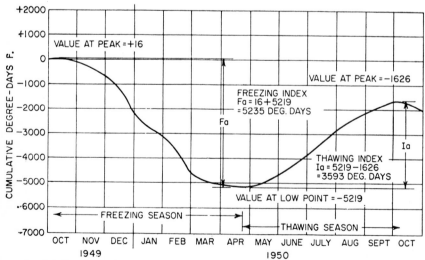

FIG. 5-1. Freezing and thawing indexes. *(U.S. Army Engineer Division, New England.)*

average daily air temperature and 32°F. The degree-days are minus when the average daily temperature is below 32°F (freezing degree-days) and plus when above (thawing degree-days). Figure 5-1 shows a typical curve for a northern location, obtained by plotting cumulative degree-days against time.

FREEZING INDEX: The number of degree-days between highest and lowest points on a curve of cumulative degree-days versus time for one freezing season. It is used as a measure of the combined duration and magnitude of below-freezing temperatures occurring during any given freezing season. The index determined for air temperatures at 4.5 ft above the ground is commonly designated as the *air freezing index*, while that determined for temperatures immediately below a surface is known as the *surface freezing index*. The freezing index is illustrated in Fig. 5-1.

MEAN FREEZING INDEX: The freezing index determined on the basis of mean temperatures. The period of record over which temperatures are averaged is usually a minimum of 10 years and preferably 30.

DESIGN FREEZING INDEX: The freezing index of the coldest winter in a 10-year period of record or the average of the three coldest winters in 30 years of record.

FREEZING SEASON: That period of time during which the average daily temperature is generally below 32°F.

FROST-MELTING PERIOD: An interval of the year during which the ice in the foundation materials is returning to a liquid state. It ends when all the ice in the ground is melted or when freezing is resumed. Although in the generalized case there is visualized only one frost-melting period, beginning during the general rise of air temperatures in the spring, one or more significant frost-melting intervals may occur during a winter season.

PERIOD OF WEAKENING: An interval of the year which starts at the beginning of the frost-melting period and ends when the subgrade strength has returned to normal period values or when refreezing starts. In seasonal frost areas, the period of weakening may be substantially longer than the frost-melting period, but in permafrost areas the periods coincide.

THAWING INDEX: The number of degree-days between the lowest and highest points on the cumulative degree-days–time curve for one thawing season. It is used as a measure of the combined duration and magnitude of above-freezing temperatures occurring during any given thawing season. The index determined for air temperatures at 4.5 ft above the ground is commonly designated as the *air thawing index*, while that determined for temperatures immediately below a surface is known as the *surface thawing index*. The thawing index is illustrated in Fig. 5-1.

MEAN THAWING INDEX: The thawing index determined on the basis of mean temperatures.

THERMAL REGIME: The temperature pattern existing in a body.

D. Terrain Terms

PATTERNED GROUND: A general term describing ground patterns resulting from frost action, such as soil polygons, stone polygons, stone circles, stone stripes, and solifluction stripes.

MUSKEG: Poorly drained organic terrain, composed of a living organic mat, overlying an extremely compressible mixture of partially decomposed peat, occurring as very deep peat bogs of varying sizes or as organic terrain a few inches to several feet in thickness, covering hundreds to thousands of square miles of arctic and subarctic terrain.

TUNDRA: A treeless region of grasses and shrubs characteristic of the Arctic.

III. NEED FOR CONSIDERING FROST ACTION IN DESIGN

Much of the United States and North America experiences at least occasional detrimental frost action in some degree. The following damaging effects to highways and airfield pavements result from frost action:

1. Seasonal frost heave and settlement
2. Surface roughness
3. Loss of compaction

4. Deterioration of the pavement surfacing
5. Loss of strength during thaw
6. Degradation through melting of permafrost
7. Restriction of subsurface drainage

The above effects may result in excessive maintenance, hazardous operational conditions, or even destruction of pavements and structures. They all are the result of the occurrence or presence of ice under freezing conditions. Roads should be designed so that they retain their stability and effectiveness in spite of these detrimental influences.

Segregated ice in soil may have three basic origins. Most commonly, ice "grows" in frost-susceptible soils in the form of lenses, veins, and masses, as the freezing level penetrates the ground, if moisture is available. As ice crystallizes within the soil, water from underlying strata is attracted strongly to the plane of freezing. The

FIG. 5-2. Ice segregation in a clay subgrade (photo). *(U.S. Army Engineer Division, New England.)*

resultant upward flow and addition of this water into the frost zone in the form of ice commonly results in an increase in volume of the soil mass. The consequent raising of the surface is called frost heave. This type of ice segregation is found in both seasonal frost and permafrost areas and is the most obvious and common source of pavement difficulties. Figure 5-2 shows typical ice segregation in a clay subgrade. Figure 5-3 shows a typical water-content increase in the frozen layer in a laboratory freezing test; note that partial reduction in moisture content is shown immediately below the plane of freezing. Figure 5-4 shows a typical record of pavement heave and penetration of 32°F temperature for a location in northern Maine; the subgrade soil in the case illustrated is believed to have frozen, not at 32°F, but at a slightly lower temperature.

A second type of ice, found only in permafrost regions, occurs as ice wedges at the margins of frost polygons. A polygonal ground pattern is formed in arctic and subarctic areas by contraction of the ground surface when it is cooled in the winter by extreme low temperatures, which results in the formation of a crack pattern. The cracks become filled with ice. These tend to grow in width each year, and in time ice wedges are formed extending to many feet in depth. In the most northern areas, these ice wedges have only a very shallow cover of earth and vegetation.

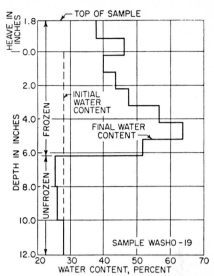

Third, ice masses in the ground may, in some cases, consist of "fossil" ice which has become buried under a protective earth covering, as by a moraine deposit or by an earth slide, and has subsequently been preserved by low ground temperatures. Such ice is, again, found only in permafrost areas.

Frost heave may be uniform or nonuniform, depending on uniformity of the soil and ground-water conditions underlying the pavement. When uniform heave occurs, adjacent areas of pavement surface are lifted by approximately equal amounts, such that the initial profile of the surface remains essentially the same. When

Fig. 5-3. Water-content increase in frozen layer. (*U.S. Army Engineer Division, New England.*)

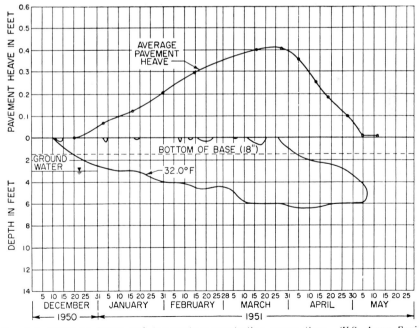

Fig. 5-4. Pavement heave and temperature penetration versus time. (*U.S. Army Engineer Division, New England.*)

heave is nonuniform, appreciable differences in the heave of adjacent areas occur, result-ing in objectionable unevenness of the pavement surface.   Heave tends to be uniform when stripping, depth of fill, soil conditions, and ground-water depth are uniform.   In uniform soil deposits, very low permeability in uniform soils also aids in producing rela-tively uniform, low heave, because moisture can move only small distances in such soils during freezing.   Nonuniform heave occurs, for example, where subgrades vary between clean sand and silty soils or at abrupt transitions from cut to fill with ground water close to the surface.   Serious differential heaving often occurs where lateral drains, culverts, or utility lines placed under pavement on frost-susceptible subgrades break the uniformity of the subgrade conditions.

High-speed traffic on modern highways and increasingly high standards of smooth-ness and safety require control of frost heave to avoid excessive roughness during the winter period.   In addition, consideration must be given to the possibility of develop-ment cf permanent and progressive roughness in the pavement as a result of differential frost loosening of the base and subgrade combined with the effects of traffic.   The accelerated deterioration of the surfacing which results from cracking produced by frost heave must also be taken into account.

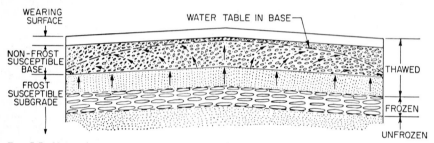

Fig. 5-5. Moisture movement upward into base course during thaw. *(U.S. Army Engi-neer Division, New England.)*

Many studies have shown that frost-susceptible soils in which ice segregation has occurred experience a loss in strength during frost-melting periods, with a correspond-ing reduction in load-supporting capacity of the pavement as the melting of the ice releases an excess of water within the soil.   This situation may occur in clay subgrades even though significant heave is not observed.   In such highly impervious soils, water for ice segregation is obtainable during freezing only from the voids of the unfrozen clay immediately below the freezing plane, and the shrinkage of the clay layers as they yield this moisture nearly balances the volume of the ice lenses formed.

The application of traffic during the period of weakening may cause remolding of the soil, with attendant further reduction in subgrade strength.   The degree of reduction in pavement-supporting capacity during a frost-melting period and the length of the period during which the supporting capacity is reduced depend on the type of soil, temperature conditions during the freezing and thawing periods, the amount and type of traffic during the frost-melting period, the availability of water during the freezing and thawing periods, and drainage conditions.

Restriction of vertical drainage is a characteristic of frozen ground in both seasonal-frost and permafrost areas.   During the frost-melting period, the excess water released by melting ice lenses cannot drain downward through the impervious, still-frozen soil below.   It therefore tends to drain upward into and saturate the base course as shown in Fig. 5-5 and may even emerge through cracks and joints in the pavement, resulting in a very unfavorable pavement condition, in which not only is the subgrade in a weakened condition, but the pavement itself may be supported in part by water under traffic loading and pumping may occur at joints, cracks, and pavement edges.   Satu-ration of the base course is made more likely by intense ice segregation in the subgrade, by thin well-graded base courses having low porosity and therefore small storage space for excess moisture, and by wide pavements requiring a long time for lateral drainage.

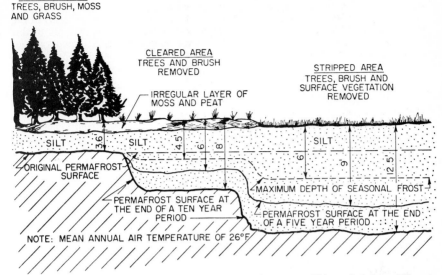

FIG. 5-6. Measured degradation of permafrost in frost-susceptible soil below different surfaces in a subarctic region. (*U.S. Army Engineer Division, New England.*)

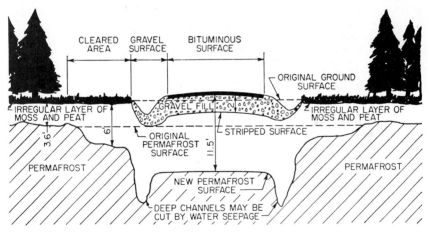

FIG. 5-7. Maximum depth to permafrost below a road after 5 years in a subarctic region. (*U.S. Army Engineer Division, New England.*)

In permafrost areas, the blockage of vertical drainage by impervious, permanently frozen ground produces a characteristic tendency toward high-water-table conditions throughout the summer in all terrain features underlain by permafrost except elevated deposits of pervious materials in favorable topographic positions.

In permafrost regions, the change of surface conditions caused by construction of the pavement may cause lowering of the permafrost level. In the more southerly permafrost regions, this lowering may be progressive, with the level continuing to lower year after year. Figures 5-6 and 5-7 illustrate typical degradation in central Alaska. When the permafrost contains lenses, veins, or masses of ice, as is common, lowering of the permafrost level will cause such ice to thaw and drain away, resulting in settlement of the overlying surfaces, usually in a very irregular manner. Under extreme conditions in the Arctic, this can make a road impassable for a jeep in a single summer.

A special road-construction problem in northern North America, not covered in the present section, is the construction of roads in muskeg terrain. Much study of this problem has been and is now being carried out in Canada, where it presents a substantial obstacle over wide areas (1).*

IV. FROST AND PERMAFROST

A. Geographical Occurrence

Figure 5-8 shows the distribution of seasonal frost and permafrost in North America based on most recent available data (2–5), and Fig. 5-9 shows the distribution of mean

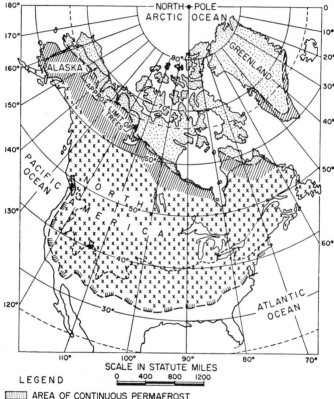

LEGEND

SCALE IN STATUTE MILES
0   400   800   1200

░░░ AREA OF CONTINUOUS PERMAFROST
▨▨▨ AREA OF DISCONTINUOUS AND SPORADIC PERMAFROST
✕✕✕ AREA OF SEASONAL FROST ONLY

ᵘᵘ—ᵘᵘ APPROXIMATE SOUTHERLY LIMIT OF AREA WHERE SEASONAL FROST
MAY BE EXPECTED TO PENETRATE PAVEMENT AND BASE TO A
DEPTH OF AT LEAST 12 INCHES 1 YEAR IN 10

----- 3 FEET COMBINED THICKNESS OF PAVEMENT AND GRANULAR, WELL-
DRAINED BASE REQUIRED TO PREVENT SUBSTANTIAL SUBGRADE
FREEZING COLDEST WINTER IN 10

FIG. 5-8. Seasonal frost and permafrost in North America. *Note:* Patches and islands of permafrost may be found in areas south of crosshatched zone, particularly in elevated mountain locations.

* Numbers in parentheses refer to corresponding items in the references at the end of this section.

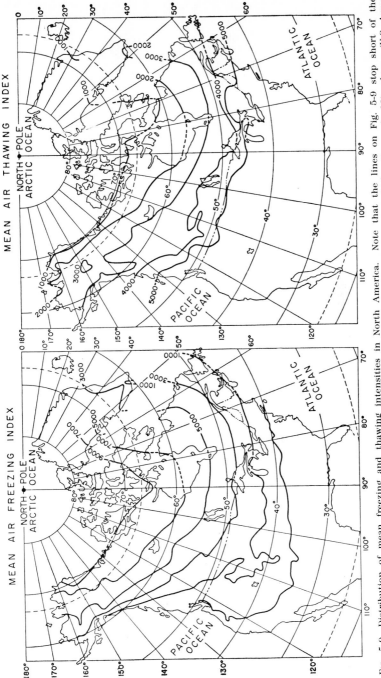

FIG. 5-9. Distribution of mean freezing and thawing intensities in North America. Note that the lines on Fig. 5-9 stop short of the southerly frost limit shown on Fig. 5-8, because the lines on Fig. 5-9 represent *mean* rather than *design* (1 year in 10) values. (*U.S. Army Engineer Division, New England.*)

air freezing indexes and mean air thawing indexes. On Fig. 5-8, the southerly limit of significant frost occurrence has been arbitrarily chosen; at the line shown, frost penetration may be expected to occur to a depth of about 12 in. on an average of once in 10 years, assuming an average pavement, kept clear of snow. As shown on Fig. 5-8, the seasonal frost areas include the major portion of the United States and the most populous areas of Canada. Depth of winter freezing increases as freezing index increases. When a point is reached where the depth of summer thaw is less than the depth of winter freezing, a condition of perennially frozen ground occurs. As freezing index increases still further and the thawing index decreases, the permafrost becomes progressively thicker and the annual frost zone at the top of the ground becomes progressively thinner. Permafrost normally does not exist unless the mean annual air temperature is at least a few degrees below 32°F; that is, the mean air freezing index must usually be somewhat greater than the mean thawing index. Actually, the temperature of the ground at its surface would be a more accurate measure than air temperature because the effects of such factors as vegetative cover, snow cover, and radiation balance are not represented in air temperatures. However, the available record of ground-surface temperature over the earth is as yet very small, and it is presently necessary to use air temperatures as the basis of computation.

B. Properties and Characteristics

1. **Seasonally Frozen Ground.** When soil freezes in winter, its strength and bearing capacity are increased because of conversion of at least a portion of the water in the soil into ice. Tests have shown that the unconfined compressive strength of frozen

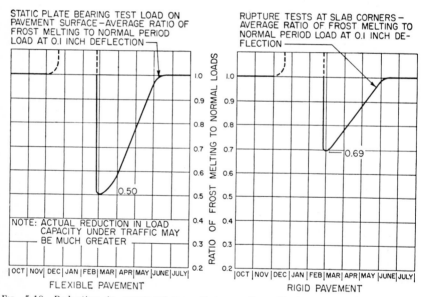

FIG. 5-10. Reduction in pavement-supporting capacity. *Note:* Curves based on static load tests from nine flexible paved test areas and from corner loading tests on six rigid paved test areas at six locations. (*U.S. Army Engineer Division, New England.*)

soils ranges from 100 to over 1,000 psi at temperatures between 25 and 32°F (6). When thawing begins in the spring, an excessively wet condition of the subgrade results if ice segregation has occurred. A lower-than-normal pavement-bearing capacity then exists until thawing stops and this water is drained away or redistributed. The degree of weakening which occurs during the thaw period may be negligible or large, depending on the characteristics and state of the soil and the general design. In seasonal frost areas, the frozen layer thaws completely during the spring and summer and there is complete return to normal strength during the summer season. This sequence of strength variation through the seasons as measured by plate-bearing tests is illustrated in Fig. 5-10. Since a surface active zone subject to seasonal freezing and thawing is present also in permafrost areas, this variation in supporting strength occurs over the entire seasonal frost and permafrost areas shown in Fig. 5-8, wherever the conditions required for frost action are present.

2. Permafrost. Depending on local conditions, permafrost may exist (Fig. 5-11):

1. As a continuous layer with its upper surface at the bottom of the annual frost zone (the most common condition in arctic regions, under natural conditions)

2. As a continuous layer with its upper surface at some depth below the annual frost zone (degrading permafrost)

3. As islands within unfrozen material

4. As layers separated by layers of unfrozen material

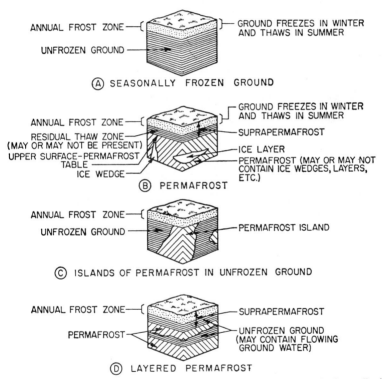

Fig. 5-11. Typical sections in seasonal-frost and permafrost areas. *(U.S. Army Engineer Division, New England.)*

Permafrost does not necessarily contain segregated ice, since by definition it is simply ground which is perennially below 32°F.

The maximum depth to the surface of the permafrost layer occurs in late summer, near the end of the thawing season. The thickness of the suprapermafrost is controlled primarily by the magnitude of the thawing index, the surface cover conditions that have existed for the previous several years, and the water content and density of the soil. Values may range from as little as 1 to 2 ft for saturated fine-grained soils under heavy moss cover, in the far-northern Arctic, to more than 20 ft in coarse, well-drained soil under a pavement, near the southern boundary of permafrost. Subsurface moisture flow, a change in elevation of the water table, or some unusual source of heat may degrade the upper layers of frozen soil. When surface conditions are altered, the depth to the permafrost table will change.

The thickness of the permafrost layer itself varies with mean annual temperature. A depth of 2,000 ft has been reported at Nordvik in Siberia. A depth of about 920 ft has been reported at Umiat, Alaska, and about 170 ft at Northway, Alaska.

Relatively clean sands and gravels located in well-drained positions generally do not present serious engineering construction problems since they do not normally contain appreciable bodies of ground ice. Permafrost consisting of fine-grained soils often contains large formations of ice in layers, wedges, or other shapes. Various patterned ground and permafrost configurations, such as polygons, pingos, stone rings, and stone stripes, are frequently indicative of possible permafrost problem areas. Permafrost containing heavy concentrations of ground ice may also occur without observable special surface configurations, particularly in the more southerly permafrost areas where trees and other vegetation hide possible surface patterns.

C. Thermal Regime in the Ground

As shown on Fig. 5-12, the amplitude of ground-temperature variations through the seasons decreases with depth. Below some depth, such as about 30 ft, the amplitude of annual temperature variation becomes so small that the temperature gradient corresponding to the normal slow flow of heat from the interior of the earth becomes discernible. When the ground-temperature curve at its warmest extreme of annual variation is below freezing over a portion of its length, as in Fig. 5-12, a permafrost condition exists. When the curve does not fall below freezing at its warmest extreme but does at its coldest extreme, a seasonal frost condition exists.

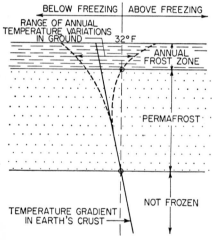

As shown on Fig. 5-12, the most extreme temperature changes are at and near the surface of the ground. The plowing of pavements in the winter accentuates the amplitude of these changes by eliminating the insulating effect of snow cover. Pavement-surface temperatures may be distinctly warmer than air temperatures on bright sunny days and distinctly colder than air temperatures on clear nights. Since ground temperatures are not the same as air temperatures, it is necessary to allow for this difference when depths

Fig. 5-12. Temperature gradients under permafrost conditions.

of freeze or thaw penetration are calculated, using air-temperature data. Such allowance is included in the computation method described later. The large surface-temperature changes through the seasons cause expansion and shrinkage of both the pavement and subgrade. In very cold climates, it has been observed that the cracks which form in the pavement surface in winter because of shrinkage may also extend downward into the underlying soils.

Transfer of heat between the earth and atmosphere is strongly affected by the type of surface. When a natural, vegetated surface is replaced by a pavement kept cleared of snow, there is usually a considerable change in temperatures at the surface and in passage of heat through this plane, which is reflected in changed conditions of seasonal frost and permafrost. In permafrost areas, the change of level of the permafrost table after construction should be considered in design.

The rate of cooling or warming of a soil under given temperature conditions is controlled by its thermal conductivity and by the amount of heat which has to be removed or added to effect a given temperature change.   The thermal conductivity of soil varies with texture, moisture content, and density; it is also dependent on whether or not the soil is frozen.   The volumetric heat capacity is determined by the volumetric specific heat and by the volumetric latent heat of fusion (latent heat of fusion per cubic foot of soil).   The amount of heat required to change ice into water at the freezing point is 144 times the amount required to raise the temperature of the same weight of water 1°F.   Thus the latent heat of fusion of the moisture phase is a major factor in controlling depths and rates of thaw and freeze in a soil.   The volumetric latent heat of fusion is a function of the moisture content of the soil.

Annual temperature fluctuations and depths of freeze and thaw are usually effective to greater depths in well-drained granular soils than in wet silts or clays.   The depth of material required to keep annual freeze-thaw within clean, non-frost-susceptible soil is a maximum approximately at the boundary between the areas of continuous and discontinuous permafrost and decreases from that line, both northward and southward.

D. Frost Susceptibility of Soils

The following criterion for frost-susceptible soils was originated by Dr. A. Casagrande (7): Inorganic soils containing 3 *per cent or more of grains finer than* 0.02 *mm* in diameter, by weight, are generally frost-susceptible.

This criterion is empirical and is based on a combination of laboratory tests and field experience.   It is not a precise criterion, but is, rather, an engineering approximation.   Materials having 3 per cent of grains finer than 0.02 mm should not be expected to be completely free from frost heave; however, any slight heave occurring will normally be within tolerable limits.   The criterion is conservative for uniform (steeply graded) soils.   Uniform sandy soils may in fact have as high as 10 per cent of grains finer than 0.02 mm by weight without being frost-susceptible.   However, their tendency to occur interbedded with other soils makes it generally impractical to consider them separately.   The criterion is less conservative for well-graded soils.   It is advisable to adhere very strictly to, or to stay well below, the 3 per cent limit when soils are extremely well graded, approaching the theoretical maximum-density grain-size distribution.   It should be carefully noted that the criterion refers to the 0.02-mm size and not the 200-mesh size.

For pavement-design purposes the U.S Army Corps of Engineers uses the following classification grouping for frost-susceptible sons (8):

Group	Description
F1	Gravelly soils containing between 3 and 20 per cent finer than 0.02 mm by weight
F2	Sands containing between 3 and 15 per cent finer than 0.02 mm by weight
F3	(a) Gravelly soils containing more than 20 per cent finer than 0.02 mm by weight; (b) sands, except very fine silty sands, containing more than 15 per cent finer than 0.02 mm by weight; (c) clays with plasticity indexes of more than 12; (d) varved clays existing with uniform subgrade conditions
F4	(a) All silts including sandy silts; (b) very fine silty sands containing more than 15 per cent finer than 0.02 mm by weight; (c) clays with plasticity indexes of less than 12; (d) varved clays existing with nonuniform subgrade conditions

Soil names are as defined in the Unified Soil Classification System (9).   The basis for division between the F1 and F2 groups is that the F1 materials may be expected to show the higher-bearing-capacity performance characteristics of gravelly, more or less well graded materials during thaw, whereas the F2 materials may be expected to have the performance characteristics of sands, under these conditions.

The groups are listed approximately in order of increasing susceptibility to frost heaving and/or weakening as a result of frost heaving.   The order of listing of the subgroups under groups F3 and F4, however, is not intended to indicate the order of susceptibility within these groups.   The soils in group F4 are of especially high frost susceptibility.

In the above frost-susceptibility grouping, varved clays have been given special consideration, as these have been found in some cases to weaken more rapidly on thawing than homogeneous clays with equal average water contents and to combine the undesirable properties of both silts and soft clays. On the other hand, other varved clays may also occur with relatively uniform subgrade conditions and give little or no trouble. Therefore, local experience and conditions must be taken into account with these soils. When subgrade conditions are uniform and there is local evidence that the degree of heave experienced in varved clays is not exceptional, then the varved clay should be assigned a group F3 frost-susceptibility classification.

It should be emphasized that the above frost-susceptibility criteria represent the potentiality of the soil alone for ice segregation and that detrimental frost action will not appear unless the other conditions necessary for ice segregation are also present. In order for ice segregation to occur, the following three conditions of soil, temperature, and water must be present simultaneously:
1. The soil must be frost-susceptible.
2. Freezing temperatures must penetrate the soil.
3. A source of water must be available.

In general, the thickness of ice layers per unit of depth is inversely proportional to the rate of penetration of freezing temperature into the soil. However, the rate of heave of the surface in inches per day is not greatly affected by the rate of frost penetration, so long as the plane of freezing is advancing. The water required for ice segregation may be obtained from an underlying ground-water table, from infiltration, from an aquifer, or from water held within voids of the soil itself. The degree of ice segregation which will occur in a given case is markedly influenced by environmental factors, such as transitions between cut and fill, lateral flow of water in cuts, and perched ground water.

V. DESIGN AND CONSTRUCTION OF HIGHWAYS IN COLD REGIONS: GENERAL CONSIDERATIONS

A. Site Investigations

Design of highways in cold regions should be preceded by field investigations which are of the same general character and scope as required in nonfrost regions, but which are carried somewhat farther. In addition to the usual data on topography, hydrology, etc., special information is needed on the climatic factors of temperature and precipitation, on the frost characteristics of soil and construction materials, and on ground-moisture conditions.

Airphoto and geological reconnaissance studies are of substantial assistance in choosing alignments which will minimize frost and permafrost problems of design, construction, and maintenance, in avoiding areas of solifluction or other ground movements, and in the locating of potential sources of non-frost-susceptible borrow and base-course materials (10, 11). Ground reconnaissance will provide information on the soils and local materials, snow cover, vegetation, ground water, surface drainage, possible icing locations, and other factors which relate to the special problems of design in cold regions.

Subsurface explorations should be of sufficient depth and extent to locate and determine the characteristics of any materials which will be subject to freeze or thaw action in the ultimate roadways. In those permafrost areas where highway construction will result in continuing degradation of the permafrost, selected holes should be carried to sufficient depth so that long-range performance at the selected alignment may be estimated. Sometimes after-the-fact explorations of this type may be needed to estimate the long-range outlook at trouble spots which develop in existing highways.

Exploration and sampling of unfrozen soils may be accomplished by any conventional methods. Frozen soils, on the other hand, may require special techniques to obtain full information on the amount of ice segregation that is present. Truck-mounted power augers are useful for exploration of frozen soils where classification and water-content information will provide sufficient basic data. Tungsten carbide cutting teeth on such augers give satisfactory service if the frozen material does not contain too many cobbles and boulders. Other techniques are available which are capable of revealing finer detail on the nature and degree of ice segregation than is attainable from the power auger. Fine-grained frozen soils can frequently be explored most effectively by drive sampling. Coarse-grained soils can be explored by means of test pits or by special core-drilling techniques (12). During the thaw season in permafrost areas, borings normally need to be cased through the soft thaw layer when drilling saturated fine-grained soils. Within the permafrost itself, casing is not needed.

When in permafrost areas it is desired to determine the upper surface of permafrost near the end of the thawing season, this can be done effectively by test pitting or by probing. Probings made near the end of the thawing season by driving rods by hand, or by use of a drill rig or other mechanical equipment, provide rapid means of outlining sporadic permafrost formations. While some studies have been made to develop geophysical exploration methods for conditions of frozen ground, further research is needed.

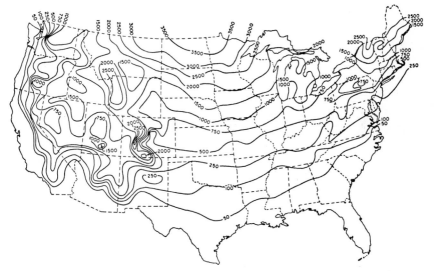

Fig. 5-13. Distribution of design-freezing-index values in continental United States. *Note:* Design-freezing-index values are cumulative degree-days of air temperature below 32°F for the coldest year in a 10-year cycle. The map is offered as a guide only. It does not attempt to show local variations, which may be substantial, particularly in mountainous areas. The actual design freezing index used should be computed for the specific project, using temperature data from stations nearest site. (*U.S. Army Engineer Division, New England.*)

The field and laboratory investigations should provide sufficient information to determine whether the combined soil, moisture, and temperature conditions beneath the pavement will result in detrimental frost action.

1. Soil. Examination of the gradation curves of the subgrade materials and comparison with the Casagrande criterion will indicate whether or not they are frost-susceptible. In borderline cases, it may be desirable to perform slow laboratory freezing tests, to measure the relative frost susceptibility. The Arctic Construction and Frost Effects Laboratory, U.S. Army Engineer Division, New England, Waltham, Mass., has developed procedures for performing such tests (13).

2. Temperature. Figure 5-13 shows distribution of design-freezing-index values in continental United States. However, the generalized plot of Fig. 5-13 will not usually be sufficiently precise to be used for specific locations, and it should be considered only as a guide. Freezing index values should therefore be computed from air temperatures obtained from weather-record stations located as close as possible to the construction locations. In areas where the design freezing index is below 1,000, differ-

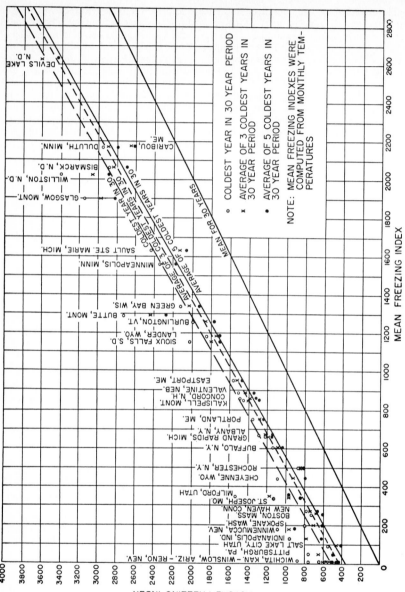

Fig. 5-14. See caption on facing page.

ences in elevation and topographical position and proximity to cities and bodies of water may cause relatively large variations of freezing index over short distances, and more care will be required in the selection of a design freezing index than in areas farther north.

Question frequently arises concerning the relationship between mean freezing conditions and colder winters occurring with various frequencies. Figure 5-14 shows the relationship, for a number of locations in the United States, between the mean and the coldest year in 30, the average of the 3 coldest years in 30 (design freezing index), and the average of the 5 coldest years in 30.

3. Depth of Frost Penetration. Penetration of freezing temperatures below a pavement kept cleared of snow may be correlated approximately with air freezing index and with water content and dry unit weight of the base and subgrade materials lying in the frost zone. Figure 5-15 and 5-16 show relationships between these factors for low and moderately high ranges of air freezing index. These charts are based upon a combination of field measurements and theoretical computation methods. However, they do not provide separate consideration of such variables as pavement type and thickness and local wind, radiation, and other effects which influence the penetration of frost. Values selected from Figs. 5-15 and 5-16 should be verified whenever possible by observations in the specific locality under consideration. When special computations are desired for a specific situation, the modified Berggren equation is considered most accurate (14, 15).

4. Water. U.S. Army Corps of Engineers studies have concluded that a potentially troublesome water supply for ice segregation is present if the highest ground water at any time of the year is within 5 ft of the proposed subgrade surface or of the top of any frost-susceptible base materials used. When the depth to the uppermost water table is in excess of 10 ft throughout the year, a source of water for *substantial* ice segregation is usually not present. However, the existence of a deep water table does not ensure that ice segregation, frost heave, pavement cracking, and subgrade weakening during thaw will be absent. Water may be available from a perched water table, by surface infiltration through pavement and shoulder areas, or by gradual accumulation of moisture content in the subgrade zone immediately beneath the pavement. The moisture content which a homogeneous clay subgrade will develop under a pavement after its construction is usually sufficient to provide water for at least limited ice segregation, even with a very deep water table.

Fig. 5-14. Relationship between mean freezing index and freezing indexes during colder years for 30 consecutive years. (*U.S. Army Engineer Division, New England.*)

Freezing Indexes Occurring with Various Frequencies in a 30-year Period at Selected Stations South of the Zero-mean-freezing-index Line

Station location	Coldest year	Average of 3 coldest years	Average of 5 coldest years
Baltimore, Md.	470	360	300
Washington, D.C.	480	380	320
Roanoke, Va.	200	140	100
Huntington, W.Va.	660	500	420
Louisville, Ky.	660	520	410
Nashville, Tenn.	330	250	180
St. Louis, Mo.	650	490	410
Ft. Smith, Ark.	310	230	180
Oklahoma City, Okla.	340	300	250
Ft. Worth, Tex.	100	90	50
Amarillo, Tex.	390	320	220
Albuquerque, N.M.	190	170	140
Roswell, N.M.	150	100	90
Eugene, Ore.	170	120	90
Seattle, Wash.	120	70	50

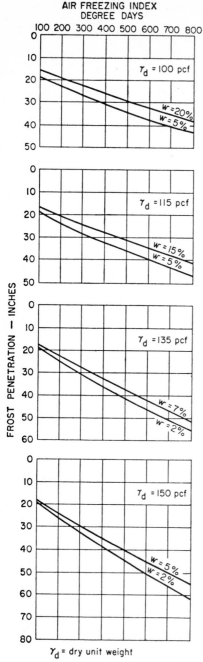

AIR FREEZING INDEX
DEGREE DAYS

γ_d = dry unit weight

w = moisture content, per cent,
based on dry unit weight

Appreciable ice segregation will not generally occur in frost-susceptible soils with a remote water table when the degree of saturation of soils in the zone subject to freezing is less than approximately 70 per cent. In fat clays, moisture moves only short distances during the freezing process, because of the low permeability of the soil, and heaving tends to be less than in lean clay and silt soils, which are more pervious. In relatively pervious but frost-susceptible sands and gravels, differential frost heave may be more intense than in clay subgrades because water may move to growing ice lenses from substantial distances, both laterally and from immediately below the lense. In all types of soils frost heave is more intense if there is a source of water closely underlying, as from a sand seam or other pervious stratum or from underlying fissured bedrock.

After the factors of *soil, temperature,* and *moisture* have been taken into account as outlined above, the engineer should, in addition, consider all reliable information concerning past frost heaving and performance of pavements in the local area under consideration; and, if necessary, the frost design requirements should be ad-

Fig. 5-15. Relationships between air freezing index and frost penetration for freezing indexes below 800 (granular, non-frost-susceptible soil beneath pavements kept free of snow and ice). (*U.S. Army Engineer Division, New England.*)

Notes: 1. Frost-penetration depths based on modified Berggren equation and computation procedures outlined in the following reports: (*a*) H. P. Aldrich and H. M. Paynter, "*Analytical Studies of Freezing and Thawing of Soils,*" U.S. Army, Corps of Engineers, Arctic Construction and Frost Effects Laboratory, June, 1953. (*b*) *Frost Penetration in Multilayer Soil Profiles,* U.S. Army, Corps of Engineers, Arctic Construction and Frost Effects Laboratory, Soil Engineering Division, MIT Dept. of Civil and Sanitary Engineering, June, 1957.

2. It was assumed in computations that all soil moisture freezes when soil is cooled below 32°F.

3. Frost-penetration depths shown are measured from pavement surface. Depths are for average thickness portland cement concrete highway pavements or for bituminous pavements over 6 to 9 in. high-quality base. For a given locality, depths may be computed with the modified Berggren equation if necessary data are available.

justed either upward or downward.   In evaluating performance records and reports, however, the engineer should always determine how the freezing and moisture conditions during the years covered by the reports compare with mean and 1-year-in-10 conditions.   *Under borderline conditions, claims of satisfactory performance should not be accepted without such comparison.*

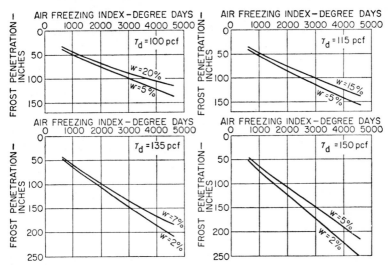

Fig. 5-16. Relationships between air freezing index and frost penetration (into granular, non-frost-susceptible soil beneath pavements kept free of snow and ice). (*U.S. Army Engineer Division, New England.*)

B. Construction on Non-frost-susceptible Materials

Clean sands and gravels under pavements do not exhibit detrimental heave and subsidence under the effects of seasonal freezing and thawing.   Sands and gravels which are slightly "dirty" or are near the border line with respect to the Casagrande frost-susceptibility criterion may, of course, show some intermediate, relatively small heave effects due to ice segregation.

The maximum heave that can occur as a result of expansion of the water phase during freezing of a saturated, completely non-frost-susceptible soil is quite small.   It may be computed by the following formula:

$$\Delta H = 0.144 w \gamma_d H \times 10^{-4} \text{ ft}$$

where w = water content, per cent of dry weight of soil
γ_d = dry unit weight of soil
H = thickness of deposit, ft

Ground in a frozen condition has a bearing strength equal to or greater than that for the unfrozen condition.   Even during thawing, clean sands and gravels remain relatively stable and maintain good bearing characteristics.   Because adverse frost effects are at a minimum or are absent in such cohesionless materials, every effort should be made to select clean, well-drained sand and gravel formations as alignment locations for planned construction and to use such soils for fill and embankments whenever possible.   While clean sands and gravels are very scarce in many geographical areas, many glaciated areas, where the higher freezing-index values are encountered, do offer substantial areas of such materials.   The tendency of sand and gravel deposits to have low ground-water levels, within limits set by surrounding terrain, contributes to their general desirability as construction sites.

Permanently frozen clean sands and gravels in permafrost regions do not usually contain troublesome bodies of ground ice. However, this cannot be made a hard and fast rule. A thin silt layer in an otherwise clean deposit can, for example, be the cause of a substantial ice layer. Usually no special measures are required where highways are constructed on clean, non-frost-susceptible deposits in permafrost areas.

Even though constructed on clean sands and gravels, it is desirable that roads in frost and permafrost regions be so designed as to ensure that the water table during the freezing period will be well below the pavement. This is in part because of the difficulty of assuring that small inclusions of fines will not occur under practical construction conditions.

Even the cleanest sands and gravels shrink considerably when cooled by low temperatures in the winter. Asphalt cement and ice have substantially greater contraction tendencies. In cold regions, not only flexible and rigid pavements, but base and subgrade materials, experience the shrinkage effects of intensive cooling in winter. At Eielson Air Force Base in Alaska, cracks totaling about $\frac{1}{2}$ in. in width per 100 ft and extending through the flexible pavement down into the underlying materials have been reported in winter, transverse to the runway. Similar cracking is experienced elsewhere in very cold regions. In the more southerly frost regions, winter cracking of flexible pavements is less intense and less obvious because even with very low surface temperatures the duration and intensity of cooling is not sufficient to cause massive cooling in depth of the underlying materials.

C. Construction on Frost-susceptible Materials

In frost and permafrost areas the possibility of detrimental consequences must always be taken into account when frost-susceptible materials may occur or be placed within the depths of annual freeze or thaw or when ice masses in permafrost may be melted by degradation of the permafrost.

Ice segregation can occur in any materials which fall within the definition of frost-susceptible materials, whether natural or artificially produced. Consideration should therefore be given to the possibility of frost action in any materials used below the ground surface, including crushed-rock base courses. While bedrock exposed in cut sections may be entirely unaffected by thaw weakening in spring, it frequently is a source of severe winter frost heave at and immediately above the surface of the bedrock. Frequently mud seams in the rock or small concentrations of fines in the immediately overlying materials will form the foci of ice segregation, which can be especially intense because of the ability of the seams in the bedrock to furnish large supplies of moisture to growing ice lenses.

Transitions should always be provided where abrupt changes in subgrade conditions occur which would otherwise result in unacceptable differential frost heave.

In permafrost areas, fine-grained frost-susceptible soils below the permafrost table usually contain ice as lenses, veins, wedges, and other forms, sometimes in very large amounts. Sometimes, on the other hand, ice may be absent or nearly absent, as near the top of a high, well-drained bluff in an arid or semiarid region. The construction of a pavement over fine-grained subgrade containing ice masses may cause thawing of these ice formations. This may result in large settlements as well as produce lower supporting capacity during the thawing period. Even in the coldest permafrost areas of northern Alaska and Greenland, a pavement which is permanently stable from the start can be obtained under these conditions only by placing at least 5 to 8 ft of non-frost-susceptible materials. In warmer permafrost zones greater thicknesses are required, and near the southerly limit of continuous permafrost as much as 20 to 25 ft of granular soil would be required under a pavement kept clear of snow to prevent thaw from reaching an underlying subgrade containing segregated ice, if the granular material were clean and well drained. For highways, it is generally economically impractical to employ base and subbase thicknesses reaching such magnitudes, except for short distances, such as adjacent to bridge abutments and similar structures.

Therefore, it is frequently necessary under these conditions to accept a possible degrading condition and provide only sufficient depth of non-frost-susceptible material to provide adequate bearing capacity for the traffic anticipated during the thaw weakening period, plus such added thickness of non-frost-susceptible base material and/or common fill as local experience may show to be needed to reduce the intensity of annual heave and settlement effects, to improve drainage, and to minimize the snow removal problem.

Highways in permafrost areas are normally elevated 2 to 3 ft above the surrounding terrain. Reliance is placed on maintenance measures to relevel the road if settlement causes the surface to become excessively uneven. Settlement from melting of ice in the subgrade tends to be most extreme in the first year following construction, but it also can be severe in following years, depending upon the chance distribution of ice formations in the subgrade. Placing of a permanent surface on a road in poor terrain is frequently undesirable. A gravel surface, treated with a dust-laying material if necessary, can be readily kept smooth through the summer with frequent passes of a blade grader; this procedure becomes impossible when a permanent surface is constructed.

D. Control of Ground Water and Drainage

1. Subsurface Moisture. Frost action could theoretically be prevented by removal of water from the frost-susceptible portions of the subgrade. However, this is not a practical method for elimination of frost action. In silts and clays, even lowering of the water table to great depth will still leave these soils nearly saturated, if there is no loss by evaporation or vegetative withdrawal, as in these fine-grained materials only a very small percentage of the total moisture, if any, can be removed by gravity drainage. Nevertheless, experience shows that pavements constructed on embankments and which thereby have superior drainage of the base course and upper portion of the embankment frequently show better frost performance than the same pavement designs in cut sections, where drainage is difficult and moisture is necessarily more readily available for frost action; where this is true it is reasonable to require somewhat less thickness of non-frost-susceptible base course in fill sections than in cut sections. Experience also shows that even in embankment sections, base courses must be drained out to the shoulders in order to obtain satisfactory performance.

Moisture conditions under the pavement should be so controlled that at no time will there be a water table in contact with the base of the pavement. It should be noted that even shallow layers of perched free water which are held at the base of the pavement by thin layers of fines produced in preparation of the subgrade and base can cause detrimental pumping of rigid pavements. The base course under a pavement should be of sufficient thickness and permeability in relation to its width between drainage points so that the base course will be able to remove the melt water produced during the thawing period without its reaching the level of the underside of the wearing surface. The U.S. Army, Corps of Engineers, specifies the general criterion (under nonfrost conditions) that base courses shall be capable of 50 per cent drainage in 10 days (16). Some details of drainage installations are illustrated on Fig. 5-17.

In permafrost areas, subsurface drainage is characteristically poor because of the underlying, impervious, frozen strata, except where clean, free-draining materials occur in favorable topographical positions. During the summer, water is supplied to the thaw zone both by infiltration of surface water and by release of water by the melting of underlying frozen soil. Under these conditions, there is a strong tendency for the water table to exist at or near the ground surface. When freezing starts in the fall, the free water tends to be drawn to the upper layers of soil, if it is frost-susceptible, forming layers of segregated ice. When freezing reaches the lower portions of the active zone later in the fall and winter, much of the available water may therefore have been already concentrated in the overlying layers of soil, if the soil is sufficiently pervious to allow movement of moisture through significant distances during the freezing process. Thus a tendency exists in permafrost areas for greater concentration of ice layers in the upper layers of the subgrade than in the lower, particularly in silt soils.

When, then, thawing of this soil occurs in the spring, the ice concentrations in the upper part of the subgrade release relatively large amounts of water during the early and most critical portion of the thawing period.

The flow of ground water is an important cause of thaw in permafrost areas, particularly when concentrated in channels. In subarctic regions, such channels tend to persist year after year when once formed and tend to be locations of progressive differential settlement during summer periods and of excessive heave during the winter,

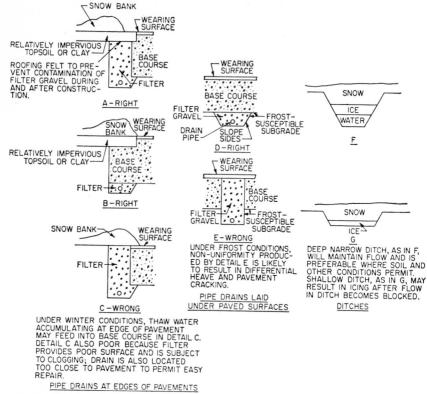

FIG. 5-17. Drainage details for cold regions. *(U.S. Army Engineer Division, New England.)*

because of the underlying source of unfrozen moisture. The possibility of formation of subsurface channels of ground-water flow should be anticipated during the planning stages in all permafrost areas. Ground-water flow beneath pavements, abutments, or other structures resting on soils containing ground ice must be prevented, if the structures are to be stable.

2. Surface Drainage. Surface drainage is an important factor in the selection of roadway alignment. Highways should be so located as to minimize the interception or blocking of existing natural drainage ways. Large cut sections should be avoided; thawed zones or water-bearing strata may be encountered and later cause serious icings in areas of high freezing index. Proper grading is essential to proper surface drainage in areas of deep frost penetration and permafrost. In permafrost areas, vegetative cover should be preserved to the maximum degree practicable. Where disturbed, it should be restored as soon as construction permits. In areas of deep seasonal frost and where the upper surface of the permafrost layer is deep, design features of a surface drainage system can be similar to those used in frost regions of the United States, if due provisions are made for the lower temperatures. In the Arctic, surface-drainage computations should assume that infiltration is zero. In the Subarctic, infiltration rates should be based on local experience. In both the Arctic and Subarctic, supplementary ponding above drain inlets should be avoided because of icing hazards.

In large areas of the Arctic and Subarctic regions, the principal fine-grained soils have silt rather than clay characteristics. Such soils not only may contain large amounts of ice lenses and wedges when frozen, but they have little cohesion on thawing and are easily eroded. Thus uncontrolled erosion is a frequent hazard of surface drainage, particularly in borderline permafrost areas. The surface of permafrost may be considerably lower beneath streams, channels, or ditches which contain flowing water than under adjacent terrain. This may result in collapse and sloughing of banks, and for this reason drainage ditches should be, under these conditions, located as far as practical from road shoulders. Locating of ditches in areas of steep slopes should be avoided if possible, as slides and sloughing may occur because of unstable conditions produced by thawing.

In subarctic regions a substantial task may be involved in freeing channels of drifted snow before spring breakup. It is helpful to design such ditches and slopes to cross sections which will accommodate heavy snow-removal equipment. On the other hand, where it is anticipated that flow will be continuous through the winter, narrow deep ditches are preferable to wide ditches, as illustrated in Fig. 5-17, in order to prevent freezing to the bottom and to avoid icings. Maintenance equipment for drainage facilities in these areas should include heavy snow-removal apparatus and a steam boiler with accessories for steam thawing of structures which have been clogged with ice. Pipes for this purpose should be fastened inside the upper portions of culverts prior to their placement. It will frequently be desirable to install culverts of much greater than normal capacity in order to allow for reduction of the cross section by ice.

Particular emphasis should be given to proper design of culverts, culvert inlets and outlets, ditches, subsurface drains, and combination drains to minimize the effects of icing. Structural design of culvert headwalls and similar structures should take into account the possibilities of frost thrust and of settlement due to degradation. Since frost thrust occurs in a direction perpendicular to the plane of freezing, the penetration of freezing temperatures through a culvert headwall into the soil behind will result in a thrust which is normal to the face of the headwall. Therefore, the fill immediately behind such structures should be clean, non-frost susceptible materials for sufficient distance back from the face so that overturning thrust will not be developed against the headwall by freezing of frost-susceptible materials.

3. Icings. Icings are irregular sheets of ice built up on the ground or pavement by thin films of water which flow out onto the surface and freeze when exposed to the cold air in winter. In northern seasonal frost areas, icings are a very serious maintenance problem, and they may completely block roads to traffic. Icings are of three general types: *river icing,* which is formed along a water course and has a river as a source of water; *ground icing,* where the source is ordinary seepage out of the ground; and *spring icing,* which has a spring as the source of water.

It is very important that road alignments in areas where icing may occur are chosen so as to minimize the problem of icing in relation to the pavement. Ground icings can be partially controlled by inducing the ice to form at a point where it does not affect the roadway. This involves bringing the ground water to the surface at a desired point by any one of various means, such as installation of impervious cutoffs or by use in permafrost areas of frost belts, which are strips from which snow is removed so as to induce rapid freezing of the underlying ground. If necessary, low barriers may be erected at the shoulder of the roadway and raised as necessary during the winter, inducing the ice to be stored vertically without drainage across the roadway itself. In case of springs, flows of water are ordinarily too large to permit use of the methods described above, and it is necessary to drain or divert the flow to a location where it will not affect the roadway. River icing may be alleviated by maintaining a sufficiently deep and narrow channel so that the river does not freeze to the bottom and thus force the flowing water to emerge onto the surface of the ice; a deep snow cover is of great assistance in maintaining the flow of water below the ice.

VI. PAVEMENT DESIGN FOR COLD REGIONS

A. Pavement Types

Both rigid and flexible types of pavements are used in highways in the seasonal frost areas of the United States. However, in permafrost areas, only flexible-type pavements are generally used, except where foundation conditions are very favorable. Flexible pavements are better able to absorb the effects of differential frost heave, although even flexible pavements may be badly damaged or destroyed if subjected to excessive differential displacement. Hot-plant-mix bituminous concrete can be placed more satisfactorily under borderline low-temperature conditions than portland-cement concrete. The latter suffers from the disadvantage that it must be protected from freezing during the placement and curing period; it must be kept sufficiently warm to gain substantial tensile strength before exposure to very low temperatures. In rigid pavements the jointing system provides a means by which shrinkage can occur without damage under very low temperatures, but for flexible pavements there is no practical means of preventing shrinkage cracking. It is unfortunate that flexible pavements are most brittle and least ductile during the period of low temperatures when the greatest shrinkage tendency and greatest heave occur. The cracking of flexible pavements is considered of structural significance only to the extent that it may provide a means for introduction of moisture into the subgrade and may offer a point where raveling may start and where freezing of moisture in and immediately below the cracks may act to further widen and intensify the cracks.

Local experience may show that greater depth of frost protection may be required under rigid pavements than under flexible, in order to assure that the pavement will not become severely cracked by differential frost heave. On the other hand, portland-cement concrete, because of its better load-distributing characteristics, shows less loss of load-bearing capacity during the frost-melting period than flexible pavement, so long as pumping conditions do not occur (Fig. 5-10). The adverse effects of frost heave on rigid pavements may also be controlled by use of relatively small slab dimensions and by use of steel reinforcement, which will hold edges of cracks tightly together. In flexible pavements, the highest penetration bitumen which will be tolerable under summer conditions should be used, in order to ensure maximum ductility under low-temperature conditions.

B. Base Course and Subgrade

Base-course materials within the depth of frost penetration should be non-frost-susceptible. If the combined thickness of pavement and base is less than the depth of frost penetration, at least the bottom 4 in. of the base course should be graded so as to provide filter action against penetration of the subgrade soil into the base course under the kneading action of traffic during and immediately following the frost-melting period.

For rigid pavements the 85 per cent size of filter or regular base-course material placed directly beneath pavements should be equal to or greater than 2.00 mm in diameter (No. 10 U.S. Sieve Series size). The 85 per cent size is that size particle for which 85 per cent of the material by weight is finer. The purpose of this requirement is to prevent loss of support by pumping of soil through the pavement joints; however, it cannot control the development of "blow holes," which occur at the outer edges of the pavement.

To prevent the movement of particles from the subgrade soil into the base course, the following conditions must be satisfied:

$$\frac{15 \text{ per cent size of base-course filter material}}{85 \text{ per cent size of subgrade soil}} \leqq 5$$

and

$$\frac{50 \text{ per cent size of base-course filter material}}{50 \text{ per cent size of subgrade soil}} \leqq 25$$

When the subgrade soil is medium to highly plastic and without sand or silt partings, the 15 per cent size of the filter material may be as great as 0.4 mm.

To permit water to drain freely in the base course, relative to the subgrade, the following criterion is specified:

$$\frac{15 \text{ per cent size of base-course filter material}}{15 \text{ per cent size of subgrade soil}} \geqq 5$$

However, the base course filter portion of the base course must not be permitted to contain more than 3 per cent of material by weight finer than 0.02 mm. Non-frost-susceptible sand is especially suitable for such a filter course.

Design should aim at maximum practical *uniformity* of base and subgrade conditions. When it is not practical to eliminate heave, it may be tolerated if it is sufficiently uniform. Much can be done during the design stage to ensure such uniformity, through specification of adequate transitions and by careful consideration of all those phases of the construction which may result in differential heave. Transitions may need to extend over as much as 100 ft under high-speed-traffic pavements. Differential heave frequently develops at abrupt changes in soil characteristics or groundwater conditions, at changes between cut and fill, and at locations of underpavement drains and culverts. At cut-fill transitions, topsoil, humus, and highly weathered surface materials may be of higher moisture-holding capacity or higher frost susceptibility than underlying subgrade soils and may serve as foci of especially strong frost action. Design should require such materials to be completely removed at cut-fill transitions, within the depth to which freezing will occur, even though stripping of the subgrade may not be generally required in fill areas.

C. Design for Frost Action

Design of pavements over frost-susceptible subgrades may be accomplished by the two methods described below (17, 18).

1. **Limited Subgrade Frost Penetration.** In this method, sufficient thickness of non-frost-susceptible base is used so that only limited, tolerable penetration of freezing temperatures into the frost-susceptible subgrade occurs. By this means, both pavement heave and subgrade weakening are reduced sufficiently in amount, frequency of occurrence, and duration so that their effects may be neglected.

The depth to which freezing will occur under a pavement kept cleared of snow should first be estimated from Fig. 5-15 or 5-16, as applicable, using the average dry unit weight of the base course and its estimated moisture content at the start of freezing and assuming the base is of unlimited depth. While this frost-penetration computation can be made for any freezing condition, such as a specific winter or the coldest winter in 30 years of record, it is recommended that the *design freezing index* be used, computed for the coldest winter in 10 years of record, or the average of the 3 coldest winters in 30 years of record. If more refined estimates of depth of frost penetration are desired than are obtainable from Figs. 5-15 and 5-16, the modified Berggren equation may be used if the necessary computational data are available (14, 15). In the absence of specific data on density and moisture content of the base and subbase material which will underlie the pavement, a relatively safe value of frost depth penetration may be obtained by using the plots for 135 or 150 lb per cu ft dry unit weight and 5 per cent moisture content on Figs. 5-15 and 5-16.

The initial frost-penetration value determined as above represents the maximum value of penetration for material of the assumed density and moisture content and of unlimited depth in the design-freezing-index year. However, it is not normally necessary to provide this full thickness of base-course material since a small amount of frost penetration into a frost-susceptible subgrade below the base course may usually be tolerated during occasional winters. Figure 5-18 shows allowable subgrade frost penetration used in this design method by the U.S. Army Corps of Engineers (17) for airfield pavements for high-speed military aircraft. The thickness of base which will result in this allowable amount of frost penetration into the subgrade in the design-

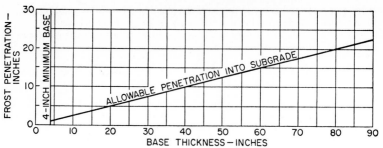

FIG. 5-18. Allowable subgrade frost penetration in design-freezing-index year (limited-subgrade-frost-penetration design). (*U.S. Army Engineer Division, New England.*)

PAVEMENT

a = COMBINED THICKNESS OF PAVEMENT AND NON FROST-SUSCEPTIBLE BASE FOR ZERO FROST PENETRATION INTO SUBGRADE (FIGS. 5-15 & 5-16)

b DESIGN DEPTH OF BASE

c = a − p

s SUBGRADE FROST PENETRATION

w_b = WATER CONTENT OF BASE

w_s = WATER CONTENT OF SUBGRADE

EXAMPLE: IF c = 60" AND r = 2.0, THEN b = 40" AND s = 10"

$$r = \frac{w_s}{w_b}$$

IF COMPUTED r ≧ 2.0, USE 2.0.

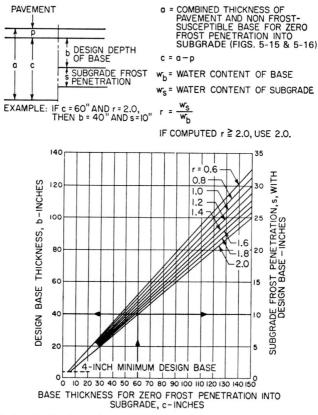

FIG. 5-19. Design depth of non-frost-susceptible base for limited subgrade frost penetration. (*U.S. Army Engineer Division, New England.*)

freezing-index winter may be determined from Fig. 5-19. This thickness will prevent penetration of frost into the subgrade in the average winter or will limit it to a relatively small and insignificant value. When the severity of the winter reaches the one-year-in-ten design condition, the computed base thickness will result in approximately the amount of frost penetration into the frost-susceptible subgrade shown on

Fig. 5-18; the resulting heave should be tolerable, even for high-speed-traffic pavements over highly frost-susceptible F3 and F4 subgrades, when considered in relation to the cost which would be required to eliminate this small and relatively infrequent heave.* For F1 and F2 subgrades, results will tend to be on the conservative side.

2. Reduction in Subgrade Strength. The second design approach is based on the reduced supporting capacity of the subgrade during the frost-melting period. This method generally permits less combined depth of pavement and base than that required under the Limited Subgrade Frost Penetration Method. The method provides sufficient combined thickness of pavement and base to prevent breakup of the pavement under *capacity* traffic during the period of minimum load-supporting capacity in the spring of the year. During all other periods of the year, then, the load-bearing capacity will normally be in excess of that required, as will be apparent from inspection of Fig. 5-10.

a. **Flexible Pavements.** For flexible pavements, the curves on Fig. 5-20 should be used to determine the combined thickness of flexible pavement and non-frost-susceptible base required. These curves are based upon traffic tests performed on pavements during the spring frost-melting period (19, 20) and upon accumulated experience and reflect the reduction in strength of the subgrade during the frost-melting period. Before entering the chart, the subgrade must be classified as to relative frost susceptibility using the classification grouping F1 through F4 shown in Art. IV.D, above.

b. **Rigid Pavements.** Under the reduction in subgrade-strength method, a non-frost-susceptible base course equal in thickness to the concrete slab should be used, but with the following specific exceptions permitted:

1. Where soils of groups F1, F2, and F3 in a subgrade occur *under very uniform conditions*† and the design freezing index is less than 1,000, the thickness of the non-frost-susceptible base may be reduced to 4 in. and it should be designed to provide filter action.

2. Where soils of groups F1, F2, and F3 occur under very uniform conditions and the depth to the uppermost water table at all periods of the year is greater than 10 ft, the thickness of the non-frost-susceptible base may be reduced to 4 in. and it should be designed as a filter. The subgrade modulus used in this determination should be obtained from Fig. 5-21 of this section; the curves of this figure, like the design curves on Fig. 5-20, are based upon traffic performance and extensive experience.

Thicknesses of pavement and base determined from Figs. 5-20 and 5-21 by the Reduction in Subgrade Strength method do not provide for control of surface roughness and cracking which may result during the winter period from frost heave. Where significant subgrade frost penetration will occur, the design studies should include consideration of frost-heaving-experience records from existing pavements in the same area under comparable conditions, and the combined thickness provided should be increased if necessary to hold heave and cracking of pavements to acceptable values under the traffic types and speeds anticipated, with the maximum value being a total thickness of cover equivalent to the Limited Subgrade Frost Penetration design-method value determined in paragraph 1 above. For F-4 subgrades thicknesses computed by the Reduction in Subgrade Strength design method should not be used unmodified, because of the severe heave which may occur; a possible exception is flexible pavement of lesser importance used for slow-speed traffic, such as a vehicle parking area where the heave and cracking may be tolerated. In some locations, up to as much as 5 ft combined thickness of pavement and base may be required over the worst subgrade conditions, though less than half this thickness may be required over the less troublesome frost materials. When thicknesses determined by the

---

* Note, however, that the weather may fail to follow the statistics in any particular year or group of years, and the coldest winter in 10, or colder, *can* occur in the first year after construction.

† In this and the following paragraph, "very uniform conditions" are intended as those conditions of soil and moisture which will result in very uniform heave, such that unacceptable differential heave and/or pavement cracking will not occur with the design thicknesses which result under this design method.

Reduction in Subgrade Strength method are greater than those obtained by the Limited Penetration method, the latter values should be used, provided they are at least equal to normal-period-thickness requirements.

The Reduction in Subgrade Strength design method presented does not make use of California Bearing Ratio, subgrade modulus, or other types of field tests performed during the frost-melting period, since most such tests do not give values which are representative of the weakening that occurs when the subgrade is thawing and is subjected to the kneading action of traffic.

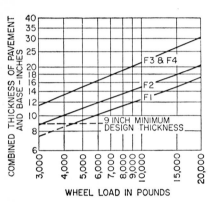

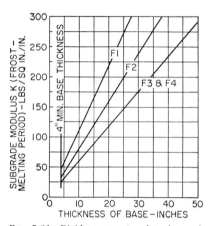

FIG. 5-20. Flexible-pavement-design curves for highways (frost condition reduced subgrade strength). (*U.S. Army Engineer Division, New England.*)

FIG. 5-21. Rigid-pavement subgrade modulus curves (frost condition reduced subgrade strength design). (*U.S. Army Engineer Division, New England.*)

D. Design Examples

1. Example 1.   Design both flexible and rigid highway pavements to carry a 16,000-lb wheel load (32,000-lb axle load) under frost conditions, using the following information:

Design freezing index.....................	600
Pavement (from normal-period design)......	3 in. bituminous concrete or 8 in. portland cement concrete (for case of protected corners)
Base material...........................	Non-frost-susceptible, dry unit weight 135 lb per cu ft, moisture content in fall 5 per cent
Subgrade..............................	Lean clay, plasticity index 14, moisture content 30 per cent; very uniform conditions; normal-period CBR 8 per cent
Highest ground water....................	3 ft below top of subgrade
Concrete flexural strength................	700 psi

The subgrade soil falls into frost group F3.

a. **Flexible Pavement.**   (1) LIMITED SUBGRADE FROST PENETRATION.   From Fig. 5-15 the estimated depth of frost penetration below the pavement surface, for base material of 135 lb per cu ft dry unit weight, 5 per cent moisture content, and unlimited depth is 43 in.   Subtracting the 3 in. of wearing surface, the penetration in base-type material would be 40 in.   From Fig. 5-19, required actual base thickness under this design method is 26 in., using a ratio of subgrade to base moisture content r of 2.0, the maximum permitted,* and about 7 in. penetration into subgrade may be expected

---

* Values of r above 2.0 are not used because not all of the moisture in fine-grained soils will actually freeze at the temperatures which will be reached in the portion of the subgrade into which freezing temperatures will penetrate.

1 year in 10. Required combined thickness of pavement and non-frost-susceptible base under the Limited Subgrade Frost Penetration Method is 26 + 3 = 29 in.

(2) REDUCTION IN SUBGRADE STRENGTH. From Fig. 5-20, 27 in. combined thickness of pavement and non-frost-susceptible base are required by the reduction in Subgrade Strength method, for this type F3 subgrade soil. This is 2 in. less than required by the Limited Subgrade Frost Penetration Method.

Since subgrade conditions are expected to produce very uniform heave, the 27-in. thickness is the logical choice. At least the bottom 4 in. of the base should be graded to provide filter action against the subgrade.

b. **Rigid Pavement.** (1) LIMITED SUBGRADE FROST PENETRATION. From Fig. 5-15, the estimated depth of frost penetration with base of unlimited depth is 43 in. Subtracting the 8-in. slab thickness applicable for normal-period design, the penetration in base materials only would be 35 in. From Fig. 5-19, the required actual base thickness is 23 in., which will allow about 6 in. frost penetration into the subgrade in 1 year in 10. Required combined thickness of pavement and non-frost-susceptible base = 23 + 8 = 31 in.

(2) REDUCTION IN SUBGRADE STRENGTH. Since design freezing index is less than 1,000 and subgrade is of a type which produces very uniform heave, exception permitting a minimum 4-in. base course to protect against loss of support by pumping is applicable. From Fig. 5-21, the reduced-strength subgrade modulus is 25 psi per in. Applying a factor of 2 to the flexural strength, the required slab thickness for the condition of protected corners is 10 in.

The selection of rigid pavement design will depend on a cost comparison between the 19 in. greater base-course thickness required by the Limited Subgrade Penetration Method and the 2 in. additional concrete thickness required by the Reduction in Subgrade Strength design method. Comparison of heave effects is not necessary as it has already been assumed that heave will be acceptably uniform.

2. Example 2. Design both flexible and rigid highway pavements to carry a 12,000-lb wheel load (24,000-lb axle load) under frost conditions, using the following information:

Design freezing index...................... 2,500
Pavement (from normal-period design)...... 3 in. bituminous concrete or 8 in. portland cement concrete
Base material......................... Non-frost-susceptible, dry unit weight 135 lb per cu ft, moisture content in fall 5 per cent
Subgrade............................. Silt, moisture content 25 per cent; normal period CBR 6 per cent
Highest ground water.................... 3 ft below top of subgrade
Concrete flexural strength............... 650 psi

The subgrade soil falls into frost group F4.

a. **Flexible Pavement.** (1) LIMITED SUBGRADE FROST PENETRATION. From Fig. 5-16, the estimated depth of frost penetration below pavement surface, for uniform material of 135 lb per cu ft dry unit weight, 7 per cent moisture content, and unlimited depth is 120 in. Subtracting 3-in. wearing surface, penetration in base-type material would be 117 in. From Fig. 5-19, required actual base thickness is 78 in., using the maximum permitted ratio of subgrade to base-course moisture content of 2.0. About 19 in. frost penetration into subgrade may be expected 1 year in 10. Required combined thickness of pavement and non-frost-susceptible base to prevent substantial freezing of subgrade is 78 + 3 = 81 in. This is not a practical thickness for most highway applications. However, it provides a guide in design studies.

(2) REDUCTION IN SUBGRADE STRENGTH. From Fig. 5-20, 23 in. combined thickness of pavement and non-frost-susceptible base is required over the type F4 subgrade soil to provide sufficient wheel-load supporting capacity in the spring. However, because of the high heave potential of the F4 subgrade and the high water table, this may be expected to result in heave, cracking, and surface roughness, which is unacceptable except possibly for vehicle-parking areas or similar pavements. Frost-heave-experience records from pavements in vicinity should therefore be studied, and a nominal, larger, combined thickness of pavement and non-frost-susceptible base between the extremes of 23 and 81 in. should be chosen which may be expected to hold

frost-heave effects within reasonable limits for the particular intended usage. Normally, such total thickness should not exceed 5 ft for highways, even for group F4 soil under very adverse moisture and freezing conditions.

b. **Rigid Pavement.** (1) LIMITED SUBGRADE FROST PENETRATION. From Figs. 5-16 and 5-19, the required combined thickness of pavement and non-frost-susceptible base is 74 + 8 = 82 in. Again, this is not a practical thickness for most highway applications (though it might be practical for an airfield).

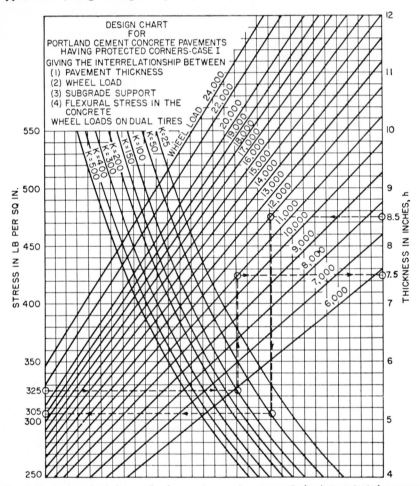

FIG. 5-22. Design chart for portland-cement-concrete pavements having protected corners, Case I. The curve $K = 25$ is intended here for frost-criteria evaluation described in this section in connection with the weakened support which occurs during spring thaw. *(Concrete Pavement Design, 1951, Portland Cement Association, fig. 4, p. 19.)*

(2) REDUCTION IN SUBGRADE STRENGTH. Because of the high heave potential of the F4 subgrade, design of rigid pavement in accordance with the Reduction in Subgrade Strength criteria is, in this case, not advisable.

Limitation of pavement to flexible type would be desirable in the situation illustrated in this example. However, if rigid type is not rejected, the use of smaller-than-normal slab sizes or use of reinforcement to hold crack edges tightly together should be considered with up to 5 ft combined thickness of pavement and base.

If, in this case, subgrade soil were of group F3 or lower classification, under very uniform conditions unlikely to produce undue cracking caused by frost heave, analysis would be made by successive approximations, as follows:

1. Assume approximate slab thickness. First approximation base thickness, then, is equal to this slab thickness.

2. Using this base thickness, determine subgrade modulus from Fig. 5-21.

3. Using this value, determine slab thickness from Figs. 5-22 and 5-23, as applicable.

4. If agreement is not obtained, repeat steps 1 to 3, using progressively more accurate slab- and base-thickness assumptions.

If, further, the subgrade soil were of group F1, F2, or F3, of a type which produces very uniform heave, and the highest ground-water level was below 10 ft, a 4-in. minimum base thickness could be used under the exception outlined under subsection, Pavement Design for Cold Regions, Rigid Pavements.   The design subgrade modulus

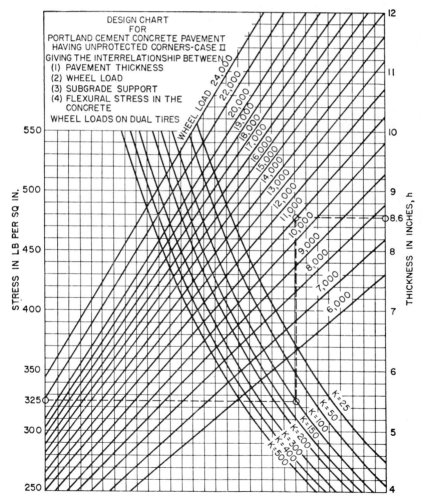

Fig. 5-23. Design chart for portland-cement-concrete pavements having unprotected corners, Case II.   The curve $K = 25$ is intended here for frost-criteria evaluation described in this section in connection with the weakened support which occurs during spring thaw. *(Concrete Pavement Design, 1951, Portland Cement Association, fig. 5, p. 20.)*

then would be determined from Fig. 5-21 and would be used to determine the required slab thickness.

VII. CONSTRUCTION OPERATIONS IN FROST AND PERMAFROST AREAS

A. Special Construction Problems of Cold Regions

Frost effects and the occurrence of low temperatures very seriously affect the scheduling of construction operations in cold regions. In both seasonal-frost and permafrost areas, effective highway construction is usually limited to the period during which air temperatures are at or above freezing, though well-drained granular materials and hot-mix bituminous surfacing are occasionally placed in below-freezing temperatures for emergency purposes, and routes through poorly drained, difficult areas are sometimes best roughed out under winter conditions when the surface is hard-frozen. Construction is also limited by the spring "mud time," during which the surface of the fine-grained soils may be practically impassable for construction equipment. In permafrost areas, cuts in undrained frozen materials are commonly made by first stripping the surface early in summer to expose the frozen material. By progressively removing thawed materials, excavation may be carried down to the elevation required. Thaw in fine-grained frozen soils may be as much as 6 in. per day following exposure. Frozen materials may also be thawed by use of cold-water or steam jets or broken up by explosives. Care must be taken to avoid permitting too great a depth of fine-grained soil to become thawed at one time, as it may become impossible for the construction equipment to cope effectively with the soft, liquefied material.

Well-drained, clean, granular materials can frequently be handled satisfactorily at very low temperatures, even though they may give trouble at temperatures which are only slightly below freezing. It has not been established whether such materials can be satisfactorily compacted in the frozen condition.

Establishment of grades when the ground is freezing or thawing is difficult in areas of frost-susceptible materials because of the rise and fall of temporary bench marks and of the pavements themselves. In permafrost regions this problem extends through the summer thawing season. In borderline permafrost areas, where freezing does not reach the permafrost surface until late in the winter, the pavement may be continuously rising and falling throughout the year.

B. Control of Subgrade Preparation

Construction operations frequently expose local adverse subsurface conditions not revealed by the design subsurface explorations. It is very important that field personnel recognize such situations as they are revealed and take action when necessary to avoid later difficulties. Construction inspectors should be aware of the design assumptions and should be alert to check their validity. Inspectors may be trained to visually identify frost-susceptible materials with the aid of comparison samples of known percentage finer than 0.02 mm. Doubtful subgrade materials encountered during the construction should be check-tested for gradation.

Since clean, granular materials are little affected by frost action, preference should be given to the cleanest available of such materials for use within the frost zone. However, it is important that these materials be uniformly clean, as small irregular inclusions of fines may cause serious differential frost heave.

Pockets of frost-susceptible soils in otherwise non-frost-susceptible subgrades or sharp variations in frost-heave potential of subgrade, within the annual frost zone, can frequently be detected only during grading operations. Such conditions can be corrected by removal and replacement or by providing transition zones between the areas of differing heave potential, so as to provide a surface which will remain acceptably smooth under the particular traffic usage the pavement will receive.

When unexpected wet areas are encountered in the subgrade which may provide a source of water for frost action, additional drainage measures should be installed as needed, such as special subsurface drains and special intercepting ditches to reduce infiltration into the ground on higher ground adjacent to the road.

Rock cuts frequently are a source of frost-heave difficulties. Each case should be studied individually. Ground-water and drainage conditions and the general nature and jointing of the rock are the key factors which must be considered. Because of the ability of bedrock to supply large quantities of water to growing ice lenses, frost heaving may frequently be actually more severe in rock cuts than in adjacent soil areas, if concentrations of fines are present at the surface of the bedrock, in mud seams, or in the base course itself. Rock cuts may therefore frequently require as much or more depth of base course than soil subgrades, to ensure pavement free of heaving and cracking. No undrained pockets should be left in the rock surface which will permit irregular reservoirs of water to accumulate in the bottom of the base course; these may produce markedly irregular raising of the surface in the winter. Filling of such drainage pockets with lean concrete may even be necessary. If emergence of seepage from the excavated rock surface cannot practically be avoided, a highly pervious drainage blanket may be used to bleed the water away laterally. If sufficiently wide and infrequent, seams in bedrock which contain frost-susceptible fines may sometimes be cleaned out and refilled with non-frost-susceptible materials. However, where this is not practical, the rock should be removed to sufficient depth to control the winter frost heave within desired limits.

When isolated ice bodies are encountered in stripping and excavation in permafrost areas, the ice should be removed and replaced with non-frost-susceptible material, if it will be within the ultimate thaw zone under the completed pavement, culvert, or other structure. If the ice is extensive, consideration should be given to relocating or to changing the design grade so as to construct the roadway entirely of fill, as nearly as possible, and of sufficient depth to protect the ice from thawing. In southern permafrost areas where summer thaw is deep or progressive degradation of permafrost cannot be prevented, it may not be possible to apply such remedial measures fully or with full effectiveness; in such a case it will be necessary to accept substantial summer maintenance of the surface as unavoidable.

C. Control of Base-course Construction

Base-course materials which are clean and unquestionably non-frost-susceptible may be placed in accordance with normal practice in nonfrost areas except that (1) scrupulous care should be exercised to avoid inclusion of small lumps, thin layers, or other small amounts of potentially troublesome fine-grained frost-susceptible materials; and (2) gradation of the material should be checked *following* placement and compaction to ensure that no untoward gradation changes have resulted during handling and rolling as a result of slaking or compaction breakdown. However, checking of materials in place does not obviate the need for inspection of materials at the pit; it is more feasible to reject unsuitable materials at the source than after they have arrived at the job. Very close checking should be done at the start of placement of material from a new borrow source. Base-course materials which are of borderline frost susceptibility (materials having close to 3 per cent, by weight, finer than 0.02 mm) need to be especially carefully checked during construction.

In order to avoid mixing of fine-grained subgrade materials with non-frost-susceptible base-course materials, the subgrade should be properly graded and compacted prior to placement of any base course, and material hauling equipment should be so routed that rutting of the subgrade by the hauling equipment is avoided. Overcompaction or insufficient thickness of the first layer of base course should also be avoided. Care must be taken that the first layer of base course provides filter action against penetration of subgrade fines into the base course under the construction traffic. Observation of the action of the base course under compaction equipment should be made to detect any areas of obviously unsuitable materials. The existence of a layer of fines on top of the base course should be avoided, since this is frequently a cause of pumping action under rigid pavements and may be a source of sufficient frost action to cause rapid deterioration of pavements, particularly after pavements have begun to crack under a number of years' service, and thus permit surface moisture to penetrate from the top of the pavement. As with subgrades, the inspection forces should be careful to provide gradual transitions in base-course thickness when this is necessary, to avoid abrupt changes in supporting conditions.

REFERENCES

1. MacFarlane, I. C.: *A Preliminary Annotated Bibliography on Muskeg*, National Research Council of Canada, Division of Building Research, Bibliography B 11, Ottawa, Ont., September, 1955.
2. Jenness, John L.: "Permafrost in Canada," *Arctic*, vol. 2, no. 1, pp. 13–27, May, 1949.
3. Pihlainen, J.: "Data on Occurrence and Distribution of Permafrost in Canada," compiled by National Research Council of Canada, Division of Building Research, Permafrost Section, Ottawa (unpublished, letter communication, December, 1956).
4. Hopkins, David M., Thor N. V. Karlstrom, and Others: "Permafrost and Ground Water in Alaska," *U.S. Geological Survey Professional Paper* 264-F, 1955.
5. U.S. Army, Corps of Engineers, Arctic Construction and Frost Effects Laboratory: Unpublished ground temperature and permafrost records from Alaska and Greenland.
6. U.S. Army, Corps of Engineers, Arctic Construction and Frost Effects Laboratory: "Investigation of Description, Classification and Strength Properties of Frozen Soils, Fiscal Year 1951," *Snow, Ice and Permafrost Research Establishment, Report* 8, Willamette, Ill., June, 1952.
7. Casagrande, A.: "Discussion on Frost Heaving," *Highway Research Board, Proceedings*, vol. 11, pt. 1, pp. 168–172, 1931.
8. Linell, Kenneth A.: "Frost Design Criteria for Pavements," in Soil Temperature and Ground Freezing, *Highway Research Board, Bulletin* 71, 1953.
9. U.S. Army, Corps of Engineers: "The Unified Soil Classification System," *Waterways Experiment Station, Technical Memorandum* 3-357, Vicksburg, Miss., March, 1953.
10. Frost, Robert E.: "Interpretation of Permafrost Features from Airphotos," in Frost Action in Soils: A Symposium, *Highway Research Board Special Report* 2, 1952.
11. Frost, Robert E.: *Evaluation of Soils and Permafrost Conditions in the Territory of Alaska by Means of Aerial Photographs*, Purdue University Engineering Experiment Station, prepared under contract with St. Paul District, Corps of Engineers, for U.S. Army, Office of the Chief of Engineers, Engineering Division, Airfields Branch Military Construction, 2 vols., September, 1950.
12. Hvorslev, M. Juul: "Interim Report on Core Drilling in Frozen Ground," paper presented before ASCE convention, Buffalo, N.Y., June, 1957, Waterways Experiment Station, Vicksburg, Miss.
13. Haley, James F.: "Cold Room Studies of Frost Action in Soils: A Progress Report," in Soil Temperature and Ground Freezing, *Highway Research Board, Bulletin* 71, 1953.
14. Aldrich, Harl P.: "Frost Penetration below Highway and Airfield Pavements," in Factors Influencing Ground Freezing, *Highway Research Board, Bulletin* 135, 1956.
15. Massachusetts Institute of Technology: "Frost Penetration in Multilayer Soil Profiles," *Arctic Construction and Frost Effects Laboratory, Technical Report* 67, Boston, Mass., June, 1957.
16. U.S. Army, Corps of Engineers: "Subsurface Drainage Facilities for Airfields," *Engineering Manual for Military Construction*, part XIII, chap. 2 (EM 110-345-282).
17. U.S. Army, Corps of Engineers: "Pavement Design for Frost Conditions," *Engineering Manual for Military Construction*, Part XII, chap. 4 (EM 1110-345-306).
18. U.S. Army, Corps of Engineers: "Roads, Walks and Open Storage Areas," *Engineering Manual for Military Construction*, Part X, chap. 1 (in revision).
19. U.S. Army, Corps of Engineers: *Report on Frost Investigations, 1944–45*, Frost Effects Laboratory, Boston, Mass., 1947.
20. U.S. Army, Corps of Engineers: *Addendum No. 1, 1945–47, to Report on Frost Investigations, 1944–45*, Frost Effects Laboratory, Boston, Mass., 1949.

BIBLIOGRAPHY

Beskow, G.: "Soil Freezing and Frost Heaving with Special Application to Roads and Railroads," *Swedish Geological Society, 26th Yearbook*, no. 3, 1935, ser. C, no. 375, with a special supplement for the English translation of progress from 1935 to 1946, Northwestern University, Technological Institute, Evanston, Ill., November, 1947 (trans. by J. O. Osterberg).

Black, Robert F.: "Permafrost—A Review," *Bull. Geological Society of America*, September, 1954.

Johnson, A. W.: "Frost Action in Roads and Airfields, A Review of the Literature," *Highway Research Board, Special Report* 1, 1952.

Muller, S. W.: *Permafrost or Permanently Frozen Ground and Related Engineering Problems*, J. W. Edwards, Publisher, Inc., Ann Arbor, Mich., 1947.

Taber, Stephen: "Frost Heaving," *Journal of Geology*, vol. 37, no. 5, pp. 428–461, 1929.

———: "Freezing and Thawing of Soils as Factors in the Destruction of Road Pavements," *Public Roads Bull.* 11, no. 6, pp. 113–132, 1930.

———: "The Mechanics of Frost Heaving," *Journal of Geology*, vol. 38, pp. 303–317, 1930.

Terzaghi, K.: "Permafrost," *Journal of the Boston Society of Civil Engineers*, vol. 39, no. 1, January, 1952.

U.S. Army, Corps of Engineers: Arctic Construction, *Engineering Manual for Military Construction*, part XV, chaps. 1–7 (EM-1110-345-370 through 376).

Section 6

EARTHWORK

L. E. GREGG, *L. E. Gregg & Associates, Consulting Engineers, Lexington, Ky.*

I. INTRODUCTION

Earthwork is that portion of highway construction required to convert the right-of-way from its natural condition and configuration to the sections and grades prescribed in the plans.   Exclusive of structures, pavements, and incidental items, it includes all phases of the actual building of the road.   In construction contracts and specifications earthwork is sometimes referred to as the grading and drainage portion of the project.

Choice of equipment and scheduling of operations are highly important to successful completion of earthwork.   Usually progress in any phase of the operation depends upon the selection and production of key machines, and scheduling the work for maximum use of the equipment is essential.   Likewise, quality of the work is dependent upon control over the manner in which selected machines are applied.

The various phases of earthwork may be generally divided into clearing and grubbing; roadway, structure, borrow, or special classes of excavation; embankment construction; and finishing or final dressing.   Included in the special classes of excavation are removal or restitution of slides, trenching for subsurface drainage or utility lines, tunneling for various purposes, and similar items.

Different construction projects involve all or most of the types of earthwork to different degrees.   Occasionally they overlap.   For example, material classed as borrow excavation usually serves the purpose of supplementing material derived from roadway excavation to provide sufficient quantities for embankment construction nearby.   However, if the roadway excavation consists of cut-through rock to a prescribed depth below subgrade elevation, some of the borrow material may be used as refill over the exposed rock.   Hence that part of the borrow excavation is replacement for material removed under the classification of roadway excavation.

Classification of the different phases has many advantages in addition to expressing the character of the work.   Types and capacities of equipment required, degrees of skill and wage rates for operators, the order in which operations should be undertaken, reasonable times for completion of the operations, and several other pertinent features are represented indirectly in the classifications.   As a consequence, practically all contracts for grading and drainage designate the different classes as separate pay items.   This is an advantage to both parties: The contractor is aided in figuring his bid, scheduling operations, realistically accounting for costs, and estimating claims for partial payment on the contract as work progresses.   On the other hand, the owner achieves flexibility for increasing or decreasing the volume of work without renegotiation, orderly supervision of the work, and a logical basis for charting progress.

II. ESTIMATING EARTHWORK QUANTITIES AND DISTRIBUTION

Estimates of the quantity and distribution of earthwork are essential prior to construction, and these are logically completed in the design stage of the project.

Many situations require only a rough estimate of earthwork quantities, in which case the crudest methods of calculation can be used.   This is usually done by approximation of some dimensions from which a gross estimate of the earthwork is obtained.

For most highway projects generalized estimates are sufficient only where there is limited reconstruction of existing roads, or in the preliminary stages of design where the object is selection of an alignment or grade for trial computation.

One method of computation particularly adaptable to borrow areas is the development of a grid with lines established at 25-, 50-, or even 100-ft intervals and elevations determined at each intersection.   After excavation has been made during construction, elevations are again determined and the volume of material borrowed is calculated by averaging the depth of cut at each of the four corners for each square defined by the grid and multiplying by the area of each grid square.

Most calculations of earthwork for highways apply to roadway excavation and embankment construction.   Both the quantity of earth to be removed from cut sections and the volume of earth necessary to form the embankment to the proper grade are involved.   Balancing of the two quantities sometimes determines the grade to which the road will be laid, but often this is offset by requirements for meeting the grade of existing roads, bridge elevations, or other controlling factors.

A. Cross Sectioning

Determination of quantities for this purpose is usually based on the cross-sectioning method since it is most practical and informative.   In addition to surface configuration, the subsurface conditions, drainage structures, utility lines, and other features can be graphically illustrated and their influence estimated.

In this operation cross sections are taken at regular intervals—usually 50 ft—and where major surface irregularities occur.   Elevations and distances from the center line are determined at sufficient points along a line at right angles to the center line to give a reasonably accurate portrait of the ground surface.   The array of points is then plotted to scale along with subsurface data and the final cross section of the proposed roadway.   Figure 6-1 illustrates a few typical examples of cross sections.

Areas inscribed by the original and final cross sections are determined usually by planimeter, differentiating areas of excavation and areas of embankment.   In areas of excavation, distinction is usually made between solid rock and soil.   These areas are computed separately, not only because of potential differences in cost of excavation, but because of their influence on the design and particularly the tendencies for "shrinkage" and "swell."   Rock will swell when excavated and broken and will occupy more space than rock in solid form.   Increases up to 25 per cent of the original volume are common.   In contrast, soils in their natural state generally occupy more volume than the same soils properly compacted in embankments—hence the term shrinkage.   Shrinkage factors of 15 to 20 per cent are typical.   Actually swell and shrinkage factors should be determined by measurements for each different type of material found on the project, but this is seldom feasible and representative values usually apply.

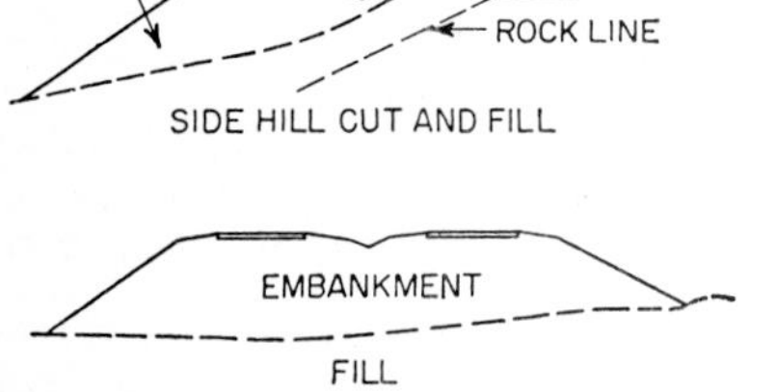

Fig. 6-1. Generalized highway cross sections.

Cross sections are not necessarily acquired by individually determining each elevation by field leveling; they may be obtained from contour maps by scaling the distance to contour lines from a precisely located center line.   The latter method is less dependable because of limitations in positioning contours even on maps having small contour intervals.

The volume of earth between each two consecutive cross sections is approximated by multiplying the average of the two cross-sectional end areas by the distance between the cross sections and expressing the quantity in terms of cubic yards. Correction factors for irregularities are sometimes applied. If the composition of material is sufficiently variable, volumes of solid rock and soil should be measured and computed separately.

B. Contour Slices

Sometimes contour maps have peculiar application to the determination of earthwork quantities, particularly where the area is large and irregular. Interchanges and

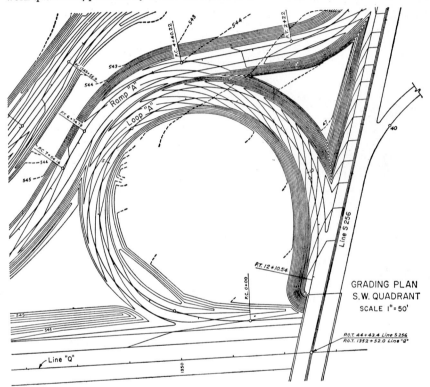

Fig. 6-2. Segment of interchange plans adaptable to earthwork calculations by contour slices.

locations of large structures are adaptable to determination of earthwork by integration of contour slices, since full contour maps are normally required in the design of drainage and several other features (Fig. 6-2).

With this approach, contours for desired grading are constructed upon the contour map of the existing ground, both in their proper horizontal location and orientation. Every finish contour must terminate coincident with the original contour of the same elevation. When all finish contours are calculated and constructed on the map, areas inscribed by each finish contour and the corresponding original contour are measured, usually by planimeter. Care must be taken to differentiate between areas of excavation and areas of embankment as outlined by the original and finish contours. Volumes are determined by summing the volumes of each contour slice limited by successive contours, the volume of each contour slice being approximated by the average of the upper and lower contour areas multiplied by the contour interval.

Volume determination by contour slices should be attempted only with maps having relatively large scales and small contour intervals.

C. Distribution of Quantities

When all volumes have been determined, an analysis of the relationships between cut and fill may be performed. Planned movement of materials from cuts to fills facilitates construction and reduces construction costs by demonstrating to the engineer and contractor how much material is to be moved and where it should be placed for maximum economy and utility.

The simplest form of planned movement of materials consists of merely summing the quantities of excavation and embankment until points are reached where the quantity of excavation, adjusted for shrinkage and swell, equals the amount of material required for embankment. These points are known as balance points, between which no borrow or waste should be necessary. Quantities between balance points must include all materials and requirements in addition to roadway earthwork. Approaches and

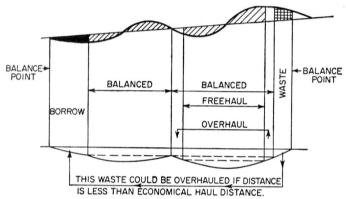

Fig. 6-3. Mass diagram.

special drainage and structure excavation must be included in the estimate. Allowance is sometimes made for subsidence of embankments since certain areas may have low supporting strength and the weight of fills upon them may exceed the supporting strength.

A more useful and reliable method of planning movement of material employs the mass diagram. As illustrated in Fig. 6-3, the mass diagram is plotted directly below the profile, the abscissa for both being stationing of the project. The ordinates of the mass curve are in cubic yards of net cut or fill plotted above or below a selected base line. Total yardages of cut and fill are not represented by the mass curve, and they must be summed between balance points selected from the mass diagram.

Distribution of quantities depends upon haul distances as well as on the materials available. To secure maximum advantage in bid prices, usually free-haul and overhaul distances are established in the specifications and the contract. Overhaul pertains to material which must be moved exceptionally long distances for which the contractor is entitled to extra pay, and overhaul quantities are normally measured in terms of yard stations, the unit being one cubic yard of material transported one station beyond the free-haul limit.

Economical haul is the distance which material may be hauled more economically than it can be wasted and borrowed. Economical haul distances vary greatly since it may be difficult to dispose of waste in some locations, and likewise borrow may not be readily available. The limit of economical haul may be expressed as

$$L = \frac{B}{O} + F$$

where L = limit of economical haul, in stations
 B = cost of borrow per cubic yard
 O = cost of overhaul per cubic yard
 F = free haul, in stations

III. SITE PREPARATION AND EXCAVATION

A. Clearing and Grubbing

Prior to the moving of earth it is essential that the site be cleared of obstructions, including structures, fences, utility lines, and major vegetation. With some exceptions, undesirable organic matter including topsoil and roots and stumps must be removed to depths necessary for satisfactory construction. Decomposition of organic matter is a continuing process after the material is covered by overlying soil or rock. As a consequence, the requirements for clearing are not limited to just those features that would obstruct the operation of equipment.

Methods used in clearing, and in turn the type of equipment and cost, are governed by the type of location, character of terrain, ground cover, requirements for disposal, and sometimes the season of the year. If the salvaging of timber is an important factor, a contract specifically for that purpose is desirable, with the work to be completed in advance of clearing and grubbing for the highway construction. Whenever conditions permit, living trees valuable in the finished landscaping of the project should be rigorously preserved.

The depth to which grubbing of roots and stumps must be carried depends mainly on whether the area will be involved in cut or fill—and more so the height of fill. In all cases, the cutting and grubbing should be carried to a depth where major remnants of stumps or root systems are at least 3 ft below final subgrade elevation. Stumps can be left protruding as much as 1 ft above the scalped ground elevation if the overlying fill will be at least 4 ft deep. Greater heights of stumps above ground would interfere with the operation of equipment in distributing and compacting soil properly.

All brush and stumps should be removed from the right-of-way to waste areas, or preferably burned in locations away from the immediate roadway site. Usually conditions for burning of brush and debris are governed by local or state statutes, and those should be consulted at the time clearing operations are first contemplated.

Unless vegetation is exceptionally heavy, both clearing and grubbing can be readily accomplished with a crawler tractor equipped with a mounted dozer or preferably a rake blade. If the vegetation is light and the land free from rock, blade graders and brush cutters can be effective. On the other hand, when large trees and stumps must be removed, large tractors up to the 16-ton class developing at least 90 hp at the drawbar are required. These conditions as well as those of extremely rough terrain call for the most experienced operators as well as equipment equal to the job.

The foregoing applies more to rural than to urban construction, where buildings, pavements, power lines, pipelines, and similar artificial obstructions must be removed, usually with greater care than that given natural growth. Maintaining utility services, protecting adjacent buildings, and numerous other considerations are of utmost importance. As a consequence, clearing in urban areas can represent a major portion of the total cost of construction, whereas on rural projects usually it is a minor cost item.

Clearing should be completed and the site inspected for satisfactory condition before earth moving is started on any major portion of the project. Otherwise organic material or debris intended for removal may become buried. In those sections where embankment is to be placed, particular care must be given to the filling and compaction of soil in depressions left by excavated trees and stumps, and the scarifying and recompaction of all soil within embankment areas. This is important from the standpoint of providing firm and stable material on which to start the embankment construction (paragraph V.A.4).

Measurement and payment are generally on the basis of the number of acres involved, although removal of existing structures is often a separate bid item on rural work. Under urban conditions, of course, several separate features may be and usually should be classified for bid items.

B. Rock Excavation—Roadway

1. General Classification. Rock is that portion of the earth's crust which underlies and sometimes emerges through the soil cover. It is variable in its origin, physical characteristics, and degree of firmness in so far as the requirements for excavation are concerned. General classes of rocks are igneous, metamorphic, and sedimentary, which terms define the mode of origin of the materials. Those of the sedimentary class, as the name implies, were formed through consolidation of sediments in the waters of ancient seas. Best-known examples are limestone, sandstone, and shales. Igneous rocks, of which granite is an example, were formed from molten lava under various environmental conditions, and metamorphic rocks are materials produced through alteration of either igneous or sedimentary rocks by processes of excessive heat, pressure, or both. Marble is a metamorphosed limestone, and slate is a metamorphosed shale.

There are almost limitless varieties of all three materials in nature, and the general character in any case determines the most desirable treatment in excavation and sometimes the usefulness of the material elsewhere on the project. Because of the variable firmness, toughness, or degree of cementation encountered in rock exposures during earth moving, it has become customary to discard classification of rock on highway projects—particularly in roadway excavation. Most specifications and bidding invitations simply combine all rock and soil under the heading of unclassified excavation. This avoids controversies over the classification of weakly cemented or highly weathered rock encountered on the project and the bid price that should be applied.

Excavation for structures is often classified into solid rock and common excavation, and provisions for separate measurement and payment are applied on the contract. This reflects the critical nature of foundation conditions, particularly for large structures such as bridges. It also stems from the fact that usually subsurface exploration and hence possibilities for estimating quantities are much more elaborate at structure locations than throughout the roadway in general.

2. Explorations. On projects of any magnitude a reasonable amount of exploration to rock and in rock is required for the roadway as well as the structures. As the size and complicity of projects increase, the need for more elaborate subsurface data increases. Quantities are large, and in the absence of qualitative information on rock and rock-soil combinations, bid prices on excavation will necessarily be high. Adequate designs of slopes, estimates of changes in volume (shrinkage or swell), location of cavities, and several other factors demand the information at the design as well as the construction stage.

While most positive information is gained through coring the rock and logging the cores, costs by this method alone often become exorbitant. Inspection of aerial photographs and geologic maps, combined with physiographic classifications of land forms in the area, are excellent reconnaissance methods. Still more definite data, particularly on depths to rock, have been obtained with great satisfaction by use of seismic methods. Reports on electrical-resistivity methods to define depths and character of rock deposits are more contradictory, with various degrees of success obtained under different conditions.

Insufficient knowledge of rock formations often results in disastrous and costly landslides during or shortly following construction (see Sec. 3). This is particularly true in regions of rough terrain. In a similar way costly changes in plans, or additions in borrow quantities, will develop in instances where mine cavities or solution channels are revealed only after construction is under way, or thick seams of coal or underclay are encountered and excessive quantities of waste result.

3. Drilling and Shooting. Most rock excavation requires blasting to loosen the material and reduce it to sizes feasible for loading and hauling. This in turn necessitates drilling into the rock formation to place the charges. Spacing of holes and the depths to which charges are carried depend on the character of the rock, the height of the face that is being worked, and the desired degree of shattering or preferable maximum size of individual pieces after the shot.

Generally, drilling is done by pneumatic equipment. Hand drills are adequate if the holes are to be less than 2 in. in diameter and no greater than 10 ft in depth. For depths running as great as 50 or 60 ft, wagon drills are efficient. Beyond that depth churn drills offer the best possibility at reasonable cost. Unless there is some special reason to the contrary, deep excavations can be worked most efficiently in stages or benches 10 to 30 ft in thickness. By this procedure each bench can be cleaned in the loading and hauling operation while drilling is progressing on top of the face that is adjacent.

Air compressors are integral parts of the drilling operation when the tools are pneumatically driven. Since line losses can increase rapidly with distance, it is desirable to have the compressor as close as possible to the drills which it serves. Size and weight of the compressor and possibilities for transporting it to the vicinity will determine the arrangement. Problems of this nature have led to the development of combined tractor-compressor units (Fig. 6-4*b*), which offer advantages of portability but sometimes at the sacrifice of compressor capacity.

Compressor requirements vary with the number and size of drills being served and the distance mentioned previously. A small compressor of 60 cubic feet per minute (cfm) capacity can supply one, or at the most two, light hand drills consistently. From that point up to heavy wagon drills the compressor capacity must be increased to as much as 315 cfm. Under some conditions compressors working in pairs are useful in serving two or more major drilling units within a limited area.

Several factors including hole spacing, character of rock, depth of face or burden, and properties of the explosive determine the amount of charge needed to dislodge a given quantity of rock. Tables for estimating particular needs are given in literature developed specifically for industries dealing with rock excavation (2).* For the purpose of generalized estimates it is considered that one pound of 40 per cent dynamite is sufficient to break up and move a yard of rock that may be termed average. Only persons trained and experienced in blasting can plan shots in a way that will move the maximum quantity, shatter the rock sufficiently, and still avoid excessive breakage beyond limits of the proposed excavation.

4. Breakage and Refill. Rock excavation in roadway cut sections will unavoidably extend beyond the backslope limits and below subgrade elevation. Specification tolerances for the backslopes must take this into account, although overbreakage need not be measured or included as a pay item in the contract. On the other hand, excessive breakage beyond plan lines should not be tolerated, and the supervising engineer should carefully determine when breakage is extending beyond reasonable limits. Excessive breakage of some rock materials in deep cut can create localized points of weakness that facilitate slides.

Excavation of rock should always carry at least 6 in. below finished subgrade elevation, unless a granular subbase overlying the rock and extending through the shoulders is regarded as refill in lieu of soil from roadway or borrow excavation elsewhere. Refill with soil or subbase material is necessary to establish a uniform surface true to section and grade from which base and pavement construction can be started. Also, the refill provides a cushion that minimizes differences in the supporting characteristics of earth embankment and solid rock in the transition from fill to cut sections.

As much as 2 and sometimes 3 ft of additional rock excavation and refill is warranted where unstable shales, coal, and similar materials are intersected at subgrade elevation. Seams of coal and associated underclays that are more than a few inches in thickness should never be closer than 1 ft below subgrade elevation, and on major highways a

* Numbers in parentheses refer to corresponding items in the references at the end of this section.

(a)

(b)

FIG. 6-4. Pneumatic-powered equipment for drilling in rock. (a) Wagon drills and compressor. (b) Self-contained unit.

minimum of 2 ft should be maintained. Similarly, shales that weather readily into active and highly plastic clays should be removed to depths at least 2 ft below subgrade elevation. On roads carrying extremely heavy traffic greater depths can be justified.

5. Loading and Hauling. Loading and hauling rock excavation normally involves a dipper shovel (Fig. 6-5) and the necessary number of trucks or rock wagons to keep the shovel occupied almost constantly. In this case the shovel is the key machine, assuming the trucks are constantly on hand. Capacity of the shovel must be fitted to the job. For excavation of rock on highway projects it is seldom that the bucket capacity will exceed 4 yd or be less than ¾ yd. Maximum sustained hourly output, assuming the most favorable conditions for digging and loading, will be about 100 times the bucket capacity, but average operations will fall well below that level.

6. Measurement—Swell. Since rock excavation for the roadway is generally unclassified, it is measured simply as a portion of all roadway excavation. Cross sections determined on the ground by the engineering party or by photogrammetric

Fig. 6-5. Excavating and loading rock.

means represent the configuration of the new ground surface after the excavation has been made. If this is done by photogrammetric means, naturally a new set of aerial photographs taken subsequent to the excavation is necessary.

In either case, the cross-section data before and after this part of the construction are the bases for computing the amount of excavation actually made by the contractor. Inasmuch as some specifications provide that certain features of rock excavation beyond plan lines will not be allowed for payment, the measurements determine only whether the required tolerances have been met. Overbreakage on backslopes is an example. Even this should be included where it can be definitely shown that the additional excavation beyond a reasonable approximation of the plan line was made as a result of the character of the rock formations and not through any fault of the contractor.

On most construction projects payment is based on the number of cubic yards of material in its original position that has been excavated in an acceptable manner. Use of that material in the embankment is subject to considerations discussed elsewhere. Under practically all conditions of rock excavation a swell factor must be applied, as noted in the discussion of earthwork-design calculations, where distribution of the rock to embankment construction is involved.

C. Soil Excavation—Roadway

1. General Classification. There are infinite varieties of soils in nature and many ways of identifying and classifying these materials. Group classifications by general properties, such as the Highway Research Board (1) or unified-system (8) designations, are highly valuable in estimating earthwork operations during construction.

With the exception of very undesirable materials such as muck or underclay, soils are usually taken as found and used to whatever advantage can be developed. Although it is desirable to separate soils from different horizons of natural formation and place those with the best properties in the most critical portions of embankments or at sub-grade elevation in rock cuts, usually this is not feasible. Considerable mixing is unavoidable.

2. Loading and Hauling. Differences in properties of soils have limited influence on equipment and methods for soil excavation. Position and land form and the length of haul are more important considerations.

Self-loading scrapers or pans are the backbone of soil excavation where the haul distances are not too great and there is sufficient space to maneuver. These units are rubber-tired and have capacities generally ranging from 10 to 25 cu yd heaped, and they can easily operate at speeds of 20 mph empty. Under favorable conditions a fully loaded scraper can operate at approximately the same speed.

Fig. 6-6. Self-loading scrapers operating with auxiliary power from dozers.

Except in cases where loading is exceptionally easy, it is desirable and is usually necessary for a pusher tractor to assist in the loading operation (Fig. 6-6). As a general rule, one pusher tractor can serve two scraper units if the loaded haul is approximately 300 ft, and an additional scraper unit can be served with each increase of 300 ft in the haul distance.

Ratings on the performance of scraper units vary with size, power, haul distance, and the terrain. As a representative estimate, an average 12-yd scraper operating over a loaded-haul distance of 500 ft should deliver approximately 90 cu yd of earth per hour. Delivery will be somewhat greater than half that yardage if the haul distance is doubled, and more than a third that yardage if the haul distance is increased to 2,000 ft. Beyond that point the delivery will reduce rapidly.

Sometimes conditions are not conducive to excavation of soil by scrapers. In rough or highly eroded terrain it may be necessary to break down or ramp over projecting ground or gullies before scrapers can operate. This and similar work is done by bull-dozers. If soil is to be moved only a short distance—less than 250 ft—most, if not all, the work will be done more efficiently by tractor dozer rather than scraper. Likewise, on hillsides that are relatively steep and where the finished section will be combined cut and fill, probably the dozer working in conjunction with compaction equipment on the fill side will carry out the operation to best advantage.

3. Grading Tolerances. Soil excavation can and should be completed to very close tolerances along both the base and the back slopes of cut sections. Deviations from plan dimensions may be as great as 1 ft in the horizontal direction, particularly in that portion of the roadway from the ditch line outward to the edge of the right-of-way. In the area between ditch lines much closer tolerances can be met and should be required.

Deviations in the vertical direction should not exceed 0.05 ft within the portion of the roadway where pavement or paved shoulders are to be placed. Preferably the variations from plan grade in these locations should be plus rather than minus, in order to facilitate fine grading of the subgrade by removal of excess material during the paving operation. Vertical deviations as great as 0.2 ft can be tolerated in the grading on earth shoulders, medians, and other portions between the ditch lines, since these too will be influenced by the final dressing, seeding, or sodding in conjunction with the later construction of the pavement.

4. Measurement—Shrinkage. Soil excavation is generally referred to in contract documents as unclassified excavation, and payment for the work is based on the number of cubic yards of material in its original position excavated and used for embankments and refill, or wasted. As in the case of rock excavation, the quantities are determined by measurements on the ground or by means of aerial photographs representing conditions before and after the grading operations.

The determination of amounts and distribution of soil thus excavated, taking into account the so-called "shrinkage" element, is carried out in accordance with procedures previously discussed in this section.

D. Structure Excavation

For purposes of earthwork, the excavation required to install or construct either pipe or cast-in-place culverts, foundations for culverts, bridges, and other similar structures is classed as structure excavation. Numerous factors in addition to the excavation of soil or rock may be involved in earthwork at structures: removing existing structures; providing temporary run-arounds; channel relocation; site protection and support; and dewatering.

1. General Classification. For reasons previously described, it is important to differentiate structure excavation into elements such as solid rock and unclassified materials—the latter being soil and loose materials that can be removed without blasting. It is also important to describe materials below and above normal levels of stream flow or ground-water table as wet and dry excavation. Because of the necessity for dewatering, shoring, or using expensive methods of removing materials from the foundation area, wet excavation will usually cost considerably more than excavation above the level of free water.

In some cases the materials to be excavated should be still more clearly defined. Where peat or muck occurs to depths that would seriously influence the foundation conditions, the limits of these materials should be carefully delineated and shown on the plans. The effect of such materials can be offset in the design or sometimes in the location of structures, such as by changes in stream channels to completely remove the structure from the area of organic deposits. Whenever this is not practicable, the design should provide not only adequate support for the structure but also practical means for dealing with the materials during the earthwork stage of construction.

2. Preparation and Protection of the Site. Preparation of structure sites prior to earthwork operations is comparable to that described for the clearing and grubbing in advance of roadway excavation. Often the removal of an existing structure is involved, and usually that requires a temporary run-around at the site if traffic is to be maintained. When realignment or relocation of a stream is required, scheduling of the different stages of earthwork construction will be necessary in the design stage to accommodate the flow of the stream, the operation of traffic (even if only construction equipment), and the immediate use of the excavated soil or rock in embankment or backfill in order to avoid double handling of the material that is excavated.

Except in those instances where lowering of the water table with well points gives adequate protection, cofferdams, cribs, or sheeting will be necessary where water must be excluded from the excavation, adjacent structures must be protected, or the slopes and depths of excavated soil are such that sides of the excavation must be supported. If exclusion of water is the object, the sheeting or cofferdam must be designed for practically watertight walls and driven or constructed before excavation is started.

The depth to which sheeting should extend will depend upon the depth to rock—and where rock is not encountered, the characteristics of the soil in the bottom of the excavation and the head created by the level of water outside the excavation will govern. If soils are fine sands and the head is sufficient to create a hydraulic gradient greater than unity, the depth of penetration of the sheeting must be increased to lengthen the path of seepage and reduce the hydraulic gradient. Otherwise, soil in the bottom of the

FIG. 6-7. Excavation at structure location by means of a clamshell bucket. *(Courtesy of Kentucky Department of Highways.)*

excavation will become "quick" or fluid, and further excavation or construction of the foundation will not be feasible. Under most conditions, penetration of sheeting to depths of 6 to 8 ft below the bottom of excavation in soil is sufficient.

3. Equipment and Methods. Choice of equipment and methods for structure excavation depends upon the type of structure and location. On the large structures and in locations where protective works such as cofferdams or sheet-pile walls create vertical lines and the bracing limits the working space, clamshell buckets (Fig. 6-7) are particularly useful. Draglines (Fig. 6-8), perhaps adaptable to the same shovel equipment, are highly productive in channel relocation or excavation to gently sloping lines.

In the cases of culverts or pipe culverts, the width and depth of excavation are usually limited. Here the backhoe is particularly useful if there is trench excavation, or the dozer if the structure is to be cast in place and water has little or no influence on the foundation excavation.

Earthwork for pipe culverts should be designed in accordance with the type of location, size of pipe, presence or absence of rock, and to some extent the height of fill that will overlie the pipe. Preferably pipe should be installed in an excavated trench, in order that there will be a prepared bedding, and side support from established walls is available in case compaction of refill and overlying materials is inadequate. If grades are such that natural stream beds do not provide for trenches, the embankment can be built to a height at least 2 ft greater than the depth of the culvert, and a trench excavated in the well-compacted soil (Fig. 6-9).

Fig. 6-8. Dragline operating in channel-change location.

Trenching and refill is particularly important in cases where the underlying material is rock. Excavation of a trench in the rock to a depth of at least 8 in. is necessary, and backfilling with granular material rather than fine-grained soil is desirable. The former will be easier to place and compact, and it offers greater possibility for distributing the load reaction uniformly from the rock to the pipe.

Similar placement of granular cushion uniformly below the pipe is essential in areas where unstable foundations prevail, such as in marshes or areas of soft marine clays. Here the cushion material should be at least 1 ft in depth and have a width at least twice the diameter of the pipe.

4. Measurement. Quantities of excavation for structures are measured in cubic yards and referenced to the excavation actually required to place the structure, revise the stream channel, or the like. Usually some excess in quantity from the required volume of the installation is given if the excavation is in soil. For example, with a bridge pier the measurement may refer to the volume of a solid bounded on the bottom by the plane of the footing; on the sides by vertical planes extending upward from the outer edges of the footing in rock but 1 ft outward from those edges if in soil; and on the top by the original ground surface.   In the case of pipe there should be tolerances also, to recognize the desirability and necessity of excavation beyond just the volume represented by the pipe itself.   Differentiation of rock and soil, or wet and dry excavation, should be in accordance with the actual depths of the two materials encountered in each instance.

Fig. 6-9. Backfilling in culvert trench excavated in compacted earth fill. *(Courtesy of Kentucky Department of Highways.)*

IV. SELECTION AND PLACEMENT OF MATERIALS IN SUBGRADES AND EMBANKMENTS

Characteristics of materials excavated in grading of the roadway, and to a lesser extent the structures, generally limit the design of embankments.   Principles of design are discussed in Sec. 3.   Utilization of the most favorable materials acquired in the excavation is basic, but separation of the most desirable soils or rock from those less desirable is seldom feasible.   However, certain general procedures can be followed in separating and distributing materials to advantage.

A. Rock

Basically, rock is undesirable in embankments since reduction of most rock materials to sizes that will provide a relatively dense fill is seldom practicable.   Large voids generally conducive to settlement caused by migration of finer materials into voids surrounding the coarser pieces are inherent in the character of materials produced by blasting.   Also, there are no compaction methods that are effective.

With the exception of micaceous shales, thinly laminated carbonaceous shales, and argillaceous shales which disintegrate and decompose rapidly through weathering, rock materials generally produce moderate to high resistance to shearing forces even when placed in a relatively loose condition.   Hence, rock fills can be made stable without exercising unusual care, but careful design and construction procedures are necessary to avoid settlement.

Sandy shales and similar materials which pulverize readily but do not decompose can be broken down, distributed, and compacted in the same manner as soils. The contrasting shales which are unstable in their composition or have a thinly laminated structure should be wasted if possible. When wasting is not feasible, they should be placed in the base of the embankment and confined by coarse broken stone such as limestone or sandstone in the outer portions of the fill.

Rock composed predominantly of large fragments or blocks of stone should be placed where it will form the base of an embankment for its full width; on the side slopes (usually to be covered to a limited depth by soil in order to support vegetation);

Fig. 6-10. Excavation of rock-soil mixture composed mainly of large fragments of rock.

or on the side slopes and top of rolled embankment made from soil. In cases where embankments are to be benched as a means of distributing the load, or at locations where loading the toe of a slope is desirable to prevent slides (see Sec. 3), rock material is particularly useful.

Rock should be placed in layers not exceeding 3 ft in depth, and any pieces having dimensions too great for placement in a layer of this thickness should be wasted or broken down to smaller size. Control of moisture has no bearing on the placement of the rock itself, but when rock forms the shell around soil in the interior of an embankment, the soil layers should always be kept higher than the adjacent rock, with adequate slope for drainage provided.

B. Rock-Soil Mixtures

Rock-soil mixtures develop frequently where thin ledges of rock are interspersed in deposits of shale or clay. Most conditions of this nature occur near the surface, where grading equipment can sometimes remove most of the soil before the drilling and shooting of rock is started. Even then some intermixing of materials cannot be avoided. Most mixtures of this description can be used satisfactorily, but careful attention must be given to their composition and placement.

When solid rock occurs in close association with highly undesirable soils—such as the relation of sandstones to coals and underclays in regions of coal seams—it will be necessary to waste the rock along with the clay and coal if separation cannot be made in the process of drilling and shooting.

Because of the uncertainty of composition and the lack of effective compaction methods, rock-soil mixtures should be distributed in much the same manner as rock. Distribution of the materials to outer and particularly lower portions of the embankment should be emphasized. Placement in layers no deeper than 3 ft is important, and if the maximum size of the rock permits, the depth of layer should be no greater than half that amount.

One feature of rock-soil mixtures to be carefully avoided is the loose placement of combinations in which unstable soils predominate but there is sufficient large rock to prevent controlled compaction. This is especially true where soft shales that readily decompose through weathering are abundant in the mixture. The moisture content in combinations of this type is potentially high, and embankments consisting of such materials are vulnerable to sliding or mud flows.

With regard to the selection of soil-aggregate mixtures for topping embankments or serving as refill in cut sections, the possibilities for detrimental moisture contents deserve attention. Selection of such materials for this purpose should be done only with full knowledge of the saturation moisture content determined by calculation and compared with the moisture contents critical in respect to plasticity characteristics of the soil portion. This problem is discussed at greater length under the heading of design considerations later in this section (see Art. V.A).

C. Soil

Soils of any description can be carefully and uniformly distributed by modern earth-moving equipment. With the exception of highly organic materials, which should never be used in embankments or as refill, soils can be closely controlled in the distribution and placement in layers conducive to densification by compaction. However, some soils are much more desirable than others, and design values for the embankment or subgrade should reflect conditions which can be reasonably achieved in placement and compaction of the soils. Absolute selection and separation of materials during construction is seldom feasible, but whenever the soil explorations indicate reasonable uniformity in the horizons of soil available from roadway or borrow excavation, different materials can be distributed and controlled separately.

The relationships outlined in Table 6-1 are a general guide to selection of soils classified in accordance with the Highway Research Board system. A similar but more extensive set of guides developed by the Highway Research Board Committee on Compaction of Subgrades and Embankments is given in Table 6-2. Both are referenced to degree of compaction or relative density, since that is the most significant property subject to field control in earthwork operations.

Selection of soils in this way will not be economically justified on rural secondary roads if considerable wasting of materials or overhaul becomes necessary. On the other hand, on urban, rural primary, and interstate highway projects the importance of using soils in this way cannot be overemphasized. Even in areas where rock excavation accounts for most of the available embankment material, upper portions of embankments (at least the top 3 to 5 ft) should consist of soils selected in this way. Requirements common in highway specification limit the top 12 in. of embankment to materials having maximum dry weights in excess of 102 lb per cu ft, and require the placement of the least desirable soils such as the expansive silts and clays at least 5 ft below subgrade elevation.

The thickness, or depth of soil spread, for each layer of embankment or subgrade is logically dependent upon the character of the soil and the equipment used to compact it. This feature is discussed more fully in portions dealing with compaction that follow. As a generalization, it can be estimated that the depth of layer after compaction

Table 6-1. General Guide to Selection of Soils on Basis of Anticipated Embankment Performance

HRB classification	Visual description	Max. dry-weight range, lb per cu ft	Optimum moisture range, %	Anticipated embankment performance
A-1-a A-1-b	Granular materials	115–142	7–15	Good to excellent
A-2-4 A-2-5 A-2-6 A-2-7	Granular materials with soil	110–135	9–18	Fair to excellent
A-3 A-3a*	Fine sand and sand	110–115	9–15	Fair to good
A-4a* A-4b*	Sandy silts and silts	95–130	10–20	Poor to good
A-5	Elastic silts and clays	85–100	20–35	Unsatisfactory
A-6a* A-6b*	Silt-clay	95–120	10–30	Poor to good
A-7-5	Elastic silty clay	85–100	20–35	Unsatisfactory
A-7-6	Clay	90–115	15–30	Poor to fair

* Ohio modification.
Source: Ohio Department of Highways.

Table 6-2. Recommended Minimum Requirements for Compaction of Embankments

Class of soil (AASHO M 145-49)	Condition of exposure					
	Condition 1 (not subject to inundation)			Condition 2 (subject to periods of inundation)		
	Height of fill, ft	Slope	Compaction (% of AASHO max. D)	Height of fill, ft	Slope	Compaction (% of AASHO max. density)
A-1 A-3 A-2-4 A-2-5	Not critical Not critical Less than 50 Less than 50	1½ to 1 1½ to 1 2 to 1	95+ 100+ 95+	Not critical Not critical Less than 10 10 to 50	2 to 1 2 to 1 3 to 1	95 100+ 95 95 to 100
A-4 A-5	Less than 50 Less than 50	2 to 1	95+	Less than 50	3 to 1	95 to 100
A-6 A-7	Less than 50	2 to 1	90–95*	Less than 50	3 to 1	95 to 100

Remarks: Recommendations for condition 2 depend upon height of fills. Higher fills of the order of 35 to 50 ft should be compacted to 100 per cent, at least for part of fills subject to periods of inundation. Unusual soils which have low resistance to shear deformation should be analyzed by soil-mechanics methods to determine permissible slopes and minimum compacted densities.
* The lower values of minimum requirements will hold only for low fills of the order of 10 to 15 ft or less and for roads not subject to inundation nor carrying large volumes of very heavy loads.
Source: Highway Research Board (4).

will be about two-thirds to three-fourths the depth of loose spread, so that if the soil to be placed is most efficiently compacted into a 6-in. layer, the loose spread should approximate 8 to 9 in.

With ordinary equipment and construction methods, fine-grained and cohesive soils should not be placed in loose depths exceeding 9 in. On the other hand, granular and cohesionless materials can be placed satisfactorily with layers up to 12 in. loose depth if pneumatic or oscillating pneumatic rollers providing a kneeding action are a part of the operation. Loose depths can be made much greater if the soil is genuinely cohesionless, vibratory compaction of favorable frequency is involved, and the layers below offer the necessary reaction or barrier to reflect the impulses.

V. COMPACTION OF SOILS IN SUBGRADES AND EMBANKMENTS

Compaction of soils has been defined as "the practice of artificially densifying and incorporating definite density into the soil mass by rolling, tamping, or other means" (4). Investigation and experience over the past 25 years have shown that physical properties of a soil mass are greatly affected by increases in unit weight or degree of density brought about by compaction under externally applied forces. In essence, the bearing value and resistance to shear are usually (though not always) increased (9); the passage of water through the voids is generally retarded, and because of reduced void volume the potential for detrimental moisture contents is decreased; tendencies for volume change (shrinkage and swell) are reduced for reasons previously cited; resistance to frost action is generally increased since heat and moisture transfer are retarded; and the mass of soil becomes more uniform and less susceptible to differential settlement.

Several factors influence the degree of density that can be achieved with a given compactive effort. These have been referred to in earlier portions of this section. The greatest influences are moisture content, type or general character of the soil, and the amount as well as the method of application of the compactive effort. To a lesser degree other factors such as temperature have a bearing on the results of compaction in earthwork operations.

A. Design Considerations—Specifications

As noted in the discussion relative to Tables 6-1 and 6-2, compaction of soils in embankments and subgrades is usually referenced to a given laboratory procedure for determining the compaction characteristics of the soil, or it is based on the use and operation of selected equipment (control of compactive effort). These in turn are related to pavement or traffic conditions, slope and height of fill, exposure to adverse conditions such as inundation, and other features pertinent to the performance of the subgrade or embankment.

Compactive effort represented in the laboratory procedure should be related in some way to the compactive effort imparted by equipment operating on the job. As shown in Fig. 6-12, variations in the effort have a marked effect on results obtained from the tests, and the degree of influence varies with general characteristics of the soils. Test methods described by AASHO Designation T 99 should apply in most cases, unless a high degree of density is the objective and large compaction equipment is available on the job.

For obvious reasons, there are wide variations in the moisture-density relations for different types of soil tested under a given compactive effort. The curves in Fig. 6-11 represent seven different soils tested in accordance with AASHO T 99. Later discussions (paragraph V.C.3) show that when the soils throughout a given region of interest fall within characteristic patterns, typical curves can be developed as a short cut to determination of moisture-density relations for a particular soil.

1. **Controlled Soil Density.** In addition to the guides contained in Tables 6-1 and 6-2, others in more abbreviated and usually more practical form have been devised for specific projects such as turnpikes or other major facilities of recent origin. Condensed versions of specification requirements widely used for design and control pur-

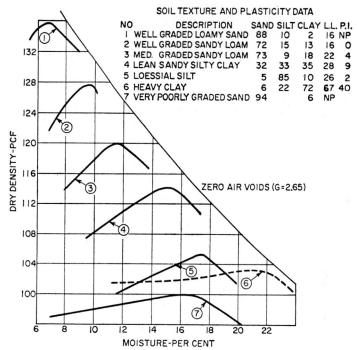

SOIL TEXTURE AND PLASTICITY DATA

NO	DESCRIPTION	SAND	SILT	CLAY	L.L.	P.I.
1	WELL GRADED LOAMY SAND	88	10	2	16	NP
2	WELL GRADED SANDY LOAM	72	15	13	16	0
3	MED. GRADED SANDY LOAM	73	9	18	22	4
4	LEAN SANDY SILTY CLAY	32	33	35	28	9
5	LOESSIAL SILT	5	85	10	26	2
6	HEAVY CLAY	6	22	72	67	40
7	VERY POORLY GRADED SAND	94		6	NP	

Fɪɢ. 6-11. Variations in moisture-density relations of different types of soils. [*Courtesy of Highway Research Board* (4)].

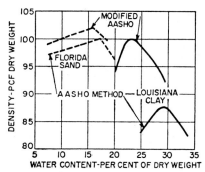

Fɪɢ. 6-12. Variations in moisture-density relations of soils resulting from different compactive efforts. [*Courtesy of Highway Research Board* (4)].

poses are given in Tables 6-3 and 6-4. Actually, these are oversimplifications of the basic principle involved, and the possibilities for differentiating the needs if the soils on a given project are distinct and can be separated. Such a condition seldom exists on the job, and recognition of this fact must be incorporated in the design requirements and specifications.

In any event, the method of specification based on controlled density has certain advantages and some disadvantages. It can be definitely related to a measurable property of the soil, both in the laboratory and in the field, and it provides broad latitude for selection of equipment most suited to the soil and the conditions under which construction will be carried out. On the other hand, application of design require-

Table 6-3. Specification Requirements for Embankment Soil Compaction

Condition 1		Condition 2	
Fills 10 ft or less in height and not subject to extensive floods		Fills exceeding 10 ft in height or fills of any height subject to long periods of flooding*	
Max laboratory dry weight, lb per cu ft	Min. field-compaction requirements, % laboratory max. dry weight	Max. laboratory dry weight, lb per cu ft	Min. field-compaction requirements, % laboratory max. dry weight
Less than 90.0	†	Less than 95.0	‡
90.0–102.9	100	95.0–102.9	102
103.0–109.9	98	103.0–109.9	100
110.0–119.9	96	110.0–119.9	98
120.0 and more	95	120.0 and more	96

* Where condition 2 applies to any portion of the embankment below a horizontal plane through subgrade elevation at pavement center line on any cross section, all portions of soil embankment throughout the total width and depth on that cross section shall be compacted in accordance with condition 2 requirements.

† Soils having maximum dry weights of less than 90.0 lb per cu ft are considered unsuitable and shall not be used in embankment.

‡ Soils having maximum dry weights of less than 95.0 lb per cu ft are considered unsuitable and shall not be used in embankment under condition 2 requirements. Soils having maximum dry weights of less than 103.0 lb per cu ft shall not be used in the top 12 in. of embankment subgrade.

SOURCE: Ohio Department of Highways.

Table 6-4. Specification Requirements for Minimum Subgrade Soil Compaction

Max. Laboratory Dry Weight, (Lb per Cu Ft)	Min. Subgrade Compaction Requirements, (% of Laboratory Max. Dry Weight)
Less than 103.0	*
103.0–109.9	102
110.0–119.9	100
120.0 and more	98

* Soils with a maximum dry weight of less than 103 lb per cu ft are considered unsuitable for use in the top 12-in. soil layer immediately below the surface of the subgrade and shall be replaced with suitable soil or granular material.

SOURCE: Ohio Department of Highways.

ments in this way can impede the construction because field control methods are time-consuming; rather careful identification of the soil is necessary as construction progresses in order to make certain which laboratory data (compaction curves) apply; and manpower requirements for sampling and testing are substantial.

2. Controlled Compactive Effort. Although in recent years the trend in specifications has been toward use of controlled density rather than controlled compactive effort, still most earthwork specifications contain references to at least the type of compaction equipment permitted. When the specification requirements are based wholly on controlled compactive effort, the equipment must be described not only with respect to type, i.e., sheepsfoot, pneumatic, vibratory, etc., but also with regard to size, weight (pressure exerted), method of operation, frequency of vibrations, or oscillations if applicable, and several other features.

Lift thickness must be specified, and the number of passes of coverages of the equipment over a given lift must be stated. To approach any semblance of control, the moisture content of the soil must be stated and carefully regulated during construction; otherwise the effort will be largely wasted if the moisture content is very far from the optimum value for compaction by the selected method.

Effectiveness of design requirements based on controlled effort is dependent on the judgment of those supervising construction, and it has no tangible connection with a characteristic of the soil which can be measured and defined in recognized terms. Also, it offers no incentive to the contractor to use resourcefulness in developing improved and more efficient operations to achieve the desired degree of density in the soil, and thus reduce construction costs.

3. Saturation of the Soil in Service. When soils are compacted to maximum unit weight at optimum moisture content with a given procedure, the difference between the moisture content at any point on the compaction curve and the moisture content on the zero air-voids curve corresponding to that same unit weight represents void space available for additional water (see dimension A, Fig. 6-13). If the service conditions are such that the necessary moisture can reach the compacted soil by percolation or capillary migration, the soil can become saturated.

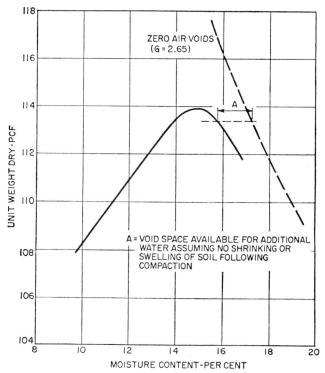

Fig. 6-13. Representation of voids remaining to be filled for complete saturation of a compacted soil.

This condition, within itself, is not cause for concern, but the characteristics of the soil to be compacted should be analyzed from this standpoint during the design stage. If the soil is of a type known to expand considerably with increases in moisture content, it should either be eliminated from consideration or singled out for placement in parts of an embankment where increases in moisture will be limited or a limited amount of expansion will have no effect on the stability of the fill.

Similar precautions apply to soil-aggregate mixtures ordinarily considered suitable for the top portions of embankments or as refill in cut sections. In either instance the mixture is positioned as subgrade material beneath the pavement. Possibilities for excessive moisture contents in the soil portion of the mixture are often overlooked, since the moisture content by usual procedures will be measured and expressed as a function of the total material.

However, distribution of moisture in the total material is not a measure of the effect on the fine-grained portion of the mixture. Through calculation of theoretical moisture-volume relationships for a representative soil-aggregate mixture (26), it has been shown that the moisture content in the soil portion (material finer than the 40-mesh sieve size) may far exceed the liquid limit even though the moisture content based on dry weight of the total mixture is much below that level.   Representative data from the calculation are given in Fig. 6-14.

Elimination of mixtures of this sort from embankments and subgrades is not feasible or desirable in many situations.   In many areas the soils generally available for construction are far inferior to these essentially granular materials.   However, the deceptive factors inherent in the situation should be recognized in advance of construction and allowance made for these possibilities in the design of overlying subbases, bases, insulation courses, and pavements.

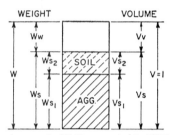

Given: Mixture of soil (20 %) and aggregate (80 %) compacted to 100 % dry unit wt. AASHO T-99
  Soil: G =   2.72
  LL = 26.8 %
  PL = 12.0 %
Aggregate: G = 2.67
  Absorption = 5 %
Mixture:
  Max. density = 130.0 pcf
  Wt % (opt) = 8.2
Find: Moisture content of total mixture and of the soil fraction (assuming only absorbed moisture distributed to the aggregate) when all voids are filled with water (saturation):

$$W_{s1} = 0.8 \times 130 = 104 \text{ lb (aggregate)}; \quad V_{s1} = \frac{104}{2.67 \times 62.4} = 0.624$$

$$W_{s2} = 0.2 \times 130 = 26 \text{ lb (soil)}; \quad V_{s2} = \frac{26}{2.72 \times 62.4} = 0.153$$

$$V_v = V - V_s = 1.0 - 0.777 = 0.223 \qquad V_s = \overline{0.777}$$

At Saturation: $Ww = V_v \times 62.4$ = 13.91 lb
  Absorption by aggregate: 0.05×104 = 5.20 lb
  Free water in voids   = $\overline{8.71}$ lb

$$\text{Wt \% (soil–aggregate mixture)} = \frac{13.91}{130} \times 100 = 10.7 \%$$

$$\text{Wt \% (soil)} = \frac{8.71}{26} \times 100 = 33.5 \%$$

FIG. 6-14. Potential moisture-volume relations in a compacted mass of soil-aggregate mixture.

4. Stabilization of Underlying Materials.   For several reasons, unstable materials below the elevations at which soils are to be compacted for subgrades or embankments must be removed or made stable by treatment.   If the embankment is high enough or of sufficient weight, the underlying material may actually fail in shear and cause the embankment to collapse; or if it is highly compressible and low in permeability, settlement under the load of the fill will be large but will occur slowly over a long period of time.

Seemingly stable soils immediately beneath subgrade material must in fact be stable; otherwise failure will occur probably under the repeated applications of load from construction traffic in placing the overlying pavement.   In any event, failure will occur under repetition of loads, and it will be regarded as subgrade failure. Actually, the subgrade soil in every respect—thickness, composition, density, and moisture content—would have performed satisfactorily had the condition of underlying soil been adequate.

Usually the unstable condition of underlying material is evident in its lack of firmness. Compaction of the first layer of embankment or subgrade is not feasible, and the specified density cannot be achieved because the underlying soil does not provide the necessary reaction. Sometimes this can be corrected by scarifying and recompacting the material below, in which case stabilization by that means alone is sufficient.

Often the cause of instability cannot be overcome by this treatment. If the water table is high and the soil has potential for rapid and relatively great migration of moisture by capillarity, sponginess of the underlying soil must be eliminated by inserting a granular layer, impervious membrane, or similar barrier to break or reduce the forces of surface tension, or the grade must be raised so that moisture cannot reach subgrade elevation (Fig. 6-15). Even then, the flabbiness of underlying material will make necessary the acceptance of degrees of density in the lower layers of embankment that are far below those that can and should be acquired in the upper portions of the fill near subgrade elevations.

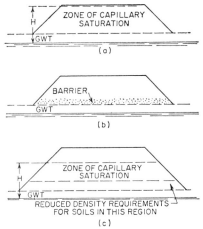

Fig. 6-15. Treatments for offsetting effects of capillary moisture in embankment construction.

Recognition of these possibilities at the design stage is important. Outstanding in this respect are beds of extinct lakes, lowlands bordering major streams, and similar locations where ground-water tables are naturally high and concentrations of silty soils prevail. However, data from soil explorations will indicate where special treatment of the type outlined is needed. Selection of the treatment will depend mainly on the influence of structure elevations, if there are any nearby, distribution of excavation quantities, availability of suitable granular materials, and the relative cost involved in each alternative.

If the need for treatment is overlooked in the design, it will become essential and usually expensive to compensate for the oversight during construction. On embankment construction it may be necessary to reduce the degree of density required even at subgrade elevation and increase the thickness of base or insulation course to provide adequate support for the loads on the pavement. In view of the sponginess of the subgrade and tendencies for rebound after the passage of conventional flat-wheel or pneumatic-tired rollers, the granular insulation or base can be compacted to best advantage by vibratory methods. This applies also to the placement of granular material for barriers at the base of fills, as shown in Fig. 6-15b.

Pronounced needs for stabilization of underlying materials are evident where embankments are placed over muck and peat deposits or in marshy areas with deep cover of organic soil over rock.   Methods of treatment for these situations can be generally placed in four categories: removal and replacement; displacement; consolidation by surcharge; and rapid consolidation by internal drainage.   All four have been used successfully.

Removal and replacement is limited to those locations where the organic deposits are shallow and construction can proceed in a manner that will permit equipment operating from positions on stable ground.   Removal can be accomplished by a dragline or clamshell, and usually the excavated organic soil can be discarded nearby. Suitable refill, granular if possible, is substituted for the soil removed.

Displacement methods usually involve jetting, blasting, trenching and blasting in combination, or loading the organic material with borrowed soil and then blasting alongside or beneath the soil.   Blasting will convert the organic soil to an essentially liquid condition, and displacement of this material by the soil is facilitated.   Planning of the operations and organization of the effort to best advantage are important (19). In cases where the organic deposit is shallow, displacement may be carried out successfully by simply loading the deposit with imported material until the organic soil fails in shear and flows from beneath the overlying soil.

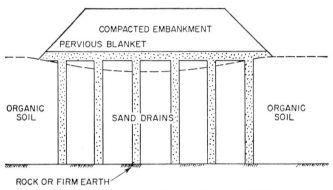

Fig. 6-16. Features of vertical-sand-drain installation.

The surcharge method is the same in operation but different in principle.   Here the imported soil is placed in layers and compacted to the extent possible, but the progress of the work is scheduled in such a way that the underlying organic material is consolidated rather than displaced.   Sometimes the compacted fill may be completed to approximate final grade, with allowance for subsequent settlement, and then additional material added as temporary load to accelerate consolidation of the organic deposit.   Later the additional material is removed and construction completed.

Extreme care must be taken in this operation, however, in order that the shearing resistance of the underlying material may not be exceeded and become displaced rather than consolidated.   The surcharge method is applicable to deposits of relatively shallow depth because of time involved in consolidation and to some extent the limited shearing resistance of the organic soils.

Where organic deposits are deep, the most feasible approach to stabilization of embankments is by means of sand drains.   In effect, this is a modification of the surcharge method.   Drainage and consequently consolidation under the weight of overlying material are expedited by vertical drainage wells filled with sand.   With such drains spaced at intervals of about 10 to 15 ft, water has only a short distance to travel from any point in the organic deposit to a medium through which it can flow rapidly and be dispelled.

The construction process involves placement of a sand blanket over the top of the organic deposit.   Working on top of this fill, equipment specifically designed for the process drills or otherwise opens holes within the muck (Fig. 6-16).   These holes are then filled with sand, and as additional fill is placed over the blanket, water is squeezed from the voids of the muck. Pressure within the entire system causes water to flow through the sand drains and be conducted away through the blanket at the surface. Flow stops and the system comes to rest after the consolidated structure of the organic soil is able to support the load entirely and there is no hydrostatic excess pressure or pressure in the voids of the organic material which would cause flow.

This treatment has been used successfully on several large highway projects throughout the United States, including the New Jersey Turnpike (15) where sand drains totaling approximately 950 miles in length were placed beneath the highway in marshy areas.

5. Permanence of Densification. For soil properly densified through compactive influences the achieved density is essentially permanent in the interior portions of embankments. Except in cases where a fill may be partially submerged such as in a causeway, or in instances where the supply of moisture from below is plentiful and frost penetration is deep, only the very outer portions of embankments are appreciably influenced by changes in density.

With subgrades the situation is different. Being at relatively shallow depth and often vulnerable to lateral seepage of water from side ditches, sloping rock strata in sidehill sections, and similar influences, soils in the subgrades undergo considerable change in density. Frost action is particularly disrupting, and there is hardly any practicable way of restoring the material at subgrade level to its original density without removing the pavement and replacing the soil or recompacting it.

Sometimes subgrade soils undergo increases in density through the influence of heavy loads operating on the pavement above. This may be to the advantage of the soil in its ultimate supporting value for the pavement, but in the process of greater densification of the soil, probably the pavement itself will become permanently deformed. Hence the problem of retaining denseness, once achieved through compaction, is largely a matter of protection of the subgrade or outer portions of embankments from adverse influences such as improper drainage or insufficient thickness of overlying pavement.

B. Methods and Equipment

Results achieved in soil-compaction operations are dependent upon selection of equipment and the methods that are used. Some equipment which is highly effective on one type of soil is only moderately effective with others. Also, compactive effort is largely wasted on most soils if the moisture content is not reasonably close to the so-called "optimum" moisture for compaction of that soil.

Developments in the size, weight, physical aspects, and operational characteristics of compaction equipment have been great during the past 20 years. Several large-scale test projects have been conducted, and reference should be made to publications dealing with those tests whenever specific information is desired (11, 20, 24). Likewise, data from manufacturers of equipment deal with performance characteristics in a way that is outside the scope of this publication.

1. Distribution-moisture Control. Distribution of soil in desired depths of loose spread is generally carried out by the earth-moving equipment. Some additional distribution may be necessary after the rough spread is accomplished, in which case the patrol grader is the most adaptable device.

If the soil is much below the optimum moisture content, or the atmospheric conditions are such that rapid drying occurs before the compaction operation is carried out, water must be added. Here the principal problem is uniform distribution of the water throughout the soil.

When the soil is taken from a borrow area it is feasible to add water either before the excavation is made or after the soil is spread loosely on the fill. Sometimes either is possible when the soil comes from roadway, structure, or channel-change excavation, but normally the moisture is added after the soil is in place. A tank truck and pressure-spray or gravity-feed sprinkling system accomplish the purpose of adding the water. Distribution should be carried out by the means most suitable to the soil.

Table 6-5 is a summary of practices as presented by the Highway Research Board Committee on Compaction of Subgrades and Embankments. Generally the addition and distribution of water throughout dry soils is more easily accomplished than the reduction of moisture in soils that are much too wet. The need for drying soils develops mostly in seasons of the year when drying is difficult or almost impossible unless artificial methods such as heating are applied. Although heating of soils in conventional aggregate-drying kilns has been done successfully, the cost is prohibitive unless there is no other alternative. Other possibilities such as the application of a direct flame from surface-heating equipment used for the restitution of bituminous pavement surfaces present themselves, but there are no known uses of this approach by which its effectiveness or relative cost can be judged.

Table 6-5. Generalized Correlation of Soil Classification and Equipment and Methods for Incorporating Water Prior to Compaction

Type of Soil	*Equipment and Methods*
Heavy clays	Difficult to work and to incorporate water uniformly. Best results usually obtained by sprinkling followed by mixing on grade. Heavy disc harrows are needed to break dry clods and to aid in cutting in water, followed by heavy-duty cultivators and rotary-speed mixers. Lift thickness in excess of 6 in. loose measure is difficult to work. Time is needed to obtain uniform moisture distribution.
Medium clayey soils	Can be worked in pit or on grade as convenience and water-hauling conditions dictate. Best results are obtained by sprinkling followed by mixing with cultivators and rotary speed mixers. Can be mixed in lifts up to 8 in. or more loose depth.
Friable silty and sandy soils	These soils take water readily. They can often be handled economically by diking and ponding or cutting contour furrows in pit and flooding until the desired depth of moisture penetration has taken place. That method requires watering a few days to 2 or 3 weeks in advance of rolling (depending on the texture and compactness of the soils) to obtain uniform moisture distribution. These soils can also be handled by sprinkling and mixing, either in-pit or on-grade, and require relatively little mixing. Mixing can be done with cultivators and rotary speed mixers to depths of 8 to 10 in. or more without difficulty.

SOURCE: Highway Research Board (4).

When atmospheric conditions are such that aeration is practicable for drying the soil, manipulation with harrows, cultivators, and particularly rotary speed mixers is highly effective. With the tail-hood section raised, a self-propelled rotary mixer can toss the soil into the air to increase the exposure and facilitate drying. Another possibility for lowering the moisture content to desired level is the combining of dry friable materials with the soil needing reduction in moisture, thereby creating a material that is near the optimum moisture content. If the two soils are not of similar composition, of course control of moisture should be based on compaction characteristics of the combined material.

In the absence of some means of accomplishing the necessary drying, overmoist soils must be wasted or placed in positions where compaction is of no real consequence. Locations for use of soils in this condition should be judged carefully, in order to avoid critical conditions for settlement or sliding of embankments. Soils with excess moisture should never be placed in subgrades.

2. Smooth-wheel Rollers. Because of its origin as the means for densifying all types of materials used in highway pavement construction, the three-wheel smooth roller has long been applied to compaction of soils. Developments over a period of many years have produced variations in sizes and characteristics. Present classification places rollers of this type in three general weight groups of 5 to 6 tons, 7 to 9 tons, and 10 to 12 tons. Compression on the drive rolls ranges from 150 to 400 lb per lineal in. of width of the drive rolls.

Effectiveness of rollers of this type in the compaction of soils is limited because of their dependence upon static compression rather than compression supplemented by kneeding, tamping, or vibratory actions which have proved so effective with soils of different characteristics. As a result of this limitation, variations of the smooth-wheel roller have been devised. Among these are the single-drum vibratory roller (towed), the self-propelled tandem roller with vibratory intermediate roll, the towed roller with gridded drum, the roller segmented in the manner shown in Fig. 6-17 or in the form of the ordinary tandem with segmented guide roll, and others of similar nature.

Experimental data on the application of rollers of this type with the various innovations are not widely developed and disseminated, and thus far manufacturers are the best sources of information on the subject. With regard to the conventional smooth three-wheel roller, the classifications in Table 6-6 are intended as a guide to their use. With limited exceptions these are intended as recommendations only when rollers better suited to the properties of the various classes of soils are not available.

FIG. 6-17. Segmented steel-wheel roller.

3. Tamping, or Sheepsfoot, Rollers. The sheepsfoot-type roller came into real being and serious use along with the awareness of compaction as an important element in soil engineering. Hence the sheepsfoot roller has developed in concept and application as the knowledge and use of compaction have developed. Although there are some soils on which compaction by tamping feet is more disturbing than beneficial, the sheepsfoot roller is still the most widely used of all devices intended for this purpose.

Rollers of this type now in widespread use may be classified in three groups as outlined in Table 6-7, along with the generalized types of soil to which they are most adaptable. For several reasons these rollers are not self-propelled, but rather are towed by other construction equipment such as a dozer or other form of tractor (Fig. 6-18).

Table 6-6. Generalized Classification of Three-wheel Rollers Most Adaptable to Compaction of Various Soils

Soil Group	Weight Group and Pressure, Wt. per Linear In. of Width of Rear Rolls
Clean, well-graded sands, uniformly graded sands (one size), and some gravelly sands having little or no silt or clay	Cannot be rolled satisfactorily with three-wheel-type rollers
Friable-silt and clay-sand soils which depend largely on their frictional qualities for developing bearing capacity	5–6 tons, 150–225 lb
Intermediate group of clayey silts and lean clayey soils of low plasticity (<10)	7–9 tons, 225–300 lb
Well-graded sand-gravels containing sufficient fines to act as filler and binder	10–12 tons, 300–400 lb
Medium to heavy clayey soils	10–12 tons, 300–400 lb

SOURCE: Highway Research Board (4).

Table 6-7. Contact Pressures and Sizes of Tamping Feet Best Suited for Compacting Different Soils with Sheepsfoot Rollers

Soil type	Contact area, sq in.	Contact pressure, psi	Remarks
Friable silty and clayey sandy soils which depend largely on their frictional qualities for developing bearing capacity	7–12	75–125	These groupings are based on stock models for use in compacting to densities of about 95% AASHO T 99 maximum density at moisture contents at or slightly below optimum when 6- to 9-in. compacted lift thicknesses are developed. It is also based on the experience that rollers are most easily towed when their weight allows them to begin to "walk up" as rolling progresses. It is realized that much heavier contact pressures may be more desirable if contact areas are increased and that such increases are necessary if higher field densities are to be produced.
Intermediate group of clayey silts, clayey sands, and lean clay soils which have low plasticity	6–10	100–200	
Medium to heavy clays	5–8	150–300	

SOURCE: Highway Research Board (4).

Effectiveness of the roller is dependent upon tamping action, and other things being equal, the greater the contact pressure the fewer the operations required to achieve desired density. Still, contact pressure cannot be carried beyond the limits of bearing offered by the soil. In essence, that is the reason for increased contact area on the tamping feet of the class of roller most applicable to the friable soils (Table 6-7). The nearer the soil approaches truly cohesionless characteristics, the less suitable the sheepsfoot roller for compaction, until with completely cohesionless soils such as clean sands, disturbance rather than compaction occurs under sheepsfoot rolling.

FIG. 6-18. Towed sheepsfoot rollers operating in tandem.

Increasing the length of tamping feet does not increase the depth of layer which can be compacted effectively. As the depth of compacted layer increases beyond 6 to 8 in. (depending on the class of roller), the greater the tendency for graded density from high values in the upper portions of the layer to much lower values at the bottom. Spacing of the feet does have some bearing on the results, but the spacing can be decreased to the point of diminishing returns from both the standpoint of actual tamping of the soil and also the amount of power required to tow the equipment.

If the project on which rollers of this description are used is of sufficient size and the soils sufficiently uniform, trial runs to establish the best combinations of equipment, thickness of layer, and number of passes can be worthwhile. Results of the trial will set the pattern for the job, enable the contractor to operate most efficiently, and make the inspection more productive. Adequate compaction will be recognized in a general way through the operational characteristics of the equipment, and control testing will be reduced.

In those cases where the roller "walks out" or the feet no longer penetrate into the soil yet the measured densities are too low, probably the lift thickness is too great, the foot pressures are too low, the soil is too dry, or the soil is greatly different from that used in determining maximum density by laboratory test. On the other hand, if the roller continues to sink into the soil and fails to walk out regardless of the number of passes, the soil may be too wet or much too dry or the foot pressures too great, or the material is not represented by the compaction test. In any event, adjustments made during the trial runs will serve as indicators of the type of adjustment to be made as difficulties arise during the course of construction.

It is desirable, also, to establish a balance between earth-moving equipment and compaction equipment that will produce adequate compaction most economically.

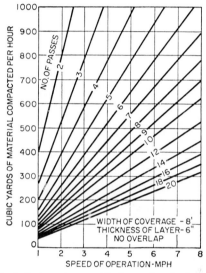

Fig. 6-19. Theoretical maximum rolling capacity of a sheepsfoot roller.

Neither equipment should be retarded or shut down in its operation because of insufficient production by the other. Conditions will vary throughout the job, with the requirements for each varying with moisture content of the soil that is brought in, temperature and humidity, and distance over which operations are being conducted.

Relationships shown in Fig. 6-19 represent the maximum possible production of a given roller having an effective width of 8 ft, producing a compacted layer 6 in. in depth, and having no overlap in its operation. If kept in operation continuously at an average speed of 3 mph for a period of 1 hr and if it was found by trial run that six passes were sufficient to accomplish the required density, such a roller would compact approximately 390 cu yd of earth per hr. This figure, when combined with that applied to a 12-yd scraper operating over a loaded haul distance of 500 ft (paragraph III.C.2), indicates that under those conditions the one roller could match the production of four scrapers with some reserve. Naturally, contingencies on the job and the assumptions that were introduced make this only an illustrative situation, and actual relationships must be worked out on the basis of job conditions and the experience of those managing the earth-moving operation.

4. Pneumatic Rollers. Rollers which depend on kneading action from passages of pneumatic tires as well as on compression from the passing load have come into widespread use within a comparatively short time. Through experiments by the U.S.

Army, Corps of Engineers (6, 14), and others, and also through use on large-scale highway projects such as freeways and turnpikes, much is known about their application and performance.

Contact pressure is important in the effectiveness of pneumatic rollers. There is a relation between contact pressure and inflation pressure, and ratings on equipment are often given in terms of inflation pressure. Increasing the total load on a roller is not a means for increasing the effectiveness unless in the process there is an increase in contact pressure or an increase in loaded area, in which case the effective depth of compaction for most soils is increased.

Cohesionless soils are affected most by compaction with relatively large contact areas, since the soil must be confined in order to become densified. Semicohesionless

Fig. 6-20. Self-propelled highly maneuverable pneumatic-tire roller. *(Courtesy of Kentucky Department of Highways.)*

soils, such as slightly dirty sands and gravels, benefit from kneeding action as well as pressure from the tires. Soils of this description, as well as more granular materials applied to subbases and even bases, account for the development of the "wobble-wheel" principle in pneumatic rolling (Fig. 6-20).

As noted in the discussion of sheepsfoot rollers, trial runs at the outset of a project may be helpful in establishing the loading of the roller, inflation pressure of the tires, and thickness of lift most advantageous to compaction of the soils available. Table 6-8 is a guide to the general selection of roller characteristics, and Fig. 6-22 is a graphical illustration of the maximum production capacity of a pneumatic roller covering a width of 7 ft, operating continuously with no overlap, and accomplishing a 6-in. depth of compacted material. Application of these data to estimates of corresponding capacity of earth-moving equipment can be made in a general way, as outlined in the discussion of sheepsfoot rollers (paragraph V.B.3).

5. Vibratory and Impact Devices. Compaction of soils by impact and vibratory forces is of comparatively recent origin. Because the effectiveness of vibrational

Fig. 6-21. Heavy pneumatic-tire roller.

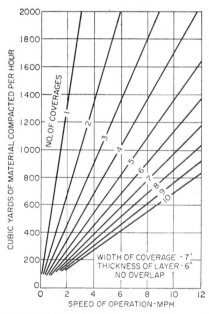

Fig. 6-22. Theoretical maximum rolling capacity of pneumatic-tire roller.

Table 6-8. Contact Pressures of Pneumatic-tire Rollers Best Suited for Compacting Different Soils

Soil Group	Contact Pressure
Clean sands and some gravelly sands	20- to 40-psi inflation pressure, the greater pressures with the large-size tires
Friable-silty and clayey sands which depend largely on their frictional qualities for developing bearing capacity	40- to 65-psi inflation pressure
Clayey soils and very gravelly soils	65-psi and up inflation pressure

SOURCE: Highway Research Board (4).

compactive effort is dependent upon the natural frequency of the soil, there is very little known about the suitability of different devices for this purpose. Since the natural frequency varies with moisture content, relative density, and other properties of the soil, no particular condition of vibration can be singled out as best for a given soil.

In general, vibration is highly effective on cohesionless soils, at least moderately effective on semicohesionless soils, and relatively ineffective on fine-grained and cohesive soils. While moisture content is not highly influential on the results of vibratory compaction for cohesionless materials, the effort is most efficient when the soil is damp and probably well on the wet side of what is generally termed "optimum."

Several types of equipment, ranging from single-shoe or single-plate machines to multiple-shoe (Fig. 6-23) devices and even vibratory smooth-wheel rollers (paragraph V.B.2), have been used for compaction under vibratory impulses. The general principle has been applied to oscillating multiple-wheel pneumatic-tire rollers. All are

FIG. 6-23. Self-propelled multishoed vibratory compactor. *(Courtesy of Kentucky Department of Highways.)*

highly successful when the soil is conducive to compaction in this manner, but conditions for greatest efficiency of each remain to be determined by experimentation in the future.

When soils are sufficiently cohesionless for compaction by this method, depth of layer is not highly critical. Some controlled tests (5) have shown effective compaction to depths twice as great as the width of loaded area through which vibrations are transmitted. Depths as great as 4 ft appear reasonable; however, in the light of present limited knowledge it is recommended that compacted depths of layer no greater than 2 ft be considered. Usually other limitations have a bearing on the practicable depth for cohesionless soils compacted for subgrades and embankments, with the possible exception of backfill behind bridge abutments where confinement at the edges may make it feasible to compact a granular soil to depths of several feet in one operation.

Compaction by impact forces has been applied mainly to bedding and backfill in trenches for drainage structures. Practical limitations on size of equipment utilizing the "rammer" principle have entered as an economic consideration. However, in addition to the manually operated portable rammer so adaptable to use in restricted areas (Fig. 6-24), pneumatic-driven pavement breakers have been successfully con-

FIG. 6-24. Portable impact or rammer compactor. *(Courtesy of Highway Research Board (4).]*

FIG. 6-25. Adaptation of a pneumatic-driven pavement breaker as equipment for compacting trench backfill. *[Courtesy of Highway Research Board (4).]*

verted to this use. Equipment of this description, mounted on mobile units and arranged for compaction of long and sonetimes moderately wide trenches, is shown in Fig. 6-25.

C. Compaction Control

The control of compaction during construction normally refers to procedures for making certain that the specified degree of density or unit weight is achieved in the embankment or subgrade. Control of the moisture content within reasonable limits of the optimum value is implied for reasons mentioned heretofore. If there is some question concerning uniformity of density throughout the depth of layer as well as uniformity over the entire area covered by the layer, the control procedure must take this into account.

Control is exercised on the basis of moisture-density tests defining that particular characteristic of the soil. If several soils that can be reasonably separated and identified under field conditions are involved, tests must be made on each.

1. Personal Examination. Sometimes control is exercised simply by personal examination by the inspector. Naturally control by this method is highly subject to human error, and it is limited by the qualifications of the inspector to whom it is entrusted.

Often experienced engineers and technicians can judge the relative moisture conditions of soils by feel and visual examination. Also, with experience based on previously determined compaction characteristics, the same individuals can estimate when a general level of densification has been reached. At best, however, these are only estimates, and their application should be limited to projects on which cost of construction, type of material (perhaps where a high percentage of excavation is rock), or similar features will not justify detailed sampling and testing.

It should be noted that specifications referring in a general way to control of moisture and expected density but prescribing no procedure for determination of those factors have little value in legal controversy. They place exceptional responsibility on the inspector or project engineer and are conducive to slipshod treatment of this important phase of earthwork. That being the case, any compromises with rigid control procedures should be made in instructions to the contractor and field personnel, waiving the specific control measures, rather than having them made as exceptions in standard specification which are applied to all segments of construction.

2. Moisture Determination and In-place Density Measurements. Rigid control can be exercised by sampling the soil in the subgrade or embankment after the compactive effort has been applied and determining the in-place moisture content and density. Obviously, this involves time, and several methods have been devised for rapidly determining or estimating the moisture content, and hence, determining the unit weight dry.

One approach is through the use of the Proctor Penetration Needle—an integral part of the original evaluation of compaction characteristics (17) but gradually having been displaced from laboratory test methods because of inherent limitations. However, if the penetration resistance was included as part of the laboratory test, sampling of soil from the compacted layer in the field will make possible its recompaction in the mold at whatever moisture content the layer possesses. Penetration of this newly compacted soil in the mold will give a rapid basis for estimating moisture content from the penetration-resistance curve. This, combined with careful measurement of the moist soil taken from a hole of measured volume, will provide the means for determining the dry unit weight and the compliance with specification requirements.

Determination of the in-place unit weight of the moist soil is dependent upon recovery of all soil, disturbed or undisturbed, from a hole of measured volume in the compacted layer. For the so-called undisturbed approach, a core of soil is taken by a drive tube designed to cut a cylindrical sample without displacing the surrounding material. After the sample is removed from the drive tube it is weighed and coated with paraffin and the volume determined by displacement (reweighing the coated sample provides a correction for volume of paraffin).

Removal of soil from the compacted layer by means of an auger or small hand tools gives a basis for determining the dry unit weight. All the material disturbed and removed in this way must be carefully recovered and weighed. Volume of the hole can be measured by means of calibrated dry sand, oil, or water contained in a rubber pouch inserted to line the hole (16). On the basis of weight of moist soil and measured volume of the hole, the dry unit weight achieved in compaction of the layer can be calculated after the soil has been dried.

If time permits, the soil should be dried to constant weight in an oven. In lieu of the use of an oven, drying can be effected with the soil in an open pan and heat supplied by a portable stove. A still more rapid method of drying is through dispersing the soil in a highly volatile fluid such as alcohol and burning the fluid to evaporate the moisture. If the amount of soil is sufficiently small and it is well dispersed in the fluid, the moisture can be essentially removed by burning the fluid three times. Several other methods for removing or evaluating the moisture in soils are known and practiced to limited extents.

All the methods of determining dry density of the soil in place have limitations. Each should be thoroughly evaluated before it is adopted for use as a routine control measure.

Inasmuch as laboratory methods normally used for determining the moisture-density relations of soil require that the test be made on only the material passing a No. 4 sieve, the effect of coarser material must be evaluated in the field if there is an appreciable amount of such aggregate contained in the soil. Methods of converting both moisture and density measurements to those of the soil actually represented have been developed (18, 23), and these should be consulted for the correction procedures.

3. Use of Typical Curves. In instances where results of a large number of laboratory tests indicate characteristic shapes of moisture-density curves, the use of typical curves for estimates and control has been feasible. On the basis of tests on 461 samples, the Ohio Department of Highways first established a set of nine curves (25) to be used for this purpose. More recently, on the basis of 10,149 tests (12), a set of 26 typical curves (Fig. 6-26) was developed for use in both the laboratory and the field.

Application of typical curves is based on the observation that results of compaction tests on soils throughout that state fitted themselves to characteristic curves. With this trait known, the amount of time and effort required to test soil and exercise the control measures could be reduced.

Two simple steps are involved in determining the typical curve applicable to a soil in question. On embankment construction, for example, the soil for the layer is compacted into the cylindrical test mold by the standard method and the wet density determined. Penetration resistance is measured by the Proctor needle, and correlation of the two values with those for the various typical curves determines which curve is applicable to the soil in question.

For example, if the data from the test were 114 lb per cu ft (wet weight) and 1,100 psi penetration resistance, the curve designation most nearly meeting this combination at a common moisture content for both describes the moisture-density relationships for the soil that was tested. It is evident that curve S (Fig. 6-26) provides the best correlation for both values at a moisture content of 19 per cent. With curve S established as the one representing the soil, the unit weight dry and the optimum moisture content are defined as 97.4 lb per cu ft and 22.7 per cent, respectively.

The use of typical curves has been extended by the Ohio Department of Highways so that both in the laboratory and in the field, the compaction characteristics and control are based on one-point tests, rather than tests involving a series of points describing the moisture-density relations.

REFERENCES

1. Allen, H.: "Classification of Soils and Control Procedures Used in Construction of Embankments," *Public Roads*, vol. 22, no. 12, 1942.
2. *Blaster's Handbook*, E. I. du Pont de Nemours & Company, Wilmington, Del.

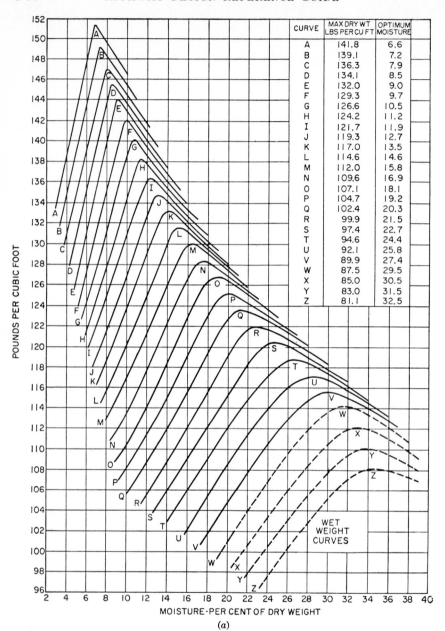

CURVE	MAX DRY WT LBS PER CU FT	OPTIMUM MOISTURE
A	141.8	6.6
B	139.1	7.2
C	136.3	7.9
D	134.1	8.5
E	132.0	9.0
F	129.3	9.7
G	126.6	10.5
H	124.2	11.2
I	121.7	11.9
J	119.3	12.7
K	117.0	13.5
L	114.6	14.6
M	112.0	15.8
N	109.6	16.9
O	107.1	18.1
P	104.7	19.2
Q	102.4	20.3
R	99.9	21.5
S	97.4	22.7
T	94.6	24.4
U	92.1	25.8
V	89.9	27.4
W	87.5	29.5
X	85.0	30.5
Y	83.0	31.5
Z	81.1	32.5

POUNDS PER CUBIC FOOT

WET WEIGHT CURVES

MOISTURE-PER CENT OF DRY WEIGHT

(a)

FIG. 6-26. Typical moisture-density curves. [*Courtesy of ASTM (12)*.]

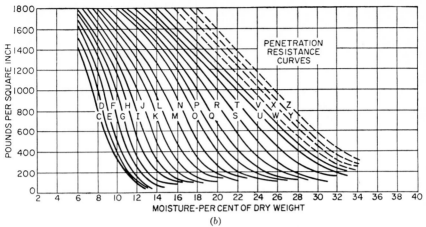

FIG. 6-26. (*Continued*).

3. "Classification of Highway Subgrade Materials," *Proceedings, Highway Research Board*, vol. 25, pp. 376–384, 1945.

4. "Compaction of Embankments, Subgrades, and Bases," *Highway Research Board Bulletin* 58, 1952.

5. Converse, Frederick J.: "Compaction of Sand at Resonant Frequency," *ASTM Special Technical Publication* 156, pp. 124–137, 1953.

6. Corps of Engineers: "Soil Compaction Investigation. Report No. 1: Compaction Study on Clayey Sands," *Waterways Experiment Station Technical Memorandum* 3-271, 1949.

7. U. S. Corps of Engineers: "Soil Compaction Investigation. Report No. 2: Compaction Studies on Silty Clay," *Waterways Experiment Station Technical Memorandum* 3-271, 1949.

8. Corps of Engineers: "Unified Soil Classification System," *Waterways Experiment Station Technical Memorandum* 3-357, 1953.

9. Foster, C. R.: "Reduction in Soil Strength with Increase in Density," *Transactions ASCE*, vol. 120, pp. 803–822, 1955.

10. Johnson, A. W.: "Practical Density Limits for Highway Subgrades and Bases," *ARBA Technical Bulletin* 203, 1954.

11. Johnson, S. J., and W. G. Shockley: "Field Penetration Tests for Selection of Sheepsfoot Rollers," *Proceedings ASCE*, vol. 79, Separate no. 363, 1954.

12. Joslin, J. G.: "Ohio's Typical Moisture-Density Curves," *ASTM Special Technical, Publication* 239, 1958.

13. Nichols, Herbert L.: *Moving the Earth*, D. Van Nostrand Company, Inc., Princeton, N.J., 1955.

14. Porter, O. J.: "The Use of Heavy Equipment for Obtaining Maximum Compaction of Soils," *ARBA Technical Bulletin* 109, 1946.

15. Porter, O. J., and L. S. Urquhart: "Sand Drains Expedited Stabilization of Marsh Section—New Jersey Turnpike," *Civil Engineering*, vol. 22, no. 1, pp. 51–55, 1952.

16. "Procedures for Soil Testing," ASTM publication sponsored by Committee D-18, 1958.

17. Proctor, R. R.: "Fundamental Principles of Soil Compaction," *Engineering News-Record*, vol. 111, August–September, 1933.

18. Shockley, W. G.: "Correction of Unit Weight and Moisture Content for Soils Containing Gravel Sizes," *U.S. Corps of Engineers, Waterways Experiment Station, Soils Division, Embankment* and *Foundation Branch, Technical Data Sheet* 2, 1948.

19. Sinacori, M. N., W. P. Hofmann, and A. H. Emery: "Treatment of Soft Foundations for Highway Embankments," *Proceedings, Highway Research Board*, vol. 31, pp. 601–621, 1952.

20. Turnbull, W. J., and G. McFadden: "Field Compaction Tests," *Proceedings Second International Conference on Soil Mechanics and Foundation Engineering*, vol. 5, pp. 235–239, 1948.

21. Turnbull, W. J., S. J. Johnson, and A. A. Maxwell: "Factors Influencing Compaction of Soils," *Highway Research Board Bulletin* 23, 1949.
22. Turnbull, W. J., and Charles R. Foster: "Stabilization of Materials by Compaction," *Transactions ASCE*, vol. 123, pp. 1–26, 1958.
23. Walker, F. C., and W. G. Holtz: "Comparison between Laboratory Test Results and Behavior of Completed Embankments and Foundations," *Proceedings ASCE*, vol. 75, Separate no. 108, 1950.
24. Williams, F. H. P., and D. J. Maclean: "The Compaction of Soil," *Road Research Technical Paper* 17, Department of Scientific and Industrial Research, Road Research Laboratory (London), 1950.
25. Woods, K. B., and R. R. Litehiser: "Soil Mechanics Applied to Highway Engineering in Ohio," *Ohio State University Engineering Experiment Station Bulletin* 99, 1938.
26. Woods, K. B., and L. E. Gregg: "Pavement Performance Related to Soil Texture and Compaction," *Proceedings, Highway Research Board*, vol. 24, pp. 426–435, 1944.
27. Yoder, E. J., and K. B. Woods: "Compaction and Strength Characteristics of Soil-Aggregate Mixtures," *Proceedings, Highway Research Board*, vol. 26, pp. 511–520, 1946.

Section 7

DISTRIBUTION, PRODUCTION, AND ENGINEERING CHARACTERISTICS OF AGGREGATES

Dr. J. F. McLAUGHLIN, *Research Engineer, Joint Highway Research Project, and Associate Professor of Civil Engineering, Purdue University, Lafayette, Ind.* (Definitions, Specifications, and Tests).

K. B. WOODS, *Head, School of Civil Engineering and Director, Joint Highway Research Project, Purdue University, Lafayette, Ind.* (Origin and Distribution).

Dr. R. C. MIELENZ, *Director of Research, Master Builders Company, Cleveland, Ohio* (Basic Properties).

NATHAN C. ROCKWOOD, *Editorial Consultant, Rock Products, Chicago (Retired)* (Production).

I. AGGREGATE TYPES AND DEFINITIONS OF TERMS RELATING TO AGGREGATES

A broad definition of the term mineral aggregate was stated by Woods (3)* as follows:

"An aggregation of sand, gravel, crushed stone, slag, or other material of mineral composition, used in combination with a binding medium to form bituminous and portland cement concrete, macadam, mastic, mortar, plaster, etc., or alone as in railroad ballast, filter beds and various manufacturing processes such as fluxing, etc."

It should be noted that the concept embodied in the use of the term "aggregate" as presented in this latter definition is much more inclusive than that of the present ASTM definition.

Materials composed of rock fragments which are used in their natural state except for such operations as crushing, washing, and sizing are usually considered natural aggregates. Some terms that are used to describe and classify the natural aggregates appear below:

1. MINERAL: A naturally occurring chemical element or compound formed as a product of inorganic processes (4). To this basic definition one should add, for the present discussion, that a mineral is a solid and usually has a definite crystalline structure and a distinctive set of physical properties (5, 6).

2. ROCK: A solid material, which is composed principally of one or more minerals, that makes up the earth's outer crust. The terms rock and stone are often used

* Numbers in parentheses refer to corresponding items in references at the end of this section.

synonymously.   *Bedrock* is solid rock still in its natural position, untouched except for geologic processes.

3. GRAVEL:† The coarse granular material, usually defined as being larger than ¼ in. in diameter, resulting from the weathering and erosion of rock by natural agencies.

4. CRUSHED GRAVEL:† The product resulting from the crushing of ordinary gravel with substantially all fragments having *one or more* faces resulting from fracture.

5. CRUSHED STONE:† The product resulting from the crushing of fragments of bedrock or large stones with *all* fragments having *all* faces resulting from fracture. Usually defined as being larger than ¼ in. in diameter.

6. SAND: The fine granular material, usually defined as being smaller than ¼ in. in diameter, that results from the natural disintegration of rock.   The foregoing material is sometimes called *natural sand,* while material made from the crushing of rock or blast-furnace slag fragments or pieces of gravel may be referred to as *stone sand, slag sand,* and *crushed-gravel sand* or *manufactured sand,* respectively.

Aggregates which result from modifications of materials which may involve physical and chemical changes are sometimes called the artificial aggregates.   These, in turn, may be of two types: (*a*) the by-product aggregates such as blast-furnace slag and cinders, or (*b*) the manufactured aggregates—those especially produced or processed from raw materials for ultimate use as aggregates.   Chief among the manufactured aggregates are the lightweight aggregates.   Some terms that may be defined with regard to artificial aggregates are:

1. BLAST-FURNACE SLAG:† The nonmetallic product, consisting essentially of silicates and aluminosilicates of lime and of other bases, which is developed simultaneously with iron in a blast furnace.

2. CINDERS: The residue from burning coal or coke in high-temperature combustion.

3. LIGHTWEIGHT AGGREGATE: A manufactured aggregate, including expanded clay or shale, processed diatomaceous earth, processed volcanic glasses, and expanded slag.

II. PROPERTIES OF AGGREGATES

Aggregate particles possess a series of physical and chemical properties which, together with grading, determine the suitability of the aggregate for an engineering application.   An understanding of these properties is essential as a basis for development of empirical tests, for establishing specification limits based upon such tests, and, in general, for evaluation and selection of aggregates for use under specific conditions of service.

The physical and chemical properties of particles of mineral aggregates arise in (1) the geologic history of the sand or gravel or of the rock formation, and (2) the processing to which the materials were subjected in the production of the finished aggregate. Under the term geologic history are included the origin or mode of formation of the rock types as well as subsequent processes like folding, faulting, jointing, recrystallization, hydrothermal alteration, weathering, erosion, deposition of secondary substances as coatings, and so on.   The geologic features of rock formations or sand and gravel deposits are important factors in the development and operation of aggregate sources, and they may determine the quality and limitations of the aggregate being produced. Sources of mineral aggregates for highway construction should be examined geologically so as to identify and determine the distribution of any unsound or otherwise deleterious materials (45–47).   The results of such an examination may indicate that certain zones or portions of the deposit or formation should be avoided or wasted in production of aggregate for use in the highway project.

† The definitions which are marked by daggers have been taken (or paraphrased from ASTM Designation: C 125-48, "Standard Definitions of Terms Relating to Concrete and Concrete Aggregates."

The reader is also referred to ASTM Designation C 294-54T, "Tentative Descriptive Nomenclature of Constituents of Natural Mineral Aggregates."

In the final analysis, however, the chemical and physical properties of aggregate particles depend upon their (1) mineralogic composition, (2) internal texture, and (3) internal structure (48, 49). Internal texture is the fabric in which the constituent grains or amorphous masses are joined, that is to say, the mutual spatial and boundary relationships among the component crystals, grains, and amorphous phases, such as volcanic glass or opal. Internal structure is the arrangement of textural units or discontinuities within the aggregate particle. Typical internal structures of rocks are stratification, schistosity, vesicularity, and various types of fractures. So, for example, a gabbro may be described as coarsely crystalline in texture, massive in structure, and composed of plagioclase and augite, together with various other minor constituents. A certain shale may be described as clastic and argillic in texture, stratified and fissile in structure, and composed of an illite-type (hydromica) clay and specific proportions of silt and sand of certain petrographic identity.

Table 7-1. Petrographic Analysis of a Gravel, near Edinburg, Tex.

Constituents	Amount, as number of particles, %, in fractions indicated			Degrees of quality	
	Plus ½ in.	½–⅜ in.	⅜–³⁄₁₆ in.	Physical quality	Chemical quality
Chalcedonic cherts................	56.0	43.3	39.1	Fair	Deleterious*
Porous cherts....................	2.7	2.7	2.3	Poor	Deleterious*
Rhyolites........................	23.3	26.7	23.0	Satisfactory	Deleterious*
Rhyolite and dacite porphyries.....	14.7	22.7	28.3	Satisfactory	Innocuous
Weathered porphyries..............	....	0.3	1.0	Fair	Innocuous
Quartzites.......................	2.0	0.7	2.7	Satisfactory	Innocuous
Sandstones.......................	0.7	0.3	0.3	Satisfactory	Innocuous
Limestones.......................	0.3	1.3	1.3	Satisfactory	Innocuous
Shales...........................	....	1.0	1.0	Poor	Innocuous
Andesites........................	0.3	1.0	....	Satisfactory	Deleterious*
Calcareous coating fragments.......	....	....	1.0	Fair	Innocuous

* Deleterious with high-alkali cements.

Table 7-2. Petrographic Analysis of a Gravel, Pearl River, La.

Constituents	Amount, as number of particles, %				Degrees of quality	
	In fractions indicated			In whole gravel	Physical quality	Chemical quality
	1½–¾ in.	¾–⅜ in.	⅜–³⁄₁₆ in.			
Dense, chalcedonic, and quartzose cherts........	35.0	22.3	11.3	20.6	Fair	Deleterious*
Sandy, dense, chalcedonic, and quartzose cherts.....	2.4	0.7	1.0	0.8	Fair	Deleterious*
Fractured, weathered, and somewhat porous, chalcedonic, and quartzose cherts................	44.5	57.0	49.0	54.6	Fair to poor	Deleterious*
Highly porous chalcedonic and quartzose cherts....	16.9	15.3	18.3	16.0	Poor	Deleterious*
Quartzites..............	1.2	4.7	20.0	7.9	Satisfactory	Innocuous
Soft, ferruginous sandstones................	....	....	0.3	0.1	Poor	Innocuous

* Deleterious with high-alkali cements.

Mineralogic composition, internal texture, and internal structure of rocks (or earth materials in general) relate specifically to the geologic conditions of origin and to subsequent geologic history. Because of this relationship, the petrographer almost always can determine the origin and geologic history of a rock from an examination of representative particles. Igneous rocks are formed by solidification of molten siliceous material intruded into the crust of the earth or extruded upon its surface. If the melt

Table 7-3. Petrographic Analysis of a Gravel, Tennessee River, Tenn.

Constituents	Amount, as number of particles, %				Degrees of quality	
	In fractions indicated			In whole aggre-gate	Physical quality	Chemical quality
	1½–¾ in.	¾–⅜ in.	⅜–³⁄₁₆ in.			
Gabbros................	....	0.3	....	0.1	Satisfactory	Innocuous
Quartzites.............	....	2.7	3.0	1.5	Satisfactory	Innocuous
Quartzose sandstones......	1.0	0.7	1.0	0.9	Satisfactory	Innocuous
Slag..................	0.3	....	....	0.1	Satisfactory	Innocuous
Calcareous shells..........	....	0.3	0.3	0.1	Fair	Innocuous
Dense brown cherts........	5.0	3.0	1.0	3.8	Fair	Deleterious*
Dense gray cherts.........	0.7	1.3	0.7	1.0	Fair	Deleterious*
Pyritiferous cherts........	....	....	0.3	Trace	Fair	Deleterious*
Fractured, weathered, or porous cherts...........	70.7	71.7	76.4	71.6	Fair	Deleterious*
Quartzose, granular cherts.	3.0	3.3	3.0	3.2	Satisfactory	Deleterious*
Highly porous cherts......	19.3	16.7	14.3	17.7	Poor	Deleterious*

* Deleterious with high-alkali cement.

Table 7-4. Petrographic Analysis of a Gravel, Northern Indiana

Constituents	Amount, as number of particles, %, in fractions indicated					Degrees of quality	
	1½–1 in.	1–¾ in.	¾–½ in.	½–⅜ in.	⅜–³⁄₁₆ in.	Physical quality	Chemical quality
Granites and gneisses...	8.8	14.8	13.5	7.3	10.0	Satisfactory	Innocuous
Diorites..............	....	....	0.7	0.3	0.3	Satisfactory	Innocuous
Diabases..............	8.8	7.4	5.0	4.7	3.3	Satisfactory	Innocuous
Weathered diabases.....	....	0.9	0.7	0.3	....	Fair	Innocuous
Schists...............	5.9	1.8	1.7	1.3	0.3	Satisfactory	Innocuous
Quartzites and quartz-ose sandstones.......	11.8	4.6	5.3	2.7	5.0	Satisfactory	Innocuous
Porous sandstones......	5.9	0.9	0.7	1.0	1.0	Fair	Innocuous
Platy sandstones.......	23.5	15.8	18.2	24.0	18.7	Fair	Innocuous
Graywackes and meta-argillites.............	5.9	6.5	6.0	5.7	5.7	Satisfactory	Innocuous
Limestones............	5.9	3.7	4.7	6.7	5.0	Satisfactory	Innocuous
Firm limestones........	2.9	....	....	0.3	0.3	Fair	Innocuous
Argillaceous limestones.	2.9	1.8	0.3	0.7	0.7	Poor	Innocuous
Dolomites.............	5.9	18.6	23.5	24.0	26.0	Satisfactory	Innocuous
Porous dolomites.......	....	2.8	7.0	7.3	5.0	Fair	Innocuous
Argillaceous dolomites..	5.9	2.8	1.3	0.3	....	Poor	Innocuous
Chalcedonic cherts and siliceous dolomites....	....	3.7	4.0	2.7	1.3	Satisfactory	Deleterious*
Porous cherts and silice-ous dolomites........	....	2.8	2.7	4.7	8.7	Fair	Deleterious*
Shales and clay stones..	5.9	6.5	4.0	4.7	6.7	Poor	Innocuous
Ferruginous concretions.	....	4.6	0.7	1.3	2.0	Poor	Innocuous

* Deleterious with high-alkali cements.

cools very slowly, the igneous rock will be wholly crystalline and a coarsely crystalline texture like that of granite may result.   If the melt cools rapidly, a portion or the entirety of the material may solidify as glass before crystallization can be completed.   Sedimentary rocks develop by grain-by-grain accumulation of particles.   The particles are joined together during geologic time by secondary processes of cementation, consolidation, or crystallization so that a coherent rock is produced.   Metamorphic rocks form by crushing, fragmentation, and recrystallization of previously existing rocks as a result of geologic processes at elevated temperature, pressure, or shearing stress.   The textures and internal structures which result reflect the nature of the original rock and the environmental conditions under which the metamorphism is consummated.

Aggregates vary widely in lithologic and mineralogic composition, both because of the nature and distribution of geologic formations and because of the selective action of weathering and erosion in destroying certain rocks and minerals and concentrating others.   Detailed petrographic analyses of gravels and sands from many parts of the United States and Canada are shown in Tables 7-1 to 7-15.

Table 7-5. Petrographic Analysis of a Gravel, Southeastern Wisconsin

Constituents	Amount, as number of particles, %, in fractions indicated			Degrees of quality	
	2-$\frac{3}{4}$ in.	$\frac{3}{4}$-$\frac{3}{8}$ in.	$\frac{3}{8}$-$\frac{3}{16}$ in.	Physical quality	Chemical quality
Granites and gneisses	4.0	3.0	3.3	Satisfactory	Innocuous
Weathered granites and gneisses	....	0.3	0.3	Fair	Innocuous
Hard schists	2.1	1.0	0.7	Satisfactory	Innocuous
Diabases and gabbros	1.1	1.7	0.7	Satisfactory	Innocuous
Rhyolite porphyries	....	1.0	0.3	Satisfactory	Innocuous
Graywackes	1.8	1.3	2.0	Satisfactory	Innocuous
Quartz and quartzites	2.9	2.3	8.0	Satisfactory	Innocuous
Quartzose sandstones	1.8	1.3	2.3	Satisfactory	Innocuous
Porous and friable sandstones	2.5	1.0	2.0	Fair	Innocuous
Argillaceous sandstones	1.4	....	....	Poor	Innocuous
Dolomites	33.1	38.0	31.4	Satisfactory	Innocuous
Porous, weathered, or fractured dolomites	35.4	23.0	10.3	Fair	Innocuous
Argillaceous dolomites	0.7	0.7	0.7	Poor	Innocuous
Sandy dolomites	0.7	0.7	1.0	Satisfactory	Innocuous
Siliceous dolomites	0.7	3.7	4.0	Satisfactory	Deleterious*
Porous siliceous dolomites	2.5	0.7	2.0	Fair	Deleterious*
Limestones	1.1	4.0	4.7	Satisfactory	Innocuous
Chalcedonic and quartzose cherts	6.8	13.3	20.0	Fair	Deleterious
Highly porous cherts	1.4	3.0	6.0	Poor	Deleterious
Feldspars	....	....	0.3	Satisfactory	Innocuous

* Deleterious with high-alkali cements.

Within each petrographic classification, such as granite, basalt, sandstone, or schist, a wide variation of composition, texture, and structure is possible.   Hence any rock types produce particles which may be either sound or unsound physically, or deleteriously reactive or innocuous chemically as a constituent of concrete aggregate.   The terms descriptive of the degrees of physical and chemical quality of the aggregate particles may be defined as follows:

SATISFACTORY: Particles are hard to firm, relatively free from fractures, and not chiplike; capillary absorption is very small or absent; and the surface texture is relatively rough.

FAIR: Particles exhibit one or two of the following qualities: firm to friable; moderately fractured; capillary absorption small to moderate; flat or chiplike; surface rela-

Table 7-6. Petrographic Analysis of a Gravel, near Jackson, Mich.

Constituents	Amount, as number of particles, %, in fractions indicated			Degrees of quality	
	1½–¾ in.	¾–⅜ in.	⅜–³⁄₁₆ in.	Physical quality	Chemical quality
Granites......................	4.4	4.0	4.5	Satisfactory	Innocuous
Anorthosites....................	0.6	....	....	Satisfactory	Innocuous
Gabbros, diabases, and greenstones..	8.2	4.3	7.7	Satisfactory	Innocuous
Quartz diorite porphyries..........	5.0	4.0	3.0	Satisfactory	Innocuous
Granite gneisses..................	8.2	6.2	5.0	Satisfactory	Innocuous
Weathered granite gneisses.........	0.3	0.4	0.6	Fair	Innocuous
Hornblende schists................	0.3	1.1	....	Satisfactory	Innocuous
Slates..........................	0.3	....	....	Fair	Innocuous
Quartzites and quartz schists.......	16.0	9.8	9.5	Satisfactory	Innocuous
Graywackes......................	10.2	8.7	10.2	Satisfactory	Innocuous
Quartzose sandstones..............	2.9	1.8	2.7	Satisfactory	Innocuous
Feldspathic sandstones............	0.3	....	....	Satisfactory	Innocuous
Argillaceous sandstones............	2.0	1.1	1.5	Poor	Innocuous
Porous sandstones.................	1.2	1.4	2.7	Fair	Innocuous
Slightly siliceous limestones.........	7.9	19.5	18.3	Satisfactory	Deleterious*
Fetid, organic limestones..........	5.3	5.1	0.3	Fair	Innocuous
Argillaceous limestones............	1.5	0.7	0.6	Poor	Innocuous
Slightly siliceous dolomites.........	13.1	14.1	17.4	Satisfactory	Deleterious*
Porous dolomites.................	0.9	....	1.5	Fair	Innocuous
Shales..........................	0.6	1.8	1.8	Poor	Innocuous
Highly siliceous limestones and dolomites......................	5.8	9.1	5.9	Satisfactory	Deleterious*
Dense chalcedonic cherts...........	2.6	3.9	3.8	Fair	Deleterious*
Porous chalcedonic cherts..........	2.4	3.0	3.0	Poor	Deleterious*

* Deleterious with high-alkali cements.

Table 7-7. Petrographic Analysis of a Gravel, near Saskatoon, Saskatchewan, Canada

Constituents	Amount, as number of particles, %, in fractions indicated			Degrees of quality	
	1½–¾ in.	¾–⅜ in.	⅜–³⁄₁₆ in.	Physical quality	Chemical quality
Granites and gneisses..............	44.6	36.8	45.4	Satisfactory	Innocuous
Weathered granites and gneisses.....	6.6	4.8	....	Fair	Innocuous
Deeply weathered granites and gneisses......................	0.1	....	....	Poor	Innocuous
Biotite schists...................	2.1	6.6	7.0	Satisfactory	Innocuous
Weathered biotite schists..........	0.7	1.8	0.9	Fair	Innocuous
Deeply weathered biotite schists....	0.3	....	....	Poor	Innocuous
Metasiltstones...................	1.3	2.3	5.1	Satisfactory	Innocuous
Quartzose sandstones and quartzites.	3.0	8.1	6.3	Satisfactory	Innocuous
Porous sandstones.................	0.4	....	....	Fair	Innocuous
Dolomites.......................	36.7	36.8	34.3	Satisfactory	Innocuous
Weathered dolomites..............	3.1	2.8	1.0	Fair	Innocuous
Deeply weathered dolomites........	0.5	....	....	Poor	Innocuous
Ferruginous siltstones..............	0.6	....	....	Poor	Innocuous

Table 7-8. Petrographic Analysis of a Gravel, St. Louis, Mo.

Constituents	Amount, as number of particles, %				Degrees of quality	
	In fractions indicated			In whole gravel	Physical quality	Chemical quality
	1½–¾ in.	¾–⅜ in.	⅜–³⁄₁₆ in.			
Gray dense cherts.........	19.0	11.3	6.2	14.9	Fair	Deleterious*
Brown dense cherts........	6.6	11.0	8.5	8.6	Fair	Deleterious*
White to tan, porous or fractured cherts.........	16.5	28.6	26.1	22.3	Fair to poor	Deleterious*
Brown, porous, or fractured cherts.................	37.2	35.7	36.4	36.7	Fair to poor	Deleterious*
Highly and finely porous cherts.................	15.7	8.0	14.0	12.2	Poor	Deleterious*
Coarsely porous quartzose cherts.................	5.0	4.7	2.9	4.8	Fair	Deleterious*
Quartzites.................	....	0.7	4.2	0.5	Satisfactory	Innocuous
Friable sandstones........	....	....	0.3	Trace	Fair	Innocuous
Coal.....................	....	....	1.0	Trace	Poor	Innocuous

* Deleterious with high-alkali cements.

Table 7-9. Petrographic Analysis of a Sand, near Kimball, Nebr.

Constituents	Amount, as number of particles, %							Degrees of quality	
	In fractions indicated						In whole sand	Physical quality	Chemical quality
	⅜–³⁄₁₆ in.	No. 4–8	No. 8–16	No. 16–30	No. 30–50	No. 50–100			
Granites.................	24.7	24.5	19.5	14.9	5.4	8.4	18.0	Satisfactory	Innocuous
Weathered granites........	1.9	....	....	....	....	....	0.6	Fair	Innocuous
Gneisses and schists.......	10.4	4.0	2.6	1.4	0.2	0.2	4.5	Satisfactory	Innocuous
Weathered gneisses and schists.................	1.4	....	....	....	....	....	0.4	Fair	Innocuous
Quartz and quartzites......	7.8	19.7	28.7	35.1	33.5	41.6	23.4	Satisfactory	Innocuous
Feldspars.................	33.0	32.1	34.8	37.9	58.7	42.5	38.5	Fair	Innocuous
Rhyolites and rhyolite tuffs.................	4.6	6.4	4.2	3.3	1.0	1.0	3.9	Satisfactory	Deleterious*
Tuffaceous sandstones.....	2.3	1.8	0.4	....	....	....	1.1	Poor	Deleterious*
Opaline and tuffaceous limestones..............	12.6	9.9	9.4	7.4	1.2	1.8	8.5	Poor	Deleterious*
Chalcedonic cherts.........	....	....	0.2	....	....	....	Trace	Fair	Deleterious*
Garnet, hornblende, biotite, etc.....................	....	....	....	....	....	4.5	0.3	Satisfactory	Innocuous
Clay balls and clay stones..	1.8	1.6	0.2	....	....	....	0.8	Poor	Innocuous

* Deleterious with high-alkali cements.

Table 7-10. Petrographic Analysis of a Gravel, South Platte River, near Denver, Colo.

Constituents	Amount, as number of particles, %, in fractions indicated		Degrees of quality	
	1½–⅜ in.	⅜–³⁄₁₆ in.	Physical quality	Chemical quality
Granites and granite gneisses......	65.7	78.0	Satisfactory	Innocuous
Weathered granites and granite gneisses......................	11.8	10.8	Fair	Innocuous
Deeply weathered granites and granite gneisses...............	5.4	3.3	Poor	Innocuous
Rhyolites......................	0.7	0.6	Satisfactory	Deleterious*
Dacite porphyries...............	0.6	0.2	Satisfactory	Innocuous
Basalts.......................	0.7	0.3	Satisfactory	Innocuous
Pumicites......................	0.3	....	Poor	Deleterious*
Biotite and sillimanite schists.....	0.8	....	Satisfactory	Innocuous
Quartz and quartzites............	8.9	4.7	Satisfactory	Innocuous
Sandstones....................	2.8	1.5	Satisfactory	Innocuous
Weathered sandstones...........	0.3	....	Fair	Innocuous
Hard siltstones.................	1.3	0.2	Satisfactory	Innocuous
Porous ferruginous siltstones......	....	0.1	Poor	Innocuous
Chalcedonic cherts..............	0.6	0.3	Satisfactory	Deleterious*
Fissile shales...................	0.1	....	Poor	Innocuous

* Deleterious with high-alkali cements.

Table 7-11. Petrographic Analysis of a Sand, Butte, Mont.

Constituents	Amount, as number of particles, %							Degrees of quality	
	In fractions indicated						In whole sand	Physical quality	Chemical quality
	No. 4–8	No. 8–16	No. 16–30	No. 30–50	No. 50–100	No. 100–200			
Granites.................	51.3	29.7	9.3	8.9	6.2	1.3	21.5	Satisfactory	Innocuous
Weathered granites.......	11.0	17.7	13.7	6.1	5.4	1.0	11.7	Fair	Innocuous
Quartz..................	1.7	13.3	22.7	39.5	34.1	29.8	21.0	Satisfactory	Innocuous
Feldspars...............	....	8.3	16.0	20.7	16.6	23.8	12.3	Satisfactory	Innocuous
Graywackes.............	15.3	6.0	6.7	4.2	1.9	....	7.0	Satisfactory	Innocuous
Calcareous coating fragments.................	1.7	0.3	0.7	0.5	4.3	5.5	1.3	Fair	Innocuous
Dacites and rhyolites.....	19.0	24.7	29.2	12.0	17.9	14.1	21.4	Satisfactory	Deleterious*
Mica (biotite)...........	....	....	1.7	5.4	6.8	5.1	2.2	Fair	Innocuous
Hornblende, magnetite, zircon, etc.............	....	....	....	2.7	6.8	19.4	1.6	Satisfactory	Innocuous

* Deleterious with high-alkali cements.

tively smooth and impermeable; very low compressibility; coefficient of thermal expansion approaching zero or being negative in one or more directions.

POOR: Particles exhibit one or more of the following qualities: friable to pulverulent; slake when wetted and dried; highly fractured; capillary absorption moderate to high; marked volume change with wetting and drying; combine three or more qualities indicated under "fair."

Table 7-12. Petrographic Analysis of a Sand, Snake River, Idaho-Oregon

Constituents	Amount, as number of particles, %					Degrees of quality	
	In fractions indicated				In whole sand	Physical quality	Chemical quality
	No. 16–30	No. 30–50	No. 50–100	No. 100–200			
Granitic rocks and gneisses.......	8.3	4.9	4.7	3.4	4.9	Satisfactory	Innocuous
Mica schists....................	3.9	0.3	....	....	0.3	Satisfactory	Innocuous
Basalts and diabases............	11.8	11.4	23.5	36.9	15.5	Satisfactory	Innocuous
Weathered basalts and diabases...	2.0	0.3	....	....	0.3	Fair	Innocuous
Andesite porphyries.............	2.0	1.5	0.5	....	1.3	Satisfactory	Innocuous
Rhyolites and andesites..........	13.8	9.3	5.2	7.1	8.2	Satisfactory	Deleterious*
Feldspars......................	16.0	26.6	17.5	10.2	23.1	Satisfactory	Innocuous
Quartz........................	40.7	45.2	35.0	21.9	41.4	Satisfactory	Innocuous
Micas.........................	1.5	0.5	0.8	2.6	0.7	Fair	Innocuous
Hornblende, sphene, magnetite, etc.	....	....	12.8	17.9	4.3	Satisfactory	Innocuous

* Deleterious with high-alkali cement.

Table 7-13. Petrographic Analysis of a Gravel, near Vancouver, British Columbia

Constituents	Amount, as number of particles, %					Degrees of quality	
	In fractions indicated				In whole aggregate	Physical quality	Chemical quality
	3– 1½ in.	1½– ¾ in.	¾– ⅜ in.	⅜– $\frac{3}{16}$ in.			
Granites and gneisses...	52.3	43.1	45.4	39.0	45.8	Satisfactory	Innocuous
Weathered granites and gneisses............	8.7	1.3	2.0	1.5	3.8	Fair	Innocuous
Schists...............	....	9.8	9.7	5.3	6.0	Satisfactory	Innocuous
Quartz and quartzites..	....	....	1.0	2.1	0.5	Satisfactory	Innocuous
Diorites..............	....	5.9	4.9	4.2	3.6	Satisfactory	Innocuous
Rhyolite porphyries....	4.3	....	0.3	0.6	1.5	Satisfactory	Innocuous
Andesites and dacites...	26.0	18.3	24.4	26.6	23.4	Satisfactory	Deleterious*
Weathered andesites and dacites..............	8.7	5.2	3.2	0.9	5.1	Fair	Deleterious*
Deeply weathered andesites and dacites.......	....	2.0	0.3	....	0.7	Poor	Deleterious*
Basaltic andesites.......	....	14.4	8.8	19.8	9.6	Satisfactory	Deleterious*

* Deleterious with high-alkali cement.

INNOCUOUS: Particles contain no constituents which will dissolve or react chemically to a significant extent with constituents of the atmosphere, water, or hydrating portland cement while enclosed in concrete or mortar under ordinary conditions.

DELETERIOUS: Particles contain one or more constituents in significant proportion which are known to react chemically under conditions ordinarily prevailing in portland

Table 7-14. Petrographic Analysis of a Gravel, near Oroville, Calif.

(Amount, as number of particles, in per cent)

Constituents	In size fractions indicated			In the whole sample					Total
				Physical quality			Chemical quality		
	1½–¾ in.	¾–⅜ in.	⅜–³⁄₁₆ in.	S*	F*	P*	I*	D*	
Granites, diorites, granodiorites, and related gneisses...	19.1	19.0	14.0	11.5	5.8	...	17.3	...	17.3
Amphibole schists............	26.3	26.9	19.3	19.9	4.0	0.3	24.2	...	24.2
Epidote-amphibole schists and fine-grained gneisses........	17.5	15.4	19.0	11.5	5.8	...	17.3	...	17.3
Meta-andesites and meta-basalts....................	18.1	23.6	35.1	21.7	3.9	...	25.6	...	25.6
Metadacites................	8.2	2.3	4.6	3.8	1.2	...	5.0	...	5.0
Serpentines................	0.3	3.6	2.0	2.0	...	...	2.0	...	2.0
Quartzites and silicified metamorphic rocks..............	5.6	4.3	2.0	4.0	...	...	4.0	...	4.0
Metagraywackes and metasubgraywackes................	0.3	2.6	3.0	2.0	...	...	2.0	...	2.0
Porous sandstones...........	1.3	....	....	....	0.4	...	0.4	...	0.4
Dacites....................	0.3	....	....	0.1	...	...	....	0.1	0.1
Basalts....................	1.3	....	....	0.4	...	...	0.4	...	0.4
Andesitic basalts.............	1.7	1.6	0.3	1.2	...	...	....	1.2	1.2
Andesites...................	....	0.7	....	0.2	...	...	....	0.2	0.2
Feldspars..................	....	....	0.7	0.2	...	...	0.2	...	0.2

* Key: S = satisfactory; F = fair; P = poor; I = innocuous;ˈ D = deleterious with high-alkali cements.

Table 7-15. Petrographic Analysis of a Gravel, San Gabriel Valley, Calif.

Constituents	Amount, as number of particles, %				Degrees of quality	
	In fractions indicated			In whole aggregate	Physical quality	Chemical quality
	1½–¾ in.	¾–⅜ in.	⅜–³⁄₁₆ in.			
Granites and gneisses.......	38.9	41.4	32.0	39.7	Satisfactory	Innocuous
Weathered granites and gneisses...............	24.1	26.0	30.1	25.2	Fair	Innocuous
Deeply weathered granites and gneisses............	1.5	0.3	....	0.9	Poor	Innocuous
Schists and schistose phases of gneisses..............	18.2	14.0	16.5	16.4	Satisfactory	Innocuous
Weathered schists and schistose phases of gneisses.....	10.8	10.0	14.3	10.7	Fair	Innocuous
Deeply weathered schists and schistose phases of gneisses................	0.5	0.3	....	0.4	Poor	Innocuous
Quartz and quartzites......	1.0	0.3	1.0	0.6	Satisfactory	Innocuous
Diorite, quartz diorite, and andesite porphyries.......	2.5	4.7	2.9	3.4	Satisfactory	Innocuous
Weathered porphyries......	0.5	0.3	1.5	0.4	Fair	Innocuous
Diabases..................	....	....	0.5	Trace	Satisfactory	Innocuous
Weathered diabases........	1.0	2.7	1.2	1.7	Fair	Innocuous
Deeply weathered diabases..	0.5	....	....	0.3	Poor	Innocuous
Weathered gabbros.........	0.5	....	....	0.3	Fair	Innocuous

cement concrete or mortar in such a manner as to produce significant volume change, interfere with the normal course of hydration of portland cement, or supply substances which might produce harmful effects upon concrete or mortar.

As a consequence of its mineralogic composition, texture, and internal structure, each rock or mineral particle in aggregate is characterized by a suite of physical and chemical properties which determines the response of the particle to environmental conditions and thus its performance in engineering service.

A. Physical Properties

The physical properties of aggregates have important engineering significance. Included are such properties as porosity, permeability, surface texture, thermal, volume change, and others.

1. Porosity, Permeability, and Absorption.   The internal pore characteristics are the most important properties of aggregates.   The size, abundance, and continuity of pores influence or control such qualities as strength, elasticity, abrasion resistance, surface texture, specific gravity, bond with cementitious binders, resistance to freezing and thawing action, and the rate and magnitude of various cement-aggregate reactions (48).   Much research remains before the significant parameters of the pore system of aggregate particles are defined and evaluated.   For example, what proportion of the total volume of a limestone, chert, or shale particle must be constituted of voids and what range of dimensions and degree of continuity must be present in the void system for the particle to produce popouts or other deterioration of concrete subjected to freezing and thawing?   This question and others like it should be answered if we are to be able to specify limits on the properties of such materials when they are constituents of aggregates.

Table 7-16. Effect of Surface Texture on Bond Strength
(Aggregate embedded in cement briquets)

Surface texture of aggregate	Bond strength in tension, psi		
	28 days in water	28 days in water, then 28 days in air	28 days in water, then 28 cycles of wetting and drying
Rough, porous	350	260	235
Rough	240	275	230
Rough	215	300	245
Fairly rough	250	185	160
Smooth	120	45	
Smooth, conchoidal	285	170	45
Very smooth	195	40	25

SOURCE: After Mather (51).

2. Surface Texture.   Surface texture of an aggregate particle is the fabric or pattern and the relative roughness or smoothness of the periphery.   Surface texture encompasses all of the physical features of the rim and their dimensions, both areally and radially, and details of the form of the surface, such as the relative planeness, curvature, and rugosity.   Clearly, surface texture depends upon the internal texture, structure, and mineralogic composition of the particle as well as the natural and artificial processes of impact and abrasion and leaching to which the particle has been subjected.

Surface texture influences the bond developed between an aggregate particle and the cementing material in concrete.   Table 7-16 shows the results of bond tests made by Goldbeck (50) and cited by Mather (51) to indicate the effect of surface texture of aggregate on bond strength in tension.   Surface texture of aggregate also influences water requirement of portland cement concrete for given consistency.

3. Volume Change with Wetting and Drying. Volume change of portland cement concrete with wetting and drying is influenced by aggregate in three general ways. First, particles containing clay minerals of the expanding lattice type, such as the montmorillonoids and some illitic (hydromica) types, expand significantly with wetting, and cyclical wetting and drying leads to progressive increase in volume of the particle at any selected level of moisture content. Second, particles of high compressibility, such as certain weak sandstones (52), fail to resist drying shrinkage of portland cement paste and so permit inordinate shrinkage of concrete (53). Third, certain aggregates, by reason of their particle shape and surface texture, increase water requirement for given consistency and thus increase drying shrinkage. In areas where aggregates contributing to drying shrinkage of concrete are likely to be proposed for use, a wetting-and-drying test should be included in the standard specifications of the highway department.

4. Thermal Properties. The thermal properties of aggregates which are of significance to the performance of portland cement concrete are (1) coefficient of expansion, (2) specific heat, and (3) conductivity (48, 54). These properties are not significant in bituminous pavement or for aggregates used as subbase materials or road metal.

In mass concrete construction, specific heat, conductivity, and diffusivity must be considered in the design of cooling systems, and these properties of the proposed aggregates may influence the selection of the portland cement for the project. In pavement concrete these properties are not of significance.

Thermal coefficient of expansion of aggregate as a whole and that of individual particles may be of importance in concrete highway structures. Of primary concern is the difference in (1) expansivity of the aggregate particle and of the concrete and (2) the expansivity of the coarse aggregate and of the mortar or cement paste. According to Callan (55), a difference of 3.0×10^{-6} per degree Fahrenheit in the thermal coefficient of expansion of the coarse aggregate and that of the mortar can reduce freezing and thawing resistance considerably. Difference in thermal coefficient of expansion between the aggregate and cement paste, low compressibility of the aggregate particles, and inferior cement-aggregate bond are thought by Rhoades and Mielenz (48, 56) to contribute to the expansion and deterioration of concrete highway pavements and other structures in Kansas, Nebraska, Iowa, and Missouri, containing the so-called sand-gravel aggregates (57, 58).

In general, however, the thermal properties of aggregates are of minimal importance as a factor in aggregate selection for concrete construction on highway projects.

Data on thermal coefficient of expansion of rocks, concrete, and portland cement paste are summarized in Table 7-17. Note that serious difference in expansivity of the cement paste and that of the aggregate is to be expected only with rocks of very

Table 7-17. Thermal Coefficient of Expansion of Some Rocks,*
Concretes, and Portland Cements

(Expansion per degree Fahrenheit, in range − 4 to 212°F)

Materials	No. of specimens	Range in mean linear thermal coefficient
Granites	27	1.0 to 6.6
Diorites and andesites	17	2.3 to 5.7
Gabbros, basalts, diabases	15	2.0 to 5.4
Sandstones	24	2.4 to 7.7
Quartzites	20	3.9 to 7.3
Dolomites	7	3.7 to 4.8
Limestones	65	0.5 to 6.8
Siliceous limestones	6	2.0 to 5.5
Cherts	49	4.1 to 7.3
Marbles	29	0.6 to 8.9
Slates and argillites	5	4.5 to 4.8
Portland cements, neat	10	5.9 to 9.0
Concretes	27	3.6 to 6.8

* All coefficients of rocks were obtained on dry specimens. After Rhoades and Mielenz (49).

low expansivity.   Rocks characterized by a linear thermal coefficient of expansion of 1.0×10^{-6} per degree Fahrenheit or less include granites, limestones, and marble.

5. Strength and Elasticity.   High strength and elasticity of the particles are desirable for road metal and ballast because the rate of disintegration is minimized and the stability of the compacted course is maximum.   However, for portland cement concrete, optimum results may be obtained by a compromise between strength and elasticity so low that the strength of the concrete suffers and drying shrinkage is excessive and level of strength and elasticity so high that adjustment of volume change of the concrete takes place primarily in the cement paste and along cement-aggregate boundaries, rather than more uniformly throughout the concrete.   Near-surface portions of pebbles and sand grains may be seriously weakened by weathering even though the interior of the grains is not modified significantly.   The strength and elasticity of aggregate are not reflected proportionately in the strength or elasticity of the concrete.

A factor in the strength and elasticity of aggregate particles is the degree to which the particles are fractured internally, either because of original fractures in the rock or because of the process of crushing by which the aggregate is manufactured.   Fractures contribute to breakdown of particles during handling and processing and to fragmentation in ballast or road metal and in the surface of bituminous pavements.   Internal fracturing of aggregate particles makes especially necessary a final screening operation just prior to use of the aggregate in order that the desired grading is maintained.

6. Density.   The density of an aggregate particle is dependent upon the true density of its mineral constituents and upon the porosity.   The density of a porous material may be defined to include in its volume all, some, or none of the volume of the pore space.   Thus there are such terms as bulk, apparent, and true specific gravity.   The bulk density (dry basis) of common rocks ranges from about 1.6 to 3.2 (48, 49).

Bulk density or bulk specific gravity of specific rock types can be correlated with approximate ranges of porosity and, commonly, also with degree of soundness, such as resistance to freezing and thawing breakdown.   Such correlations are empirical but, with proper control and evaluation, are of great value in aggregate selection in specific areas.

7. Hardness.   The resistance of an aggregate to abrasion and degradation is controlled by the hardness of the mineral constituents, the firmness with which the individual grains are cemented or interlocked together, and the frequency of fracturing (see above).   Particles composed of minerals of a low degree of hardness may be designated as "soft"; those which are easily disintegrated because of poor cementation or intensity of fracturing may be designated as "weak."   Weak and soft particles are objectionable in aggregate because they break down during processing, changing the gradation of the aggregate from that contemplated, and are susceptible to continuing disintegration during service as a result of impact, abrasion, and weathering.

8. Particle Shape.   The shape of aggregate particles may significantly affect the engineering performance of the aggregate.   The workability of concrete, the strength, asphalt demand, and workability of asphaltic mixtures, and the frictional characteristics of graded aggregate mixtures are but a few examples of the influence of particle shape on the quality of aggregate.

The size and shape of rock and mineral particles in aggregate depend to a considerable extent upon the presence and spacing of natural partings and cleavage in the parent formation.   It is for this reason that certain formations characteristically produce more or less equidimensional, slabby, or elongated forms in certain size fractions and why it is not possible to obtain substantial production of large-size aggregate from certain sources.   The shape of particles depends also upon the relative strength, elasticity, and abrasion resistance of the rock or mineral and upon the natural or artificial processes whereby the aggregate is produced.   It is common to find certain rock and mineral types in an aggregate represented by subangular or even well-rounded particles, whereas other rock and mineral particles in the same sample are angular.

Mather has discussed particle shape of aggregate in some detail (51).   Two relatively independent properties, sphericity and roundness, control particle shape. Sphericity describes the relation of the surface area of the particle to its volume or the relative volume of the particle and that of the circumscribing sphere.   Roundness measures the relative sharpness or angularity of the edges and corners of the particle.

9. Coatings. A coating is a layer of substance covering a part or the entirety of the surface of an aggregate particle. The coating may be of natural origin, such as mineral deposits formed in sand and gravel by ground water, or artificial such as stone dust resulting from crushing and handling. Natural coatings usually do not cover the entire surface of pebbles and sand grains; rather, they tend to be concentrated at the bottom of the particles as they lie in the deposit. Coatings usually are composed of silt, clay, and calcium carbonate; organic matter, iron oxides, opal, manganese oxides, alkali and alkali-earth sulfates, and soluble phosphates have been identified.

The coating materials may be essentially inert chemically, or they may be potentially deleterious. Many coatings are physically weak, porous, and absorptive, and poorly bonded to the aggregate particles. In such instances they may seriously impair the quality of the aggregate for the intended use.

B. Chemical Properties

The chemical properties of aggregates are frequently of great engineering significance.

1. Solubility. Few rocks or gravels and sand that would be considered seriously for use as aggregate contain a sufficient proportion of water-soluble substances to affect the quality of the aggregate in service. Nevertheless, occasional formations and deposits of sand and gravel contain concentrations of water-soluble substances, such as gypsum, in the form of coatings or seam fillings, sufficient to cause difficulty in portland cement concrete. Such substances can be detected by petrographic examination or by qualitative chemical tests; they may be revealed by efflorescence at natural exposures of the formation or deposit.

2. Oxidation, Hydration, and Carbonation. Various unstable minerals are susceptible to oxidation, hydration, or carbonation if exposed to the atmosphere. These effects are insignificant for road metal, ballast, or bituminous construction, if ordinary caution is used in selection of the materials. However, portland cement concrete may become unsightly or distressed by these actions. Susceptible substances include iron sulfides (marcasite, pyrite, and pyrrhotite), ferric and ferrous oxides in clay-ironstone particles, free lime (CaO) and free magnesia (MgO) in industrial products and wastes, and certain zeolites.

3. Reactions with Portland Cement. There are several types of substances occurring in mineral aggregates which can participate in reactions with hydrating portland cement. These include:

1. Soluble sulfates, such as gypsum, oxidation products of iron sulfides, and alunite-jarosite, which may produce expansion and disintegration of cement paste as a result of sulfate attack. Except for the products of oxidation of iron sulfides, these substances are rarely significant.

2. Miscellaneous inorganic compounds, such as water-soluble salts of zinc and borates, which may cause serious retardation of hardening and strength development of portland cement concrete. These substances are rarely significant in aggregates.

3. Various organic substances, such as carbohydrates, certain hydrocarbons, or natural surface-active residues occurring within or on the surface of aggregate particles, which may retard hardening, reduce strength development, produce staining, or cause excessive air entrainment in portland cement concrete. Among these substances, only those producing undesirable air entrainment are of relatively common occurrence (59).

4. Certain forms of silica and siliceous materials which may produce deleterious expansion and cracking of concrete by interaction with alkalies released during the hydration of portland cement (60). Alkali-aggregate reaction is a serious problem in highway pavements and structures in many parts of the United States, including not only most of the western states but also areas in the Middle West, South, Atlantic Seaboard, and portions of upper New York State, as well as in parts of Ontario, Canada. Test methods in common use to identify and evaluate the potential alkali reactivity of aggregate and of specific cement-aggregate combinations are described elsewhere in this chapter.

III. USES OF AND GENERAL REQUIREMENTS FOR AGGREGATES IN HIGHWAY CONSTRUCTION

Mineral aggregates are materials which are basic to each element in the pavement structure, whether it be a low-cost road or a high-type flexible or rigid pavement. It is the purpose of this portion of the section to review briefly the scope of usage of aggregates in the various types of pavement layers and to indicate, in a general way, the requirements for the aggregates in these various applications. Article IV of this section on "Specifications" treats in greater detail the specific requirements for aggregates for particular uses.

A. Aggregate Used without the Addition of a Cementing Material

Aggregates are used by themselves, that is, without the addition of a cementing material, in such applications as bases or subbases for flexible and rigid pavements. In the case of a base for a rigid pavement, the purpose of the layer may be to prevent pumping, to cover a frost-susceptible material, or to improve the general drainage characteristics of the section. Load-carrying capacity of the base layer is not a primary consideration in the choice of a material. Rather, gradation is the primary consideration (61). In the case of bases and subbases for flexible pavements, however, load-carrying capacity is a primary factor in the selection of materials. Here, too, gradation plays an important role, but in addition, the material used in any particular layer of the flexible-pavement system must reinforce those layers beneath it and must be capable of withstanding and transmitting the load to which it will be subjected. Strength tests such as the triaxial compression test and the California Bearing Ratio Test (CBR) are commonly used to evaluate such materials (62).

Aggregates by themselves frequently make up the entire pavement structure. This type of pavement, sometimes referred to as the low-cost road, is one "in which a layer of predominantly granular material is placed on the natural soil to serve as a wearing course or a base course for a relatively thin flexible surface" (63). Again, the requirements for these materials are mainly gradational requirements, with perhaps a restriction placed on the plasticity of the material passing the No. 200 sieve. The emphasis here is on the maximum utilization of local materials.

Aggregates are also used without the addition of a cement for shoulder material. The main requirements in this application are again gradation and stability. Naturally, all shoulders are not composed of uncemented graded aggregates. The trend toward paved shoulders has progressed rapidly.

B. Aggregates for Bituminous-pavement Layers

There are many types of bituminous-pavement layers used by highway agencies throughout the country. The situation with regard to bituminous mixtures is not so simple as it may appear with portland cement concrete mixtures. In the latter case there is one basic type of mixture for slab construction, whereas there is a multiplicity of types of mixtures in the former.

In illustration of this, Table 7-18 shows a brief outline of types of bituminous layers classified on the basis of type of construction. This classification does not take into account local terminology, which, if added, could expand the table tenfold.

Table 7-18. Outline of Bituminous-pavement Layers on the Basis of Type of Construction

Application Types	Mixture Types
Penetration macadam Surface treatments	Road mixes
Single Multiple	Traveling plant mixes
Prime coats ⎫ No aggregate Tack coats ⎬	Stationary plant mixes

In spite of the many types of bituminous layers which are used, Pauls and Carpenter (64) state:

"The ideal aggregate for bituminous construction regardless of class or type would have the following characteristics:

(1) Strength and toughness.

(2) Ability to crush into chunky particles, free from flakes, slivers, and pieces that are unduly thin and elongated.

(3) Low porosity (however, it should not be completely lacking in porosity).

(4) Hydrophobic characteristics.

(5) Particle size and gradation appropriate to the type of construction."

Several of the items listed above have been discussed in the part of the text dealing with "Properties of Aggregate Particles." The details on specifications and tests are discussed in a later section.

C. Aggregates for Portland Cement Concrete

A major use of aggregates in the highway industry is, of course, in portland cement concrete for rigid-pavement slabs, bridges, and other structures.

Aggregates for concrete should be physically and chemically stable. It was earlier stated that the properties of an aggregate were primarily dependent upon its mineralogic composition, internal texture, and internal structure. These factors then influence the concrete-making quality of an aggregate to a major degree.

Blanks and Kennedy (2) summarize the general requirements for aggregates for portland cement concrete as follows:

" . . . Controlling factors include: size, distribution and interconnection of voids; surface character and texture; gradation; internal texture and structure; mineral composition; particle shape."

They also point out that:

"Probably no other physical characteristic of an aggregate has such an important effect on the water requirements and workability of fresh concrete mixtures as the gradation or particle size distribution. . . . These factors profoundly influence a large number of other important concrete properties. Particle shape also affects concrete workability, segregation, and bleeding."

IV. SPECIFICATIONS FOR AGGREGATES

"A specification may be defined as a concise description, preferably in measurable terms, of the significant characteristics of a material. The determination of such needs in terms of characteristics that can be set up easily in a specification requires logical thought based on sound chemical and engineering principles.

"Requirements should be set forth in a clear, detailed but concise manner and should be quantitative rather than qualitative to reduce to a minimum decisions based on personal opinion.

"There are five important requisites of a specification:

1. Accuracy and precision
2. Workability
3. Suitability
4. Flexibility
5. Acceptability

"A specification should never be considered as final nor complete. Frequently an outmoded specification will prove a greater deterrent to progress than no specification at all. Any specification should be examined periodically, in view of technological advances in manufacture, testing, and use requirements" (71).

Specifications for any material represent the purchaser's or user's conception of those characteristics of the material which are necessary for the successful use of the product in his application. Preparing a specification for a material is not a simple matter. It is frequently very difficult to put into specific quantitative terms the properties which the material should and should not possess. There is an ever-present danger of specifying too rigidly, the consequences of which are higher prices. In addition, one may know well what specific characteristics of an end product are desirable but, in preparing specifications for components of that end product, may not know with the same certainty the properties of the components that determine the ultimate success of the end product.

This is sometimes the case with mineral aggregates. Take, for example, aggregates for portland cement concrete for a highway pavement. A desirable aggregate for this application is one that will make workable, then strong, durable concrete. However, present-day methods are not always adequate to perform this evaluation. The relatively recent discovery of concrete failures caused by the presence of certain forms of silica support the contention that there may still be present in aggregates certain very important characteristics hitherto ignored and unknown.

This should not imply, however, that specifications for aggregates are at the level of guesswork. There are many, perhaps most, aggregate properties which can be specified on a firm scientific basis. Aggregate specifications are dynamic. They are constantly being refined and improved as new research and further experience is gained.

A. Local Specifications and Their Importance

Many of the examples of specifications for aggregates are those of national rather than local character. Such specifications must necessarily be broad. Hence they should be regarded primarily as guides rather than as absolute standard. There are many mistakes which may be made by using or imposing a general specification in all areas of the country. First of all, the aggregates differ. Secondly, the climate may differ, so a material which may be extremely deleterious in one area may be completely innocuous in another.

Each of the state highway departments and several other agencies dealing with road building have standard specifications for aggregates for one or several uses. These specifications are collections of experience and research effort which, in their present form, represent years of work. They are usually, for any given area, the best specifications to be found even though changes are bound to be made as experience and research data become available. For the most part the state highway specifications are extremely valuable documents.

B. Gradation Requirements

The gradation or particle-size distribution of an aggregate is usually specified to be within certain limits for various types of construction. There is a great difference between what is considered an acceptable grading for aggregate for portland cement concrete, for the various bituminous mixtures, or for base layers. Again, the local conditions and background are important factors in the determination of specific construction types and grading requirements for the materials used in the construction. In spite of these local factors, however, there are a sufficient number of common elements found in each broad class of construction type so that general limits for gradation may be drawn.

References 63, 72, and 73 present excellent summaries of gradings used by various agencies for the three classes of aggregate use in highway construction. Tables 7-19 to 7-22 are taken from these references and will illustrate gradation requirements which are commonly specified for some types of construction. Reference to ASTM Designation D 1241 will provide a summary of gradation requirements for soil-aggregate materials.

C. Quality Requirements for Aggregates

Under "quality requirements" one can list all usual specification provisions other than those dealing with gradation. Quality requirements can be divided into five distinct groups:
1. General quality requirements
2. Abrasion resistance

3. Soundness
4. Restrictions on "deleterious" constituents
5. Special requirements

It is the purpose of this section to discuss each of these five types of quality requirements, giving reasons for each.

Table 7-19. Typical Grading Requirements for Aggregate for Portland Cement Concrete

Sieve size	Cumulative percentage passing				
	A*	B	B	A	B
3 in.				100	
2½ in.				95–100	
2 in.				80–95	95–100
1½ in.	95–100	95–100	100	65–87	85–100
1 in.	70–86			50–75	
¾ in.	55–75	80–100	95–100	45–66	60–85
⅜ in.	40–55			38–55	
No. 4	30–45	35–85	35–90	30–45	40–70
No. 8	23–35			23–35	
No. 16	17–27			17–27	
No. 30	10–17	15–22	15–25	10–17	12–20
No. 50	4–9	3–10	5–12	4–9	3–7
No. 100	0–3			0–3	
No. 200	0–2			0–2	

* Letters denote different specifying agencies.
SOURCE: After Price (73).

Table 7-20. Typical Grading Requirements for Dense-graded Bituminous Concrete

Sieve size	Cumulative percentage passing					
	A*	B	C	D	E	F
1 in.				100	95–100	
¾ in.	100	100		95–100		100
½ in.			100		60–80	95–100
⅜ in.		70–85	80–100	50–75		
No. 4	50–83		45–70		40–55	50–60
No. 8			30–55	30–50		40–50
No. 10	25–60	40–55			28–38	
No. 16				20–30		
No. 20			20–40			30–38
No. 30	15–42					
No. 40		20–35			20–27	
No. 50	10–28		5–25	10–20		
No. 80					13–20	
No. 100	5–17		2–12	7–10		15–23
No. 200	2–8	2–10	2–5	5–7	5–11	6–10

* Letters denote different specifying agencies.
SOURCE: After Benson (72).

1. General.   Most specifications for aggregates for use in portland cement concrete, bituminous mixtures, or in similar classes of use begin with a paragraph with a heading such as "Description," "General Characteristics," "General Requirements," or something similar.   It is the purpose of the paragraph first to describe in general terms the type of material which is considered to be acceptable, and second, to collect in one place those rather nebulous requirements which undoubtedly influence the acceptability of the material but for which a good quantitative method of test does not exist.

For example, the following is an excerpt from the first paragraph of a specification dealing with coarse aggregates called "Description":

"Coarse aggregates . . . shall be composed of clean, tough, durable fragments of crushed limestone or dolomite, crushed or uncrushed gravel, or slag, free from an excess of flat, elongated, thinly laminated, soft or disintegrated pieces and free from fragments coated with dirt or other objectionable matter" (74).

Table 7-21. Typical Grading Requirements for Open-graded, Cold-mix Bituminous-surface-course Mixtures

Sieve size	Cumulative percentage passing			
	A*	B	C	D
1½ in.	100	100		
1 in.	90–100	90–100		
¾ in.	40–75			
½ in.	15–35	0–15		100
⅜ in.	0–15		100	
No. 4	0–5		20–45	20–45
No. 8				10–30
No. 10			0–5	
No. 50				0–8
No. 200		0–2		0–2

* Letters denote different specifying agencies.
Note: All mixtures require seal courses, and, in addition, agency B requires a keystone application prior to sealing.
SOURCE: After Benson (72).

Table 7-22. Summary of Grading Requirements for Water-bound Macadam
(Average of state highway specifications in round numbers)

Sieve size	Cumulative percentage passing		
	Coarse aggregate	Medium aggregate	Screenings
4 in.	100		
2½ in.	65–75	100	
1½ in.	30–70	35–70	
½ in.		0–5	100
No. 4			90–100
No. 100			5–25

SOURCE: After Willis and Kelly (63).

2. Abrasion Resistance. The qualities of a material known as *hardness* and *toughness* have historically been regarded as essential to good aggregate. Shelburne states that "The properties of hardness and toughness are very closely related." He indicates that hardness is made up, in part, by abrasion resistance and that "Toughness is generally understood to mean the power possessed by a material to resist fracture under impact" (75).

3. Soundness. The soundness of aggregates or their resistance to the forces of weathering is undoubtedly one of the most important considerations in the selection of a material for highway construction. The primary exposure that one is concerned with is alternate freezing and thawing. Somewhat less frequently one may be concerned with resistance of materials to alternate heating and cooling, wetting and drying, or the action of aggressive waters. Most specifications for aggregates from northern areas, however, include a provision for soundness which is designed to ensure the selection of a material which is durable in freezing and thawing.

4. Restrictions on Deleterious Constituents. It is generally recognized that the presence of certain substances in aggregates for portland cement concrete is undesirable. These substances, if present, would tend to decrease the durability of the concrete, make it less abrasion-resistant than would otherwise be the case, cause pop-outs, or inhibit strength gain. Specifications for such aggregates, therefore, normally contain a section in which the deleterious materials are named and a limit placed on the amount of each that is allowable.

Obviously, the same deleterious substances do not occur in all locations, and even materials which are called by the same name may differ in their effect on concrete from one area of the country to another. Hence one cannot list all deleterious materials with fixed allowable percentages of each. The most commonly recognized deleterious materials are clay lumps, coal and lignite, soft particles, shale, chert, materials reactive with cement alkalies, organic impurities, lightweight particles, and material finer than the No. 200 sieve. Some specifications list in detail each type of substance with a corresponding limiting percentage. Others may set up a general category based on a physical property rather than on specific names.

Table 7-23. Limits for Deleterious Substances in Coarse Aggregates for Concrete

Item	Maximum Per Cent by Weight of Total Sample
Clay lumps..	0.25
Soft particles...	5.0
Chert that will readily disintegrate (soundness test, 5 cycles)........................	1.0
Material finer than No. 200 sieve...	1.0*
Oven-dry material floating on a liquid having a specific gravity of 2.0...............	1.0†

* In the case of crushed aggregates, if the material finer than the No. 200 sieve consists of the dust of fracture, essentially free from clay or shale, this percentage may be increased to 1.5.

† This requirement does not apply to blast-furnace-slag coarse aggregate.

It should be stated, however, that some agencies specify aggregates on the basis of class rather than on the basis of use and may employ the same class of aggregate for portland cement concrete and some high types of bituminous construction. In this case, specification provisions that are primarily intended for the control of quality of concrete aggregates may be imposed upon aggregates for bituminous construction, and vice versa.

5. Special Requirements. Under this heading may be listed a number of items that are sometimes included in specifications, often simply as a general statement, but sometimes with quantitative limits attached. Some of the most common of these are:

1. A statement limiting the presence of flat and elongated pieces in coarse aggregate.
2. A "service-requirement" provision which may say, in effect, that even if the aggregate conforms to all other requirements of the specification, if it has been shown to give poor field performance it can be rejected. On the other hand, it may be a permissive statement such as one saying that even if the aggregate fails to meet a certain requirement, it may be acceptable if it has had a satisfactory service record in the past.
3. Aggregates for use in bituminous mixtures may be required to pass some stripping test.
4. Aggregates for use in bituminous mixtures may have to meet a specification requirement on minimum percentage of crushed pieces.
5. Blast-furnace slag may have to meet a requirement for minimum unit weight.

REFERENCES

1. American Society for Testing Materials: *ASTM Standards*, part 3, Philadelphia Society for Testing Materials, 1955.
2. Blanks, R. F., and H. L. Kennedy: *The Technology of Cement and Concrete*, vol. 1, John Wiley & Sons, Inc., New York, 1955.
3. Woods, K. B.: "Introduction to Symposium on Mineral Aggregates," *Symposium on Mineral Aggregates*, *ASTM Special Technical Publication*, 83, 1948.

4. Dana, E. S., and C. S. Hurlbut, Jr.: *Manual of Mineralogy*, 15th ed., John Wiley & Sons, Inc., New York, 1946.
5. Pirsson, L. V., and Adolph Knopf: *Rocks and Minerals*, John Wiley & Sons, Inc., New York, 1949.
6. Longwell, C. R., et al.: *Outlines of Geology*, John Wiley & Sons, Inc., New York, 1937.
7. *The Profitable Use of Testing Sieves*, The W. S. Tyler Co., Catalogue 53, 1955.
8. Lambe, T. William: *Soil Testing for Engineers*, John Wiley & Sons, Inc., New York, 1951.
9. Talbot, A. N., and F. E. Richart: "The Strength of Concrete: Its Relation to the Cement, Aggregate and Water," *Engineering Experiment Station Bulletin* 137, University of Illinois, October, 1923.
10. Fuller, W. B., and S. E. Thompson: "The Laws of Proportioning Concrete, *Transactions, ASCE*, vol. 59, 1907.
11. Thwaites, F. T.: *Outline of Glacial Geology*, University of Wisconsin, Madison, 1937.
12. Runner, D. G.: *Geology for Civil Engineers as Related to Highway Engineering*, Gillette Publishing Company, Chicago, 1939.
13. Jenkins, D. S., D. J. Belcher, L. E. Gregg, and K. B. Woods: "The Origin, Distribution, and Airphoto Identification of United States Soils," *Technical Development Report 52*, Civil Aeronautics Administration, May, 1946.
14. Flint, R. F.: "Glacial Geology of Connecticut," *State Geological and Natural History Survey Bulletin* 47, 1930.
15. Burchard, E. F.: "Stone Resources East of the Mississippi River," *Mineral Resources of the United States*, part II, pp. 782–834, U.S. Geological Survey, 1911.
16. Dale, T. N.: "The Chief Commercial Granites of Massachusetts, New Hampshire, and Rhode Island," *Bulletin 354, U.S. Geological Survey*, p. 228, 1908.
17. Dale, T. N.: "The Granites of Vermont," U.S. Geological Survey, *Bulletin* 404, 1909.
18. Loughlin, G. F.: *Mineral Resources of the United States*, part II, pp. 761–842, U.S. Geological Survey, 1915.
19. Calvin, Samuel: "The Aftonian Gravels and Their Relation to the Drift Sheets in the Region about Afton Junction and Thayer, Iowa," *Proceedings, Davenport Academy of Science*, vol. 10, pp. 18–31, 1907.
20. Calvin, Samuel: "The Buchanan Gravels: An Interglacial Deposit in Buchanan County, Iowa," *American Geologist*, vol. 17, pp. 76–78, 1896.
21. "Geologic Map of the United States," U.S. Geological Survey, 1932.
22. Blatchley, W. S., et al.: "The Roads and Road Materials of Indiana," *Thirtieth Annual Report*, U.S. Department of Geology and Natural Resources, pp. 17–1007, 1905.
23. Merrill, F. J. H.: *Road Materials and Road Building in New York*, vol. 4, no. 17, University of the State of New York, 1897.
24. Stout, W.: "Dolomites and Limestones of Western Ohio," *Geological Survey of Ohio Bulletin* 42, 4th ser., 1941.
25. Hotchkiss, W. O., and E. Steidtmann: "Limestone Road Materials of Wisconsin," *Wisconsin Geological and Natural History Survey Bulletin* 34, 1914.
26. Burchard, E. F.: Stone Resources in Great Plains and Rocky Mountain States, *Mineral Resources of the United States*, part II, pp. 764–818, U.S. Geological Survey, 1912.
27. Fenneman, Nevin M.: *Physiography of Western United States*, McGraw-Hill Book Company, Inc., New York, 1931.
28. Culver, H. E.: "Geology of Washington, part I, General Features of Washington Geology," *Department of Conservation and Development Bulletin* 32, 1936.
29. Gilbert, G. K.: "Lake Bonneville," *Monographs, U.S. Geological Survey*, vol. I, 1890.
30. Russell, I. C.: "Geological History of Lake Lahontan," *U.S. Geological Survey Monographs* XI, 1885.
31. Burchard, E. F.: "Stone Resources in the States West of the Rocky Mountains," *Mineral Resources of the United States*, part II, pp. 1335–1410, U.S. Geological Survey, 1913.
32. Darton, N. H.: "Preliminary Report on the Geology and Underground Water Resources of the Central Great Plains," *U.S. Geological Survey Professional Paper 32*, 1905.
33. Nash, J. P.: "Road-building Materials in Texas," *University of Texas Bulletin* 1839, July, 1918.
34. Cooke, C. W.: "Geology of Florida," *Florida Geological Survey, Geology Bulletin* 29, 1945.
35. Furcron, A. S.: "Dolomites and Magnesian Limestones in Georgia," *Department of Mines, Mining, and Geology, Information Circular* 14, 1942.
36. Swanson, H. E., and N. C. Rockwood: "Louisiana—A Sand and Gravel Producers Legacy," *Rock Products*, July, 1946, p. 82.
37. Woodward, T. P., and A. J. Gueno, Jr.: "The Sand and Gravel Deposits of Louisiana,"

Louisiana Geological Survey Bulletin 19, 1941. Frink, J. W.: "Subsurface Pleistocene of Louisiana," *ibid.*

38. Darton, N. H.: "Gravel and Sand Deposits of Eastern Maryland Adjacent to Washington and Baltimore," *U.S. Geological Survey Bulletin* 906-A, 1939.
39. Adams, G. I., C. Butts, L. W. Stephenson, and W. Cooke: "Geology of Alabama," *Geological Survey of Alabama Special Report* 14, University of Alabama, 1926.
40. Spain, E. L., Jr., and M. A. Rose: "Geological Study of Gravel Concrete Aggregates of the Tennessee River," *American Institute of Mechanical Engineers Technical Paper* 840.
41. Crider, A. F.: "Geology and Mineral Resources of Mississippi," *U.S. Geological Survey Bulletin* 283, 1906.
42. Cantrill, Curtis, and Louis Campbell: "Selection of Aggregates for Concrete Pavement Based on Service Records," *Proceedings, Highway Research Board*, vol. 39, p. 937, 1939.
43. Dumble, E. T.: "The Geology of East Texas," *University of Texas Bulletin* 1869, 1918.
44. "Mineral Industry of California in 1947," *California Journal of Mines and Geology*, California State Division of Mines, vol. 43, no. 4, October, 1947.
45. Mielenz, Richard C.: "Petrographic Examination of Concrete Aggregates," *ASTM Special Technical Publication* 169, p. 253, 1955.
46. Spencer, J. W., and O. K. Dart, Jr.: "Locating Gravel Sources for Highway Use," *Public Works*, vol. 88, no. 7, p. 94, 1957.
47. Rhoades, Roger: "Influence of Sedimentation on Concrete Aggregate," *Applied Sedimentation*, John Wiley & Sons, Inc., New York, chap. 24, p. 437, 1950.
48. Rhoades, R. F., and R. C. Mielenz: "Petrographic and Mineralogic Characteristics of Aggregates," *ASTM Special Technical Publication* 83, p. 20, 1948. Rockwood, N. C.: "Production and Manufacture of Fine and Coarse Aggregates," *ibid.*, pp. 88–116.
49. Rhoades, Roger, and R. C. Mielenz: "Petrography of Concrete," *Proceedings, American Concrete Institute*, vol. 42, p. 581, 1958.
50. Goldbeck, A. T.: *Crushed Stone Journal*, vol. 14, no. 5, p. 8, 1939.
51. Mather, Bryant: "Shape, Surface Texture, and Coatings on Concrete Aggregates," *ASTM Special Technical Publication* 169, p. 284, 1955.
52. Stutterheim, N.: "Excessive Shrinkage of Aggregates as a Cause of Deterioration of Concrete Structures in South Africa," *South African Institute of Civil Engineers Transactions*, vol. 5, p. 20, 1954.
53. Carlson, R. W.: "Drying Shrinkage of Concrete as Affected by Many Factors," *Proceedings, ASTM*, vol. 38, part II, p. 419, 1938.
54. Cook, H. K.: "Thermal Properties of Concrete Aggregates," *ASTM Special Technical Publication* 169, p. 325, 1955.
55. Callan, E. J.: "Thermal Expansion of Aggregates and Concrete Durability," *Proceedings, American Concrete Institute*, vol. 48, p. 485, 1952.
56. Mielenz, R. C.: "Potential Reactivity of Aggregate in Concrete and Mortar," *ASTM Bulletin*, 193, p. 41, 1953.
57. Gibson, W. E.: "A Study of Map-cracking of Sand-gravel Concrete Pavements," *Proceedings, Highway Research Board*, vol, 18, part I, p. 227, 1938.
58. Conrow, A. D.: "Studies of Abnormal Expansion of Portland-cement Concrete," *Proceedings, ASTM*, vol. 52, p. 1205, 1952.
59. Macnaughton, M. F., and J. B. Herbech: "Accidental Air in Concrete," *Proceedings, American Concrete Institute*, vol. 27, p. 273, 1954.
60. Lerch, William: "Chemical Reactions of Concrete Aggregates," *ASTM Special Technical Publication* 169, p. 334, 1955.
61. Allen, C. W., and L. D. Childs: "Report on Pavement Research Project in Ohio," *Highway Research Board, Bulletin* 116, 1956.
62. "Design of Flexible Pavement," *Highway Research Board, Research Report* 16-B, 1954.
63. Willis, E. A., and J. A. Kelley, Jr.: "Mineral Aggregates for Low-cost Roads and Water-bound Macadams," *Symposium on Mineral Aggregates, ASTM Special Technical Publication* 83, 1948.
64. Pauls, J. T., and C. A. Carpenter: "Mineral Aggregates for Bituminous Construction," *Symposium on Mineral Aggregates, ASTM Special Technical Publication* 83, 1948.
65. *Blaster's Handbook*, E. I. Du Pont de Nemours & Co., Wilmington, Del.
66. Nichols, H. L., Jr.: *Moving the Earth: The Work Book of Excavation*, North Castle Books, Greenwich Conn., 1955.
67. Taggart, A. F.: *Handbook of Mineral Dressing*, John Wiley & Sons, Inc., New York, 1945.
68. See technical literature of manufacturing of air-separating equipment such as that of the Fuller Co., Catasauqua, Pa., and the Kennedy-Van Soun Co., New York.

69. Addison, Herbert: *A Treatise on Applied Hydraulics*, John Wiley & Sons, Inc., New York, 1954.
70. Price, W. L.: "Ten Years of Progress in Gravel Beneficiation—1948–1958," *National Sand and Gravel Association, Circular 71*, Washington, D.C., 1958.
71. American Society for Testing Materials: *Proposed Recommendations on Form of ASTM Specifications*, Philadelphia, 1957.
72. Benson, Jewell R.: "The Grading of Aggregates for Bituminous Construction," Symposium on Mineral Aggregates, *ASTM Special Technical Publication 83*, 1948.
73. Price, Walter H.: "Grading of Mineral Aggregates for Portland-cement Concrete and Mortars," Symposium on Mineral Aggregates, *ASTM Special Technical Publication 83*, 1948.
74. State Highway Commission of Indiana: *Standard Specifications for Road and Bridge Construction and Maintenance*, 1957.
75. Shelburne, T. E.: "Crushing Resistance of Surface Treatment Aggregates," *Engineering Bulletin of Purdue University, Research Series 76*, vol. 24, no. 5, 1940.
76. Woolf, D. O., and D. G. Runner: "The Los Angeles Abrasion Machine for Determining the Quality of Coarse Aggregates," *Proceedings, ASTM*, vol. 35, part II, pp. 511–532, 1935.
77. Woolf, D. O.: "The Relation between Los Angeles Abrasion Test Results and the Service Records of Coarse Aggregates," *Proceedings, Highway Research Board*, vol. 17, pp. 350–359, December, 1937.
78. Brard, *Annales de Chimie et de Physique*, vol. 38, p. 160, 1828.
79. Garrity, L. V., and H. F. Kriege: "Studies of Accelerated Soundness Tests," *Proceedings, Highway Research Board*, vol. 15, pp. 237–260, 1935.
80. Woolf, D. O.: "An Improved Sulfate Soundness Test for Aggregates," *ASTM Bulletin*, April, 1956.
81. Bloem, D. L.: "Soundness and Deleterious Substances," Significance of Tests and Properties of Concrete and Concrete Aggregates, *ASTM Special Technical Publication 169*, 1956.
82. Powers, T. C.: "Basic Considerations Pertaining to Freezing-and-thawing Tests," *Proceedings, ASTM*, vol. 55, 1955.
83. Proudley, C. E.: "Sampling of Mineral Aggregates," Symposium on Mineral Aggregates, *ASTM Special Technical Publication 83*, 1948.
84. Woolf, D. O.: "Toughness, Hardness, Abrasion, Strength and Elastic Properties," Symposium on Mineral Aggregates, *ASTM Special Technical Publication 169*, 1956.
85. Kriege, H. F.: "The Stability of Chert," *Rock Products*, Apr. 27, 1929.
86. Sweet, H. S., and K. B. Woods: "A Study of Chert as a Deleterious Constituent in Aggregates," *Engineering Bulletin of Purdue University*, vol. 26, no. 5, September, 1942.
87. Walker, R. D., and J. F. McLaughlin: "Effect of Heavy Media Separation on Durability of Concrete Made with Indiana Gravels," *Highway Research Board Bulletin 143*, pp. 14–26, 1956.
88. Mielenz, R. C., and E. J. Benton: "Evaluation of the Quick Chemical Test for Alkali Reactivity of Concrete Aggregate," *Highway Research Board Bulletin 171*, 1958.
89. Ricketts, W. C., John C. Sprague, D. D. Tabb, and J. C. McRae: "An Evaluation of the Specific Gravity of Aggregates for Use in Bituminous Mixtures," *Proceedings, ASTM*, vol. 54, pp. 1246–1257, 1954.
90. "Symposium on Specific Gravity of Bituminous Coated Aggregates," *ASTM Special Technical Publication 191*, 1956.

Section 8

PORTLAND CEMENT CONCRETE— MATERIALS AND MIX DESIGN

LEO H. CORNING, *Chief Consulting Structural Engineer, Portland Cement Association, Chicago* (Mix Design and Properties of Portland Cement Concrete).

H. L. FLODIN, *Retired, Portland Cement Association, Chicago* (Mix Design and Properties of Portland Cement Concrete).

W. J. McCOY, *Director of Research, Lehigh Portland Cement Company, Research Laboratory, Coplay, Pa.* (Composition, Manufacture, and Properties of Portland Cement).

I. INTRODUCTION

A. Historical

Cement is an age-old product, but portland cement is relatively new, dating only from 1824, when the first patent was granted. Credit for its invention goes to an Englishman, Joseph Aspdin.

The new cement was named "portland" both to set it apart from its more primitive ancestors and specifically because the color of concrete made from it resembled the color of natural stone quarried on the Isle of Portland, off the southern coast of England.

Portland cement was introduced to the United States shortly after the Civil War, being shipped over as ballast about 1870. The first portland cement manufactured in this country was produced about four years later at Coplay, Pa. Some of the old vertical kilns used in those days have been preserved there for their historical value.

Production from one of these kilns was about 100 to 150 bbl a week. In view of this fact, it is readily understandable that only a limited amount of portland cement was produced until the advent of the rotary kiln about 1890.

B. Growth and Development

From that date on, the cement industry experienced a very rapid growth because portland cement could now be made in a continuous process. For example, during the next ten years, the yearly production of portland cement increased from 335,000 to 8,500,000 bbl—a tremendous gain, which accurately reflects the importance of the rotary kiln to the portland cement industry.

A rotary kiln is a cylindrical metal shell with a refractory lining, rotating on an axis inclined slightly from the horizontal. The raw material is fed in the upper end and, because of the kiln's rotating action, gradually passes through to the mouth of the kiln, flowing countercurrent to the heat, which is furnished by a flame utilizing oil, gas, or pulverized coal.

Modern rotary kilns are among the largest pieces of moving equipment or machinery in the world. Some are 10 or 12 ft in diameter and 500 ft long and can produce 5,000 bbl of cement clinker a day.

Today there are approximately 50 different independent companies manufacturing portland cement in the United States. Together they operate about 300 plants and produce over 500 million bbl of cement per year.

We have just mentioned both barrels and bags of cement. Except for a very small amount of export, no cement has been shipped or sold in barrels for many years. At one time, however, much of it was shipped in big wooden barrels, each containing about 376 lb. This tradition of measure is still in use today even though most cement is now shipped in bulk, with lesser quantities in paper bags containing 94 lb, or $\frac{1}{4}$ bbl. The bag is a convenient measure for proportioning concrete mixes, since one bag of cement contains approximately a cubic foot.

Before concluding this very brief review of the history of portland cement, it seems appropriate to point out the tremendous increase in the use of concrete. Other construction materials, such as lumber, stone, brick, and tile, have been used and improved upon for thousands of years. The same is true of the ferrous metals, such as iron and steel, and the nonferrous metals, such as copper, bronze, and brass. Portland cement concrete, scarcely 100 years old, is truly the "baby" of the construction-materials field, yet today the tonnage of concrete used exceeds that of all these others combined.

II. MANUFACTURE OF PORTLAND CEMENT

A. Raw Materials

Since portland cement is a finely pulverized material consisting principally of certain definite compounds of lime or calcium oxide, silica, alumina, and ferric oxide, all in combined form, perhaps you can readily guess the type of raw materials from which it is made. Some of these are limestone, marl, chalk, and marine shells. They provide the calcareous or lime components. Shale, slate, clay, and sand provide the siliceous or argillaceous components. Such materials as slag and cement rock provide a source of both calcareous and argillaceous components.

The number of raw materials needed for the production of portland cement depends entirely on their composition. They must be such that when blended together and burned, they are capable of producing a clinker with 60 to 65 per cent of lime, 19 to 23 per cent silica, and 6 to 9 per cent alumina and ferric oxide. Usually only two or three materials are required, but in some cases four or five may be necessary.

B. Blending Raw Materials

For an example of the blending process, let us consider a plant using three raw materials, namely, limestone, shale, and sand.

The plant chemist first determines the composition of each material. From this knowledge, he adjusts the proportion of each material to provide a raw mixture of the right composition to produce good-quality cement clinker. These materials are then ground together to a high degree of fineness approximating that of finished portland cement, to ensure intimate contact during the burning process.

The raw mix is sampled as it is transferred from the grinding mills to large storage bins, known as blending bins. This sample is carefully analyzed. In all probability, this single proportioning of raw materials will not result in quite the desired composition. For example, if the chemist tries for a lime content, on an ignited basis, of 63 per cent, he may find that the analyzed raw mix has only 61 per cent. Another bin is then filled with a raw mix with more lime-bearing material that will provide a lime content of about 65 per cent.   In this manner several bins are filled with raw mix, each closely approximating the composition desired by the chemist.

Based on the analysis of the samples, he then blends the material from three, four, or more bins to obtain a raw mix of the exact desired composition.   This final raw-mix blend is referred to as the kiln feed and is ready for the burning operation.

The foregoing is true of a dry-process plant.   In a wet-process plant, water is added during the grinding operation and the ground raw mix is in slurry form, containing 30 to 40 per cent water.   The slurry is stored in tanks, analyzed, and blended to provide a kiln feed mix with the right composition.

C. Burning the Raw Mix

The burning operation is generally considered the most important part of the entire manufacturing process.   The finely ground, blended raw materials are fed into the upper end of the inclined rotary kiln, which is fired at the lower end.

In a wet-process plant the first thing that happens to the raw mix is that the water is driven off.   From this point on there is no difference between the wet and the dry process, as these terms refer only to the preparation of the raw materials prior to burning.   No matter how carefully you might examine the clinker or cement, it would be impossible to tell by which process it was produced.

After all moisture is driven off, the limestone component, upon further heating, is converted to calcium oxide or quicklime.   As the raw mix passes on through the kiln, it continues to increase in temperature until it reaches approximately 2700°F.   At this point the mix partially fuses into small lumps called clinker.

D. Portland Cement Clinker

As the raw mix clinkers, the lime chemically combines with the silica and alumina to form tricalcium silicate (C_3S), dicalcium silicate (C_2S), tricalcium aluminate (C_3A), together with other minor compounds.   These are new compounds, which have been synthesized during the burning operation.   Portland cement clinker, therefore, has entirely different properties from the materials from which it is produced.

Neither the blended raw mix nor any of its raw materials have any cementitious properties, but these raw materials during the burning process form into an entirely new product—portland cement clinker—which consists primarily of hydraulic calcium silicates and aluminates.   It is these new compounds that develop cementitious properties when mixed with water.

E. The Final Process

If portland cement were made just from the clinker, it would in most cases be extremely quick setting.   In fact, if mixed with water, it sometimes would set so quickly you would have great difficulty in getting it out of the container.   Thus it is necessary to add a material to retard and control the set.   That material is gypsum, which not only retards the set but also helps in the development of strength.   Gypsum in the amount of 4 to 5 per cent is interground with the clinker in the final operation to produce portland cement.

III. COMPOSITION AND PROPERTIES

A. Hydration Reactions and Their Significance

When cement is mixed with water, certain basic reactions occur.   In this connection it is well to remember that the three important compounds in cement are C_3S, C_2S, and C_3A.   The percentage of these compounds in all portland cement usually totals more than 80 per cent.   Another compound in most cements is tetracalciumaluminoferrite (C_4AF), the percentage of which is normally 7 to 14 per cent.   In addition, small amounts of magnesium oxide and alkalies are to be found in practically all cements.   However, these compounds have little cementing action; so, for the sake of simplification, we shall concern ourselves here with the hydration characteristics of C_3S, C_2S, and C_3A.

When cement is mixed with water, these compounds immediately begin to react. This reaction is referred to as hydration and is accompanied by the evolution of heat. In general, the cementing compounds are attacked or decomposed by water with the formation of hydrated compounds and supersaturated solutions from which the excess solids precipitate as a complex combination of gels and crystalline hydrates. These products of hydration provide the cementing or gluing action which is the basis for the setting and strength-gaining properties of portland cement.

At this point it would be well to consider the difference between setting time and strength. *Set* is the state of rigidity achieved by concrete, after which it is impossible to work or finish it. *Strength* can be roughly defined as the ability to withstand applied loads or stresses. The fact that a cement sets fast does not necessarily mean that it will gain strength fast. Set and strength are two separate and distinct properties of portland cement.

The three compounds under consideration hydrate at different rates, the C_3A being the fastest-acting. In fact, if gypsum were not added to the clinker in the final grinding operation, the C_3A would hydrate in a few moments and most likely produce a flash set, with considerable heat being evolved. If this happened, the aluminate hydrate would establish the structure of the hydrated paste and handicap the development of strength by the subsequent hydration of the other compounds.

In preventing the C_3A from flash setting, gypsum reacts with it to form a relatively insoluble compound of calcium sulfoaluminate. This reaction continues until all the gypsum is used. If there are enough of the unhydrated aluminates remaining at this point, they will hydrate rapidly and produce a final set. In cements with low C_3A content, there may not be enough unhydrated C_3A at this point to produce final set. In this case final set will be brought about by the hydration of the calcium silicates. Besides helping to control the rate of reaction and the set, gypsum has a significant effect on early-strength and volume-change characteristics of the hydrated cement.

It is obvious that portland cement is a well-engineered construction material because its three basic compounds hydrate and contribute to strength gain at different rates and ages. The C_3A starts to hydrate immediately after the cement is mixed with water, but as explained above, the gypsum prevents this action from proceeding too fast. Within the first day or so the C_3A has essentially completed its hydration, but in the meantime the C_3S has started to hydrate and will be the chief contributor to strength up to ages of 28 days. From this point on the C_2S begins to take over and is thought to be responsible for most of the strength gain after 6 to 9 months; in fact, it continues to hydrate and provides for continuing strength gain even after 10 years.

B. Types of Cement

1. Portland Cement Types I to V. Today there are five basic types of portland cement:

Type I, the most direct descendant of Joseph Aspdin's product, is generally referred to as "regular," or "normal," cement. This is the cement that is used in most ordinary concrete work. Type I cement has a C_3S content of about 45 per cent, C_2S of 28 per cent, and C_3A of 12 per cent.

Type II is often called "modified" cement and was developed for use where moderate sulfate resistance or moderate heat of hydration is desired. Some waters and soils contain a relatively high percentage of sulfate, which can cause structural disintegration by attacking the C_3A component of cement. Because Type II cement is intended to lessen this risk, Federal and ASTM specifications require that it contain less than 8 per cent C_3A. Since Type II releases heat more slowly during hydration, it is also valuable for use in moderately massive sections of concrete. This is because heat released too rapidly in mass concrete will not escape fast enough, causing volume changes and thus cracks. To achieve this moderate heat of hydration the proportion of faster-acting C_3A and C_3S is usually reduced, the slower-acting C_2S being increased. However, for normal conditions Type I will produce as durable concrete as Type II and will in most cases have a somewhat higher early strength.

Type III is known as high-early-strength cement. The rapid strength-gaining property of this cement is of increasing importance in modern construction, where speed is so often a vital factor. For example, service strengths ordinarily requiring 7 days can generally be obtained in 1 to 3 days. For this reason the use of Type III can often save time and money, even though there is usually a small premium of the order of 50 cents a barrel charged for this type as compared with Types I and II. As would be expected, the percentage of C_3S and sometimes C_3A is increased in Type III cement, while C_2S is decreased. Another distinguishing characteristic is its high degree of fineness, which accelerates the rate of hydration for the same reason that powdered sugar dissolves faster than granulated. Because of this fineness, and the higher percentages of faster-acting compounds, the percentage of gypsum required for Type III cement is greater than for Types I and II.

Type IV, commonly referred to as low-heat cement, is generally not stocked but is manufactured on special order. The heat of hydration produced by Type IV is less than Type II, so its primary use is in gigantic dams, where the heat problem cannot be solved by the use of Type II. As would be expected, the amount of C_3S and C_3A in Type IV is considerably less than in Type I and somewhat less than in Type II.

Type V is sulfate-resistant cement and is valuable where exposure to sulfate is so great as to threaten the durability of concrete made with Type II cement. Since C_3A is the compound susceptible to sulfate attack, ASTM and Federal Specifications require that the C_3A content of Type V cement be limited to 5 per cent.

Because of their chemical composition, the rate of strength gain of Types IV and V is less than that of Types I and II and considerably less than that of Type III.

2. Air-entraining, Low-alkali, and White Portland Cements. *Air-entraining Portland Cements.* These cements are produced by intergrinding with the gypsum and clinker a minute amount of soluble resin soap, or other air-entraining agent, usually not more than a few hundredths of 1 per cent. Such cements when used in making concrete will in most instances entrain from 3 to 5 per cent of air. The entrainment of this amount of air markedly improves the workability, durability, and salt resistance of concrete.

In plastic concrete the ratio of water to cement is usually about 50 per cent. If cement could be completely hydrated, it would require only about 25 per cent water by weight, indicating that even after some of the uncombined water is lost by evaporation an appreciable quantity still remains in the pores of the concrete. When concrete is exposed to severe frost action, this free or uncombined water freezes, with a resulting expansion. Upon the first freezing cycle the effect is not as dramatic as when the water pipes in a house freeze, but the same potential forces are at work. The pore structure of the concrete can be broken down gradually by successive cycles of freezing and thawing, first evidenced by surface scaling, which can lead to progressive deterioration.

In some respects the action of salt crystallizing in the pores of concrete is not unlike the formation of ice crystals and has somewhat the same damaging effect.

Air-entraining cement prevents this type of deterioration by incorporating literally millions of tiny, discrete air bubbles in the concrete. They act as small safety chambers to absorb the expansive effects of the ice and salt crystals without damaging the concrete. Thus one of the principal uses of air-entrained concrete is in highway paving, where both the freezing-thawing action and salt used for ice and snow removal are a hazard to durability.

Low-alkali Cement. In some cases it has been shown that the alkalies in portland cement react with certain aggregates, and excessive expansion and subsequent deterioration of the concrete are associated with this reaction. In view of this the U.S. Government Federal Specification SS-C-192b and the American Association of State Highway Officials Specifications have an optional requirement of a maximum limit of 0.6 per cent total alkali (per cent Na_2O + 0.658 per cent K_2O) for cement when alkali-reactive aggregates are to be used in the concrete.

White Portland Cement. The gray color of ordinary portland cement is due primarily to the iron oxide (Fe_2O_3). The essential difference between the gray and white portland cement is that the iron oxide content of white cement is usually in the range of 0.3 to 0.4 per cent, a maximum amount being of the order of 0.5 per cent. Even small amounts of manganese oxide have a marked effect on the color of white cements, so special care is exerted during manufacture to avoid this material.

3. Natural, Portland-slag, Slag, and Pozzolan Cements. *Natural Cement.* Natural cement is obtained by finely pulverizing a naturally occurring mixture of calcareous and argillaceous materials to which nondeleterious materials not to exceed 5 per cent may be added, and then calcining at a temperature not significantly higher than necessary to drive off carbon dioxide and in all cases below that at which sintering takes place. Since the composition is largely determined by the naturally occurring raw material, it is usually more variable than portland cements. Today natural cements are very seldom used in concrete except when used with portland cement in the ratio of 1 bag of natural to 4 or 5 bags of portland.

Portland Blast-furnace Slag Cement. Portland blast-furnace slag cement is an intimately interground mixture of portland cement clinker, granulated blast-furnace slag, and in most cases a small amount of gypsum to control the set.

Slag Cement. Slag cement is an interground mixture of water-quenched granulated blast-furnace slag, hydrated lime, and usually a small amount of gypsum. As compared with portland cement, slag cements are lighter in color and have a considerably lower strength. As a result of their low-strength characteristic, their use is in most cases limited to masonry work or, as in the case of natural cement, used in concrete when 1 bag of slag cement is used with 4 or 5 bags of portland cement.

Portland Pozzolan Cement. Portland pozzolan cement is an intimate mixture of portland cement and pozzolan obtained by either intergrinding or blending of the materials subsequent to their having been pulverized. A pozzolan is usually defined as a siliceous or siliceous and aluminous material which in itself possesses little or no cementitious value but will in finely divided form and in the presence of moisture react with calcium hydroxide at ordinary temperatures to form compounds possessing cementitious properties.

In general, portland pozzolan cements have a lower early strength for all types of concrete mixes as compared with portland cement; however, in some cases, depending on the nature of the pozzolan, higher compressive strengths may be obtained at later ages for lean mixes.

IV. AGGREGATES FOR PORTLAND CEMENT CONCRETE

A. Types

Aggregates commonly used in concrete may be divided into the two broad classifications of natural and artificial materials. In the first group are natural sands and gravels, stone sand (stone crushed to fineness of sand) and crushed rock, and a few lightweight materials such as pumice, scoria, and volcanic cinders. In the second classification are most lightweight aggregates produced by crushing air-cooled blast-furnace slag or by expanding the slag by special cooling processes; by expanding and crushing clays and shales or by pelletizing, expanding, and burning them.

Because aggregates constitute about 75 per cent of the volume of concrete, they must meet certain requirements and their characteristics influence to a large degree the workability, strength, durability, wear resistance, and economy of the concrete. Aggregates must be hard, to resist abrasion; sound, to withstand repeated cycles of freezing and thawing and other types of weathering; strong, to produce concrete strong in compression and flexure; tough, to resist impact and wear; and they must be inert to chemical reaction with the alkalis in cement.

Aggregates containing weak, friable, or laminated particles are undesirable, and shale, stone laminated with shale, and most cherts should be avoided. Visual inspection will usually reveal unsatisfactory material in coarse aggregate, but for more important work both fine and coarse aggregates should be tested. Aggregates should meet the requirements of the "Specifications for Concrete Aggregates" (ASTM), which limit the permissible amounts of deleterious materials and also give the requirements for grading, strength, and soundness. When durability is important only aggregate having a record of proved resistance to the particular exposure to which the concrete will be subjected should be used, irrespective of test results.

B. Physical Characteristics

1. Particle Shape. Particle shape is important in both fine and coarse aggregates. Long, flat particles or those that are very angular require a higher percentage of fine material and more cement to produce a workable concrete than do aggregates that are round or nearly cubical. Concretes made with aggregates containing in excess of 15 per cent of long, flat particles may be weak in flexure and not economical because richer mixes are required for workability and to provide adequate durability and strength to meet service conditions.

Crushed rock and other crushed materials make excellent concrete, but the particles should be more or less cubical in shape. The type of crusher as well as the nature of the material itself has considerable influence on the shape of crushed particles. Gyratory crushers or cone crushers with curved breaking plates produce the larger size of coarse aggregate with the least amount of flat and elongated pieces. For the smaller sizes, corrugated roll crushers produce best results. Some rocks break into long, slivery, or flat pieces, regardless of the type of crusher used. Such material should not be used.

Natural sands and gravels usually consist of rounded particles. Stone sand made by crushing stone is more angular than natural sand, but if the particles are generally cubical, not thin and sharp, satisfactory concrete can be made with only slightly higher cement content than with natural sand. If stone sand made by crushing limestone is to be used in pavement, it should be investigated to be sure it does not result in a slippery surface under traffic.

2. Maximum Size. Maximum-size aggregate consistent with the job conditions should be used. In structures the largest practicable size depends on the size and shape of members and the spacing of reinforcing bars. Generally, the maximum size should not exceed one-fifth the minimum dimensions of the member, nor three-fourths of the clear spacing between bars.

For pavements, well-graded coarse aggregate up to 3 in. maximum size has been used successfully, but $2\frac{1}{2}$ in. top size is more common. A maximum size of $1\frac{1}{2}$ in. is also entirely satisfactory.

For practically all types of work it is desirable to furnish coarse aggregate in at least two separate sizes. It is particularly important to do so when the maximum size is $1\frac{1}{2}$ in. or larger in order to prevent segregation of sizes in handling and thereby to maintain better control of the quality of the mix. Material graded from No. 4 to $1\frac{1}{2}$ in. should be divided at the $\frac{3}{4}$-in. sieve, and at the 1-in. sieve for material graded up to 2 or $2\frac{1}{2}$ in.

3. Gradation. Gradation, or particle-size distribution, of aggregate is important because it affects cost, workability, segregation, density, shrinkage, and durability. Generally, aggregates that give a smooth grading curve, that is, where there is neither a large excess nor deficiency of any size, produce best results. Very fine sands are uneconomical, and very coarse sands produce harsh unworkable mixes, so they should be avoided if possible.

A sieve analysis is made of both fine and coarse aggregates to determine their gradation. For fine aggregate, the standard sieves are Nos. 4, 8, 16, 30, 50, and 100; for coarse aggregates they are 6 in., 3 in., $1\frac{1}{2}$ in., $\frac{3}{4}$ in., $\frac{3}{8}$ in., and No. 4. These sizes designate square openings.

4. Fineness Modulus. This is a term widely used as an index to the fineness or coarseness of aggregate. It is the summation of the cumulative percentages of the aggregates retained on the standard sieves divided by 100. The fineness modulus does not indicate the gradation because many different gradations will give the same value. The higher the value, the coarser the material.

Sands falling within the range of requirements in the ASTM specifications may be quite different in degree of fineness, as indicated by the fineness modulus. The choice of sand A or sand B would depend upon the type of mix best suited to the job, the richness of mix, and the size and grading of the coarse aggregate. When a relatively small size coarse aggregate, $\frac{3}{4}$ in. maximum size or less, is used or for lean mixes, sand B or a grading approaching the maximum percentage passing each sieve is desirable. With large coarse aggregate, such as 2½ in. maximum size, frequently used in pavements or for rich mixes, sand A would be preferable.

5. Effect of Gradation. The workability of a mix is particularly sensitive to the amount of material passing the Nos. 50 and 100 sieves. A deficiency in these sizes may result in excessive water gain, or "bleeding," which causes so-called "sand streaks" on the surface of concrete in contact with forms as the water and fine particles in the mix separate and rise to the surface. There should be not less than 15 to 18 per cent of fine aggregate passing the No. 50 sieve and 3 to 5 per cent passing the No. 100 sieve to produce smooth surfaces and for good consolidation of the concrete in constricted places. An adequate amount of fines is more important in wetter mixes than in stiffer mixes, and in leaner mixes than in rich ones, to minimize bleeding and to provide satisfactory workability and optimum economy.

Coarse aggregate should be graded from No. 4 size up to the largest size. In some localities large-size coarse aggregate is not economically available. Under such circumstances, the largest available size should be used. The larger the maximum size of well-graded coarse aggregate, the less mortar and cement paste is required to fill the voids; consequently less water and cement are necessary to produce concrete of the desired quality for the service to which it will be subjected.

The water required in a mix of given consistency decreases with increases in maximum size of aggregate. For non-air-entrained concrete having a 3-in. slump, the water content per cubic yard varies from 39 gal for 1-in. maximum-size aggregate to about 34 gal for 2-in. aggregate, as shown in Fig. 8-1. The water content remains essentially constant for a given consistency and given aggregates regardless of cement content or proportions of water to cement. The cement content per volume of concrete also decreases as the maximum size of aggregate increases, as shown in Fig. 8-2.

The grading of coarse aggregate of a given top size may vary considerably without having an appreciable effect on the cement content, provided a near-optimum amount of sand for good workability is used. If the amount of sand is held constant, however, the cement content of the mix will vary considerably, depending upon the coarse-aggregate gradation, and may result in an uneconomical mix. Table 8-1 shows several gradations of coarse aggregate and the optimum amount of sand required to produce concrete of best workability. The water content in each mix is 6.3 gal per

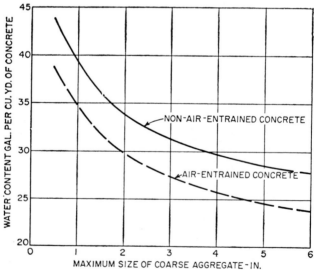

FIG. 8-1. Water requirements for concrete of 3 to 4 in. slump and different maximum sizes of aggregate.

sack of cement. For the optimum amount of sand, the cement content is the same, although the gradation of the coarse aggregate varies widely. For a uniform sand content, however, the amount of cement varies as much as 23 per cent.

It is evident that the sieve-analysis curve for coarse aggregate need not be as smooth as that for sand, provided the fine and coarse aggregates are proportioned for good workability. The curve, however, should always fall within the specified limits.

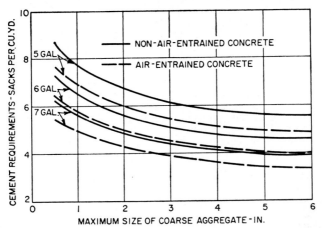

Fɪɢ. 8-2. Cement requirements for concrete of 3 to 4 in. slump and different maximum sizes of aggregate.

Table 8-1. Effect of Gradation of Coarse Aggregate on Cement Requirement

Grading of coarse aggregate, % by weight			Optimum* amount of sand, %	Cement required at % of sand indicated, sacks per cu yd	
No. 4–⅜ in.	⅜–¾ in.	¾–1½ in.		Optimum	35%
35.0	00.0	65.0	40	5.4	5.7
30.0	17.5	52.5	41	5.4	5.8
25.0	30.0	45.0	41	5.4	6.2
20.0	48.0	32.0	41	5.4	6.0
00.0	40.0	60.0	46	5.4	7.0

* Amount giving best workability with aggregates used. Water content 6.3 gal per sack of cement.

C. Other Factors Affecting Aggregate Quality

1. Deleterious Substances. In aggregates these may cause unsoundness, decrease durability and strength, and mar the appearance of concrete. Contaminating materials may be certain chemical substances, silt or clay, soft particles, salts, humus or other organic materials, wood fragments, and coal, lignite, or other combustible or volatile materials.

Silt or clay may be in the form of finely divided particles which tend to rise to the surface while concrete is being placed, resulting in a layer of weak material. When in excessive amounts, additional cement paste is required to coat the greater surface area of the fine material. If the silt or clay occurs as a coating on the aggregate particles, it is particularly objectionable because it breaks the bond between the cement paste and the aggregates and thereby reduces the strength of the concrete.

Organic substances often affect the setting of the cement, delay or even prevent the hardening of the concrete, and may lead to deterioration and loss of strength. Sugar, even in very small amounts, prevents setting of portland cement. Tannic acid sometimes present in sand because of humus or other organic material in the aggregate also tends to inhibit setting of the cement and full development of strength in the concrete.

Coal, lignite, and other combustible materials in excessive amounts affect the strength and durability of the concrete. Particles near the surface not only disintegrate but often pop out and spall the concrete.

2. Alkali Reactivity. Aggregates are usually considered to be inert filler materials, but it is now recognized that certain aggregates display deleterious chemical reaction under some conditions when used in concrete. The result of such reaction is excessive expansion, causing deterioration. It has been demonstrated that certain silica minerals such as opal, chalcedony, tridymite, glass, rhyolites, and others can react with the alkalis in cement if present in sufficient quantity. Pure limestones and dolomites are not deleteriously reactive, but if they contain reactive silica minerals they may contribute to deterioration as a result of alkali reactivity. Likewise, igneous rocks normally entirely satisfactory for concrete aggregates can be deleteriously reactive if impregnated with reactive substances.

Field service records of concrete structures and pavements provide the best information as to the soundness and nonreactivity of aggregates. Aggregates that have a good service record when used with cements of different compositions can be assumed to be free from reactive materials detrimental to concrete. Many Federal, state, and private agencies have kept performance records involving a great variety of aggregate materials and cements.

D. Acceptance Tests

It is very important when performing acceptance tests either in the laboratory or in the field that each step of the standard procedure be followed meticulously. Failure to do so, even for any one step, may give results not truly representative of the qualities of the aggregate. For this reason, copies of the most recent applicable ASTM standard specifications and methods of tests on concrete and concrete materials should be obtained from the American Society for Testing Materials, 1916 Race Street, Philadelphia, Pa. 19103.

V. DESIGN OF CONCRETE MIXTURES

A. Basic Considerations

1. Cement, Aggregates, Water, and Admixtures. It has been indicated in the previous section that the aggregates constitute something like two-thirds to three-fourths of the volume of concrete. The balance is paste, a mixture of cement, water, a small amount of air, and in some cases small amounts of materials added to affect certain properties of the concrete. The paste is the cementing medium that binds the aggregate particles into a solid mass. Thus the quality of the concrete is largely dependent on the quality of the paste.

The hardening of concrete is due to chemical reactions between the cement and water which take place rapidly in the early stages but continue for a long time under favorable conditions. Water must be present to continue the reactions, and the rate of reaction is affected by the temperature, being much slower below 40°F than above this temperature and practically ceasing at freezing temperature or below.

Tests have shown that the most significant factor affecting the quality of the paste, and hence the quality of the concrete, is the amount of mixing water used relative to the amount of cement. The less water used, so long as the mixture is placeable, the stronger and more durable will be the concrete. Only a relatively small amount of water can combine with the compounds of the cement. Water in excess of this amount dilutes the paste and reduces its durability, strength, and watertightness. Some excess water is generally used to produce more economical concrete, but the exact amount that can be used will depend on the properties of the concrete required for the job for which it is intended.

The relation of the amount of water to the amount of cement is referred to as the water-cement ratio and is generally expressed in gallons of water per sack of cement, but sometimes it is expressed on a weight basis. Modern specifications include maximum water-cement ratios permitted for the different classes of concrete.

2. Effect of Water-Cement Ratio on Strength. Many tests have been made by a large number of investigators which prove the water-cement-ratio principle. The exact strength values obtained have varied with the materials used and other conditions, but in all cases they have a direct relation to the water-cement ratio. Table 8-2 gives values for compressive strength at 28 days for Type I portland cement. These are conservative values, reported as minimum, but they show the marked effect of the water-cement ratio. Similar relations exist for flexural strength, tensile strength, and bond strength.

Table 8-2. Compressive Strength of Concrete for Various Water-Cement Ratios*

Water-cement ratio, gal per bag of cement	Probable compressive strength at 28 days, psi	
	Non-air-entrained concrete	Air-entrained concrete
4	6,000	4,800
5	5,000	4,000
6	4,000	3,200
7	3,200	2,600
8	2,500	2,000
9	2,000	1,600

* These average strengths are for concretes containing not more than the percentages of entrained and/or entrapped air shown in Table 8-4. For a constant water-cement ratio, the strength of the concrete is reduced as the air content is increased. For air contents higher than those listed in Table 8-4, the strengths will be proportionally less than those listed in this table.

Strengths are based on 6- × 12-in. cylinders moist-cured under standard conditions for 28 days. See Method of Making and Curing Concrete Compression and Flexure Test Specimens in the Field (ASTM Designation C 31).

SOURCE: Report of Committee 613, "Recommended Practice for Selecting Proportions for Concrete," *Journal of the American Concrete Institute*, September, 1954, vol. 51, table 4, Proceedings.

Compressive and *flexural strengths* obtained for a range of water contents and different ages from 1 day to 28 days for Type I and Type III portland cements are shown in Fig. 8-3 and Fig. 8-4. Because ASTM specifications permit certain tolerances in the requirements for portland cement and concrete aggregates and because the care with which standard testing procedures are followed affects test results, the relation between water-cement ratio and compressive or flexural strength is shown by bands rather than single curves. The most accurate and reliable values are obtained by making preliminary tests, using the cement, aggregates, and other materials selected for the job.

3. Effect of Water-Cement Ratio on Durability. The most destructive force of natural weathering on concrete is the freezing and thawing of moisture in the concrete. On pavements and highway structures there is the additional action of salt used in the winter months to remove ice from pavements. Tests have shown that resistance of concrete to freezing and thawing and to salt action is greatly influenced by the water-cement ratio in the same manner as it affects strength. Table 8-3 may be used as a guide in selecting maximum water-cement ratios permissible for different exposure conditions.

4. Effect of Air Entrainment on Strength and Durability. Extensive laboratory tests and experience have proved conclusively that relatively small percentages of entrained air per volume of concrete produced concrete highly resistant to severe frost action and to effects of applications of salt to pavements for snow and ice removal. Air-entrained concrete is also more cohesive and workable than non-air-entrained concrete and has less tendency to segregate and bleed.

When air is introduced into a concrete mixture there is some reduction in strength if no changes are made in the mix proportions. For the amount of air recommended, the reduction in compressive, flexural, and bond strengths may be up to 10 or 15 per cent. The entrained air increases the volume of concrete produced per unit volume of cement and increases the proportion of mortar. In designing air-entrained concrete mixtures, therefore, consideration should be given to this increased volume of mortar, and advantage may also be taken of the inherent workability and cohesiveness

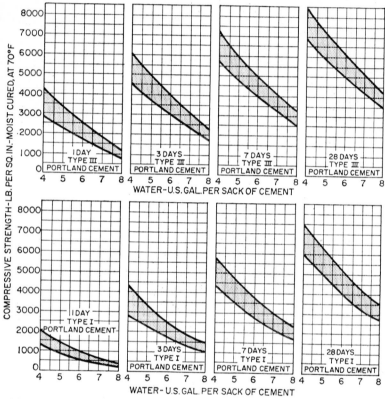

Fig. 8-3. Age-compressive-strength relation for Types I and III portland cements. A large majority of the tests for compressive strength made by many laboratories using a variety of materials complying with the specifications of the ASTM is in the area within the band curves.

of the mortar. These factors permit some reduction in the quantities of fine aggregate and water commonly used in mixes designed for concrete containing no air-entraining materials.

The mixing times of 1 to 2 min usually specified for normal concrete are adequate for air-entrained concrete. Inadequate mixing may result in entrainment of insufficient air; conversely, prolonged mixing under certain conditions tends to entrain air beyond the limits desirable for satisfactory durability and minimum strength reduction. Because of the cohesiveness and freedom from bleeding of water to the surface, air-entrained concrete should be spread, puddled, or vibrated and finished as soon as possible after placing.

Where air-entraining admixtures are used at the mixer, the amount of air entrained will be affected by the amount and nature of the admixture and will also vary with the cement used and the job conditions. In using admixtures for this purpose, it is essential that tests be made of the job-mixed concrete to determine the amount of admixture required. Only by making such tests can there be assurance that sufficient air is entrained to secure the desired results and not so much that the strengths are reduced below the values required.

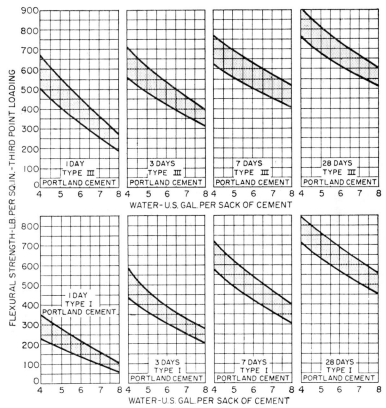

Fig. 8-4. Age-flexural-strength relation for Types I and III portland cements. A large majority of the tests for flexural strength made by several laboratories using a variety of materials complying with the specifications of the ASTM is in the area within the band curves.

The amount of entrained air which will give best results varies somewhat with the size of aggregate. Tests show that in concretes containing aggregates having maximum size of 2½ in. or 1½ in., the air content should be about 4½ per cent; for ¾-in. aggregate, 5½ per cent; and for ⅜-in. aggregate, 7 per cent. A tolerance of plus or minus 1½ per cent in these percentages is generally allowed in specifications. These tests also show that with air-entraining cements these larger amounts of air are usually obtained with the smaller sizes of aggregate because of the larger proportion of mortar in the concrete. The strength reduction for each 1 per cent entrained air with the small aggregate is less than that with the larger aggregate; consequently the total strength reduction may be no greater. This is largely because the reduction in water requirement which accompanies entrained air is considerably greater with small aggregate than with large aggregate.

Table 8-3. Maximum Permissible Water-Cement Ratios for Different Types of Structures and Degrees of Exposure

Type of structure	Exposure conditions*					
	Severe wide range in temperature or frequent alternations of freezing and thawing (air-entrained concrete only)			Mild temperature rarely below freezing, or rainy, or arid		
	In air	At the water line or within the range of fluctuating water level or spray		In air	At the water line or within the range of fluctuating water level or spray	
		In fresh water	In sea water or in contact with sulfates†		In fresh water	In sea water or in contact with sulfates†
Thin sections, such as railings, curbs, sills, ledges, ornamental or architectural concrete, reinforced piles, pipe, and all sections with less than 1 in. concrete cover over reinforcing....	5.5	5.0	4.5‡	6	5.5	4.5‡
Moderate sections, such as retaining walls, abutments, piers, girders, beams..............	6.0	5.5	5.0‡	§	6.0	5.0‡
Exterior portions of heavy (mass) sections.....................	6.5	5.5	5.0‡	§	6.0	5.0‡
Concrete deposited by tremie under water.................	...	5.0	5.0	...	5.0	5.0
Concrete slabs laid on the ground..	6.0	...	....	§	...	....
Concrete protected from the weather, interiors of buildings, concrete below ground.........	§	...	....	§	...	....
Concrete which will later be protected by enclosure or backfill but which may be exposed to freezing and thawing for several years before such protection is offered......................	6.0	...	....	§	...	....

* Air-entrained concrete should be used under all conditions involving severe exposure and may be used under mild exposure conditions to improve workability of the mixture.

† Soil or ground water containing sulfate concentrations of more than 0.2 per cent.

‡ When sulfate-resisting cement is used, maximum water-cement ratio may be increased by 0.5 gal per bag.

§ Water-cement ratio should be selected on the basis of strength and workability requirements.

SOURCE: Report of Committee 613, *Journal of the American Concrete Institute*, September, 1954, table 4, vol. 51, Proceedings.

B. Selecting Proportions for Concrete

After the materials have been selected and the water-cement ratio has been determined, the problem of the engineer is to proportion the materials to produce a concrete mixture that will be economical and readily placeable and a hardened concrete that will have the specified properties. All aggregate particles must be completely coated with cement-water paste, and all voids in the aggregate must be completely filled. The mixture should be such that it can be placed in the forms by the means at hand without undue effort and without segregation of the ingredients and such that the concrete may be finished as specified.

Stiff mixes require less paste than wetter ones. Large, well-graded aggregates having rounded or cubical-shaped particles require less paste than smaller, poorly graded aggregates having flat, elongated particles. Hence, for greatest economy, the mixture should be only wet enough for proper handling and placing and the aggregates should be well graded to the largest size practical for the job and made up of rounded or cubical-shaped particles.

Many methods of proportioning have been used. One that is now widely accepted is the trial-batch method in which several batches are made, either small ones preliminary to job operations or full-size ones on the job. The batch that best meets the requirements of the job is selected. A fairly close estimate of the best proportions may be made from experience with similar materials or by reference to literature on the subject.

1. Water Requirements. In addition to the wetness of the mix and the size, grading, and particle shape of the aggregate, the quantity of water required per unit volume of concrete is influenced by the amount of entrained air. Within the normal range of mixes, the water requirement is not affected by the quantity of cement. Table 8-4 shows approximate water requirements for concrete made with natural sand, average coarse aggregate of different sizes, and concrete having slumps of 1 to 7 in. The values are based on many tests and experience on a wide variety of jobs and are accurate enough for preliminary estimates of proportions.

2. Cement and Aggregate Requirements. The amount of cement per unit volume of concrete for a given set of conditions is estimated by dividing the water requirement listed in Table 8-4 by the water-cement ratio (gallons per sack) in Table 8-3. The volume of paste is determined by adding the volume of water and the solid volume of cement. The solid volume of aggregate may be found by subtracting the volume of paste from the unit volume of concrete less the volume of entrained air. Estimates

Table 8-4. Approximate Mixing-water Requirements for Different Slumps and Maximum Sizes of Aggregate*

Slump, in.	Water, gal per cu yd of concrete, for indicated maximum sizes of aggregate							
	⅜ in.	½ in.	¾ in.	1 in.	1½ in.	2 in.	3 in.	6 in.
	Non-air-entrained concrete							
1–2	42	40	37	36	33	31	29	25
3–4	46	44	41	39	36	34	32	28
6–7	49	46	43	41	38	36	34	30
Approximate amount of entrapped air in non-air-entrained concrete, %......	3	2.5	2	1.5	1	0.5	0.3	0.2
	Air-entrained concrete							
1–2	37	36	33	31	29	27	25	22
3–4	41	39	36	34	32	30	28	24
6–7	43	41	38	36	34	32	30	26
Recommended average total air content, %..........	8	7	6	5	4.5	4	3.5	3

* These quantities of mixing water are for use in computing cement factors for trial batches. They are maxima for reasonably well shaped angular coarse aggregates graded within limits of accepted specifications. If *more* water is required than shown, the cement factor, estimated from these quantities, *should* be increased to maintain desired water-cement ratio, except as otherwise indicated by laboratory tests for strength. If *less* water is required than shown, the cement factor, estimated from these quantities, should *not* be decreased except as indicated by laboratory tests for strength.

SOURCE: Report of Committee 613, *Journal of the American Concrete Institute*, September, 1954. table 3, vol. 51, Proceedings.

of the amounts of fine and coarse aggregates per unit volume of concrete are then made, using the data in Table 8-5.

The solid volume of a granular material, such as cement or aggregate, usually referred to as the absolute volume, is obtained by dividing the weight of the material by its specific gravity times the unit weight of water. Portland cement has an average specific gravity of 3.15, average sand and gravel about 2.65, limestone 2.60, granite 2.70, and traprock 2.90.

Table 8-5. Volume of Coarse Aggregate per Unit of Volume of Cement*

Maximum size of aggregate, in.	Volume of dry-rodded coarse aggregate per unit volume of concrete for different fineness moduli of sand			
	2.40	2.60	2.80	3.00
⅜	0.46	0.44	0.42	0.40
½	0.55	0.53	0.51	0.49
¾	0.65	0.63	0.61	0.59
1	0.70	0.68	0.66	0.64
1½	0.76	0.74	0.72	0.70
2	0.79	0.77	0.75	0.73
3	0.84	0.82	0.80	0.78
6	0.90	0.88	0.86	0.84

* Volumes are based on aggregates in dry-rodded condition as described in Method of Test for Unit Weight of Aggregate (ASTM). These volumes are selected from empirical relationships to produce concrete with a degree of workability suitable for usual reinforced construction. For less workable concrete such as required for concrete pavement construction they may be increased about 10 per cent.

SOURCE: Report of Committee 613. *Journal of the American Concrete Institute*, September, 1954, table 6, vol. 51. Proceedings.

C. Calculation of Mix Proportions

The calculations involved in proportioning a mixture may best be illustrated by examples. Assume that for a bridge to be constructed in a northern climate the concrete is to have 28-day compressive strength of 3,000 psi, which could be obtained with 6¼ gal of water per sack of cement, but because of the severe exposure conditions the water-cement ratio is not to exceed 5½ gal per sack. Coarse aggregate for this job has specific gravity of 2.70, maximum size 1½ in., and dry-rodded weight 108 lb per cu ft. The volume of coarse aggregate per unit volume of concrete may be obtained from Table 8-5. Fine aggregate has a specific gravity of 2.62 and fineness modulus of 3.00. Slump of concrete is to be 3 to 4 in., and air content 4½ per cent. Then, for a cubic yard of concrete:

$$\text{Water, estimated from Table 8-4} = 32 \text{ gal} = \frac{32}{7.5} = 4.27 \text{ cu ft}$$

$$\text{Cement} = \frac{32}{5.5} = 5.82 \text{ sacks} = \frac{5.82 \times 94}{3.15 \times 62.3} = 2.79 \text{ cu ft}$$

$$\text{Air} = 4\tfrac{1}{2} \text{ per cent} = 0.045 \times 27 = 1.21 \text{ cu ft}$$

$$\text{Coarse aggregate} = \frac{108 \times 0.70 \times 27}{2.70 \times 62.3} = 12.13 \text{ cu ft}$$

Vol. of water, cement, air, and coarse aggregate = 20.40 cu ft

$$\text{Sand} = 27.00 - 20.40 = 6.60 \text{ cu ft} = 6.60 \times 2.62 \times 62.3 = 1,077 \text{ lb}$$

Assuming the fine aggregate contains 3 per cent free moisture and the coarse aggregate 1 per cent, the weight of free water will be $1,077 \times 0.03 = 32.3$ lb in the fine aggregate,

27 × 0.70 × 108 × 0.01 = 20.4 lb in coarse aggregate, or a total of 52.7 lb. Then, for a cubic yard of concrete, the field weights will be:

$$\text{Cement} = 5.82 \times 94 = \quad 547 \text{ lb}$$
$$\text{Fine aggregate} = 1{,}077 + 32 = 1{,}109 \text{ lb}$$
$$\text{Coarse aggregate} = 2{,}041 + 20 = 2{,}061 \text{ lb}$$
$$\text{Water} = 32 - \frac{52.7}{8.33} = 25.67 \text{ gal} = \quad 214 \text{ lb}$$

For a second example, assume that a concrete pavement is to be constructed where specifications require mixing water not to exceed 5½ gal per sack of cement, 4 ± 1 per cent entrained air, 2 in. maximum size of aggregate, and slump 1 to 2 in. Coarse aggregate is found to weigh 102 lb per cu ft with specific gravity 2.65. Fine aggregate has specific gravity of 2.60 and fineness modulus of 2.80. For a cubic yard of concrete:

$$\text{Water} = 27 \text{ gal} = \frac{27}{7.5} = \quad 3.60 \text{ cu ft}$$
$$\text{Cement} = \frac{27}{5.5} = 4.91 \text{ sacks} = \frac{4.91 \times 94}{3.15 \times 62.3} = \quad 2.35 \text{ cu ft}$$
$$\text{Air} = 4 \text{ per cent} = 0.04 \times 27 = \quad 1.08 \text{ cu ft}$$
$$\text{Coarse aggregate} = \frac{102 \times 0.75 \times 27}{2.65 \times 62.3} = \underline{12.52 \text{ cu ft}}$$

$$\text{Vol. of water, cement, air, and coarse aggregate} = 19.55 \text{ cu ft}$$
$$\text{Sand} = 27.00 - 19.55 = 7.45 \text{ cu ft} = 7.45 \times 2.60 \times 62.3 = 1{,}207 \text{ lb}$$

Assuming fine aggregate contains 2 per cent free moisture and coarse aggregate contains ½ per cent free moisture, the weight of free water in the fine aggregate would be 1,207 × 0.02 = 24.1 lb and in coarse aggregate 2,065 × 0.005 = 10.3 lb. Total moisture in aggregate = 24.1 + 10.3 = 34.4 lb = 4 gal. Then, for a cubic yard of concrete, the field weights would be:

$$\text{Cement} = 4.91 \times 94 = \quad 462 \text{ lb}$$
$$\text{Sand} = 1{,}207 + 24 = 1{,}231 \text{ lb}$$
$$\text{Coarse aggregate} = 2{,}065 + 10 = 2{,}075 \text{ lb}$$
$$\text{Water} = 27 - 4 = 23 \text{ gal} \times 8.33 = \quad 192 \text{ lb}$$

In these examples the proportions obtained by the calculations may be assumed for the first trial batch. The batch should be carefully observed as to the placeability of the concrete and the finish obtained. Adjustments should then be made in succeeding batches until a satisfactory mixture is obtained. On many paving jobs the mixture is established by the state highway department laboratory based on tests made with the materials to be used and on past experience with the same materials. Final adjustments are then made on the job if this is found necessary.

D. Job Practices

The methods of mixing, handling, placing, finishing, and curing affect the final quality of the concrete. If proper methods are not used in each of these operations, much of the potential strength, durability, and wear resistance of the concrete which could be obtained from the mixture chosen in the design may be sacrificed.

1. Mixing. It is essential that each batch of concrete is mixed thoroughly until it is uniform in appearance, with all the ingredients uniformly distributed. The time of mixing required to accomplish this will vary with the efficiency of the mixer and the nature of the concrete. Specifications generally require a minimum of 1 min mixing for mixers up to 1 cu yd capacity, with an increase of 15 sec mixing for each $\frac{1}{2}$ cu yd, or fraction thereof, additional capacity. A few state highway departments require $1\frac{1}{4}$ min mixing for either 27E or 34E mixers, and two or three require 2 min for both sizes of mixers. The mixing period is measured from the time all solid materials are in the mixer drum, provided all the water is added before one-fourth of the mixing time has elapsed. Some mixers, particularly paving mixers, are equipped with timing devices which may be locked so that the batch cannot be discharged until the designated time has elapsed.

Concrete may be mixed on the job or in a central plant and transported to the job. Portable plants with overhead bins for storage of the materials and scales underneath for weighing the materials are often set up on the job or at a nearby railway siding. The materials may be dry-batched, and three or four batches loaded into a compartmented truck and hauled to the mixer. Central mix plants may deliver completely mixed concrete in agitator trucks, partially mixed concrete in mixer trucks to complete the mixing during transit, or the dry materials may be placed in a transit mixer truck with water in a separate tank. The water may be released into the drum and mixing may proceed during transit, or the complete mixing may be done at the job site after the truck arrives. Truck mixers have capacities up to 5 cu yd.

2. Handling. At the job, concrete is transported from the mixer to the point of placing, using buggies operated on runways or buckets moved by cranes or cableways. Buggies may be powered or hand-operated. They should be moved on smooth runways and have rubber tires to reduce jolting to a minimum. When buggies or buckets are used to haul concrete from ready-mix trucks to the forms, the concrete should be run from the truck into a hopper and from the hopper into the buggy or bucket. Otherwise uniformity of the concrete will not be maintained (Fig. 8-5).

Sometimes concrete is pumped through a 7- or 8-in. steel pipe, using a heavy-duty, single-acting, horizontal piston-type pump. The method is favored in tunnel and other construction where space is limited. Only concrete of good proportions which will not segregate can be successfully pumped. A single pump will move the concrete up to 1,000 ft horizontally or 125 ft vertically. Bends in the pipe reduce these distances. Where concrete is placed directly from the truck into the forms, short chutes may be used. These should be of metal, round-bottomed, and of ample size to avoid overflow. The slope of the chute should be such that segregation of the concrete does not occur as it flows in the chute or when it is discharged into the forms.

3. Placing. In placing concrete on a subgrade or in vertical forms, considerable care is required to avoid segregation, stone pockets, and water bleeding (Fig. 8-6). On subgrades each load of concrete should be dumped into the previously placed concrete. Dumping loads of fresh concrete on the subgrade or other flat surface some distance apart and then working them together results in stone pockets and segregation. The concrete should drop vertically onto the subgrade or into deep forms, and in some cases hoppers and drop chutes are necessary to accomplish this. Drop chutes also help to avoid segregation in deep forms and also incrustations of hardened mortar on forms and reinforcement above the point of placement.

In walls and similar types of work the concrete should be placed in horizontal layers 12 to 20 in. deep, each layer thoroughly consolidated, and the next layer placed while the one below is still soft. The upper portions of walls and columns should be cast with concrete of the lowest slump possible. The concrete should be thoroughly vibrated and then allowed to settle several hours before slabs and beams the columns support are cast. Concrete in the slabs and beams should be placed, however, while the vibrators will still penetrate of their own weight into the concrete in the columns and walls below.

Where concrete is to be placed on hardened concrete or on rock, a layer of mortar $\frac{1}{2}$ to 1 in. thick should be well worked into the irregularities of the hardened surface. This provides a cushion against which the new concrete can be placed, prevents stone pockets, and secures a tight joint. The mortar should be made with the same water content as the concrete.

Nearly all concrete in both pavements and structures is consolidated by some form of vibration. This permits the use of less water for a given cement content, thereby improving the strength, durability, and other properties of the hardened concrete, or it permits the use of less cement to produce a given quality concrete, thus resulting in better economy. Labor is also saved, resulting in further economy. Vibration should

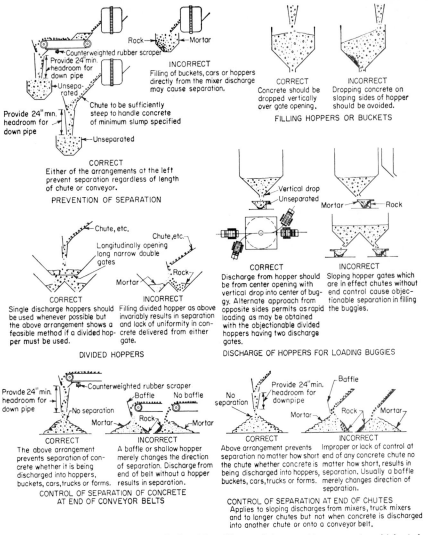

Fig. 8-5. Correct and incorrect methods of handling and transporting concrete. [*Adapted from "Recommended Practice for Measuring, Mixing and Placing Concrete" (ACI 614-42) of the American Concrete Institute.*]

be uniform and thorough, but should not result in bringing excess water or mortar to the surface. There is little chance of overvibration of a mix designed for vibration but more likelihood of undervibration. The inspector must be constantly on the alert to see that the vibration is sufficient to consolidate the concrete. Adequate

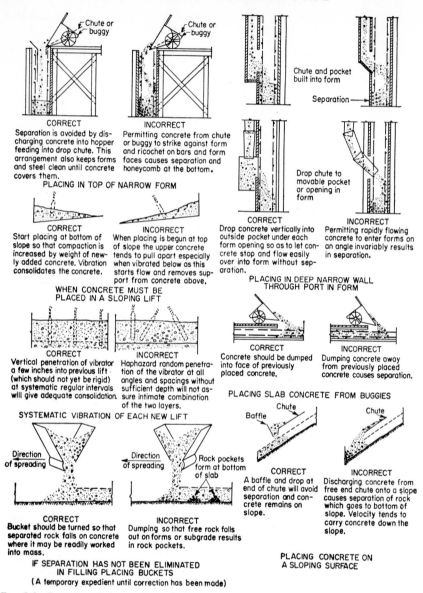

CORRECT
Separation is avoided by discharging concrete into hopper feeding into drop chute. This arrangement also keeps forms and steel clean until concrete covers them.

INCORRECT
Permitting concrete from chute or buggy to strike against form and ricochet on bars and form faces causes separation and honeycomb at the bottom.

PLACING IN TOP OF NARROW FORM

CORRECT
Start placing at bottom of slope so that compaction is increased by weight of newly added concrete. Vibration consolidates the concrete.

INCORRECT
When placing is begun at top of slope the upper concrete tends to pull apart especially when vibrated below as this starts flow and removes support from concrete above.

WHEN CONCRETE MUST BE PLACED IN A SLOPING LIFT

CORRECT
Vertical penetration of vibrator a few inches into previous lift (which should not yet be rigid) at systematic regular intervals will give adequate consolidation.

INCORRECT
Haphazard random penetration of the vibrator at all angles and spacings without sufficient depth will not assure intimate combination of the two layers.

SYSTEMATIC VIBRATION OF EACH NEW LIFT

CORRECT
Bucket should be turned so that separated rock falls on concrete where it may be readily worked into mass.

INCORRECT
Dumping so that free rock falls out on forms or subgrade results in rock pockets.

IF SEPARATION HAS NOT BEEN ELIMINATED IN FILLING PLACING BUCKETS
(A temporary expedient until correction has been made)

Chute and pocket built into form

Separation —

Drop chute to movable pocket or opening in form

CORRECT
Drop concrete vertically into outside pocket under each form opening so as to let concrete stop and flow easily over into form without separation.

INCORRECT
Permitting rapidly flowing concrete to enter forms on an angle invariably results in separation.

PLACING IN DEEP NARROW WALL THROUGH PORT IN FORM

CORRECT
Concrete should be dumped into face of previously placed concrete.

INCORRECT
Dumping concrete away from previously placed concrete causes separation.

PLACING SLAB CONCRETE FROM BUGGIES

CORRECT
A baffle and drop at end of chute will avoid separation and concrete remains on slope.

INCORRECT
Discharging concrete from free end chute onto a slope causes separation of rock which goes to bottom of slope. Velocity tends to carry concrete down the slope.

PLACING CONCRETE ON A SLOPING SURFACE

Fig. 8-6. Correct and incorrect methods of placing concrete. [*Adapted from "Recommended Practice for Measuring, Mixing and Placing Concrete" (ACI 614-42) of the American Concrete Institute.*]

consolidation is indicated when a thin line of mortar appears at the form surface and the top surface of the concrete is level but outlines of coarse-aggregate pieces still show although covered with mortar. The rate of placing should be adjusted to the rate the vibrator crew can properly consolidate the concrete. Placing of concrete should be continuous in the designated section or until a construction joint can be provided at a suitable location. The location and manner of making construction joints are usually covered by the specifications. It is important that the surface of the hard concrete at the joint be clean, rough, and damp when the new concrete is placed. Forms should be retightened against the hard concrete to avoid leakage and the formation of an offset at this location. At horizontal joints a layer of mortar on the hard concrete should be provided as described above. The lower area of the new concrete should be carefully spaded or vibrated to reduce the possibility of stone pockets.

4. Finishing. Both unformed and formed surfaces of concrete may require finishing. Highway pavements are usually finished by large machines supported on the side forms. They consolidate and strike off the concrete to proper level and belt the surface to the desired finish. Several machines having different functions in the finishing operation may be operated in tandem. A "bridge" over the pavement is provided on which men can work in tooling the edges of joints. Outside edges of the pavement are also tooled with an edging tool. Bull floats are sometimes used transversely across the pavement to assist in consolidating and smoothing the surface and to provide a smoother-riding pavement. Highway specifications usually permit a tolerance in surface smoothness of $\frac{1}{8}$ in. in 10 ft, and the pavement is carefully tested with a straightedge or a measuring device to check the surface.

On smaller jobs such as driveways and sidewalks the unformed surface may be screeded to proper level and then floated, leaving a level but gritty, nonslip surface. Tops of walls, parapets, and platforms are often finished in this manner also. Floating should be done only after the concrete has stiffened somewhat and the water sheen has disappeared. It should provide a surface of uniform texture free of marks from the strike-off board.

Brooming is sometimes done to provide a gritty, nonslip surface of uniform texture. It usually follows floating, except that where a fine texture is desired it may follow a single troweling. Rough scoring is produced by using a steel-wire or a coarse-fiber broom. Finer textures are produced with hair brooms. Brooming should be done when the concrete is hard enough to retain the scoring.

Where a smooth, dense surface is desired, floating is followed by steel troweling. The operation should be delayed as long as possible, at least until the concrete is hard enough so that water and excess fine material are not brought to the surface. The water sheen should have disappeared from the surface before it is troweled. Spreading dry cement or a mixture of cement and sand on the surface to blot up excess water should not be permitted. The water is an indication that too much water or improper proportions of the other materials in the mix have been used, and the mixture should be corrected.

At one time formed surfaces on bridge abutments and in other locations were finished by rubbing the concrete with carborundum stones. A grout was worked up and spread over the surface, providing a uniform, smooth, temporarily attractive appearance. Often this surface had little resistance to weather, disintegrated, and pulled away from the concrete. It was also quite costly. Consequently, many states and others have discontinued the practice and now use other methods, the most popular being the grout-cleaning method. After forms have been removed and the tieholes have been filled with mortar, the surface is saturated and a grout of 1 part cement to $1\frac{1}{2}$ to 2 parts of fine sand is brushed on the surface, completely filling air bubbles and holes. White portland cement is sometimes used for part or all of the cement to give a lighter color. The surface is then floated with a wood or cork float, and the wall is scoured thoroughly. When the grout has hardened sufficiently so that it will not be pulled out of the holes, the excess grout is removed with a sponge-rubber float. After the surface has dried thoroughly, it is rubbed with dry burlap to remove any dry grout. No visible film of grout appears on the surface after the rubbing.

Sometimes the grout is applied by rubbing it on with burlap, completely filling all pits and holes. When it has dried enough so it will not smear, most of the excess grout is removed by rubbing with dry burlap. This is followed by 2 days of damp-curing, and excess mortar is then removed with coarse sandpaper.

5. Curing. As stated previously, concrete continues to gain strength so long as conditions are favorable. Moisture must be present and temperature above freezing to continue the hydration of the cement. The chemical reactions occur most rapidly in the early stages, and it is this period that is the most important in developing the greater portion of the potential strength, watertightness, durability, and wear resistance. The fresh concrete contains more than enough water for complete hydration of the cement, but much of this water is quickly lost by evaporation unless certain precautions are taken. Water is retained by covering the surfaces with moisture-retaining covers such as burlap or cotton quilts, watertight paper, or wet sand or by sealing the surface with a curing compound. Such compounds are available in black, colorless, or white. Important precautions in curing are to apply the coverings early enough after finishing the surface and keeping the concrete constantly moist. Alternate cycles of wetting and drying of green concrete may result in crazing or cracking of the surface. The curing should start as soon as the covering can be placed without marring the surface. Sealing compounds should be applied uniformly and in the specified amounts to produce a complete seal over the surface. On pavement construction it is important that coverings are brought out over the edges of the pavement. In structural work, leaving the forms in place is of assistance in retaining moisture for curing, but only if the forms are kept wet.

In cold weather it is often necessary to cover the concrete with straw and tarpaulins or other insulating materials to protect the concrete against low temperatures. Hydration of cement develops some heat, and if this is retained, it usually keeps the concrete at a satisfactory temperature. In structural work the concrete may be covered with tarpaulins or may be housed in and the enclosure heated with salamanders. The heating devices should not be too close to the concrete or the forms because the concrete may become overheated and dry out or the forms may catch fire. It is just as important that concrete be kept wet in cold-weather as in warm-weather construction. It is advisable to provide thermometers in enclosures and under insulating covers as a check on the curing temperatures.

Leaving forms in place to aid in curing is effective only when they are kept wet so that the concrete surfaces do not dry. Generally it is better to remove the forms fairly early so that the regular curing may be started without delay. Patching and repairing of formed surfaces should be done as early as possible, which, of course, requires removal of the forms. If a rubbed finish is required, early removal of the forms is necessary. On massive work in normal weather it is desirable to remove heavy wood forms to cool the concrete. Then, too, the contractor often requires early re-use of the forms.

6. Removal of Forms. Forms should not be removed until the concrete has attained sufficient strength to ensure structural stability. There should be no measurable deflection or distortion and no damage to the corners and edges of the hardened concrete upon stripping of the forms. Beams, girders, floors, and other supporting members should not be stripped nor the shoring removed until they are strong enough to support their own weight and any construction loads. The time when forms can be safely removed will depend on many factors such as temperature and the materials used. The best guide is provided by test specimens of the concrete made and cured under job conditions. The strength is usually considered sufficient when the tests show a safety factor of 2 for the stresses to be sustained.

Rapid cooling of the concrete should be avoided in cold weather since it may cause surface cracks. Enclosures and forms should be removed in a manner and at a time to permit the concrete to cool slowly.

7. Placing of Reinforcement. The exact locations, sizes, and depth of embedment of reinforcement are shown on the plans, and it is important that the requirements be met. The stability of the structure is dependent on the amount and location of the steel, and adequate cover of concrete over the steel is necessary to protect it against corrosion and the effects of fire. After steel is in place it should be carefully inspected to check positions, spacing, and splices and to make sure that it is supported properly so that it will not be displaced during concrete placing. Rust on the bars is not objectionable unless it is in the form of a loose scale that can be easily removed.

VI. PROPERTIES OF FRESH CONCRETE

A. Workability and Consistency

1. Workability. This term refers to the ease with which concrete can be mixed, handled, and placed without segregation. A workable mixture is usually a plastic mixture, but the degree of workability may vary widely, depending on the conditions of placing and the means used for placing. A mixture that has suitable workability for pavement and other flat work would not have the workability required for placing in narrow, deep forms with reinforcement present. Workability is not subject to measurement by test and must be judged by the behavior of the fresh concrete while it is being placed.

2. Consistency. This term refers to the fluidity or degree of stiffness of the fresh concrete and is a component of workability. It is a property of fresh concrete that is measured by the slump test in which a truncated cone form is used. The cone is filled with concrete following a prescribed procedure. It is then withdrawn, and the fall or slump of the concrete is measured. Concrete of stiff consistency has less slump than the more fluid mixtures. The workability and consistency are influenced by the constituents in the mixture and their proportions. Under conditions of uniform operation, changes in consistency, as indicated by the slump test, are useful in indicating changes in the character of the aggregate, the proportions, or the water content. The test is therefore often used in the field as one means of controlling the quality of concrete.

3. Influence of Constituents. With a given amount of cement paste, more aggregate is used in stiff mixes than in more fluid mixes, so that stiff mixes are more economical in cost of materials. The stiff mix requires more labor in placing, and this may offset the saving in cost of materials. The addition of water to a mix makes it more fluid, but as shown previously, this makes a weaker, less durable, and less watertight concrete. Adding aggregate to a proportioned mix stiffens it; the finer the aggregate, the greater is the effect. Thus sand has about twice the effect as the same amount of coarse aggregate graded up to about 1½ in. Sands containing a large proportion of fine material passing Nos. 50 and 100 sieves have a much greater effect in stiffening a mix than sands containing small amounts of these sizes. Yet aggregates must provide enough of these fine sizes to produce concrete of good workability that can be handled and placed without segregation and finished with smooth surfaces. The shape of aggregate pieces has its influence also. Aggregates made up of rough angular pieces or flat pieces require more of the finer materials—sand and cement—to give good workability than aggregates made up of rounded pieces.

Entrained air has an effect on workability similar to that obtained by adding very fine sand. It increases the volume of paste and makes the mixture more cohesive and less apt to segregate. Additions of fine materials such as lime or similar fine admixtures have similar effects on workability.

VII. PROPERTIES OF HARDENED CONCRETE

A. Strength

As stated in Art. V.A, the first step in concrete-mix design is to select relative amounts of water and cement to be used based on the conditions to which the concrete is to be exposed and on the strength desired. Certain material-acceptance tests and other field control tests on the fresh concrete are required to be made to assure production of concrete of the desired quality and strength. Finally, follow-up tests are made to determine the effectiveness of the field control and the properties of the hardened concrete.

The strength of concrete for structures is commonly measured in compression. For pavements, however, it is the usual practice to measure the strength in flexure because the flexural strength governs in pavement design. Some highway departments also require that compressive-strength tests be made. All such tests should be made in strict accordance with standard methods established by the American Society for Testing Materials.

B. Elastic Properties and Durability

Procedures for determining the elastic properties of hardened concrete and its durability are closely related. Reduction in the modulus of elasticity is considered to be an indication of the potential durability of laboratory-concrete specimens or of concrete in structures. Other tests are also made to determine the durability of concrete under certain conditions of exposure, such as freezing and thawing, which is generally considered to be the most severe weathering condition. Concrete shown to be durable under repeated cycles of freezing and thawing will be resistant to the less severe exposure of alternate wetting and drying.

REFERENCES

1. American Society for Testing Materials: *Book of Standards*, part 4, 1958, pp. 1–292.
2. Bogue, R. H.: *The Chemistry of Portland Cement*, 2d ed., Reinhold Publishing Corporation, New York, 1955.
3. Blanks, R. F., and H. L. Kennedy: *The Technology of Cement and Concrete*, John Wiley & Sons, Inc., 1955, vol. I, pp. 1–247.
4. Lea, F. M., and C. H. Desch: *The Chemistry of Cement and Concrete*, rev. ed., Edward Arnold & Co., London, 1956; also available St. Martins Press, New York, 1956.
5. Rader, L. F.: *Materials of Construction*, 6th ed., John Wiley & Sons, Inc., 1955, pp. 327–350.
6. U.S. Department of Interior, Bureau of Reclamation: *Concrete Manual*, Denver, Colo., 1956.
7. Goldbeck, A. T.: *The Proportioning of Concrete*, National Ready-Mixed Concrete Association, Washington, D.C., June, 1949.
8. National Ready-Mixed Concrete Association: *Control of Ready-Mixed Concrete*, Washington, D.C., June, 1953.
9. Portland Cement Association: *Design and Control of Concrete Mixtures*, tenth edition, Chicago, Ill., 1924.
10. Portland Cement Association: *Influence of Aggregate Characteristics in Proportioning Concrete*, Chicago, Ill., 1937.
11. Portland Cement Association: *Effect of Aggregate Characteristics on Quality of Concrete*, Chicago, Ill., 1955.
12. Portland Cement Association: *Field Tests for Determination of Air Content of Fresh Concrete*, Chicago, Ill., 1948.
13. Portland Cement Association: *Prevention of Plastic Cracking in Concrete*, Chicago, Ill., 1955.
14. Portland Cement Association: *Concrete Pavement Inspectors Manual*, Chicago, Ill., 1948.
15. Portland Cement Association: *Concrete Pavement Manual*, Chicago, Ill., 1928.
16. "The Strength of Concrete; Its Relation to the Cement, Aggregates and Waters," *Univ. Illinois Bull. Eng. Expt. Sta. Bull. Ser.* 137, 1923.
17. American Concrete Institute Committee 613: "Recommended Practice for Selecting Proportions for Concrete," *Proc. Am. Concrete Inst.*, vol. 51, p. 49, 1954.

Section 9

BITUMINOUS MATERIALS AND MIXTURES

W. H. GOETZ, *Professor of Highway Engineering, Research Engineer, Joint Highway Research Project, Purdue University, Lafayette, Ind.* (Bituminous Materials and Tests).

Dr. L. E. WOOD, *Associate Professor of Civil Engineering, Purdue University, Lafayette, Ind.* (Bituminous-aggregate Mixtures).

I. INTRODUCTION

According to nomenclature commonly in use in the United States, the term "bituminous material" is used to denote substances in which bitumen is present or from which it can be derived. Bitumen is hydrocarbon material of either natural or pyrogenous origin, gaseous, liquid, semisolid, or solid, which is completely soluble in carbon disulfide (ASTM).

With respect to use in highway construction, the term bituminous material is used to include both natural and manufactured materials regardless of origin, but is restricted to those hydrocarbon materials which are cementitious in character or from which a residue of this character will develop.

Figure 9-1 outlines the principal bituminous materials used in road construction, classifying them as to type and origin and relating these factors to the principal marketed products: asphalt cements, cutback asphalts, slow-curing or road oils, asphalt emulsions, and road tars. Other bituminous materials such as joint and crack fillers, asphalts and pitches for waterproofing, asphalt mastics, etc., are outside the scope of this section.

The first part of this section is devoted to bituminous materials, both asphalts and tars, their manufacture, properties and specifications, and applicable methods of test and their significance. Aggregates for bituminous mixes are discussed, but for a complete treatment of aggregate properties, production, tests, and their significance, the reader is referred to Sec. 7. Finally, the last portion of this section is devoted to the design of bituminous mixes, including specific mix-design methods. The following definitions are pertinent:

A. Terms Relating to Asphalt

ASPHALTS: Black to dark brown semisolid to solid cementitious materials consisting principally of bitumen that gradually liquefy when heated and which occur in nature as such or are obtained as a residuum in the refining of petroleum.

ASPHALT CEMENT, OR PAVING ASPHALT: An asphalt specially prepared as to quality and consistency for direct use in paving. It has a normal penetration between 40 and 300 and must be used hot.

NATIVE, OR NATURAL ASPHALT: One occurring as such in nature. It may be of the

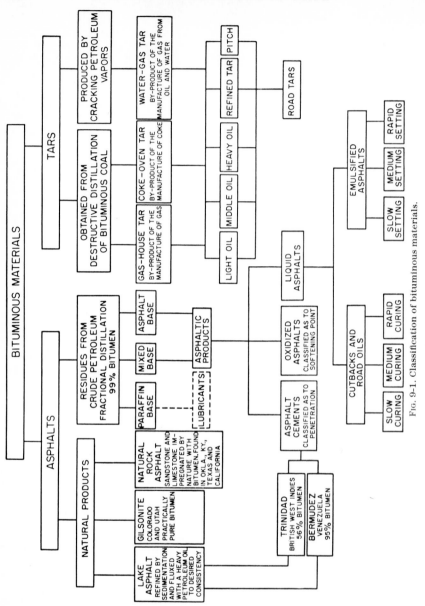

Fig. 9-1. Classification of bituminous materials.

lake, rock, or vein variety, may be essentially pure bitumen or contain a large amount of mineral matter, and the asphalt may vary from hard to soft.

PETROLEUM ASPHALT: Asphalt produced by the refining of petroleum.

STEAM-REFINED ASPHALT: Petroleum asphalt which has been produced in the presence of steam in the distillation process.

OXIDIZED ASPHALT: Asphalt produced in part by the blowing of air through it at high temperature.

BLOWN ASPHALT: The same as oxidized asphalt except that if a catalyst is used in the air-blowing operation, the product is known as a catalytically blown asphalt.

CUTBACK ASPHALT: Asphalt dissolved in naphtha (gasoline) or kerosene to render it temporarily fluid for use. If the solvent is naphtha and the dissolved asphalt relatively hard, the cutback is known as a rapid-curing one. If the solvent is kerosene and the dissolved asphalt relatively soft, the cutback is designated as a medium-curing one.

SLOW-CURING ASPHALT: A soft residuum obtained by distilling off the lighter fractions from petroleum or by blending a residual material with heavy distillates. This material is sometimes referred to as a cutback, but strictly speaking, it is not, even when produced by blending. This material is sometimes called a slow-curing oil or a road oil.

EMULSIFIED ASPHALT: A mixture in which an asphalt cement in a finely dispersed state is suspended in chemically treated water. An inverted asphalt emulsion is one in which asphalt is the continuous phase with water dispersed in it. Inverted asphalt emulsions used in road construction usually are made from liquid asphaltic materials.

LIQUID ASPHALTS: Include rapid- and medium-curing cutbacks, slow-curing asphalts, and emulsified asphalts.

FLUX, OR FLUX OIL: A bituminous material, generally liquid, used for softening other bituminous materials.

B. Terms Relating to Tar

TAR: A black or dark brown bituminous material obtained as a condensate in the destructive distillation of organic substances such as bituminous coal, petroleum, and wood. It is principally bitumen of liquid to semisolid consistency which yields pitch as a residue when fractionally distilled.

COAL TAR: Tar produced by the destructive distillation of bituminous coal.

COKE-OVEN TAR: A variety of coal tar obtained as a by-product from the destructive distillation of bituminous coal in the production of coke.

OIL-GAS TAR: A petroleum tar produced by cracking oils at high temperature in the production of oil gas.

WATER-GAS TAR: A petroleum tar produced by cracking oils at high temperature in the production of carbureted water gas.

REFINED TAR: Produced from crude tar by distillation to remove water and to produce a residue of the desired consistency. Refined tar may also be produced by blending tar residuum with a tar distillate. A tar refined in quality and consistency for use in paving is a *road tar*.

PITCHES: Black or dark brown solid cementitious residues which gradually liquefy when heated and which are produced by distilling off the volatile constituents from tar.

C. Terms Relating to Bituminous Pavements

BITUMINOUS PAVEMENTS: Highway or similar structures, placed to carry traffic, which consist of a mineral-aggregate surface cemented together with bituminous material. The bituminous portion may be only a thin surface, or it may be 8 in. or more in thickness, consisting of mixtures of aggregate and bituminous material more or less carefully proportioned and compacted upon a foundation soil, gravel or stone base, or old bituminous, cement concrete, brick, or block pavement.

PRIME COAT: Liquid bituminous material applied to penetrate the surface of a base on which a bituminous pavement is to be placed. The prime coat is applied to bond together loose particles on the surface, to protect the surface from weather, traffic, and construction equipment, and to prepare the surface to receive the bituminous pavement.

TACK COAT: A bituminous material applied to provide bond between a new or old surface and the mixture with which it is being covered. It may be used in addition to a prime coat and may be applied as a hot asphalt cement or as a liquid bituminous material.

SEAL COAT: An application of bituminous material, usually with aggregate cover and seldom over $\frac{1}{2}$ in. in thickness, which is applied for one or more of the following reasons:

1. To prevent the entrance of air or moisture

2. To rejuvenate an old, dry, or weathered surface
3. To provide a nonskid surface texture
4. To change surface color for visibility or for demarcation purposes
5. To supply additional asphalt on the surface for more effective sealing by traffic

BITUMINOUS BASE COURSE: Consists of mineral aggregate such as stone, gravel, or sand bonded together by a bituminous material and used as a foundation upon which to place a binder or surface course.

BITUMINOUS BINDER COURSE: A bituminous-aggregate mixture used as an intermediate course between the base and surface courses or as the first bituminous layer in two-course bituminous resurfacing. It is sometimes called a leveling course, if that is its primary purpose.

BITUMINOUS SURFACE COURSE: The uppermost course of a bituminous pavement.

BITUMINOUS SURFACE TREATMENT: The application of mineral aggregate and bituminous material to a road or pavement surface, either with or without a mixing operation, which results in a surface course not substantially greater than 1 in. in thickness. Double- or triple-surface treatments are constructed with multiple applications of bituminous material and aggregate.

PENETRATION MACADAM: A type of bituminous-pavement construction in which relatively coarse and uniform-size stone aggregate is placed in layers and stabilized by rolling sufficiently to cause an interlocking of the aggregate, after which the course is penetrated by bituminous material. The bituminous material may be asphalt cement or a heavy tar applied hot or a rapid-curing cutback or rapid-setting emulsion.

MIXED-IN-PLACE, OR ROAD MIX: A method of mixing mineral aggregate and bituminous material on the road surface by means of blade graders, drags, farm implements, or special road-mixing equipment.

TRAVEL-PLANT MIX: Refers to a method of mixing aggregate and bituminous material in which a traveling mechanical mixer is used. The aggregate may be in place in a windrow or may be dumped into the hopper of the traveling mixer by trucks.

PLANT MIX: A term which refers to the mixing of mineral aggregate and bituminous material in a mechanical mixer into which the aggregates and bituminous material are carefully proportioned. The mixture may be hot or cold, and cold mixes may or may not be produced with dried aggregate. The resulting mix is transported to the road and laid either hot or cold.

COLD-LAID MIXTURES: Plant mixes made with bituminous materials of the liquid type or with solvents or flux oils of a type that will permit the mixture to be spread and compacted at atmospheric temperature. Some cold-laid mixtures are produced hot.

ASPHALTIC CONCRETE: A plant mix of closely graded mineral aggregate and asphalt, designed and controlled to produce a mixture of high quality from the standpoint of both stability and durability. Such mixtures are usually produced hot with asphalt cement, but other types may be used as long as the resulting mixture is as described. It may be produced as base, binder, or surface courses.

SHEET ASPHALT: A hot plant mixture of sand, filler, and asphalt cement carefully proportioned and controlled to produce a sand-type mixture of high quality. It is usually used for surfacing courses.

SAND ASPHALT: A mixture of sand and asphalt that may be made with asphalt cement, emulsified asphalt, or other liquid type, mixed in a plant or by road-mix methods.

II. ORIGIN AND USE OF BITUMINOUS MATERIALS

Bituminous materials have been used by man since the dawn of history. Fossilized remains of plants and animals, preserved by being embedded in asphalt, such as those found in the Rancho-la-Brea asphalt pits of California, attest to the fact that asphalt was available on the earth long before the advent of man.

With respect to the use of asphalt and tar in pavement construction in more modern times, it has been established that the Incas of Peru in constructing an elaborate system of highways paved at least a portion of them with a type similar to modern penetration macadam. This occurred about 1500. Following the discovery of Val de Travers (Switzerland), Limmer (Germany), and Seyssel (France) asphalt deposits, these materials were used for foot pavements, first in Europe and then in the United

States.   Seyssel asphalt was used first for a foot pavement in Paris in 1835 and in London in 1836.   Coal tar was first used for constructing a tar-macadam pavement in Gloucester, England, between 1832 and 1838.   The first report of an asphalt deposit in the United States was made in 1837 and concerned solid and semisolid bitumens in Connecticut.   However, the first asphalt applied to paving in the United States was Seyssel rock asphalt, which was used for sidewalks in Philadelphia in 1838.

The first asphaltic road in modern times was constructed in 1852 from Paris to Perpignan, France, using Val de Travers rock asphalt in a macadam type of construction.   A short stretch of asphalt pavement was laid in Paris in 1854 by compressing rock asphalt.   This was the first construction in which a mixture requiring compaction was used.   Following this, the first large area of modern pavement of the sheet-asphalt type was laid in Paris, using a mastic of Val de Travers rock asphalt compressed to 2 in. in thickness and placed over 6 in. of concrete.   A similar pavement was laid in London in 1869.

The first asphalt roadway pavement in the United States apparently was a short experimental section composed of rock asphalt and laid in Newark, N.J., in 1870.   In 1871, pavements were laid in Washington, D.C., which were composed of crushed rock, sand, coal-tar pitch, and creosote oil.   Also, in 1871, 1872, and 1873, Val de Travers rock-asphalt pavements were laid in New York City.   In 1876, Congress passed an act authorizing the paving of Pennsylvania Avenue in Washington, D.C.   Part was paved with Val de Travers rock asphalt, and the remainder with sheet asphalt made with Trinidad asphalt.   Bermudez asphalt (Venezuela) was first used extensively for paving in Detroit, Mich., in 1892 and the following year in Washington, D.C.

Tank trucks or transports also are preferably insulated, but these are usually used for shorter and more direct hauls than railroad tank cars.   In this case, the transport is equipped with an oil burner for heating, which may be in operation during transit.

Asphalts are shipped by water in barges or tankers, frequently with a capacity of several million gallons.   This may also be true for tar, especially in the case where the tar is being imported.   Hot materials in such large quantities retain their heat over long periods of time.   Such asphalt shipments are made from Texas gulf ports to eastern Atlantic ports, including New England.   Asphalt is transported in this way from gulf ports via the Mississippi, Ohio, and Missouri Rivers to terminals located along these navigable streams.

Bituminous materials are also shipped in drums, usually of 55 gal capacity.   Liquid materials can be shipped this way easily, but semisolid materials present a problem of removal from the drum at the destination.   Some type of drum or barrel heater must be used.   Hard asphalts or pitches may be shipped in paper or cardboard containers.   In this case, the paper or cardboard is usually treated with clay, powder, etc., to prevent the asphalt from sticking and thereby facilitate the stripping of the container from the bituminous material.

III. MANUFACTURE OF ASPHALT

In the strictest sense of the word, asphalt is not manufactured, since the petroleum refiner uses principally physical means to separate asphalt from the crude oil in which it exists.   The essential process is one of fractional distillation wherein heat is used to boil off from the crude oil such distillates as naphtha or gasoline, kerosene, diesel oil, etc., and to leave behind as a residue various grades of road oil or asphalt, depending upon the degree to which distillates are removed as determined by the conditions of distillation.   In any event, the asphalt is always the residual material, and the terminology "residual asphalt" is used.   These residual asphalts of various grades may be further processed by air blowing, solvent extraction, blending, compounding, and admixing with other ingredients to make a great variety of asphalt products used in paving, roofing, waterproofing, paints, coating and sealing materials, and materials for hundreds of industrial applications.

The character of asphalt is dependent, to some degree, upon the nature of the petroleum or crude oil from which it is made.   Historically, the most desirable asphalts for paving purposes have been straight-run materials produced by the vacuum and steam reduction of asphaltic-base crudes.   However, modern refinery practice is a very versatile operation in which the refiner can subject materials to a variety of processes economically. As a result, asphalts can be produced from other than strictly asphaltic-base crudes, and residual materials from vacuum reduction may be improved by further processing. In the modern refinery a variety of blending operations is possible, and such processing as solvent extraction and air blowing may be used. The net result is the ability of the refiner to "tailor-make" an asphalt with specific properties and relative independence from the source of crude oil for this control.

With respect to paving asphalts, however, the user of these materials has not yet learned to take full advantage of the possibilities open to the refiner, primarily because of his inability to define in specific terms those properties which make for quality in the product. As a result, many engineers of the road fraternity, being unable to evaluate the effects of various refining methods, take the attitude that any processing that a refiner practices beyond simple recovery by fractional distillation is detrimental to the asphalt. It follows that these individuals are of the opinion that satisfactory paving asphalt can be made only from specific crude sources. Many of these individuals feel, also, that present-day paving asphalts are inferior to those produced in the past because of the modern refiner's ability to break down, blend, recombine, and to otherwise modify asphaltic materials. This situation results from rapid advances being made in the petroleum industry and the difficulty of finding a satisfactory substitute for the test of time.

This problem has been further complicated in recent years by promotional efforts to incorporate various rubbers, both natural and synthetic, and other elastomers into asphalts for paving purposes. It is certainly true that many of these materials will modify asphalt properties markedly, particularly with respect to temperature susceptibility, low-temperature ductility, and cohesive strength. Such additions, with the resulting asphalt modification they provide, probably have merit for certain uses, but like the processing available to the asphalt refiner, proof of this merit, particularly the economic justification of their use, is dependent upon a more precise definition of those asphalt qualities that are desired. Current technical literature attests to the fact that research is making rapid strides in providing answers to these problems.

This discussion of asphalt manufacture is presented because it is felt that a basic

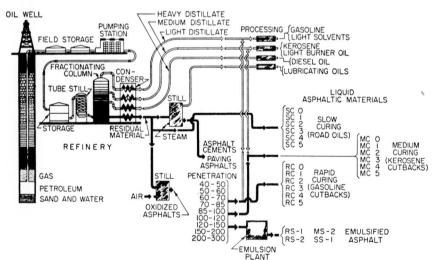

Fig. 9-2. Simplified flow chart showing recovery and refining of petroleum-asphaltic materials. [*Courtesy of The Asphalt Institute* (20).*]

---

* Numbers in parentheses refer to corresponding item in the references at the end of this section.

knowledge of the engineering processes used in asphalt manufacture is necessary to the understanding of asphalt as a material. While the subject is much too complex to expect the highway engineer to possess the knowledge of the petroleum engineer, nevertheless, the highway engineer can be helped greatly in his selection of materials for the job and in understanding their behavior in use by a rudimentary knowledge of petroleum processing. Also, it is only by this means that the highway engineer will be able to advance his understanding of asphalt as a material to the point where he will be better able to define quality in his specifications.

Figure 9-2 shows a simplified flow chart for the recovery and refining of asphaltic materials from petroleum or crude oil.

IV. ASPHALT CEMENTS

Asphalts are complicated colloidal systems of hydrocarbon material which from a chemical point of view are largely unknown. They have been arbitrarily divided into so-called asphaltenes, resins, and oils. These fractions can be separated by selective-solvent and absorption techniques. Asphaltenes are large, high-molecular-weight hydrocarbon molecules with a carbon-hydrogen ratio of more than 0.8. Resins, sometimes referred to as "cyclics," are hydrocarbon molecules with a carbon-hydrogen ratio of less than 0.8 and more than 0.6. Oils are hydrocarbon molecules with a carbon-hydrogen ratio of less than 0.4 (1).

It is generally considered that asphalt is a colloidal suspension of asphaltenes in an oily medium in which the resins act as peptizing agents to prevent coagulation of the asphaltenes. Thus the continuous medium is oil, with the properties of the asphalt dependent upon its nature, as well as upon the concentration of the disperse phase and its degree of dispersion. There is evidence that the amount and character of the cyclics or resins may be one of the most important factors. From the practical point of view, these factors are determined by crude source and kind of processing.

A. Character of Asphalts

Asphalts are thermoplastic materials since they gradually liquefy when heated. They are characterized by their hardness or viscosity and by the character of their rheologic, or flow, properties. Thus their properties may vary from brittle to rubber-like through all forms intermediate to these.

The hardness, or viscosity, of an asphalt is a function of its temperature, but the change in viscosity with temperature is a function of its character. In basic terms, an asphalt is characterized by the relationship between rate of shear, shearing stress, and time. In this way the material may be classified as to its shear susceptibility, which indicates the state of colloidal structure it possesses.

In the testing of asphalt cements, the industry has standardized on two convenient methods of test to characterize the material. These are the penetration and softening-point tests. The penetration is the depth that a standard needle penetrates the asphalt under standard conditions of load, temperature, and time. It measures hardness of the material and is used to designate grade. The softening point is the temperature at which the asphalt has a fixed consistency as determined by the conditions of the test.

1. **Viscosity-Temperature Relationships.** When the logarithm of the viscosity of an asphalt is plotted against temperature, an approximate straight line results (8). Figure 9-3 shows such a relationship for asphalts of various penetration grades made from one crude source (11).

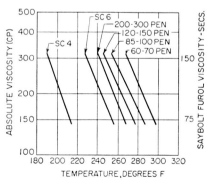

FIG. 9-3. Temperature-viscosity variations of asphalts. [*After Vallerga* (11).]

2. Temperature Susceptibility. An asphalt cement is characterized by the extent to which its consistency changes with temperature. This characterizing factor is called temperature susceptibility. It may be measured by comparing consistency measurements such as penetration or viscosity at two or more temperatures or by determining the penetration-softening-point relationship. The following relationships have been used:

1. Susceptibility factor $= \dfrac{\text{penetration } 77°F, \ 100 \ g, \ 5 \ sec}{\text{penetration } 32°F, \ 100 \ g, \ 5 \ sec}$

2. Susceptibility factor

$$= \frac{\text{penetration } 115°F, \ 50 \ g, \ 5 \ sec \ - \ \text{penetration } 32°F, \ 200 \ g, \ 60 \ sec}{\text{penetration } 77°F, \ 100 \ g, \ 5 \ sec}$$

3. $M = \dfrac{\log p_2 \ - \ \log p_1}{t_2 \ - \ t_1}$

where M = temperature susceptibility = slope of straight line expressing logarithm penetration vs. temperature

p_1 = penetration 100 g, 5 sec, at temperature t_1
p_2 = penetration 100 g, 5 sec, at temperature t_2

4. The straight-line relationship between the logarithm of penetration and temperature also has been used to determine a penetration index by including the softening-point temperature.

According to Pfeiffer and Van Doormaal (9), there is a close relationship between their penetration index and type of asphalt:

a. PI $-$ 2 to $+$2, normal residual-type asphalts.
b. PI greater than 2, blown type with low temperature susceptibility.
c. PI less than $-$2, pitch type with high temperature susceptibility like coal-tar pitches.

Where temperature susceptibility has been included in specifications, the method of control has been to specify penetration requirements at 32, 77, and 115°F. Low temperature susceptibility is desirable if it is not gained at the expense of other characteristics, since it means the asphalt will be softer at low temperature and harder at high temperature than otherwise would be the case.

3. Effect of Crude Source. The source of the crude oil from which a residual asphalt is made determines in part the temperature susceptibility of the product.

B. Specifications for Asphalt Cements

Almost all state highway departments and other user agencies have specifications for asphalt cements.

It will be noted that the standard specifications for asphalt cements do not provide for control of temperature susceptibility. In addition, some technologists feel that the requirements for loss on heating, including penetration loss, provide only for control of the asphalt as affected by heating in bulk form whereas, from a quality point of view, the critical factor in durability is the ability to resist hardening when heated in thin films. As a consequence, some consumer agencies have developed specifications they consider to be of an improved type in which these deficiencies of the standard specifications are eliminated. The State of California is one of these.

It will be noted that the California specifications provide control of temperature susceptibility by specifying a minimum penetration ratio and a range for Furol viscosity at 275°F for each penetration grade. An attempt is made to secure a quality product from the durability standpoint by use of a modified loss-on-heating test. Although time and temperature are the same in this test as in the standard one, the asphalt is spread out over a larger area in the California test so that the film thickness is about ⅛ in. Thus the modified test is more severe than the standard one. This test method was developed by the U. S. Bureau of Public Roads (10). In addition, the ductility requirement is placed on the asphalt after loss on heating rather than on the original material.

V. CUTBACK ASPHALTS AND ROAD OILS

Cutback asphalts and road oils are types of liquid asphaltic materials that are produced in various grades. They may be solutions of asphalt cement in a solvent (rapid-curing and medium-curing) or may be produced by straight-run distillation or blending (slow-curing).

These materials are widely used in bituminous construction of both the application and mixture type. They have the advantage, as compared with asphalt cements, that they can be used with cold aggregates and with a minimum of heat.

A. Types and Grades

Cutback asphalts and road oils are designated by type as rapid-curing (RC), medium-curing (MC), and slow-curing (SC). The rate of curing of these materials, or the type, is controlled by the nature of the solvent used to render the base asphalt liquid. In the case of the cutback materials, RC and MC, in particular, the liquefaction is intended to be more or less temporary. The object is to render the asphalt fluid for use without resorting to high temperatures, but to have the solvent leave more or less rapidly after the asphalt cement has been deposited in place.

A similar situation exists with regard to the SC, or road-oil, type of material, but in this case the asphalt deposited initially is very soft and cannot be classed as a cement. Curing is much slower than is the case for RC and MC materials. Whereas RC and MC materials cure primarily by volatilization of solvent, it is probable that SC materials cure mostly by oxidation, perhaps by a photo-oxidation process.

While cutback asphalts and road oils are typed on the basis of curing rate, it is equally or more important to recognize that these types of materials also differ markedly in the base asphalt they contain. Thus RC materials, in addition to being rapid-curing, contain a relatively hard base asphalt. The asphalt contained in MC materials is intermediate in hardness between that of RC and SC, but is more nearly like that of the RC material and is classed as an asphalt cement. These differences in base asphalt frequently are of more importance than curing rate in selecting a type of liquid asphalt for use.

These three types of liquid materials are made in various numbered grades that are distinguished on the basis of viscosity. In standard specifications there are six grades of RC and MC materials numbered from 0 to 5 and seven grades of SC material numbered from 0 to 6. Within each type of material the grades are numbered consistently with viscosity, the 0 grade being the lowest viscosity and the 5 or 6 grade the highest. Also, like-numbered grades have the same viscosity for all three types (RC, MC, and SC).

Viscosity is controlled by composition; the more asphalt, the greater the viscosity. Therefore, the lower-numbered grades contain the least asphalt and the most solvent, and the higher-numbered grades contain the most asphalt and the least solvent. In specifications, for purposes of practicability and accuracy in testing, the viscosity ranges of the various grades are specified at different temperatures. Although exact relationships are dependent upon specific materials because of temperature-susceptibility factors, the approximate viscosity ranges at 140°F for the various grades of liquid asphaltic materials are as shown in Fig. 9-4.

It will be noted that the upper limit of viscosity is always twice the lower limit for each grade and that a definite gap exists between grades. While it may seem that the viscosity ranges of the lower-numbered grades are not consistent with the viscosity ranges of the more viscous grades, this situation will be seen to be realistic if it is recognized that it takes a greater percentage of solvent to lower the viscosity of the 0 grade from 30 to 15 sec than it does to lower the viscosity of the 5 grade from 3,000 to 1,500 sec.

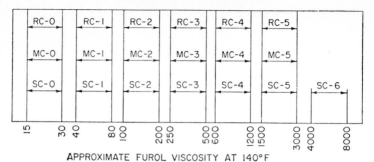

APPROXIMATE FUROL VISCOSITY AT 140°F

Fig. 9-4. Viscosity comparison of RC, MC, and SC liquid asphalts at 140°F.

VI. EMULSIFIED ASPHALTS

An emulsion is an intimate mixture of two immiscible liquids, one dispersed in the other in the form of minute droplets. The dispersed liquid is called the disperse phase, and the other the continuous phase. In the case of emulsified asphalts for road-building purposes, the asphalt is normally the disperse phase and water the continuous phase. Some asphalt emulsions are used in which asphalt is the continuous phase and water the disperse phase. The latter materials are referred to as inverted emulsions, and the asphaltic material they contain is usually a rapid-curing cutback. Asphalt emulsions may be simple or complex systems.

Asphalt emulsification is a means of liquefying an asphalt cement for use. It has advantages over hot asphalt in that it can be used with cold as well as with hot aggregate and with aggregate that is dry, damp, or wet. The ability to be used with wet aggregate in general is an advantage for this material over a cutback, asphalt as well. In addition, the emulsification process completely eliminates the fire and toxicity hazards which cutback asphalts possess. Asphalt emulsions may be formulated to deposit an asphalt film with better physical properties, such as lower temperature susceptibility (less brittle at cold temperatures and less fluid at high temperatures), as compared with the original asphalt cement. For this reason, among others, emulsified asphalts are sometimes used in hot- or semihot-mixture types (24).

Asphalt emulsions do not depend upon evaporation of solvent for their setting characteristic and therefore may set faster and more completely than cutback materials that cure out over a long period of time. However, the successful use of asphalt emulsions in most road-building applications does depend upon sufficient drying weather for removal of water. These materials also may be subject to washing by sudden rains if they are not properly formulated. With respect to washing by rain, however, emulsified asphalts can be formulated to be resistant to this action and to be weatherproof in this sense (25).

A. Manufacture of Emulsified Asphalts

Emulsified asphalts for use in road construction are made by mechanically dispersing an asphalt cement in liquid form (hot) in water which has been treated with an emulsifying agent. Dispersion of the asphalt is normally accomplished with some type of colloid or emulsifying mill that provides the necessary shearing action.

The emulsifying agent is a material of such molecular structure, usually containing polar and nonpolar groups, that the molecules will concentrate at the interfacial layer between the two liquids and orient themselves to give the required action. In the case of materials which ionize, the long-chain portion of the molecule providing the emulsifying action may be either negative or positive, and the resulting emulsion is then of either the anionic or cationic type, respectively. If the emulsifying agent does not ionize, the resulting emulsion is a nonionic one.

Common emulsifying materials are soaps, of which there are a great variety. Many water-soluble surface-active chemicals may be used as well as protein materials. Finely dispersed solids such as clay also are emulsifying agents. The stability of asphalt emulsions, either with regard to their ability to mix with aggregate or their storage stability, may be increased with the addition of substances called stabilizers. These may be water-soluble chemicals or dispersed solids.

Figure 9-5 shows a simplified flow diagram for emulsified asphalt manufacture.

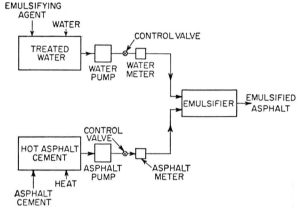

FIG. 9-5. Simplified flow diagram for manufacture of emulsified asphalt.

B. Types and Grades of Emulsified Asphalts

Emulsified asphalts are made in three basic types: rapid-setting (RS), medium-setting (MS), and slow-setting (SS). In manufacture, the type of emulsion produced is controlled principally by the type and amount of emulsifying agent used. The type and amount of stabilizing agent used also may have some bearing, as does the asphalt from which the emulsion is made and the particular processing used.

The grade of emulsified asphalt, within any given type, may be based upon variations in mixing stability, emulsion viscosity, or the hardness of the asphalt from which the emulsion is made. Mixing stability is controlled by choice of the quantity or type of emulsifying agent, and viscosity is controlled principally by the proportions between asphalt and water. Generally speaking, any asphalt cement with a hardness of 40–50 penetration or above may be emulsified, but the harder materials may be more difficult to process.

Rapid-setting types of asphalt emulsions are used in application types of construction such as penetration macadam, seal coats, and surface treatments of the application type. They are made in two grades which differ principally in viscosity. Commonly only one grade of SS emulsion is specified, although some agencies include a second grade which differs in that a harder base asphalt is required. These SS materials are used wherever their high mixing stability is required, as with aggregate containing a large amount of fine dust, in soil stabilization, or in special applications such as slurry seals.

Medium-setting (MS) types of asphalt emulsions characteristically are used in mixture types employing relatively clean coarse aggregate. In some areas they are used in open or graded hot-mix types. They also are used in application types such as seal coats, mat coats, retreads, and surface treatments. Many current specifications do not delineate this material from an SS type. In those states where the MS types of asphalt emulsion are adequately specified, such materials are widely used (24).

VII. ROAD TARS

Tars used in road building are materials refined by fractional distillation of crude tars that are produced from the destructive distillation of bituminous coal or by the cracking of petroleum. In this country in the past, and in other countries currently, bituminous coals were processed to produce both coke and gas as primary products. The equipment used and the processing involved varied, depending upon whether coke or gas was the chief product. The tars produced varied as well. However, with the widespread distribution of natural gas, local coal-gas plants (and the gas-oven tars they produced) have practically disappeared in this country.

A. Manufacture of Crude Tar (1, 27)

Practically all the bituminous-coal distillation carried out in the United States today is for the purpose of producing coke for the steel industry and other uses. Crude coke-oven tars are by-products of this processing. The other main source of crude tar is from the cracking of petroleum in the carbureted-water-gas process.

1. Coke-oven Tar. Coke-oven tar is produced when bituminous coal is heated in the absence of air so that it decomposes into volatile substances and coke. The volatile substances include gases, light oils, tars, and chemicals. The processing is done continuously by the use of a battery of high, narrow, brick-lined ovens, each holding a charge of 10 tons or more of coal which is heated to temperatures of 2500°F or higher. Properties of the tar vary with the coal processed, design of the oven, temperatures used, and pressure in the system as it affects the time volatile gases are exposed to incandescent coke.

Tar, being part of the volatile product, is collected with the gases that go to the by-product plant associated with the coke ovens. Here, the tar is removed from the gases by cooling and by scrubbing with ammonia liquor, itself a by-product of the operation. The tar and ammonia liquor are separated by decantation.

2. Water-gas Tar. In some areas not supplied by natural gas, gas is manufactured for municipal and industrial heating by the carbureted-water-gas process. In this process, carbon from coke used as fuel is combined chemically with hydrogen and oxygen formed when water, as steam, is decomposed by passing it through a bed of incandescent coke. Carbon monoxide and hydrogen gases are formed which are known as water gas. However, these gases are too low in Btu value for heating purposes, and hydrocarbon gases are added for enrichment.

The hydrocarbon gases are formed from the cracking of petroleum oils as they are passed over hot firebrick in a so-called carburetor. The cracking of the petroleum oils produces, in addition to hydrocarbon gases, a residue of heavy hydrocarbons which condenses to form tar. This tar is called water-gas tar if a light-fuel or gas-oil petroleum fraction is used, but if a heavier residue is used, the tar is referred to as a residuum tar. In either case, the crude tar must be refined for use.

3. Comparison of Coal and Water-gas Tars. Coal and water-gas tars are quite different materials, both from the standpoint of their chemical make-up and their physical characteristics. As compared with coal tars, water-gas tars are thinner in consistency and contain much smaller amounts of free carbon. The water associated with water-gas tar is practically free from ammonia compounds. Water-gas tars contain substantially more of the aromatic compounds benzene, toluene, and xylene than coal tars, while coal tars usually contain more pitch.

VIII. AGGREGATES FOR BITUMINOUS MIXES

The mineral aggregates most widely used in bituminous mixes are crushed stone, slag, crushed and uncrushed gravel, sand, and mineral fillers such as portland cement, stone dust, hydrated lime, and some suitable soils (28). Since the mineral aggregates usually constitute 88 to 96 per cent by weight, or approximately 80 per cent by volume of the total mix, their influence upon the final characteristics of bituminous mixes is very great. In this section aggregates and aggregate properties are treated with specific reference to bituminous mixes.   For a more complete coverage on aggregates, the reader is directed to Sec. 7.

A. Desirable Aggregate Characteristics

The choice of an aggregate for use in bituminous construction depends upon the aggregates available, their cost, and the type of construction in which they are to be used—which may range from a low-cost surface treatment or road mix to a high-type bituminous concrete.   However, regardless of the construction type, an ideal aggregate for use in bituminous construction should have the following characteristics (29):

1. Gradation and size appropriate to type of construction
2. Strength and toughness
3. Cubical shape
4. Low porosity
5. Proper surface texture
6. Hydrophobic characteristics

1. Gradation and Size.   One of the most important characteristics of an aggregate affecting the stability and working properties of a mix is the gradation.   Maximum aggregate size also has a great effect upon workability and density of bituminous mixtures.   It has been observed that the use of a maximum aggregate size greater than 1 in. in graded mixtures often results in harsh or unworkable bituminous mixtures that tend to segregate in the handling operations.   This results in pavement surfaces that have objectionable surface voids which are unsightly and may lead to raveling.

Aggregates may be divided on the basis of gradation as follows: (1) dense-graded, (2) open-graded, and (3) one-sized.   The dense-graded materials include appropriate amounts of all sizes from coarse to fine, including dust or material passing the No. 200 sieve.   They are used in hot-mix and other dense-graded types.

The open-graded materials may have an incomplete grading, or they may differ from dense-graded materials in that they contain much less material passing the No. 200 sieve.   These materials are generally used for road or plant mixes.

One-sized materials are, as the name implies, essentially one size.   These materials are generally used in macadams, surface treatments, and seal coats.

Dense-graded mixes tend to have a large number of points of contact between individual aggregate pieces resulting in high frictional resistance.   The increased number of contact points as compared with poorly graded materials also results in a greater area for load transfer from one aggregate to another.   This decreases the possibility of crushing of the individual aggregate piece by point loadings.

Logically, it might seem that the best method of attaining high stability in a bituminous mixture would be to use the densest aggregate gradation possible—to approach a solid as closely as possible by packing the maximum amount of aggregate into the given volume with just enough bituminous material present to bind the aggregate together.   The disadvantage of this concept is that it does not provide sufficient space for the bitumen necessary for durability of the mixture.   Numerous investigators have proposed aggregate gradations for maximum density.   One of the best known of these gradations is Fuller's curve.

The equation for Fuller's maximum density curve is

$$P = 100 \sqrt{d/D}$$

where P = total percentage passing
d = size of sieve opening considered
D = largest size (sieve opening) in gradation

The basic idea of the theory is that the amount of material of a given size should be just sufficient to fill the voids between aggregate of larger sizes.   In actual practice, the smaller particles tend to wedge between the larger ones, increasing the voids that must be filled by the smaller sizes.   As a result, maximum densities are actually produced by gradations having an excess of the small sizes as compared with the theoretical amounts.

Figure 9-6 was prepared to compare a typical gradation suggested by the Asphalt Institute (20) with Fuller's theoretical maximum-density curve. This figure shows that the recommended gradations for dense-graded mixes (asphaltic concrete) correspond quite well with the theoretical curve. The largest deviation is in the greater amounts of the smaller sizes, which are a necessity for maximum density in actual practice.

Aggregate gradation specifications for bituminous mixtures have been developed through actual field experiences which show that certain gradations are more satisfactory than others when construction types and thicknesses, traffic demands, subgrade conditions, durability, types of aggregate used, and economic factors are considered (30). In many cases aggregate gradation specifications have been the result of trial and error.

Laboratory testing has shown rather conclusively that aggregate gradation is a very important factor in the stability of bituminous mixes and that those gradations which

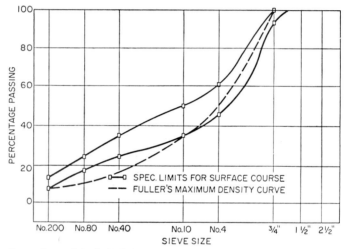

FIG. 9-6. Comparison of Asphalt Institute recommended gradation for asphaltic concrete with Fuller's curve.

produce the best laboratory strength approach the theoretical gradation for maximum density (31). However, field tests do not always support this concept, and recent research has suggested that laboratory strength measurements may not always give realistic values (32). An important contribution to aggregate gradation has been the modification of the theoretical gradation curves to fit actual aggregate characteristics (33, 34).

Aggregate gradation specifications have been developed because of the need to control construction in order to get a desired end product, to obtain the best utilization possible of locally available aggregate, and to effect reduction in cost through the standardization of sizes.

2. Strength and Toughness. The aggregate in a bituminous mixture supplies most of the mechanical stability. It must support the loads imposed by traffic and at the same time distribute these loads to the subbase at a reduced intensity. The aggregates used in bituminous mixes tend to be broken or degraded by the loads imposed upon them, both during construction and later by the action of traffic. Degradation may take place by compressive failure from concentrated loads at points of contact between aggregate particles and by abrasive action when the individual pieces move with respect to the others. The amount of degradation is affected by both the magnitude of the applied loads and the resistance to crushing or abrasion of the aggregate.

One major cause of degradation results from the compaction operation during construction. This is especially true in the construction of surface treatments and seal coats. Aggregate breakdown in macadam can cause overchoking of the voids, which leads to a loss of stability. During construction, the problem of degradation under steel rollers has been overcome to some extent by the use of pneumatic-tired rollers.

In an extensive investigation it was found that degradation of aggregates was a function of the original grading (35). It has been recommended that grading limits should depend upon the crushing characteristics of the aggregates (36). In dense-graded mixes, if the aggregate used has a reasonable crushing strength, little degradation should occur. However, in mixture design, an allowance for some degradation should be made in those cases where it is necessary to use softer materials. Open gradations are more subject to degradation than dense ones. A danger from degradation in open mixes is the possibility of increasing the surface area to the point where insufficient bituminous material is present for proper coating. To prevent excessive degradation, the aggregates used for open-graded mixes should have a greater crushing resistance than aggregates used for dense-graded mixes.

Both crushed and uncrushed gravels have been used successfully in bituminous mixtures. When uncrushed gravels are used, extra care must be taken to control the gradation and the bitumen content.

B. Specific Gravity of Aggregates

The specific gravity of aggregates is quite important from the standpoint of mixture calculations for the determination of void contents in compacted bituminous mixtures. There are three generally accepted values of specific gravity for aggregates: (1) apparent, (2) bulk, and (3) effective.

1. Apparent Specific Gravity. By referring to Fig. 9-7, apparent specific gravity of an aggregate as defined by ASTM (12) equals the weight of the solids divided by the sum of the volume of the solids and the volume of the impermeable voids and by the unit weight of water (γ_ω):

$$\text{App. sp. gr.} = \frac{W_s}{(V_s + V_i)\gamma_w}$$

Those voids that undergo essentially no change in moisture content during either prolonged soaking in water or drying are referred to as impermeable voids. Permeable voids refer to those voids whose moisture content changes readily with soaking in water or with drying.

2. Bulk Specific Gravity. By referring to Fig. 9-7, bulk specific gravity of an aggregate as defined by ASTM equals the weight of the solids divided by the sum of the volume of the solids and the volume of the permeable and impermeable voids and by the unit weight of water (γ_ω):

$$\text{Bulk sp. gr.} = \frac{W_s}{(V_s + V_i + V_p)\gamma_w}$$

The choice of aggregate specific gravity used in mixture calculations has an appreciable effect upon the calculated amount of air voids in a compacted bituminous mixture, unless the aggregate is nonabsorptive, in which case the apparent and bulk specific gravity have the same value. The actual specific gravity of the aggregate in the mixture depends upon the manner in which a bitumen will penetrate the permeable voids. If the apparent specific gravity is used, this means that the permeable voids are assumed to be filled with bitumen to the same extent as they are with water in 24-hr soaking. If the bulk specific gravity is used, this means that the asphalt is assumed not to penetrate into the permeable voids. The actual case of an aggregate that exhibits the normal degree of bitumen absorption is the case where the permeable voids are filled to some extent with bitumen. This gives rise to the term "effective specific gravity."

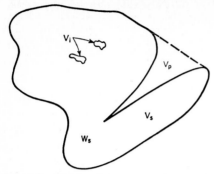
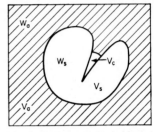

FIG. 9-7. Sketch of an aggregate particle showing distribution of various weights and volumes.

W_s = weight of solids; V_s = volume of solids; V_p = volume of permeable voids; V_i = volume of impermeable voids; γ_w = unit weight of water.

FIG. 9-8. A graphical representation of an aggregate particle submerged in asphalt.

W_a = weight of asphalt; V_a = volume of asphalt; W_s = weight of solids; V_s = volume of solids; V_c = volume of voids not filled with asphalt; S_m = specific gravity of mass; S_a = specific gravity of asphalt; γ_w = unit weight of water.

3. Effective Specific Gravity. By referring to Fig. 9-8, the concept of effective specific gravity can be represented as

$$S_e = \text{effective specific gravity} = \frac{W_s}{(V_s + V_c)\gamma_w}$$

or

$$S_e \gamma_w = \frac{W_s}{V_s + V_c} \tag{9-1}$$

$$S_m = \frac{W_a + W_s}{(V_a + V_s + V_c)\gamma_w} \tag{9-2}$$

$$S_a = \frac{W_a}{V_a \gamma_v} \tag{9-3}$$

If W_a and W_s are weight percentages such that

$$W_a + W_s = 100\% \tag{9-4}$$

substituting Eqs. (9-1), (9-3), and (9-4) into Eq. (9-2) gives

$$S_m = \frac{100}{\dfrac{100 - W_a}{S_e} + \dfrac{W_a}{S_a}}$$

and solving,

$$S_e = \frac{100 - W_a}{100/S_m - W_a/S_a}$$

C. Methods of Combining Aggregates to Meet Grading Requirements

The combining of two or more aggregates of different gradings to produce a blended aggregate that will meet certain grading specifications is a common problem in construction. The determination of the relative amounts of the various aggregates required to meet a definite gradation specification is a problem in proportioning, and it can be solved by a number of methods.

The simplest method of determining proportions is generally the trial-and-error method. The individual aggregate gradations are examined, and a certain percentage of each aggregate is chosen. A few simple calculations will soon determine if this choice of percentages was suitable.

The second method of combining aggregates to produce a certain gradation is the graphical method, in which the gradations of the aggregates are plotted. This is a little more time-consuming than the first method, but it has the added advantage of picturing all of the possible combinations of the different aggregates. It is very simple to pick the most satisfactory combinations with a minimum of calculations.

1. **Trial-and-error Method of Blending Aggregates.** Table 9-1 gives the sieve analysis of each of three aggregates, along with the gradation specifications.

In trying to meet a specification range, start with the average of the limits as listed in column 5. Upon examination of this particular situation it can be seen that all material coarser than the No. 4 sieve must come from aggregate A. In 100 lb of blended aggregate, 53 lb must be coarser than the No. 4 sieve. Checking aggregate A, it can be seen that 81 per cent is retained on the No. 4 sieve. Thus, in order to obtain 53 lb of material coarser than the No. 4, take 53/0.81 = 66 lb of aggregate A.

Now examine the gradings with respect to the material passing the No. 200 sieve. In the 100 lb of blend, 7 lb of material passing the No. 200 sieve is needed. There is some −200 material in all three aggregates. In the 66 lb of aggregate A there is 66 × 0.02 = 1.2 lb of material passing the No. 200 sieve. Examine aggregate B and assume 0.5 lb of −200 material would come from this source. This leaves 7 − 1.2 − 0.5 = 5.3 lb of −200 material which must come from aggregate C. Only 88 per cent of aggregate C passes the No. 200 sieve, so 5.3/0.88 = 6 lb of aggregate C that must be taken. The amounts of aggregate A and C have now been established. By difference it is possible to establish the amount of aggregate B. Since 100 lb of blend was desired initially, 100 lb − 6 lb aggregate C − 66 lb aggregate A = 28 lb aggregate B.

Combining 66 lb of aggregate A, 28 lb of aggregate B, and 6 lb of aggregate C would give a blend having the following gradation:

Table 9-1. Sieve Analysis of Three Aggregate Samples Used in Blending Example

(Total percentage passing)

Sieve size	(1) Agg. A	(2) Agg. B	(3) Agg. C	(4) Specification range	(5) Specification average
1 in.	100	100	100	95–100	98
½ in.	63	100	100	70–85	77
No. 4	19	100	100	40–55	47
No. 10	8	93	100	30–42	36
No. 40	5	55	100	20–30	25
No. 80	3	36	97	12–22	17
No. 200	2	3	88	5–10	7

(Total percentage passing)

Sieve size	Blend	Specification average
1 in.	100	98
½ in.	75.6	77
No. 4	46.5	47
No. 10	37.3	36
No. 40	24.7	25
No. 80	17.9	17
No. 200	7.4	7

The normal state of affairs would require two or more trial solutions in order to obtain the desired grading. This particular example was simplified due to the fact that all material retained on the No. 4 sieve had to come from one source. However, the procedure for attacking a more complicated problem is exactly the same as for the example given. There are generally one or two sieve-size fractions in the gradations that will give an excellent clue as to amounts of various aggregates to choose. It usually pays to examine the gradation closely before starting calculations.

2. Graphical Method of Blending Aggregates. When it is required to blend three aggregates to obtain a specified gradation, the problem may sometimes be attacked best by a graphical solution. One type of graphical solution which consists of two sets of coordinate graphs with appropriate scales is outlined as follows (refer to Table 9-1 for aggregate gradations and specifications limits):

PART I. Plot the gradation (percentage passing) of aggregate A on scale IV and the gradation (percentage passing) of aggregate B on scale III in Fig. 9-9.

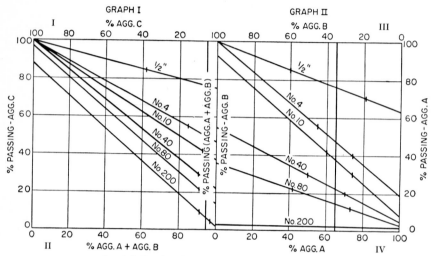

FIG. 9-9. Graphical method of blending aggregates.

PART II. Connect by a straight line drawn between scales III and IV the percentage passing for each respective sieve size. Observe now that any vertical line through graph II in Fig. 9-9 will define the gradation of the blend composed of the relative amounts of aggregate A and aggregate B as shown on the horizontal scales (top and bottom).

PART III. In graph II, place tic marks on each sieve line to indicate the specification limits for that particular sieve.

PART IV. Now choose a vertical line that will strike the best average between the specification limits indicated by the tic marks on each individual sieve line. In this example, the vertical line selected represents 65 per cent aggregate A and 35 per cent aggregate B.

PART V. Project horizontally the intersections of each sieve line with the selected vertical line (65 per cent aggregate A and 35 per cent aggregate B) to scale II on graph I. The values projected on scale II represent the percentages passing the respective sieves for an aggregate blend composed of 65 per cent aggregate A and 35 per cent aggregate B.

PART VI. Plot the percentage passing of aggregate C on scale I and repeat the process as outlined in Parts II to IV to determine the final proportions for blending aggregate C with the combination of aggregates A and B. In this example, the vertical line chosen represents 5 per cent aggregate C and 95 per cent of aggregates A and B, or

$$
\begin{array}{llll}
\text{Agg. C} & & = & 5\% \\
\text{Agg. B} & = 0.95 \times 35\% & = & 33\% \\
\text{Agg. A} & = 0.95 \times 65\% & = & \underline{62\%} \\
& & & 100\%
\end{array}
$$

PART VII. Project horizontally the intersections of each sieve line with the selected vertical line (5 per cent aggregate C and 95 per cent aggregate A plus aggregate B) to scale I. The numerical values obtained in this manner represent the sieve analysis of an aggregate blend composed of 62 per cent aggregate A, 33 per cent aggregate B, and 5 per cent aggregate C. Reading from scale I, the following gradation for the blend was determined:

Sieve size	Percentage passing	Specification range
1 in.	100	95–100
½ in.	77	70–85
No. 4	50	40–55
No. 10	41	30–42
No. 40	27	20–30
No. 80	19	12–22
No. 200	7	5–10

It can be seen that the gradation of the blend falls within the required specification range. However, the material passing the No. 10 sieve is approaching the upper limit. From this initial plot it would be a simple matter to alter the various amounts of the different aggregates and to determine a blend whose gradation is closer to the average values of the specification ranges.

A large number of different types of blending problems could be presented. Their solutions, however, would, in general, be approached by one of the methods previously outlined.

IX. DESIGN OF BITUMINOUS MIXTURES

In considering the design of bituminous mixtures, it is necessary to be aware of the desirable properties that should be incorporated in a paving mixture. The final design should result in a mixture with satisfactory workability and one with the best possible combination of the following properties: stability, durability, flexibility, and skid resistance.

Stability refers to the ability of the paving mixture to offer resistance to deformation under sustained or repeated loads. This resistance to deformation generally manifests itself in two forms: frictional or interlocking resistance and cohesive resistance. Frictional or interlocking resistance develops in the aggregate framework. The magnitude of *frictional resistance* varies directly with the area of aggregate contact and applied pressure and with the surface roughness of the aggregate particles. Frictional resistance is not dependent upon loaded area, rate of load application, or temperature. *Interlocking resistance* is dependent upon aggregate size and shape. Cohesive resistance generally develops in the bituminous-binder portion of the paving mixture. Its magnitude varies directly with rate of loading, loaded area, and viscosity of bitumen. Cohesive resistance varies inversely with temperature.

Durability refers to the resistance of a paving mixture to disintegration by weathering or the abrasive forces of traffic. The disintegration from weathering manifests itself in two main forms: raveling, pitting or potholing, and cracking of the paving surface. The most important factor with respect to durability is the amount of bitumen.

In order to prevent a mixture from raveling and potholing, it must be formulated for resistance to the action of moisture. For such to be the case, the bitumen must show a high degree of adhesion for the aggregate surface. The aggregate-bitumen combination should be such that the aggregate is easily wetted by the bitumen. Asphalt viscosity and film thickness have a major effect upon the difficulty of displacing a bituminous film with water. In general, the more viscous the bitumen and the thicker the film, the more difficult it is for water to displace it. The matter of water resistance, use of additives to improve adhesion, and the measurement of stripping resistance are discussed in more detail in Art. X, dealing with additives. Raveling also may occur because of the bitumen becoming hard and losing its cementing qualities through weathering.   Resistance to such action is a function of both quantity and quality of the bitumen used.

Cracking of the paving surface may be due to the bituminous mixture being brittle. This brittleness occurs when the bituminous material used in the mixture has insufficient ductility at low temperatures either initially or because of weathering.   Brittleness also may result from insufficient bitumen in the mixture.   Cracking may occur because of a combination of factors: the shrinkage which takes place during a volume change because of a lowering of temperature and the low tensile strength of the binder. The volume of a bituminous material decreases with a decrease in temperature. Brittleness in a paving mixture may develop because of steric hardening of the binder or the oxidation or polymerization of the bituminous film from exposure to air and sunlight.

To resist the abrasive action of traffic, a bituminous mixture must have sufficient bitumen to hold aggregate particles securely.   The bitumen must have sufficient ductility, and the aggregate itself should have adequate toughness and resistance to abrasion.

Flexibility refers to the ability of a bituminous mixture to bend repeatedly without cracking and to conform to changes in shape of the base course.   In order for a paving mixture to have the desired flexibility, it is necessary that the proper amount and grade of asphalt be incorporated.   In general, open aggregate gradings produce more flexible mixtures than do dense gradings.   As the thickness of the paving layer increases, the flexibility decreases.   A paving mixture becomes more flexible as the temperature increases.   The flexibility required in a paving mixture is a function of the resiliency of the structure of which it is a part and the number and weight of applied loads.

Skid resistance refers to the ability of the paving surface to offer resistance to slipping or skidding of automobile tires.   Excess asphalt, which results in a surface condition called "bleeding," decreases skid resistance.   The aggregate used in a paving mixture plays the dominant role in skid resistance.   The surface texture of the pavement also determines skid-resistant properties.   Dirt and other debris on a paving surface lower the skid resistance.   For a discussion of skid resistance, the reader is directed to Sec. 10.

A. Function of Bitumen and Aggregate

The bitumen is used in a bituminous paving mixture principally as a binding and a waterproofing ingredient.   When an excessive amount of bitumen is present, it may enhance the waterproofing function, but it then ceases to act as a binder and tends to lubricate the aggregate mass in such a manner that stability is seriously lowered.   In acting as a binder, the bitumen tends to keep the aggregate particles in position as they were distributed during construction.   This is one form of offering resistance to displacement under the action of traffic.   The bitumen present in the mixture allows the deformed paving layer to rebound upon removal of the load and to once again offer resistance to deformation from additional loads.   The bitumen can offer considerable shear resistance if loads are applied rapidly.

Under normal conditions, a bitumen is quite plastic in character and will flow under a small load.   Therefore, the major share of the resistance to deformation must be borne by the aggregate framework present in the bituminous paving mixture.   The aggregate framework has a great effect on the flexibility of a bituminous paving mixture.   A weak aggregate framework will deform easily under load, while a rigid aggregate framework may result in brittleness and weakness under impact loads.

The aggregate framework can be varied considerably by altering the gradation (see Sec. 7 and the part of this section dealing with aggregate grading).   One form of aggregate framework results when the aggregate is uniformly graded.   This means the aggregate particles are of a uniformly decreasing size from coarse down through fine and including dust.   Under these conditions, a highly stable aggregate framework may be developed because of the close packing of the various sizes (Fig. 9-10*a*).

Aggregates may be gap-graded, which means some part of the total aggregate grading is entirely missing or seriously deficient. One form of gap grading results when the coarse aggregate is present in such quantity that the total aggregate framework developed has a stability similar to that of the coarse aggregate alone (see Fig. 9-10b). A second form of gap grading results when not enough coarse aggregate is present to form a framework. In this case, the coarse-aggregate pieces tend to float in a matrix composed of bitumen and fine aggregate, and the stability of the bituminous mixture takes on the properties of the fine aggregate (see Fig. 9-10c).

B. Determination of Design Bitumen Content

In most cases, the most important factor in bituminous-mixture design is the determination of the design bitumen content. In the over-all analysis this requires a compromise between the stability and durability requirements of the bituminous mixture. A bituminous mixture is resistant to the action of air and water in direct proportion to the degree that they are kept out of the mixture. If the voids in the aggregate mass are completely filled with bitumen, the entrance of water and air is reduced to a minimum. A mixture of this type, even though it might provide the ultimate in durability, would be quite undesirable from the standpoint of stability. When placed in the field, the stability would be low and rutting and shoving would occur. Under the action of traffic, bleeding or flushing of bitumen to the pavement surface would take place. This would result in an unsightly appearance and a surface with low skid resistance.

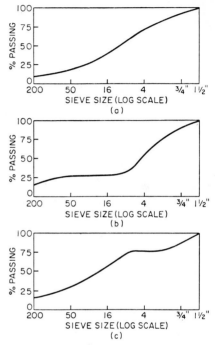

FIG. 9-10. Dense and skip-graded aggregate curves.

Maximum stability is not reached in an aggregate mass until the film thickness of bitumen reaches some critical value. Additional bitumen then tends to act as a lubricant rather than a binder, and the stability of the mixture is reduced. It is good practice, however, to keep the bitumen content as high as possible without sacrificing needed stability. Under these circumstances, the best compromise between durability and stability requirements is reached.

1. Aggregate-surface-area Method. The starting place in the aggregate-surface-area method of mixture design is with the concept that the bitumen content of a bituminous mixture is a function of the surface area of the aggregate to be covered. Following this concept, a number of simple, empirical expressions were developed by numerous state highway departments from their experience.

One of the early bituminous-mixture designs based on aggregate surface area was established by McKesson and Frickstad of the California Highway Department. This method is as follows:

$$P = 0.015a + 0.03b + 0.17c$$

where P = percentage of bitumen in the mix by weight
a = percentage of aggregate retained on No. 10 sieve
b = percentage of aggregate passing No. 10 and retained on No. 200 sieve.
c = percentage of aggregate passing No. 200 sieve

The numerical values preceding a, b, and c are a measure of the surface capacity of the various aggregate sizes. Since this expression uses only two sizes of sieves, it is possible to have a wide variation in surface area within the same gradation. The formula was established, though, for local conditions in California and for materials that were generally well graded. The bitumen used was a narrow range of slow-curing oils. This formula was varied widely in applying it to other aggregates and other conditions.

There are a number of factors that could affect surface-area formulas. They are listed as follows: (1) specific gravity of the aggregate, (2) specific gravity of the bituminous material, (3) absorption of the aggregate, (4) surface texture and shape of the aggregate, (5) character of the material passing the No. 200 sieve, (6) type of bituminous material, and (7) amount of voids in the aggregate structure. A formula devised by the State Highway Department of Nebraska attempted to take into account a number of these factors. The Nebraska formula is as follows:

$$P = AG(0.02a) + 0.06b + 0.10c + Sd$$

where P = percentage of bitumen residue by weight of mixture at time bituminous mixture is laid

A = absorption factor for aggregate retained on No. 50 sieve

G = specific-gravity correction factor of aggregate retained on No. 50 sieve; $G = 2.62$/apparent specific gravity of aggregate

a = percentage of aggregate retained on No. 50 sieve

b = percentage of aggregate passing No. 50 and retained on No. 100 sieve

c = percentage of aggregate passing No. 100 and retained on No. 200 sieve

d = percentage of aggregate passing No. 200 sieve

S = factor which depends upon character of material passing No. 200 sieve

This expression is more complicated than the McKesson-Frickstad formula, but it does have much wider application since it uses an additional sieve. The accuracy of the surface-area determination is a function of the number of sieves used. The bitumen demand of the material passing the No. 200 sieve is quite high. Therefore any method that includes the character of this fraction is an improvement. Present-day uses of surface-area formulas are for preliminary estimates of amount of bitumen needed in high-type bituminous paving mixtures and for determining amounts of bitumen to be used in low-cost roads. In general, these formulas should be used only in those cases where facilities for application of better design procedures are not available.

2. Voids Considerations for Determining Design Bitumen Content. According to the voids theory of bituminous-mixture design, the bitumen content is a function of the available space in the compacted aggregate framework or of the voids in the aggregate structure. Based on this concept, the design bitumen content is that which permits filling of the voids, when the aggregate is compacted, to such an extent that there will be space remaining for expansion of the bitumen on hot summer days and for the bitumen if further densification is expected under traffic. It should be evident that certain factors such as aggregate size, shape, and gradation will have quite an effect on the amount of voids developed in the aggregate mass and therefore on the amount of bitumen that the mixture can carry.

If the voids concept of design is applied to the determination of a satisfactory asphalt content for a very-open-graded mixture, the asphalt content so determined would tend to be too large because such aggregate gradations cannot carry in film form enough asphalt to fill the available void space. Such mixtures gain durability from thick asphalt films and do not depend entirely upon imperviousness. However, the axiom that trouble will result if available void space is overfilled with asphalt is applicable to mixes of all types. Voids concepts of design can best be used in the area of dense-graded mixtures, where asphalt cement is to be used as the binder.

Considering the concepts presented, it is apparent that an upper and lower limit of criteria for voids in the bituminous mixture must exist. For dense mixtures, the upper limit generally is set at 5 per cent and is established on the basis of durability considerations. Holding the upper limit to 5 per cent keeps the porosity to a reasonable limit such that the action of air and water will not be objectionable. The lower limit, generally set at 3 per cent, is established on the basis of allowing space for expansion of the binder on hot summer days and for some additional compaction of the paving mixture under the action of traffic.

The voids theory is not used to any great extent by itself as a means of determining bitumen content. Its major use is in providing for the durability factor in connection with various design procedures that utilize a stability test of some type. This should be apparent from a study of Arts. III and V in this section.

C. Volume-Weight Relationships in Bituminous Mixtures

It is necessary to have a working knowledge of the fundamentals involved in volume-weight relationships in order to handle the required mix calculations in the various bituminous-mixture design procedures. In this discussion a minimum of equations are presented since it is the feeling of the authors that it is much better to understand the basic concepts and to remember a few definitions of terms rather than to try to remember a large number of special equations. The same concepts presented here are used in the fields of soil mechanics and portland cement concrete mixes.

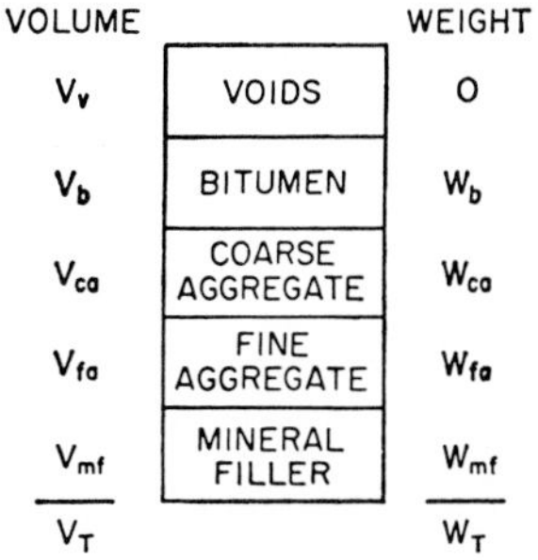

VOLUME		WEIGHT
V_v	VOIDS	O
V_b	BITUMEN	W_b
V_{ca}	COARSE AGGREGATE	W_{ca}
V_{fa}	FINE AGGREGATE	W_{fa}
V_{mf}	MINERAL FILLER	W_{mf}
V_T		W_T

FIG. 9-11. A graphical division of weights and volumes in a compacted bituminous specimen.

In order to picture volume-weight relationships, it is recommended that a block diagram be constructed for each set of calculations such as that shown in Fig. 9-11. Consider the following mixture:

	Specific gravity	Mixture composition, by weight, %
Bitumen....................	0.989	3.9
Coarse aggregate.............	2.63	72.1
Fine aggregate..............	2.72	17.0
Mineral filler...............	2.90	7.0
Total......................		100.0

A sample of this mixture is compacted into a specimen, and the following bulk-density data obtained:

Plain uncoated specimen:

Height	=	6.35 cm
Diameter	=	10.16 cm
Weight of compacted specimen in air	=	1,220 g
Weight of compacted specimen in water	=	705 g

Paraffin-coated specimen:

Weight of compacted, coated specimen in air	=	1,268 g
Weight of compacted, coated specimen in water	=	700 g
Specific gravity of paraffin	=	0.90

In the following presentation, the V and W refer to volume and weight, S refers to specific gravity. The subscripts identifying each V, W, and S refer to the specific ingredient used in the mixture as indicated in Fig. 9-11. γ_w is the unit weight of water, which in the foot-pound-second system is 62.4 lb per cu ft. In the centimeter-gram-second system γ_w equals unity. Thus specific gravity $S = W/V\gamma_w$.

The total weight of the compacted sample:

$$W_T = W_{mf} + W_{fa} + W_{ca} + W_b$$

The total volume of the compacted sample:

$$V_T = V_{mf} + V_{fa} + V_{ca} + V_b + V_v$$

or expressing the volumes of the individual ingredients in terms of their respective weights (W) and specific gravities (S), one obtains

$$V_T = V_v + \frac{W_b}{S_b \gamma_w} + \frac{W_{ca}}{S_{ca} \gamma_w} + \frac{W_{fa}}{S_{fa} \gamma_w} + \frac{W_{mf}}{S_{mf} \gamma_w}$$

1. Bulk-density Determinations. The bulk density of a compacted bituminous specimen γ_m is calculated by determining the ratio between its weight in air W_T and its total volume V_T and multiplying this quotient by the unit weight of water γ_w. The bulk density of a specimen may be determined by three different procedures. The procedure chosen depends upon the shape and surface texture of the specimen.

A specimen that has a dense, smooth surface may have its bulk density γ_m evaluated by method I. The total volume of the specimen is evaluated as the difference between its weight in air and its weight (uncoated) in water.

For uncoated specimens:

$$\gamma_m = \frac{W_T}{V_T} = \frac{W_{SA}}{W_{SA} - W_{SW}}$$

where γ_m = bulk density
W_{SA} = weight of specimen in air, g
W_{SW} = weight of specimen in water, g
Calculation:

$$\gamma_m = \frac{1,220}{1,220 - 705} = 2.37 \text{ g per cc, or } 147.89 \text{ lb per cu ft}$$

Method II is used on specimens whose surface is quite open and porous. The volume of the specimen is the difference between its weight in air and its paraffin-coated weight in water.

For paraffin-coated specimens:

$$\gamma_m = \frac{W_T}{V_T} = \frac{W'_{SA}}{W'_{SA} - W'_{SW} - \dfrac{(W'_{SA} - W_{SA})}{S_p}}$$

where W'_{SA} = weight of specimen plus paraffin coating in air, g
W'_{SW} = weight of specimen plus paraffin coating in water, g
S_p = specific gravity of paraffin

Calculation:

$$\gamma_m = \frac{1,220}{1,268 - 700 - \dfrac{(1,268 - 1,220)}{0.9}} = 2.376 \text{ g per cc, or } 148.26 \text{ lb per cu ft}$$

Method III may be used on specimens that have a dense surface and whose dimensions are such that they can be accurately measured. The volume of the specimen is determined from the height and diameter of the specimen.

For dense specimens of exact dimensions:

$$\gamma_m = \frac{W_T}{V_T} = \frac{W_{SA}}{\pi d^2 h / 4}$$

where d = measured diameter, c
h = measured height, c
W_{SA} = weight of specimen in air, g

Calculation:

$$\gamma_m = \frac{1,220}{\dfrac{\pi(10.16)^2}{4} \times 6.35} = 2.372 \text{ g per cc, or } 148.01 \text{ lb per cu ft}$$

2. Maximum Theoretical Density. The maximum theoretical density of a compacted bituminous specimen can be depicted as the density value resulting from a specimen that is compacted such that no voids remain in the aggregate-bitumen mixture. From Fig. 9-11 it can be seen

$$\text{Maximum theoretical density} = \gamma_m' = \frac{W_T}{V_T - V_v} = \frac{W_T}{V_b + V_{ca} + V_{fa} + V_{mf}}$$

or

$$\gamma_m' = \frac{W_T \gamma_w}{\dfrac{W_b}{S_b \gamma_w} + \dfrac{W_{ca}}{S_{ca} \gamma_w} + \dfrac{W_{fa}}{S_{fa} \gamma_w} + \dfrac{W_{mf}}{S_{mf} \gamma_w}}$$

Calculation:

$$\gamma_m' = \frac{1,220}{\dfrac{0.039 \times 1,220}{0.989} + \dfrac{0.721 \times 1,220}{2.63} + \dfrac{0.17 \times 1,220}{2.72} + \dfrac{0.07 \times 1,220}{2.9}}$$

$$= 2.50 \text{ g per cc, or } 156 \text{ lb per cu ft}$$

3. Percentage Air Voids in Compacted Mixture. The volume of air voids or the void content of a compacted bituminous mixture is the difference between the total volume V_T of the specimen and the volumes of the ingredients, or from Fig. 9-11,

$$V_v = V_T - V_b - V_{ca} - V_{fa} - V_{mf}$$

Calculation:

$$V_v = \frac{1,220}{2.37} - \frac{0.039 \times 1,220}{0.989} - \frac{0.721 \times 1,220}{2.65} - \frac{0.17 \times 1,220}{2.72} - \frac{0.07 \times 1,220}{2.90}$$

$$= 27 \text{ cc}$$

$$\% \text{ air voids} = \frac{V_v}{V_T} \times 100 = \frac{27}{515} \times 100 = 5.2\%$$

4. Aggregate Voids. The voids in the aggregate mass are those voids in the compacted aggregate framework that are available for filling with bitumen. The volume of these voids is determined by subtracting the volume of the aggregate fractions from the total volume. From Fig. 9-11, it follows:

Volume of voids in the aggregate mass: $V_{vam} = V_T - V_{ca} - V_{fa} - V_{mf}$. A measure of the degree to which bitumen fills the voids in the aggregate mass is the expression $\dfrac{V_b}{V_{vam}} \times 100$.

Note: $V_{vam} = V_v + V_b$.

Calculation:

$$V_{vam} = \frac{1,220}{2.37} - \frac{0.74 \times 1,220}{2.65} - \frac{0.17 \times 1,220}{2.72} - \frac{0.07 \times 1,220}{2.90}$$

$$= 75 \text{ cc}$$

$$\% V_{vam} = \frac{V_{vam}}{V_T} \times 100 = \frac{75 \times 100}{1,220/2.37} = 14.6\%$$

$$\% \text{ voids in aggregate mass filled with asphalt} = \frac{V_b \times 100}{V_{vam}} = \frac{\dfrac{0.039 \times 1,220 \times 100}{0.989}}{75}$$

$$= 63.5\%$$

X. USE OF ADDITIVES AND RUBBER

In the bituminous field as a whole, both asphalts and tars are modified by adding to them a great variety of materials to form compounds for special uses. These include inorganic fillers and fibers, resins, oils, fatty acids, fatty-acid pitches, pigments, etc., to name only a few (1). However, in the paving industry, asphalts and tars used as road binders generally are modified only in two ways. One of these is by treatment with so-called "additives" to promote coating and adhesion; and the other is with additions of rubber.

A. Use of Additives

Additives are used in bituminous materials to aid the coating of aggregate, particularly when in a damp or wet condition, and to increase the resistance of the bituminous film to stripping by the action of water. In this connection, a discussion of factors affecting the coating and adhesion phenomena is pertinent. This is necessary to an understanding of the function of additives and to an appreciation of test conditions applied to the measurement of water resistance.

1. Coating and Adhesion (91). In order for a bituminous material to adhere to an aggregate, it is first necessary that the aggregate be wetted by the bituminous material. This can be accomplished only by bringing the two in close proximity where the forces of wetting can act. These distances are extremely small because the forces acting are molecular ones and their sphere of influence is of the order or magnitude of the size of the molecule itself. Films of dust or of moisture can effectively prevent this required closeness of contact, if the bituminous material is not capable of penetrating the dust film or of displacing the film of water.

It is possible for a bituminous material to coat an aggregate and yet not to wet it and therefore not to adhere to it. In this connection the physical and chemical properties of both the asphalt and the aggregate are of importance. Viscosity of the bituminous material is an important factor because low-viscosity materials coat and wet better than high-viscosity ones. However, materials with high viscosity are more difficult to displace by water than those with low viscosity because of their greater resistance to movement. Low surface tension of the liquid is an aid to wetting, but solids with high surface tension are more easily wetted than those with lower surface tension. Smooth, rounded aggregate is coated and wetted more easily, but angular, rough-textured aggregate is apt to retain a coating better once it is wetted and adhesion established. However, except for porous aggregates where the bond may be principally a mechanical one, the surface-chemical characteristics of the aggregate and the chemical nature of the hydrocarbon molecules play the dominant role in determining how tenaciously the bituminous film will be held. The aggregate has polarity in a surface-chemical sense, and that liquid which satisfies the energy demand of the surface to the greatest degree will show the best adhesion.

Aggregates vary greatly in their ability to hold a bituminous film in the presence of water, depending upon the type of aggregate or mineral and depending upon how they have been processed. Siliceous aggregates such as quartz and granites, characterized as being acidic in nature, show relatively poorer adhesion in the presence of water than those basic materials such as limestone and dolomite. Aggregate surfaces which have weathered may show adhesion characteristics that are quite different from the freshly crushed material. Adsorbed moisture films may be a determining factor in adhesion (92). With respect to the bituminous materials, these, too, vary in their ability to adhere to aggregate and, more importantly, vary in adhesiveness depending upon the particular aggregate with which they are used. Interfacial tension characteristics of the oil-water-aggregate system are most important. Of these factors. aggregate variables are often the controlling ones.

2. Additive Materials. The adhesion characteristics of bituminous materials are dependent upon their chemical make-up, and particularly upon the presence of so-called "dipole" molecules. These are materials which act as though they have a positive and negative end and therefore are able to satisfy the surface-energy demands of the aggregate. It follows that adhesion may be improved by introducing this type of molecule into the system. Another possibility is to change the surface-chemical characteristics of the aggregate.

Materials which have been used for improving adhesion include fatty acids, various soaps, particularly soaps of the heavy metals, basic dyes, fatty amines (high molecular weight), amine acetates and hydrochlorides, and amine soaps. The heavy-metal soaps have been applied to the aggregate by treatment first with a heavy-metal salt followed by a soluble soap to form a coating of water-insoluble soap on the aggregate which is better wetted by bitumen than by water. Basic dyes have been used for aggregate treatment as well.

It is generally considered that the practical approach to the problem is to treat the asphalt in those cases where the aggregate is to be coated wet or where resistance to stripping by water needs to be improved. Most materials being used are some type of cationic additive, added to the asphalt in amounts of 0.2 to 2.0 per cent by weight. These are frequently an amine or amine soap with a long-chain molecule containing a polar and a nonpolar end (93). Such materials orient themselves at the interface because one end is attracted to the nonpolar bituminous substance and the other to the polar surface. Materials of this kind are added to cutback asphalt to promote wet coating and to improve adhesion. Treatment of tar for adhesion purposes generally is not considered necessary in this country, but additive treatments are used in Great Britain to improve wet-coating and initial adhesion in tar-surface-treatment construction (94).

In recent years, additive treatments of asphalt cements have been used to improve the resistance of hot mixtures to stripping by the action of water. For this purpose a heat-stable additive is required. These are usually cationic materials formulated to be chemically stable so that they will not be destroyed by breakdown with heat or by reaction with components of the asphalt (93, 95).

B. Use of Rubber

High polymers or copolymers commonly referred to as rubbers have been used with bituminous materials to improve their properties as cementing materials for aggregate. It has been demonstrated by many studies that additions of these materials in amounts as little as 0.1 per cent, but more normally in amounts of the order of 1 to 5 per cent, materially affect the properties of bitumen (98–101). In general, the addition of rubber to bitumen increases the viscosity and softening point and decreases penetration. The resistance to flow is increased, and temperature susceptibility is decreased. Low-temperature ductility may be very materially increased by such additions. Effectively rubberized bituminous materials also show properties of great extensibility at high rates of movement and high resistance to such movement (99).

Natural, synthetic, and reclaimed rubber materials have been used (102). Early work was with crumbs and powders of these materials. However, since a major consideration affecting the use of such materials is the ease with which they may be incorporated, other means have been used. These include the formation of high-rubber-content pellets with a bituminous material; mineral powder, such as barytes (103), coated with rubber; and rubber latex. It has been found that adding rubber in latex or emulsion form is a very effective method (104). If added to hot asphalt cement, tars, or to cutback asphalt by this means, provision must be made for removing water. Asphalt emulsions may be rubberized by the direct addition of latex if the two emulsion systems are formulated to be compatible.

Much of the synthetic material used for rubberizing asphalt has been of the styrene-butadiene copolymer type formerly called GR-S, but now designated by the letters SBR. There are many different materials of this type, and their effects on asphalt vary with the particular molecular structure involved. Other synthetic types, including butyl, nitrile, and neoprene, have been used also (101).

Field trials of rubberized asphalt, until comparatively recently, were confined to mixtures of the bituminous-concrete type. This was because rubber additives raise the cost of the mixture substantially and it was felt that the economic justification would be found in the more expensive types of construction. However, since well-designed bituminous concrete has a life expectancy of many years, improvements effected by the addition of rubber may take many years to be demonstrated. With this realization, more recent field demonstrations have been with rubberized asphalts (cements, cutbacks, and emulsions) in seal coats and surface treatments.

In the surfacing of airfields for jet-aircraft operation, the bituminous mixture must be resistant to both the solvating action of jet fuel and the high temperature and erosion of the jet blast. Tars have been used in such construction because of their resistance to petroleum solvents. In order to improve the characteristic low temperature susceptibility of these materials, they have been rubberized for this use, usually with a rubber of the oil-resistant type (105).

REFERENCES

1. Abraham, Herbert: *Asphalts and Allied Substances*, 5th ed., vols. 1 and 2, D. Van Nostrand Company, Inc., Princeton, N.J., 1945.
2. McGovern, E. W.: "Accelerated Weathering of Road Tars," Symposium on Accelerated Durability Testing of Bituminous Materials, *ASTM, Special Technical Publication* 94, 1950.
3. Lee, A. R., and E. J. Dickinson: "The Durability of Road Tar," *Road Research Technical Paper* 31, Department of Scientific and Industrial Research Road Research Laboratory (London), 1954.
4. Thurston, R. R., and E. C. Knowles: "Asphalt and Its Constituents: Oxidation at Service Temperatures," *Industrial and Engineering Chemistry*, vol. 33, no. 3, 1941.
5. Tyler, O. R., W. H. Goetz, and C. Slesser: "Natural Sandstone Rock Asphalt," *Engineering Bulletin of Purdue University*, vol. 25, no. 1, 1941.
6. Vokac, Roland: "Petroleum Processing and Asphalt Manufacture," a seminar paper presented at Purdue University, Lafayette, Ind., June, 1956. (Unpublished.)
7. Benson, J. R., and C. J. Becker: "Exploratory Research in Bituminous Soil Stabilization," *Proceedings, Association of Asphalt Paving Technologists*, vol. 13, 1942.
8. Lewis, R. H., and J. Y. Welborn: "Report on the Physical and Chemical Properties of Petroleum Asphalts of the 50–60 and 85–100 Penetration Grades," *Proceedings, Association of Asphalt Paving Technologists*, vol. 11, 1940.
9. Pfeiffer, J. P.: *The Properties of Asphaltic Bitumen*, Elsevier Press, Inc., Houston, Tex., 1950.
10. Lewis, R. H., and J. Y. Welborn: "Report on the Properties of the Residues of 50–60 and 85–100 Penetration Asphalts from Oven Tests and Exposure," *Proceedings, Association of Asphalt Paving Technologists*, vol. 12, 1940.
11. Vallerga, B. A.: "Notes on the Design, Preparation, and Performance of Asphalt Pavements," University of California, Institute of Transportation and Traffic Engineering, Berkeley, Calif., 1953.
12. American Society for Testing Materials: *1958 Book of ASTM Standards*.
13. American Association of State Highway Officials: *Standard Specifications for Highway Materials and Methods of Sampling and Testing*, parts I and II, 1955.
14. Brown, Marshall, and Fred J. Benson: "Significance of Tests for Asphaltic Materials," *Texas Engineering Experiment Station Bulletin* 119, Agricultural and Mechanical College of Texas, 1950.
15. "Significance of Tests for Highway Materials—Basic Tests: Progress Report of the Committee on Significance of Tests for Highway Materials of the Highway Division," *Journal of the Highway Division of the American Society of Civil Engineers, Paper* 1385, vol. 83, no. HW4, September, 1957.
16. American Society for Testing Materials: "Symposium on the Durability Testing of Bituminous Materials," *ASTM Special Technical Publication* 94, 1950.
17. Oliensis, G. L.: "The Oliensis Spot Test as a Quality Test," *Proceedings, Association of Asphalt Paving Technologists*, vol. 26, 1957.
18. Benedict, A. H.: "Development of the Modified Quantitative Oliensis Spot Test," *Proceedings, Association of Asphalt Paving Technologists*, vol. 11, 1940.
19. Lewis, R. H., and W. J. Halstead: "Behavior of Asphalts in Thin-film Oven Test," *Public Roads*, vol. 24, no. 8, 1956.
20. The Asphalt Institute: *Asphalt Handbook*, Construction Series 81, 1958.

21. The Asphalt Institute, Pacific Coast Division: *Manual on Design and Construction of Asphaltic Roads and Streets*, 1952.
22. Griffin, R. L., T. K. Miles, and W. C. Simpson: "A Curing Rate Test for Cutback Asphalts Using the Sliding Plate Microviscometer," *Proceedings, Association of Asphalt Paving Technologists*, vol. 26, 1957.
23. Hoiberg, A. J.: "Residues of 100 Penetration from Slow Curing Road Oils by Vacuum Distillation and ASTM Procedures," *Proceedings, Association of Asphalt Paving Technologists*, vol. 17, 1948.
24. State Highway Department of Indiana: *Standard Specifications for Road and Bridge Construction and Maintenance*, 1957.
25. *Weather-proof Emulsified Asphalt*, K. E. McConnaughay Asphalt Plants and Processes, Lafayette, Ind., 1958.
26. Tyler, O. R.: "A New Method of Recovering Bitumen from Bituminous Emulsions," *Proceedings, Association of Asphalt Paving Technologists*, vol. 13, 1942.
27. *Where Tars Come From*, Koppers Company, Inc., Pittsburgh, Pa., 1949.
28. *Principles of Highway Construction as Applied to Airports Flight Strips and Other Landing Areas for Aircraft*, Federal Works Agency, Public Roads Administration, 1943.
29. Pauls, J. T., and C. A. Carpenter: "Mineral Aggregates for Bituminous Construction," Symposium on Mineral Aggregates, *ASTM Special Technical Publication* 83, 1948.
30. Benson, J. R.: "The Grading of Aggregates for Bituminous Construction," Symposium on Mineral Aggregate, *ASTM Special Technical Publication* 83, 1948.
31. Vokac, R.: "A Type of Aggregate Gradation for Most Suitable Asphaltic Mixtures," *Proceedings, Association of Asphalt Paving Technologists*, vol. 10, 1939.
32. Oppenlander, J. C., and W. H. Goetz: "Triaxial Testing of Open-type Bituminous Mixtures," *Proceedings, Association of Asphalt Paving Technologists*, vol. 27, 1958.
33. Hveem, F. N.: "Gradation of Mineral Aggregates for Dense-graded Bituminous Mixtures," *Proceedings, Association of Asphalt Paving Technologists*, vol. 11, 1940.
34. Campen, W. H.: "The Development of a Maximum Density Curve and Its Application to the Grading of Aggregates for Bituminous Mixtures," *Proceedings, Association of Asphalt Paving Technologists*, vol. 11, 1940.
35. Shelburne, T. E.: "Crushing Resistance of Surface Treatment Aggregates," *Engineering Bulletin* 76, vol. 24, no. 5, Purdue University, Lafayette, Ind., 1940.
36. Howe, H. L., and H. W. Hughes: "Cold Plant Mix or Hot Plant Mix," *Proceedings, National Asphalt Conference*, December, 1940.
37. MacNaughton, M. F.: "Physical Changes in Aggregates in Bituminous Mixes under Compaction," *Proceedings, Association of Asphalt Paving Technologists*, January, 1937.
38. Herrin, M., and W. H. Goetz: "Effect of Aggregate Shape on Stability of Bituminous Mixes," *Proceedings, Highway Research Board*, vol. 33, 1954.
39. Lottman, R. P., and W. H. Goetz: "Effect of Crushed-gravel Fine Aggregate on the Strength of Asphaltic Surfacing Mixtures," *National Sand and Gravel Association Bulletin* 63, March, 1956.
40. Traxler, R. N., F. R. Olmstead, and R. E. Bollen: "Comparison of the Various Tests Used to Evaluate Mineral Fillers," *Proceedings, Association of Asphalt Paving Technologists*, vol. 9, 1937.
41. Traxler, R. N., and J. S. Miller: "Mineral Powders: Their Physical Properties and Stabilizing Effects," *Proceedings, Association of Asphalt Paving Technologists*, vol. 7, 1936.
42. Traxler, R. N.: "The Evaluation of Mineral Powders as Fillers for Asphalt," *Proceedings, Association of Asphalt Paving Technologists*, vol. 8, 1937.
43. *Investigation of the Design and Control of Asphalt Paving Mixtures*, 3 vols., U.S. Army, Corps of Engineers, Waterways Experiment Station, Vicksburg, Miss., May, 1948.
44. Zapata, J., and R. S. von Hazmburg: "The Effect of Fillers on the Durability of Asphalt," *Proceedings, Association of Asphalt Paving Technologists*, vol. 18, 1949.
45. Martin, J., and A. Layman: "Hot-mix Asphalt Design Studies," *Oklahoma A and M Engineering Experiment Station Bulletin* 75, 1950.
46. Ricketts, W. C., J. C. Sprague, D. D. Tabb, and J. L. McRae: "An Evaluation of the Specific Gravity of Aggregates for Use in Bituminous Mixtures," *Proceedings, American Society for Testing Materials*, vol. 54, 1954.
47. Serafin, J.: "Measurement of Maximum Theoretical Specific Gravity of a Bituminous Mixture by Solvent Immersion," *Proceedings, Association of Asphalt Paving Technologists*, vol. 23, 1954.
48. Rice, J. M.: "New Test Method for Direct Measurement of Maximum Density of Bituminous Mixtures," *The Crushed Stone Journal*, September, 1953.

49. Craig, H. R., and F. W. Kimble: "Determination of Specific Gravity of Bituminous Concrete Mix for Field Density Control," *Highway Research Board Bulletin* 105, 1955.
50. McLeod, N. W.: "Selecting the Aggregate Specific Gravity for Bituminous Paving Mixtures," *Proceedings, Highway Research Board*, vol. 36, 1957.
51. Vokac, R.: "Compression Testing of Asphalt Paving Mixtures," *Proceedings, American Society for Testing Materials*, vol. 36, part 2, 1936.
52. Hillman, W. O'B.: "Bending Tests on Bituminous Mixes," *Public Roads*, vol. 21, no. 4, June, 1940.
53. Rader, L. F.: "Investigation of Physical Properties of Asphalt Mixtures at Low Temperatures," *Proceedings, American Society for Testing Materials*, vol. 35, part 2, 1935.
54. Eriksson, R.: "Tension Tests on Sheet Asphalt," *Statens Vaginstitute*, Stockholm, 1954.
55. Pfeiffer, J. P.: "Observations on the Mechanical Testing of Bituminous Road Materials," *Journal, Society of Chemical Industry*, vol. 57, 1939.
56. Vokac, R.: "Correlation of Physical Tests with Service Behavior of Asphaltic Mixtures," *Proceedings, Association of Asphalt Paving Technologists*, vol. 8, 1937.
57. Wood, L. E., and W. H. Goetz: "Strength of Bituminous Mixtures and Their Behavior under Repeated Loads," *Proceedings, Highway Research Board*, vol. 35, 1956.
58. Pauls, J. T., and J. F. Goode: "Application and Present Status of the Immersion-Compression Test," *Proceedings, Association of Asphalt Paving Technologists*, vol. 16, 1947.
59. Mack, C.: "The Deformation Mechanism and Bearing Strength of Bituminous Pavements," *Proceedings, Association of Asphalt Paving Technologists*, vol. 23, 1954.
60. McLeod, N. W.: "Application of Triaxial Testing to the Design of Bituminous Pavements," Triaxial Testing of Soils and Bituminous Mixtures, *ASTM Special Technical Publication* 106, 1951.
61. Goetz, W. H., J. F. McLaughlin, and L. E. Wood: "Load-deformation Characteristics of Bituminous Mixtures under Various Conditions of Loading," *Proceedings, Association of Asphalt Paving Technologists*, vol. 26, 1957.
62. Endersby, V. A.: "The History and Theory of Triaxial Testing, and the Preparation of Realistic Test Specimens: A Report of the Triaxial Institute," Triaxial Testing of Soils and Bituminous Mixtures, *ASTM Special Technical Publication* 106, 1951.
63. Nijboer, L. W.: "Mechanical Stability of Bituminous Aggregate Mixtures," *Journal, Society of Chemical Industry*, vol. 67, 1948.
64. Smith, V. R.: "Triaxial Stability Method for Flexible Pavement Design," *Proceedings, Association of Asphalt Paving Technologists*, vol. 18, 1949.
65. Worley, H. E.: "Triaxial Test Methods Used in Flexible Pavement Design," *Proceedings, Highway Research Board*, vol. 23, 1943.
66. Hubbard, P., and F. C. Field: "The Rational Design of Asphalt Paving Mixtures," *Asphalt Institute Research Series* 1, 1935.
67. Vallerga, B. A.: "Recent Laboratory Compaction Studies of Bituminous Paving Mixtures," *Proceedings, Association of Asphalt Paving Technologists*, vol. 20, 1951.
68. Phillippi, O. A.: "Molding Specimens of Bituminous Paving Mixtures," *Proceedings, Highway Research Board*, vol. 31, 1952.
69. Hughes, E. C., and R. B. Farris: "Low Temperature Maximum Deformability of Asphalts," *Proceedings, Association of Asphalt Paving Technologists*, vol. 19, 1950.
70. Nijboer, L. W.: "The Determination of the Plastic Properties of Bitumen-Aggregate Mixtures and the Influence of Variations in the Composition of the Mix," *Proceedings, Association of Asphalt Paving Technologists*, vol. 16, 1947.
71. Mack, C.: "A Quantitative Approach to the Measurement of the Bearing Strength of Road Surfaces," *Proceedings, Association of Asphalt Paving Technologists*, vol. 17, 1947.
72. Eriksson, R.: "Deformation and Strength of Asphalts at Slow and Rapid Loading," *Statens Vaginstitute*, Stockholm, 1951.
73. Vokac, R.: "An Impact Test for Studying the Characteristics of Asphalt Paving Mixtures," *Proceedings, Association of Asphalt Paving Technologists*, vol. 6, 1935.
74. Itakura, C., and T. Sugaware: "Dynamic Tests on the Stability of Bituminous Mixtures for Pavements at Low Temperature," *Memoirs of Faculty of Engineering, Hokkaido University*, vol. 9, no. 4 (no. 42), November, 1954.
75. Skidmore, H. W.: "Some Low Temperature Characteristics of Bituminous Paving Compositions," *Transactions, American Society of Civil Engineers*, no. 101, 1936.
76. Rader, L. F.: "Correlation of Low Temperature Tests with Resistance to Cracking of

Sheet Asphalt Pavements," *Proceedings, Association of Asphalt Paving Technologists,* vol. 7, 1936.

77. Nijboer, L. W.: "Mechanical Properties of Asphalt Materials and Structural Design of Asphalt Roads," *Proceedings, Highway Research Board,* vol. 33, 1954.

78. Monismith, C. L.: "Flexibility Characteristics of Asphaltic Paving Mixtures," *Proceedings, Association of Asphalt Paving Technologists,* vol. 27, 1958.

79. Mack, C.: "Rheology of Bituminous Mixtures Relative to the Properties of Asphalts," *Proceedings, Association of Asphalt Paving Technologists,* vol. 13, 1942.

80. Lee, A. R., and A. H. D. Markwick: *Journal, Society of Chemical Industry,* vol. 56, 1937.

81. The Asphalt Institute: "Mix Design Methods for Hot-mix Asphalt Paving," *Manual Series* 2, April, 1956.

82. Stanton, T. E., and F. N. Hveem: "Role of the Laboratory in the Preliminary Investigation and Control of Materials for Low Cost Bituminous Pavements," *Proceedings, Highway Research Board,* vol. 14, 1935.

83. California Division of Highways, Department of Public Works: "*Materials Manual: Testing and Control Procedures,*" vol. 1, 1956.

84. Hveem, F. N.: "Use of the Centrifuge Kerosene Equivalent as Applied to Determine the Required Oil Content for Dense-graded Bituminous Mixtures," *Proceedings, Association of Asphalt Paving Technologists,* vol. 13, 1942.

85. McFadden, G., and W. C. Ricketts: "Design and Field Control of Asphalt Paving Mixtures for Military Installations," *Proceedings, Association of Asphalt Paving Technologists,* vol. 17, 1948.

86. U.S. Army, Corps of Engineers: "Airfield Flexible Pavement Design Procedure," *Technical Report* 3-475, February, 1958.

87. Smith. V. R.: "Triaxial Stability Method for Flexible Pavement Design," *Proceedings, Association of Asphalt Paving Technologists,* vol. 18, 1949.

88. Nadai, A.: "*Plasticity: A Mechanics of the Plastic State of Matter,*" McGraw-Hill Book Company, Inc., New York, 1931.

89. Barber, E. S., and C. E. Mershon: "Graphical Analysis of the Stability of Soil," *Public Roads,* vol. 21, no. 8, October, 1940.

90. Love, A. E. H.: "The Stress Produced in a Semi-infinite Solid by Pressure on Part of the Boundary," *Philosophical Transactions, Royal Society of London, series A,* vol. 228, 1928.

91. Hubbard, P.: "Adhesion of Asphalt to Aggregate in the Presence of Water," *Proceedings, Highway Research Board,* part I, 1938.

92. Rice, J. M.: "Relationship of Aggregate Characteristics to the Effect of Water on Bituminous Paving Mixtures," Symposium on Effect of Water on Bituminous Paving Mixtures, *ASTM Special Technical Publication* 240, 1958.

93. Huber, C. F., and P. F. Thompson: "The Function and Chemistry of Asphalt Additives," *Proceedings, Association of Asphalt Paving Technologists,* vol. 24, 1955.

94. "Adhesion," *Road Tar Bulletin* 4, British Road Tar Association, London, 1948.

95. Critz, P. F.: "Laboratory Study of Anti-stripping Additives for Bituminous Materials," Symposium on Effect of Water on Bituminous Paving Mixtures, *ASTM Special Technical Publication* 240, 1958.

96. Goetz, W. H.: "Methods of Testing for Water Resistance of Bituminous Paving Mixtures," Symposium on Effect of Water on Bituminous Paving Mixtures, *ASTM Special Technical Publication* 240, 1958.

97. American Society for Testing Materials: "Stripping Tests for Bitumen-Aggregate Mixtures," *ASTM Standards on Bituminous Materials for Highway Construction, Waterproofing, and Roofing,* Appendix, December, 1955.

98. Lewis, R. H., and J. Y. Welborn: "The Effect of Various Rubbers on the Properties of Petroleum Asphalt," *Public Roads,* vol. 28, no. 4, October, 1954.

99. Benson, J. R.: "New Concepts for Rubberized Asphalts," *Roads and Streets,* vol. 98, no. 4, April, 1955.

100. Wood, P. R.: "Rheology of Asphalts and the Relation to the Behavior of Paving Mixtures," Rheological and Adhesion Characteristics of Asphalt, *Highway Research Board Bulletin* 192, 1958.

101. Thompson, D. C., and J. F. Hagman: "The Modification of Asphalt with Neoprene," *Proceedings, Association of Asphalt Paving Technologists,* vol. 27, 1958.

102. Rex, H. M., and R. A. Peck: "A Laboratory Study of Rubber Asphalt Paving Mixtures," *Public Roads,* vol. 28, no. 4, October, 1954.

103. Winters, W. F.: "Barytes in Rubber-Asphalt Mixtures," *Proceedings, Association of Asphalt Paving Technologists,* vol. 25, 1956.

104. Gregg, L. E., and W. H. Alcoke: "Investigations of Rubber Additives in Asphalt Paving Mixtures," *Proceedings, Association of Asphalt Paving Technologists*, vol. 23, 1954.
105. U.S. Army, Corps of Engineers, Waterways Experiment Station: "Laboratory Investigation of the Use of Various Elastomers with Tar as a Binding Agent for Jet-fuel- and Jet-blast-resistant Pavements," *Miscellaneous Paper* 4-245, 1957.
106. Neppe, S. L.: "The Influence of Rheological Characteristics of the Binder on Certain Mechanical Properties of Bitumen-Aggregate Mixes," *Proceedings, Association of Asphalt Paving Technologists*, vol. 22, 1953.

Section 10

PAVEMENT SLIPPERINESS

DR. JOHN W. SHUPE, *Associate Professor, Applied Mechanics Department, Kansas State College, Manhattan, Kans.*

I. INTRODUCTION

A. Relation between Highway Safety and Pavement Slipperiness

Not all accidents in which skidding occurs can be attributed to a slippery pavement. It is impossible to construct highways which will completely eliminate skidding of the modern vehicle, with its potential for high speeds and extreme rates of acceleration and deceleration. A vehicle traveling at 70 mph possesses sufficient kinetic energy to cause it to hurtle over 160 ft in a vertical direction, and even the best type of pavement surface cannot control this tremendous quantity of kinetic energy for all possible driving maneuvers.

There are many types of pavements, however, which for certain surface conditions may be extremely slippery. Analyses of accident records indicate that the incidence of total accidents, as well as accidents directly involving skidding, increases significantly when the coefficient of friction between the sliding tire and pavement surface is less than 0.40. Currently there is an appreciable mileage of highways in the United States which cannot develop this friction value for all surface conditions, and as the polishing effect of traffic becomes increasingly intensified the mileage of slippery pavements will tend to increase. This presents a challenge to the highway engineer, for, by designing, constructing, and maintaining highways which possess adequate skid resistance, he can help to minimize the occurrence of skidding accidents and contribute to over-all driving safety more effectively than by acknowledging and identifying specific sections of highways as "slippery when wet."

B. Surface Condition

All highway surfaces can develop adequate skid resistance in the dry state. In fact, some surfaces which are extremely slippery when wet exhibit excellent antiskid characteristics in the dry condition. Conversely, any surface which becomes glazed with ice or packed snow constitutes an extreme skidding hazard. This is a special condition which will not be discussed in this section.

The majority of the pavement surfaces, however, will frequently be exposed to rainfall, and it is in this wet state that a tremendous variability in the skid resistance of pavements occurs. Some surface types exhibit nearly as good antiskid characteristics when wet as when dry, while other surfaces can develop less than one-fifth of the skid resistance in the wet state as compared with the dry. Consequently, the critical condition from skid-resistance considerations is when the surface is wet, and the major portion of this section will pertain to the skidding characteristics of pavement surfaces in the wet condition only.

C. Scope of Section

Discussion of the following topics is included in this section: (1) skidding phenomena for a rubber tire sliding on a pavement surface; (2) measurement of skid resistance by different field and laboratory methods with recommendations as to preferred procedures for various testing programs; (3) polishing characteristics of mineral aggregates due to the action of traffic; (4) building skid resistance into new construction; and (5) improving skid resistance of existing pavements.

II. SKIDDING PHENOMENA

A. General

A brief consideration of the nature of the forces developed between a sliding rubber tire and a pavement surface illustrates some of the factors contributing to pavement slipperiness and suggests possible remedies for minimizing the deleterious effects of these factors. The phenomenon of a rubber tire sliding on a dry surface is considered in this section, along with a discussion of how the forces generated between the tire and the dry pavement are modified by the presence of a water film and also by the debris that accumulates on the surface during certain seasons of the year.

B. A Rubber Tire Sliding on a Dry Surface

The total force resisting skidding of a rubber tire on a dry pavement is composed of two related effects. The first of these is due to mechanical interlock and results from the envelopment of some component of the pavement structure by the sliding tire. For a freshly broomed portland cement concrete pavement the tire "gears" into the grooves left by the brooming, while for an open-graded bituminous surface the tire envelopes the individual aggregate particles. The second effect is similar to mechanical interlock, but on a microscopic scale. This is the so-called true friction in which the minute surface irregularities of the tire and pavement mesh and resist sliding.

For a dry pavement the true-friction component contributes significantly to the total skid resistance. Dry-skidding tests indicate that the highest skid resistance is obtained on pavements that provide a maximum of contact area between the surface and the sliding tire (19).* This condition is typified by a smooth-tread tire skidding on a tight, dense surface.

C. The Effect of Water on the Surface

The combination of a smooth-tread tire and a dense surface which possesses excellent skid resistance in the dry state may result in an extremely slippery condition when wet. The minute irregularities of the tire and surface become filled with water, and the tire in effect glides on an unbroken film of water. For the extreme condition a polished wet surface may exhibit only one-tenth of the skid resistance at normal driving speeds that it is capable of developing when dry (31).

The presence of water on the pavement will also adversely affect the antiskid properties of an open-textured surface, i.e., one in which the size of the individual surface voids is appreciable. Mechanical interlock may still take place, but the majority of the "true frictional" resistance will be eliminated since the water film prevents the small irregularities of the tire and surface from meshing. For many of the aggregates commonly used in highway construction a typical water-film thickness will vary from 0.01 to 0.02 in. (13). Most of this film can be dissipated rather easily by a sliding rubber tire, but the last 0.001 in. or so is very tenacious and, under certain conditions, may

---

* Numbers in parentheses refer to corresponding items in the references at the end of this section.

greatly reduce the skid-resisting force otherwise developed. A major problem of the highway engineer in preventing pavement slipperiness is to construct highways which will permit a rubber tire to penetrate this boundary layer of water and generate positive interaction with the pavement surface.

D. Minimizing the Water-film Effect

1. Surface Drainage. The water which falls upon a pavement should be removed as quickly as possible. This may be accomplished by an adequate crown and an absence of a curb or lip at the pavement edge. An excess of water increases the likelihood of a water film becoming entrapped between the tire and surface.

The degree of openness of the pavement surface will dictate the amount of water required for complete saturation. A "polished" portland-cement-concrete pavement or a "bleeding" asphalt surface requires only a very small amount of water to arrive at its most slippery condition, while a fine-grained, open-graded surface, such as a silica sand or a Kentucky rock-asphalt pavement, requires a greater amount of water for saturation. An open-textured surface, as typified by a bituminous seal coat composed of one-size coarse aggregate, will seldom receive sufficient water to bridge across each projecting piece of aggregate. However, a pavement of this nature may be dangerously slippery when only a small amount of water is present on the surface, since a film of water may exist on each piece of aggregate long before complete saturation of the surface occurs; and, as previously discussed, this film may result in a slippery pavement.

The existence of a film of water between the tire and the surface, and the degree to which it affects skid resistance, is a function of the time. At low vehicle speeds the tire has sufficient time to dissipate the water film and develop a positive skidding force, while at high speeds the entrapped water may prevent this action. This is particularly true for dense surfaces with a tight texture, since they do not possess the drainage channels that an open-textured surface possesses and require a greater period of time for the water to drain from beneath the tire. Pavements with dense, tight surfaces experience a greater loss in wet-skid resistance at high speeds than do open-textured surfaces (14, 19).

A final item pertaining to drainage relates to the degree of porosity of the aggregate. An appreciable hydraulic pressure is generated in a water film entrapped between a tire and an aggregate particle. A porous aggregate can relieve this excess hydraulic pressure and help to develop positive interaction between the tire and the surface.

2. High Unit Pressure between Tire and Surface. A necessary condition for penetrating the water film is to maintain a high unit pressure between the tire and the pavement. Any surface which wears uniformly or contains exposed aggregate that is rounded and polished will be slippery at normal driving speeds when wet. There must be sufficient surface angularity or differential wear so that the area of contact between the tire and surface is low.

It is not essential for the individual surface voids to be large in order for a pavement to develop adequate skid resistance. Wet-skidding tests indicate that two of the surface types which exhibit excellent antiskid properties are an angular silica-sand surface treatment (8) in which the majority of the quartz particles pass a No. 30 mesh sieve and a granite-chip seal coat (20), with the aggregate ranging from $\frac{1}{4}$- to $\frac{3}{8}$-in. in size. Although each of the silica sand grains can generate only a small fraction of the frictional force developed by a $\frac{3}{8}$-in. granite chip, the accumulated skid resistance generated over the total contact area for these two dissimilar surfaces is nearly identical, since in both cases the angular aggregates slice through the water film and actively engage the tire.

3. Tire Effect. There are a few harsh, abrasive surfaces for which a smooth-tread tire will generate equal, or even greater, wet-skid resistance than a tire with a conventional tread pattern. For the majority of surface types, however, a tire with a good tread pattern will exhibit better resistance to skidding than one with a smooth tread, and for some dangerously slippery pavements, this advantage is appreciable. With a surface containing highly polished aggregate, for which a water film tends to develop between the tire and the aggregate, the tread pattern supplies built-in drainage channels which facilitate removal of the water. In addition, the grooves of the tread pattern permit greater aggregate envelopment and supply a more positive "wiping" action than can be accomplished with a continuous smooth-tread tire. Test results indicate that some surfaces exhibit over twice as great wet-skid resistance with a tire having a normal tread pattern as compared with a tire in which the tread has been worn smooth (31).

The composition of the rubber and the configuration of the tread pattern also have some effect on the skid-resisting capabilities of a tire. These factors deserve, and receive, consideration from the tire industry, but will not be included in this discussion of pavement surfaces. An excellent coverage of this subject was presented at the First International Skid Prevention Conference (25).

E. Traffic Film

An accumulation of oil, worn rubber, and dust particles on a pavement has a significant effect on the antiskid characteristics of the surface (15, 19). By filling in the minute surface irregularities, this debris may decrease the wet-skid resistance by as much as 50 per cent, as compared with the wet-skid resistance of the pavement when clean. A tight, dense surface is more sensitive to the traffic film than an open-textured surface and will show a greater variation in skid resistance because of the presence of this "road scum."

The traffic film forms during a dry period and is fairly well dispersed by a driving rain. Consequently, the exposure of a pavement to moisture following a prolonged dry spell will result in the most slippery surface condition. This factor should be considered in any series of field tests in which different surface types are evaluated on a relative basis. The amount of wetting that a pavement receives immediately prior to testing is usually insufficient to leave the surface in a "scrubbed" condition, so an effort should be made to assure some consistency of the traffic film if a comparison of the different surface types is to be made.

The degree to which the traffic film accumulates and lowers the antiskid properties of a pavement depends upon the length of time between rains. The wet-skid resistance of pavement surfaces is the poorest during the dry season, which normally corresponds to the summer months, so the term "seasonal effect" appears in the literature to describe this phenomenon. In addition to the traffic film, the amount of polish on the individual pieces of aggregate contributes to the seasonal variation in skid resistance. During a dry period the size of dust particles present on the highway and adhering to the tire are very fine and put a high polish on the aggregate. During a rainy spell these fine particles are washed from the pavement and relatively coarser particles are embedded in the tire. They tend to scour and roughen the polished aggregate somewhat, increasing the skid resistance. This condition has been simulated in the laboratory (33).

A final factor, which is related to the traffic film, may occur when appreciable amounts of silt or clay are present on the pavement surface, as can happen when a vehicle returns to a highway from a muddy shoulder. Not only do these fine particles tend to fill the surface depressions, but they also supply additional lubrication through a rolling or "ball-bearing" action and may result in a very slippery condition.

The accident potential of this situation is aggravated by the fact that there may be a differential braking effort developed by each pair of tires on the right and left sides of the sliding vehicle. If the two outside tires are skidding on slick area of wet clay while the inside pair is generating a reasonable skidding effort, this unbalanced torque will cause the vehicle to spin, with the direction of spin toward opposing traffic. This condition may also occur if one pair of tires is sliding in an oil slick between wheel paths or if one wheel track of a bituminous pavement has an excess of asphalt at the surface. A pavement surface which permits a differential braking effort to be developed between the right and left pair of tires is more dangerous, in some respects, than a surface which is uniformly slick; and every effort should be made to eliminate these surfaces from our highway systems.

III. MEASUREMENT OF SKID RESISTANCE

A. General

As a motor vehicle travels over a highway, the condition of maximum stability, during which the driver has complete control of the vehicle, occurs when the wheels are rolling without slipping, i.e., when there is no relative motion between the tire and pavement. There are driving maneuvers which result in simultaneous rolling and sliding. Rounding curves at excessive speeds, endeavoring to achieve extreme rates of acceleration, and braking the vehicle to the point of impending or incipient skidding, all will result in relative motion between the tire and surface in addition to the rolling action. When the wheels are locked, which normally occurs during an emergency stop on a wet pavement, there is sliding without rolling, and the driver retains very little steering control.

A difference of opinion as to which of the above conditions is the most significant with regard to evaluating the skid resistance of a pavement surface is one reason for the wide variety of skid-test equipment developed. In Europe the viewpoint is that "cornering" represents the critical condition and the emphasis has been on measuring the sideways skid resistance. In the United States, however, the majority of the field testing programs have evaluated the skid resistance of surfaces by a straight-ahead skid along the length of the pavement. Most of the field testing in this country has stemmed from the work of Agg and Moyer at Iowa State College (1, 18). One of the conclusions from their original series of tests, which included both sideways and straight-ahead skid-test measurements, was that the straight-ahead method gave results which evaluated the surfaces on the same relative basis as the sideways method and required somewhat less complicated instrumentation.

An interesting phenomenon noted in field testing with some of the towed-vehicle units is that at the point of impending or incipient skidding, i.e., when the tire is still rolling but just on the verge of locking, the skid resistance developed may be 50 to 100 per cent higher than that generated in the wheels-locked condition. This should present a challenge to the automotive industry, since by designing a braking system which could develop the maximum skid-resistance potential of a surface without permitting the wheel to lock, an appreciable contribution to driving safety would be made. Currently, however, it is merely a matter of academic interest to the design engineer, for in an emergency stop a driver will generally lock the wheels, particularly on a wet slippery pavement. Consequently, the wheels-locked testing condition will receive major consideration in this section.

IV. POLISHING CHARACTERISTICS OF MINERAL AGGREGATES

A. General

The antiskid properties of a well-designed pavement surface are dependent to a large degree upon the polishing characteristics of the mineral aggregate or aggregates of which the paving mixture is composed. Skid resistance will be dependent to a lesser degree upon the type of cementing agent, the gradation, and the openness of the mixture; but the ultimate state of pavement slipperiness will be dictated by the nature of wear of the pavement, which, in turn, is directly related to the resistance of the surface aggregate to polishing.

This is true both for portland cement and bituminous paving mixtures. However, it usually requires less traffic to define the polishing characteristics of aggregates in bituminous mixtures, since the degree of exposure is greater for the individual pieces of aggregate than with portland cement surfaces. The extreme condition is represented by a bituminous surface treatment, consisting of a uniform coarse-aggregate cover one particle thick, in which case the tire has the opportunity to envelop and subject the aggregate to an intense polishing action as soon as the highway is open to traffic. For portland cement surfaces the coarse aggregate is protected by the mortar and, for low-volume highways, may never be directly exposed to the polishing effects of traffic.   Even for this condition, however, the fine aggregate in the mortar continues to play an important role in the skid resistance of portland cement surfaces, and for heavily traveled highways, the coarse aggregate will ultimately be exposed.

Although polishing of aggregates was noted occasionally in the initial highway skid-resistance studies, only within the last decade has pavement slipperiness due to polished aggregate become a problem of major importance to the highway engineer. Factors contributing to this recent development would include: (1) Both the number of vehicles and travel per vehicle have increased tremendously since World War II. (2) The flexing action of low-pressure passenger-car tires traveling at high speeds exerts a "squee-gee" polishing effort on the pavement surface, in addition to the relative sliding between tire and pavement due to braking and acceleration.   (3) High-pressure truck tires place an increasingly severe unit load on the components of the pavement surface.   (4) The sand and calcium or sodium chloride, frequently used in many localities for de-icing, tends to multiply the abrasive effects of the factors previously enumerated.

Recent studies in Indiana have shown that for moderate traffic both bituminous and portland cement surfaces may lose as much as 50 per cent of their initial skid resistance within two years (32).   If a reasonable degree of skid resistance is to be maintained on our highway system for anticipated high-volume traffic, it becomes increasingly important for the design engineer to give some thought to the polishing characteristics of mineral aggregates and to their effect upon pavement slipperiness.

B. Limestone

Limestones, as a group, have developed a poor reputation with regard to susceptibility to polishing.   Moyer, in his initial work (18), reported that for bituminous pavements "the use of limestone dust and limestone aggregate exclusively resulted in surfaces with low coefficients."   Studies in Virginia (23), Tennessee (39), and Kentucky (38) indicated that both bituminous and portland cement surfaces, in which the aggregate consists entirely of limestone, may become slippery under the action of traffic, and measures have been taken to limit the use of limestone as a surface material in these three states.   In many sections of the United States, limestones have been the primary source, and in some areas essentially the only source, of highway aggregate.   Economic considerations, as well as the many desirable qualities of limestone aggregate in paving mixtures, warrant maximum utilization of this material without, of course, sacrificing highway safety.

There is some variability in the resistance to polishing of different types of limestone.   For example, in the report of the Tennessee field investigation (39), it was indicated that "several surfaces containing limestone aggregates have been found to be relatively skid resistant even after the passage of a rather considerable amount of traffic."   Laboratory studies have been directed toward an evaluation of the properties of highway materials which contribute to pavement slipperiness, so a differentiation can be made between those aggregates which definitely should be eliminated as the total surface material and other aggregates which, for certain traffic conditions, may be entirely satisfactory.

The results of a laboratory investigation at Purdue University (33), in which the resistance to polishing of 22 mineral aggregates from five different states were studied, are summarized in Fig. 10-1.   An open-graded asphaltic concrete was used in evaluating the aggregates.   In order to facilitate a comparison of the various aggregates, the asphalt content was kept low, and each test specimen was composed entirely of one aggregate type.   The skid resistance of each test specimen was obtained at the completion of a laboratory wear and polish procedure, which was intended to simulate the polishing effect of traffic.   The relative resistance value (RRV), plotted as the ordinate to indicate the skid resistance of the various specimens, was based on a value of unity for a Kentucky rock-asphalt specimen; i.e., a core taken from a Kentucky rock-asphalt pavement, which from field studies has consistently exhibited excellent antiskid properties, gave a relative resistance value of 1.00 as evaluated in the laboratory skid-test apparatus.

The results for test specimens made with the 12 limestones included in this study, which are identified as L-1 to L-12, are plotted to the left in Fig. 10-1 with values for mixtures containing the other 10 aggregates listed to the right. The 12 limestone specimens, with an average RRV of 0.45, possessed less resistance to polishing, as a group, than mixtures containing the other 10 aggregates, which averaged 0.63. There was also appreciable variability in the polishing characteristics of the different limestones, with specimens containing L-4 exhibiting 75 per cent greater skid resistance than those made with L-7.

Petrographic and chemical analyses were performed on the 12 limestones used in the study. Those limestones which readily polished possessed either a very-fine-grained crystalline structure or consisted of rounded oolitic grains supported in a calcite matrix of similar hardness. For both cases, however, the limestones were almost pure calcium

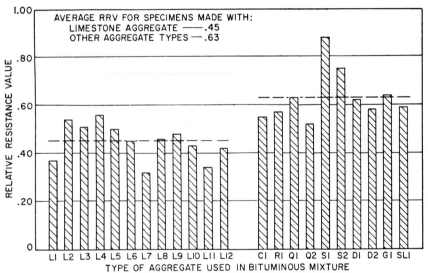

Fɪɢ. 10-1. Resistance to polishing of bituminous mixtures composed of different aggregate types.

carbonate, and the nature of the wear was very uniform, resulting in a highly polished surface that was slippery when wet.

The limestones which exhibited the best antiskid characteristics were crystalline granular limestones composed of angular interlocking grains. For the 12 limestones under consideration, this structure occurred with those limestones relatively low in calcium carbonate and high in "impurities," which consisted primarily of magnesium carbonate with a slight amount of silica. Consequently, for this investigation the highly dolomitic limestones exhibited the best resistance to polishing. The highest relative resistance value for a bituminous mixture containing any of the 12 limestones was obtained with a dolomitic limestone composed of angular interlocking crystals with some porosity present.

C. Sandstone

Both in laboratory and field investigations, sandstone, in a variety of forms, has consistently exhibited excellent antiskid characteristics. Kentucky rock asphalt, previously discussed as the reference skid-resistant material in the Purdue study, occurs as a Mississippian-age sandstone which has been impregnated with asphalt; the antiskid surface treatments developed in Virginia are composed of crushed sandstone (8); sandstones when used as a conventional aggregate in paving mixtures are skid-resistant (39); and laboratory investigations at both Kentucky (38) and Purdue (33) have established the high resistance of sandstone to polishing. Referring to Fig. 10-1, the two aggregates identified as S-1 and S-2, which show appreciably greater skid resistance than the next-best aggregate, were both sandstones.

In general, sandstones are composed of small angular quartz grains supported by a somewhat weaker cementing matrix. This combination results in excellent antiskid characteristics. The individual quartz grains are highly resistant to polishing, since quartz is No. 7 on Moh's relative-hardness scale, and the only material present on the highway capable of polishing it is more quartz. Calcite, one of the major constituents of limestone, is No. 3 in the hardness scale. In addition, there is a particle-by-particle type of wear, so that before an exposed quartz grain has the opportunity to become highly polished, it is dislodged from the weak-cementing matrix and a fresh, harsh particle appears at the surface.

This phenomenon was noted in the Purdue study on tests performed with a calcareous sandstone composed of 60 to 70 per cent of angular quartz grains supported in a matrix of calcite. During the laboratory wear and polish procedure, the weak calcite matrix gradually wore away from the resistant quartz exposed at the surface of the test specimen, so that each quartz particle lost its support and was dislodged before it had the opportunity to polish excessively. A similar type of wear occurs with the skid-resistant Kentucky-rock-asphalt and silica-sand surfaces. Individual sand particles are dislodged from the pavement before polishing, thereby continuously rejuvenating the riding surface.

D. Other Aggregate Types

Although all aggregate types other than limestone and sandstone are grouped under this general heading, there is a tremendous amount of variability in their relative resistance to polishing; however, their respective characteristics have not been too well defined as yet. Some of the coarse-grained granites and traprocks exhibit wearing characteristics similar to those of sandstone and, consequently, possess excellent antiskid properties. Such a granite was noted in California (20) in which the varying degrees of hardness of the quartz, feldspar, and micaceous minerals promoted differential wear of the various components and resulted in high resistance to skidding.

Conversely, some of the fine-grained uniform aggregates possess polishing characteristics similar to those of the more resistant of the limestones. This is illustrated in Fig. 10-1 by the results for the chert (C-1) and rhyolite (R-1) aggregates. Initially, bituminous mixtures containing these two harsh aggregates exhibited high resistance to skidding, but during the simulated wear and polish procedure the uniform nature of wear lowered the skid resistance to values comparable with those of specimens containing highly dolomitic limestone. Another factor contributing to the decrease in skid resistance of bituminous mixtures composed of these two aggregates was the particle orientation that occurred during the rolling cycle. This operation, which was intended to simulate the rolling effect of traffic, lowered the effective angularity of the surface aggregate by causing any flat faces present on the aggregate to become oriented parallel to the surface.

Aggregate polishing is generally considered as a function of the traffic. Some gravels, however, are highly polished in their natural state and, if used in an uncrushed condition, may constitute an extreme skidding hazard. A seal coat constructed in California with polished beach gravel exhibited very little resistance to skidding when wet (20). Both the accident records and the coefficient-of-friction values, which were less than 0.2 at 50 mph, identified this section as slippery when wet. This condition is shown to a lesser degree in Fig. 10-1. Aggregates Q-1 and Q-2 were both high-quartz gravels. All the aggregate used in the Q-1 test specimens was crushed, but for the Q-2 specimens the coarse-aggregate fraction consisted of the natural rounded gravel. As illustrated, the Q-1 specimens, with all aggregate crushed, exhibited over 20 per cent greater resistance to skidding than the Q-2 specimens containing rounded coarse aggregate.

The four other aggregates of Fig. 10-1, which have not yet been discussed, consist of two diabases (D-1 and D-2), a granite (G-1), and a blast-furnace slag (SL-1). These aggregates all imparted adequate skid resistance to their respective bituminous mixtures, with the relative resistance values falling slightly above those of the best limestones but appreciably below those of the sandstones. The blast-furnace slag was composed essentially of a vitrophyric glass which appeared as though it might polish excessively. The vesicular nature of the slag, however, broke up the continuity of the structure and resulted in reasonable antiskid properties.

E. Summary of Aggregate Polishing

The following remarks with regard to the suitability of various types of aggregates as highway surface material apply primarily when the total surface aggregate is composed of one aggregate type. Subsequent discussion will consider mixtures composed of more than one type of aggregate.

Some varieties of limestone, particularly those consisting essentially of pure calcium carbonate, should not compose the total surface aggregate except for very-low-traffic roadways. Surfaces of this nature possess only fair skid resistance when new and may become dangerously slippery after a moderate amount of traffic.

Other types of limestone—and initial studies indicate that some highly dolomitic limestones fall in this category—result in paving mixtures which may be expected to exhibit adequate antiskid characteristics for a reasonable period of time if severe traffic is not anticipated. Pavements composed of these mixtures may ultimately polish and, if traffic is excessive, may do so in a relatively short period. However, this same observation can also be made of surfaces containing some of the so-called resistant aggregates, such as the fine-grained basalts, chert, and high-quartz gravel. These resistant aggregates may retain their initial angularity for an appreciable period but, because of the uniform nature of wear of the fine-grained structure, can ultimately polish excessively.

In order for a pavement constructed with conventional highway materials to retain a nonskid surface under prolonged action of heavy traffic, some type of coarse particle-by-particle wear is necessary. This condition is typified by a paving mixture composed of sandstone or granite in which the highly resistant quartz particles are dislodged from the wearing surface before they have the opportunity to polish appreciably.

F. Predicting Aggregate Polishing Characteristics

Valuable information with regard to the polishing characteristics of specific aggregates may be obtained from their field performance. The identity of potentially slippery aggregates may result from correlating a field skid-test program or a skidding-accident analysis with specific aggregate sources. Frequently, certain geologic formations or strata result in aggregates that have uniquely good or poor antiskid characteristics. However, information of this nature is generally lacking, and there also may be appreciable variability between the polishing characteristics of aggregates from a given source; so it becomes desirable to have a method of evaluating the polish susceptibility of an aggregate considered for use in a highway paving mixture.

Unfortunately, no correlation has been established between the polishing characteristics of an aggregate and such physical tests as specific gravity, degree of water absorption, or abrasion loss. In the Tennessee field studies of 50 pavement surfaces, aggregate samples were taken from each of the test sections and subjected to standard physical tests (39). No correlation was noted between these properties and pavement slipperiness. Similarly, in the Purdue laboratory investigation, there was no relation between the antiskid characteristics of the 22 mineral aggregates and their respective physical-test results.

Currently, the best method of predicting the polishing characteristics of an aggregate in a specific paving mixture is to duplicate this mixture in the laboratory, subject it to an accelerated wear-and-polish procedure, and evaluate the change in skid resistance due to simulated wear. Laboratory equipment of the type developed at Purdue University and the University of Tennessee lends itself to a study of this nature. A more fundamental approach, and one which shows appreciable promise, is to relate such basic aggregate properties as those obtained from petrographic and chemical analyses to the polishing characteristics of the aggregate. Additional work in this area may result in a better understanding of the phenomenon of aggregate polishing.

V. MINIMIZING PAVEMENT SLIPPERINESS IN NEW CONSTRUCTION

A. General

Until quite recently, very little consideration was given in the design of paving mixtures to the antiskid characteristics of the resulting pavement surface. In the early days of highway design and use, when traffic volume, weight, and speed were all relatively low, this was not too serious an omission. With the exception of a surface containing an excess of asphalt or one composed of polished uncrushed gravel, the majority of new construction exhibited adequate skid resistance for a reasonable period of wear.

Currently, however, and with increasing importance as the polishing effect of traffic becomes more intensified, it is essential for the highway engineer to consider the antiskid properties of a paving mixture in addition to the other design parameters. An additional initial expense may be economically justified in order to construct a pavement which will retain a safe driving surface after an appreciable amount of polishing by traffic, whereas ignoring the antiskid characteristics of a paving mixture may result in the need for a relatively expensive deslicking treatment after only a short period of wear. The following discussion and recommendations for building in skid resistance and minimizing pavement slipperiness are based exclusively on skid-resistance considerations, with the realization that in actual practice other factors may modify these proposals.

B. Portland Cement Concrete

Historically, portland cement pavements have not exhibited the extremes in skid resistance that have been noted with bituminous surfaces. In portland cement concrete, the individual pieces of aggregate are supported by the mortar and are not exposed to traffic to as great a degree as with some types of bituminous surfaces. Consequently, during a skid the amount of aggregate envelopment, and the related resisting force, are less than would occur with, for instance, an open-graded granite-chip seal coat. Conversely, the aggregate in portland cement concrete is not exposed to as severe a wearing action and usually will not exhibit the high polish that may occur with some bituminous surfaces.

1. **Finishing Procedure.** The initial skid resistance of a portland cement pavement will be dependent upon the nature of the finish that the surface receives and the type of fine aggregate in the mortar. A well-designed concrete mix which permits good workability with a minimum of segregation should result in a pavement with high initial skid resistance. If an excess of finishing is required, however, the surface may consist essentially of cement paste which exhibits poor antiskid properties.

The final texture may be supplied to the pavement with a burlap drag or with a longitudinal or transverse broom drag. By any of these methods, it is possible to arrive at a surface with a slightly roughened coarse sandy texture possessing good skid resistance. It is not necessary to broom deep grooves in the pavement in order to achieve a skid-resistant surface. In fact, the individual ridges may polish and result in a slicker pavement than would occur with a denser surface. It is necessary, however, to use a delay-finish procedure, for if the drag is applied while the mortar is too soft, a dense smeary surface is obtained rather than the desired coarse sandy texture (20).

2. **Concrete Aggregates.** For a properly finished portland cement surface, the initial skid resistance will depend upon the angularity and hardness of the fine aggregate contained in the mortar. A harsh resistant sand will result in somewhat greater skid resistance than either a rounded natural sand or a crushed softer aggregate. As traffic begins to polish the surface, it is imperative that some type of differential wear of the various components occur. If the cement paste and the fine aggregate both possess essentially the same resistance to wear, a uniformly polished surface may result

that is slippery when wet. If the concrete also contains a coarse aggregate with similar wear characteristics, the uniformity of polish increases as wear progresses. The major occurrence of slipperiness for portland-cement-concrete pavements has been with mixes consisting entirely of crushed-limestone fine and coarse aggregate. Consequently, if at all possible, a fine aggregate, such as natural sand, with wear characteristics dissimilar to those of the cement paste should be used in concrete mixes, particularly if the coarse aggregate is limestone.

There is evidence that the mortar, which derives its characteristics from the fine aggregate, has a dominant role in establishing the antiskid properties of a concrete pavement even after an appreciable period of wear. In field studies in Indiana (17, 35) in which over 60 concrete pavements were tested for skid resistance, not a single surface was found to be dangerously slippery when wet. Some of these sections were over 30 years old and showed considerable wear. However, all the mixes had been made with natural-sand fine aggregate, and the differential nature of the wear had prevented the surfaces from polishing excessively. A number of concrete pavements have been constructed in Indiana containing 100 per cent limestone, but it was impossible to test these sections since, without exception, they have been resurfaced because of their slippery condition.

Some of the concrete pavements tested in Indiana were constructed with limestone coarse aggregates, while other sections contained the more wear-resistant gravel coarse aggregates. After an appreciable amount of wear, the limestone coarse-aggregate sections exhibited slightly better skid resistance than the gravel sections. The limestone coarse aggregate possessed less resistance to wear than the supporting mortar and consequently had worn slightly below the datum level of the mortar. Although the limestone aggregate contributed little to the skid resistance, the nature of the wear was such that the natural sand mortar and the sliding tire were in intimate contact and adequate skid resistance resulted.

With the gravel sections, however, the coarse aggregate was more resistant than the mortar, so projected slightly above the mortar datum. This decreased the contact pressure, and related skid-resistance component, as developed between the tire and mortar for the limestone sections. Some of the natural gravels were initially rounded, and others, because of their fine-grained structure, tended to polish under the action of traffic, so although the skid-resistance component of the gravel coarse aggregate was appreciably greater than that of the recessed pieces of limestone, the total skid resistance was somewhat less, on an average, than for the limestone coarse-aggregate sections.

This again emphasizes the need for a high-quality fine aggregate in concrete, particularly if a coarse aggregate with low wear resistance is present in the mix. It also indicates that if the concrete contains a wear-resistant coarse aggregate which will ultimately project above the mortar datum, then this aggregate should be crushed. A naturally rounded pebble will polish more readily and to a higher degree when exposed to the action of traffic than will a harsh angular piece of aggregate.

C. Bituminous Surfaces

As indicated in the previous section, the high degree of aggregate envelopment permitted by some types of bituminous surfaces may result in a wide range of antiskid properties, depending upon the nature of the aggregates composing the surface. In addition, the presence of an excess of bitumen at the surface may cause an otherwise adequate pavement to be a potential skidding hazard when wet. Thirty or forty years ago, this was by far the most common cause of pavement slipperiness and is still one of the major contributors today, even though improved design and construction practices have lessened the frequency with which a flushed or "bleeding" asphalt surface occurs.

The major problems, therefore, in constructing skid-resistant bituminous surfaces are: (1) Eliminate polished aggregate, which may arrive at this polished condition either naturally or because of the action of traffic, from the pavement surface. (2) Prevent an excess of asphalt from accumulating at the surface. In discussing methods of minimizing slipperiness with bituminous surfaces, only the high-type plant-mix asphaltic concrete and, at the other end of the economic range, the bituminous surface treatments will be considered. The recommendations presented for these two surface types may be extended to include the intermediate range of bituminous surfacing.

1. Asphaltic Concrete. With the control that is inherent in a plant-mix operation, there should be no occasion for an asphaltic concrete surface constructed by this process to become flushed with asphalt. Some design procedures advocate the use of a high asphalt content in order to promote good durability; and if this is carried to the extreme, an excess of asphalt may appear at the surface after the additional compaction due to traffic. However, a reasonable asphalt content will seldom result in a bleeding pavement and may contribute to better antiskid properties than if too little asphalt is used. With a lean mixture, i.e., one in which the asphalt content is quite low, the fine-aggregate and asphalt matrix may be readily displaced by traffic, resulting in a concentrated polishing effort on the coarse aggregate.

A related problem pertains to the degree of openness of the asphalt concrete. Some engineers advocate an open-graded mixture containing coarse aggregate in which the large surface voids facilitate drainage from between the sliding tire and surface and promote better skid resistance. To a certain extent this is true, particularly for a smooth-tread tire. However, the degree of harshness and the shape of the individual pieces of aggregate are more important than the openness of the surface in eliminating the water film, and there is evidence to the effect that greater skid resistance can be developed by loading the tire uniformly over its entire area, rather than by having voids of appreciable individual magnitude in the surface. The sand-asphalt type of pavement of the Coastal Plain area of Virginia is an example of a nonskid surface in which the individual voids are small (30).

Laboratory studies at Purdue (33) also indicate that if there is no excess of asphalt present, dense-graded mixtures exhibit better resistance to skidding than open-graded mixtures. Therefore, although open-graded mixtures possess a higher void content than dense-graded mixtures and are less likely to become flushed with asphalt because of compaction of traffic, they do expose the aggregate to a greater polishing effort, and so may experience a larger decrease in skid resistance because of traffic than dense-graded mixtures. Currently, there is some disagreement as to whether an open- or a dense-graded asphaltic concrete does possess the better antiskid properties. However, as improved design and control procedures tend to eliminate the possibility of a bleeding asphalt surface, dense-graded mixtures will probably emerge as the more skid-resistant of the two.

Whereas the fine aggregate in the mortar may play the more dominant role in establishing the antiskid characteristics of portland cement concrete, it is generally the coarse-aggregate ingredient which dictates the skid resistance of asphaltic-concrete surfaces, particularly after an appreciable amount of wear. The fine aggregate and asphalt matrix in which the coarse aggregate is supported is not as wear-resistant as the portland cement mortar, so even the weaker aggregates remain level with, or project somewhat above, the surrounding matrix and have a significant effect on the skid resistance. Of course, the fine aggregate will have some effect on the skid resistance of the mixture, but both field studies in Virginia (8) and laboratory work at Purdue (36) indicate that very sizable quantities of a resistant silica sand are required to appreciably increase the skid resistance of a bituminous mixture composed of a polish-susceptible coarse aggregate.

Asphaltic-concrete pavements should not be constructed with rounded uncrushed gravel. Field skid tests have identified surfaces of this nature as potential skidding hazards immediately following the final rolling operation (31). Specifications requiring a certain percentage of gravel with one or more crushed faces improve the above situation, but in a bituminous concrete surface the slick faces have a tendency to orient themselves parallel to the surface under the kneading action of traffic. A reasonable requirement if a skid-resistant surface is desired is to specify 100 per cent crushed material in the coarse-aggregate fraction.

An aggregate which is quite angular or harsh will impart good initial antiskid characteristics to an asphaltic-concrete surface. However, there is no guarantee that the pavement will retain its early skid resistance after an appreciable period of wear. The aggregate is directly exposed to the polishing effects of traffic, and the remarks pertaining to aggregate selection in Art. IV on "Polishing Characteristics of Mineral Aggregates" would apply.

2. Bituminous Surface Treatments. The following discussion will pertain to the relatively inexpensive type of surface treatment in which separate applications of asphalt and aggregate are applied to a prepared base. An objective of this treatment is to provide a wearing surface in which the tire rides almost completely on the exposed aggregate. Unfortunately, this treatment frequently results in an excess of asphalt at the surface and a pavement which is dangerously slippery when wet.

There are many ways in which a bleeding asphalt surface treatment may result. Too much asphalt on the pavement, whether through design or construction error, will "drown" the aggregate, while too little asphalt will result in inadequate aggregate retention and ultimate whip-off by traffic. Whip-off may also occur with certain types of surface treatments if the aggregate is too cold or wet when applied; if the asphalt distributor is poorly adjusted; if the period between shooting the surface with asphalt and the aggregate application is too great; if the temperature is cold or damp; if inadequate speed control is maintained when traffic is first admitted to the surface; or if dusty or coated aggregate is used (3).

A porous base will blot up the asphalt and prevent adequate aggregate retention, whereas if the treatment is placed on an old bituminous surface, rich in asphalt, bleeding may progress up through the surface treatment. An extremely hard base can result in fracture and degradation of certain aggregates, while a soft base may permit the aggregate to punch down into it, leaving a surplus of asphalt at the surface. Considering the tremendous varieties of ways in which a bleeding surface treatment may occur, a higher percentage of the total construction cost should be spent for engineering and inspection on surfaces of this nature than on any other type of pavement construction. Frequently, the opposite is true.

When bituminous surface treatments are placed on roadways carrying light to moderate traffic, which generally is the case, the tendency is to select the most convenient available aggregate. Although some of the less polish-resistant aggregates listed in Art. IV may be considered for use, some caution should be exercised in the aggregate selection. This type of surface permits maximum exposure of the aggregate to the polishing effect of traffic, and the pavement may exhibit a very open texture even though the individual pieces of aggregate are highly polished. As pointed out in a British study (29), this condition has a particularly high accident potential, since by presenting an open texture in which the individual pieces of aggregate "stand proud" even when wet, the surface gives the driver a false sense of security and encourages excessive speeds, whereas, in reality, the pavement may be dangerously slippery.

Another situation to be avoided is the use of rounded natural gravels as the cover aggregate. A seal coat of this nature constructed in California with "highly polished water-worn beach gravel" exhibited very poor antiskid characteristics when wet (20). A wearing surface consisting of highly polished aggregate, no matter whether the polish occurred naturally over a period of centuries or because of the action of traffic shortly after the surface was placed, will be dangerously slippery when wet.

The previous discussion has enumerated many of the difficulties encountered with bituminous surface treatments. However, by adhering to a design and construction procedure such as that developed in Texas (4) and selecting a suitable aggregate, it is possible to construct surface treatments which are completely adequate for their intended use. Many miles of satisfactory bituminous surface treatments have been and will continue to be built, but not without appreciable engineering effort in all phases of design and construction.

VI. IMPROVING THE ANTISKID PROPERTIES OF SLIPPERY PAVEMENTS

A. Deslicking Existing Surfaces

A number of methods for improving the skid resistance of slippery pavements by modifying the existing surface have been developed. Although most of these procedures received only limited application at the time of their development, and none of them is widely used today, a brief summary of the methods will indicate the variety of approaches attempted through the years and may tend to lend support to nonskid surface treatments, which are recommended in the following discussion as the best current remedy for a slippery pavement. The main reason that these deslicking methods are not particularly successful is that they do not correct the initial cause of pavement slipperiness, and although temporary improvement in skid resistance is frequently realized, the surface will generally return to a slippery condition after a relatively short period of wear.

1. **Portland Cement Concrete.** One of the early methods of deslicking polished portland cement concrete involved the direct application on the pavement surface of a dilute hydrochloric acid (9). The acid reacts with the cement paste, as well as with any calcareous surface aggregate, and etches the smooth surface so that the resulting texture shows an increase in skid resistance over the polished condition.

The other general procedure for improving the antiskid properties of existing concrete pavements is to mechanically roughen the surface. The surface may be sandblasted, chipped, or ground with rotary drills, with as much as $\frac{1}{16}$ in. of the concrete being removed (10, 11). A related procedure is the Kogel process, in which the surface is heated by directing flames onto the pavement from a hand-propelled generator (12). The high temperature attained causes the surface to spall, resulting in a roughened texture similar to that achieved by a mechanical process.

2. **Bituminous Surfaces.** Deslicking procedures for bituminous surfaces generally involve the elimination of slippery sections of bleeding asphalt pavements. It is frustrating to the engineer to observe a surplus of a perfectly good binder material on the roadway and yet to be unable to incorporate sufficient cover aggregate with it to provide a reasonably good riding surface. However, when considering the difficulties encountered in obtaining adequate aggregate retention with new construction, as enumerated in Art. V, it is not surprising that the degree of success is rather low in accomplishing aggregate retention with the high-viscosity asphalt occurring on the roadway.

One of the common methods of "blotting" an excess of asphalt at the surface is to spread a surplus of sand over the bleeding area and permit traffic to roll the sand into the asphalt. This may provide temporary improvement in skid resistance, but generally the degree of aggregate retention is small, so that the asphalt may drown the retained aggregate and again present a flushed surface after a short period of wear. Efforts have been directed toward heating either the sand or the surface immediately prior to aggregate application, and this has resulted in somewhat better retention, although it is debatable whether the degree of improvement justified the additional expense. A related procedure was developed in Texas in which crushed stone was treated with kerosene, placed on the slippery areas, and rolled into the surface (6). The kerosene reduced the softening point of the asphalt and promoted retention of the crushed stone.

A treatment developed in California for roughening slick seal coats involves softening the binder by means of heat and grooving the surface with a rake-type drag (2). This results in an improvement in skid resistance, but leaves an unsightly riding surface and one which gives the driver a feeling of pavement instability. The longitudinal grooves tend to guide the vehicle, so that the operator does not retain complete steering control.

B. Nonskid Surface Treatments

The surface types discussed under this general heading differ from the bituminous surface treatments of Art. V in a number of respects. Rather than consisting of separate applications of bituminous material and cover aggregate, they are generally placed in relatively thin layers in one operation, although a tack coat is frequently required. They are composed for the most part of fine-grained abrasive materials and are somewhat more expensive than the conventional bituminous surface treatment. However, they do possess excellent antiskid characteristics, both initially and after appreciable amounts of wear.

1. Kentucky Rock Asphalt. This material occurs as a Mississipian sandstone naturally impregnated with asphalt. The individual particles consist of hard angular quartz grains, most of which will pass a No. 30 and be retained upon a No. 100 sieve. Kentucky rock asphalt is normally placed with an initial compacted thickness of approximately ½ in., but has been applied as a deslicking treatment at rates as low as 8 lb per sq yd.

A Kentucky-rock-asphalt surface typifies the conditions necessary for permanent skid resistance. The individual quartz grains are highly resistant to polishing, and before they have the opportunity to polish excessively, they are dislodged by traffic and replaced at the surface by fresh harsh particles. This continuous rejuvenation results in excellent antiskid characteristics of the pavement during the entire life of the rock-asphalt surface, but in so doing limits the effective life of the treatment. For high-traffic conditions a ½-in. layer of Kentucky rock asphalt may be completely displaced from the wheel-track portion of the roadway in somewhat less than 10 years.

Recently, Kentucky-rock-asphalt surfaces have shown some evidence of excessive stripping and, because of the geographic limitation of source location, will never be too widely used. However, endeavors to simulate the excellent antiskid characteristics of this fine-grained material have resulted in the development of some very effective nonskid treatments, a discussion of which follows.

2. Silica-sand Surface Treatments. Different types of sand-asphalt mixtures have been used with varying degrees of success as deslicking treatments for a number of years. However, it was a concentrated effort in 1955 on the part of the Virginia Council of Highway Investigation and Research to find an inexpensive substitute for Kentucky rock asphalt that resulted in a type of nonskid surface treatment that appears to be a major contribution in preventing and correcting pavement slipperiness (8, 23). This treatment has received wide acceptance and has been placed on some of the most heavily traveled highways in the United States, including the elevated Pulaski Skyway in New Jersey, with its 60,000 vehicles per day, and on many of the major bridges in the New York metropolitan area.

The major constituent of this surface treatment is a harsh silica sand, most of which falls between the No. 30- and No. 100-mesh sieves. Approximately 8 per cent of a penetration-grade asphalt cement is used in the bituminous mixture, along with a small amount of hydrated lime filler to prevent stripping. The mixture may be applied to the surface either with a conventional Barber-Green paver or with a spin-spreader. The paver has been used successfully for application rates as low as 25 lb per sq yd, or somewhat less than a ¼-in. compacted thickness. The spin-spreader distributes from 6 to 10 lb per sq yd per pass and is used for the lighter deslicking treatments where low traffic counts are anticipated. A tack coat is required to promote bonding to the existing surface.

The nature of wear for the silica-sand surface treatment is identical with that of Kentucky rock asphalt and results in excellent antiskid characteristics during the effective life of the treatment. On the basis of experience with Kentucky rock asphalt and with the initial silica-sand test sections, it is estimated that for the traffic encountered on Virginia highways a 15 lb per sq yd application placed on a smooth surface will provide adequate skid resistance for at least 6 years (8). The Port of New York Authority estimates the expected life of a ½-in. silica-sand surface to be from 7 to 10 years with traffic of 5,000 to 7,000 vehicles per day and 4 to 5 years with 50,000 to 60,000 vehicles per day (28).

3. Additional Nonskid Surface Treatments. There are many other varieties of nonskid treatments which have not been used as extensively as the two previously considered. A brief summary of some of these methods will serve to illustrate the variety of materials used in certain applications and may suggest to the reader possibilities of using local materials in developing a satisfactory nonskid treatment for his particular locality.

a. **Rockite.** Rockite consists of a crushed diabase traprock to which is added a powdered asphalt and an asphalt fluxing oil. The resulting material has the appearance of a wet black sand and may be stockpiled for a reasonable period of time. It is placed in thicknesses of from ⅛ to ½ in. on any hard surface, with the greater thickness recommended for high traffic. The nature of wear is similar to that of the silica-sand surface treatment, and the antiskid characteristics are good.

b. **Vultite.** Vultite is composed of an asphalt emulsion, cement, sand, and water. The materials are mixed, spread, and screeded to a thickness of approximately ½ in. The surface is then finished with a mechanical float and rolled with a 1-ton roller.

c. **Resinous Surface Treatments.** The recent application of a resin binder to surface treatments has introduced a new cementing material to highway construction. The resin binder is placed on a prepared surface in a liquid state, and a sharp abrasive grit such as crushed quartz, emery, or alundum is distributed over the surface. As the resin sets chemically to form a tough plastic binder, the grit is permanently fixed in place. The first surface treatment of this nature to be described in the technical literature carried the trade name of Relcote (7, 21, 22).

This treatment is rather expensive and as yet has received only limited application. Initial installations were in toll-booth areas, which receive a tremendous polishing effort from traffic, and early reports on the test sections indicate good results. There is no particle-by-particle wear with this type of surface, and the antiskid characteristics will be determined entirely by the initial shape and the resistance to polishing of the abrasive material. Aluminum oxide, with a hardness of 9 on Moh's scale, may cost 30 to 40 times as much as a silica sand with a hardness of 7. However, for high anticipated traffic the more resistant material may be necessary. It will be interesting to observe the results with this new highway surfacing material and to determine if the high initial expense is economically justified in obtaining skid resistance.

4. Summary on Nonskid Surface Treatments. As traffic continues to increase, some form of thin nonskid surface treatment appears to be the most logical and economical method of eliminating paving slipperiness in those areas for which the major portion of the aggregate is polish-susceptible. The type of treatment will depend upon the availability of material.

A nonskid silica-sand surface treatment, such as that developed by Virginia, will probably be the most generally accepted method of combating pavement slipperiness in the future. In most of the areas in which polish-susceptible aggregates exist, there are also sources of silica sand, frequently occurring as sandstone which can be crushed and used in deslicking operations. Such nonskid surface treatments, when placed on existing slippery pavements that are structurally adequate or used as a preventive measure in new construction with polish-susceptible aggregates, can make a significant contribution to driving safety.

UPDATE

Designers experimented with grooved pavements in the 1970s. The grooves are typically 1 to 2 in. apart, ¼-in. wide, and cut into existing asphalt pavement with diamond-edge circular blades. The grooves do improve control on curves in wet weather. However, they also decrease driver wheel control slightly. They are useless under icy conditions. In fact, grooved pavements have increased accident frequencies in some parts of the country; apparently, driver confidence increases under rainy conditions to the point where drivers are less cautious with ice. The bottom line: Grooved pavements should be reserved only as a last resort, when nothing else works. Pavement grooving should not be designed into new roadways.

Vehicle handling has improved greatly in recent years, thanks to radial tires on autos and antiskid brakes on semitrailers. Nevertheless, such advances are not universal; it will be well into the twenty-first century before all vehicles are equipped with antiskid equipment, for instance. Roadways must be designed for the least-capable vehicles using them.

REFERENCES

1. **Agg, T. R.:** "Tractive Resistance and Related Characteristics on Roadway Surfaces," *Iowa Engineering Experiment Station, Bulletin* 67, 1924.

2. Bangert, N. R.: "Roughening Slick Seal Coats," *California Highways and Public Works,* vol. 31, no. 2-3, March-April, 1952.
3. Benson, F. J., and B. M. Gallaway: "Retention of Cover Aggregate by Asphalt Surface Treatments," *Texas Engineering Experiment Station, Bulletin* 133, 1953.
4. Benson, F. J.: "Seal Coats and Surface Treatments," *Proceedings, 44th Annual Road School,* Purdue University, Lafayette, Ind., 1958.
5. Bressot: *Chairman's Report,* Committee on Slipperiness, presented at the 10th Congress in Istanbul, Permanent International Association of Road Congresses, Paris, 1955.
6. Burton, Joe: "Slick When Wet," *Construction and Maintenance Bulletin 21,* Texas Highway Department, March, 1953.
7. Creamer, W. M., and R. E. Brown: "Application of a New Non-skid Surface Treatment on Connecticut State Highways," *Highway Research Board, Bulletin* 184, 1957.
8. Dillard, J. H., and R. L. Alwood: "Providing Skid-resistant Roads in Virginia," *Proceedings, Association of Asphalt Paving Technologists,* vol. 26, pp. 1–22, 1957.
9. Discussion on "Acid Treatment of Concrete Pavements," *Science News Letter,* vol. 27, no. 14, Apr. 6, 1940.
10. Discussion on "Roughening Treatment for Concrete, Granite and Other Surfaces," *Engineering* (Great Britain), vol. 171, no. 4445, Apr. 6, 1951.
11. Discussion on "Road Roughening," *Surveyor* (Great Britain), vol. 111, no. 3169, Nov. 29, 1952.
12. Discussion on "Skid Proofing with the Kogel Process," *Highways. Bridges, and Engineering Works* (Great Britain), vol. 20, no. 1023, Feb. 17, 1954.
13. Giles, C. G., and A. R. Lee: "Non-skid Roads," *Proceedings, The Public Health and Municipal Engineering Congress,* London, 1948.
14. Giles, C. G., and F. T. W. Lander: "The Skid-resisting Properties of Wet Surfaces at High Speeds," *Journal, Royal Aeronautical Society* (London), February, 1956, pp. 83–94.
15. Giles, C. G.: "The Skidding Resistance of Roads and the Requirements of Modern Traffic," *Proceedings, Institute of Civil Engineers* (London), vol. 6, pp. 216–249, 1957.
16. Goodwin, W. A., and E. A. Whitehurst: "A Device for the Determination of Relative Potential Slipperiness of Pavement Mixtures," paper presented at the 37th Annual Meeting of the Highway Research Board, 1958.
17. Grunau, D. L., and H. L. Michael: "Skid Characteristics of Pavement Surfaces in Indiana," *Highway Research Board, Bulletin* 139, 1956.
18. Moyer, R. A.: "Skidding Characteristics of Automobile Tires on Roadway Surfaces and Their Relation to Highway Safety," *Iowa Engineering Experiment Station, Bulletin* 120, Ames, Iowa, 1934.
19. Moyer, R. A., and J. W. Shupe; "Roughness and Skid Resistance Measurements of Pavements in California," *Highway Research Board, Bulletin* 37, 1951.
20. Moyer, R. A.: "Building Safety into California Highways," *Proceedings, Tenth California Street and Highway Conference,* University of California, Berkeley, Calif., 1958.
21. Nagin, H. S., T. G. Nock, and C. V. Wittenwyler: "The Development of Resinous Skid-resistant Surfaces for Highways," *Highway Research Board, Bulletin* 184, 1957.
22. Nagin, H. S., and Others: "The Development of Techniques for Applying Resinous Skid-resistant Surfaces to Highways," paper presented at the 37th Highway Research Board Meeting, 1958.
23. Nichols, F. P., J. H. Dillard, and R. L. Alwood: "Skid Resistant Pavements in Virginia," *Highway Research Board, Bulletin* 139, 1956.
24. Normann, O. K.: "Braking Distances of Vehicles from High Speed," *Public Roads,* vol. 27, pp. 159–169, June, 1953.
25. Papers of Committee C on "The Relationship of Tire Design and Composition to Skidding," *Proceedings, First International Skid Prevention Conference,* University of Virginia, Charlottesville, Va., 1958.
26. Rice, J. M.: "Highlights of Recent Activities at NCSA Laboratory," *Journal, The National Crushed Stone Association,* December, 1955.
27. Rice, J. M.: Discussion of paper "Providing Skid-resistant Roads in Virginia," *Proceedings, The Association of Asphalt Paving Technologists,* vol. 26, pp. 19–21, 1957.
28. Ruefer, A. L.: "Test Installations of Thin Silica Sand Asphalt Resurfacing," paper presented to the 37th Annual Meeting of the Highway Research Board, 1958.
29. Sabey, B. E.: "Accident Reports as a Guide to Slippery Lengths of Road," *Roads and Road Construction* (Great Britain), vol. 34, pp. 203–206, July, 1956.
30. Shelburne, T. E., and R. L. Sheppe: "Skid Resistance Measurements of Virginia Pavements," *Highway Research Board, Research Report* 5-B, 1948.
31. Shupe, J. W.: "Factors Affecting the Skidding Resistance of Bituminous Pavements," thesis submitted to the University of California, Berkeley, for the degree of Master of Science, 1951.

32. Shupe, J. W.: "Progress Report No. 5 on Skid Resistance Study for U.S. 31 Test Road," *Joint Highway Research Project, Report* 36, Purdue University, Lafayette, Ind., 1956. (Unpublished)

33. Shupe, J. W., and R. W. Lounsbury: "Polishing Characteristics of Mineral Aggregates," *Proceedings, First International Skid Prevention Conference,* University of Virginia, Charlottesville, Va., 1958.

34. Shupe, J. W., and W. H. Goetz: "A Laboratory Method for Determining the Skidding Resistance of Bituminous Paving Mixtures," paper presented to the 61st Annual Meeting of the American Society for Testing Materials, 1958.

35. Shupe, J. W., and W. H. Goetz: "A Laboratory Method for Evaluating Slipperiness," *Proceedings, First International Skid Prevention Conference,* University of Virginia, Charlottesville, Va., 1958.

36. Shupe, J. W.: "A Laboratory Investigation of Factors Affecting the Slipperiness of Bituminous Paving Mixtures," thesis submitted to Purdue University, Lafayette, Ind., for the degree of Doctor of Philosophy, 1958.

37. Skeels, P. C.: "Measurement of Pavement Skidding Resistance by Means of a Simple 2-wheel Trailer," paper presented at the 37th Annual Meeting of the Highway Research Board, 1958.

38. Stuzenberger, W. J., and J. H. Havens: "A Study of the Polishing Characteristics of Limestone and Sandstone Aggregates in Regard to Pavement Slipperiness," paper presented at the 37th Annual Meeting of the Highway Research Board, 1958.

39. Whitehurst, E. A., and W. A. Goodwin: "Pavement Slipperiness in Tennessee," *Proceedings, Highway Research Board,* vol. 34, pp. 194–209, 1955.

Section 11

SOIL STABILIZATION

A. W. JOHNSON, *Engineer of Soils and Foundations, Highway Research Board, Washington, D.C.* (Cement-treated Soil Mixtures).

DR. MORELAND HERRIN, *Associate Professor of Civil Engineering, University of Illinois, Urbana, Ill.* (Bituminous Soil and Aggregate Stabilization).

DR. D. T. DAVIDSON, *Professor of Civil Engineering, Iowa Experiment Station, Iowa State University of Science and Technology, Ames, Iowa* (Soil Stabilization with Lime, Lime-pozzolan, Chlorides, Lignin Derivatives, and other Chemicals; Construction).

DR. R. L. HANDY, *Assistant Professor of Civil Engineering, Iowa Experiment Station, Iowa State University of Science and Technology, Ames, Iowa* (Soil Stabilization with Lime, Lime-pozzolan, Chlorides, Lignin Derivatives, and other Chemicals; Construction).

I. INTRODUCTION

When a less stable soil is treated to improve its strength and its resistance to change, it is said to be "stabilized." Thus stabilization infers improvement in both strength and durability. In its earlier usage, the term stabilization signified improvement in a qualitative sense only. More recently stabilization has become associated with quantitative values of strength and durability which are related to performance. Those quantitative values are expressed in terms of compressive strength, shearing strength, or some measure of bearing value or load deflection to indicate the load-bearing quality; and in terms of absorption, softening, and reduction in strength, or in terms of direct resistance to freezing and thawing, and wetting and drying to indicate the durability of the stabilized construction.

II. CEMENT-TREATED SOIL MIXTURES

A. Introduction

Cement stabilization consists of a mixture of pulverized soil and measured amounts of portland cement and water, compacted to a high unit weight and protected against moisture loss during a specified curing period.

Cement-treated soil mixtures differ somewhat for the two principal types of soils. In the fine-grain silty and clayey soils, the cement, on hydration, develops strong linkages among and between the mineral aggregates and the soil aggregates to form a matrix which effectively encases the soil aggregates. The matrix forms a honeycomb type of structure on which the strength of the mixture depends, since the clay aggregations within the matrix have little strength and contribute little to the strength of the soil cement. The matrix is effective in fixing the particles so they can no longer slide over each other. Thus the cement not only destroys the plasticity but also provides increased shear strength. The surface chemical effect of the cement reduces the water affinity and thus the water-holding capacity of clayey soils. The combination of reduced water affinity and water-holding capacity and a strong matrix provides an encasement of the larger unpulverized raw soil aggregates. Because of its reduced water affinity and strength, this serves not only to protect them but also to prevent them from swelling and softening from absorption of moisture and from suffering detrimental effects from freezing and thawing.

In the more granular soils the cementing action approaches that in concrete except that the cement paste does not fill the voids in the aggregate. In sands, the aggregates become cemented only at points of contact. The more densely graded the soil, the smaller the voids, the more numerous and greater the contact areas, and the stronger the cementing action. Uniformly graded ("one-size") sand, which has a minimum of contact area between grains, requires fairly high cement contents for stabilization. Since well-graded granular soils generally also have a low swell potential and low frost susceptibility, it is possible to stabilize them with lesser cement contents than are needed for the uniformly graded sands and the more frost-susceptible silts and the higher swelling- and frost-susceptible clayey soils. For any type of soil, the cementing process is given the maximum opportunity to develop when the mixture is highly compacted at a moisture content that facilitates both the densification of mix and the hydration of the cement.

B. Types of Cement-treated Soil Mixtures

Four major variables control the degree of stabilization of soils with cement. They are (1) the nature of the soil, (2) the proportion of cement in the mix, (3) the moisture content at the time of compaction, and (4) the degree of densification attained in compaction. It is possible, simply by varying the cement content, to produce mixes that, after hydration of the cement, may range from those which result in only a slight modification of the compacted soil to the product known as soil cement, which must meet certain minimum strength and durability requirements. When moisture is increased sufficiently to produce a plastic mix and the cement content adjusted to meet strength and durability requirements for the plastic condition, the product becomes plastic soil cement.

The possibility of controlling the properties of the mix to suit the construction and the degree of stabilization to satisfy the strength and durability requirements have resulted in the development of four principal types of cement-soil mixtures, as follows:

1. Soil Cement. Soil-cement mixtures are designed to satisfy stated criteria determined by Standard AASHO and ASTM tests. This material is commonly termed soil cement.*

2. Cement-modified Granular Soil Mixtures. Cement is used here principally to reduce plasticity and swell characteristics and thus to improve the bearing value of marginal or substandard granular materials to make them acceptable base or subbase materials for both rigid and flexible pavements. The cement contents may range from about 1 per cent by weight upward, but is always less than that required for soil cement.†

---

* Material specified in California and some other Western states as cement-treated base, Class A (and Class B Mixtures), satisfies criteria for soil cement.

† This type is equivalent to some Class C "cement-treated base" specified in some Western states.

3. Cement-modified Silt-clay Soil Mixtures. In cement-modified mixtures the cement is used to control the swell-shrink characteristics of the soil. This degree of stabilization may also be used to strengthen abnormally weak soils or wet-soil areas. This type always contains less cement than is required for soil cement.

4. Plastic Soil Cement. This is a cement-soil mixture that can be placed in a plastic state, yet hardens into a material that meets the strength and durability requirements set for soil cement. It is usually made from the lighter-textured, usually sandy soils.

5. Cement-treated Soil Slurries and Grouts. These are used for mudjacking pavements and for stabilizing railroad ballast.

C. Properties

1. Optimum Moisture Content and Maximum Density of Cement-treated Soil Mixtures. The optimum moisture contents and maximum densities of compacted cement-treated soil mixtures are approximately those of the raw soil for a large proportion of the soils tested when normal mixing times are observed. Some soils do exhibit marked departures in optimum moisture content and maximum density, but they are limited in number. Most departures are of the order of 1 to 3 lb per cu ft. Increases in density usually occur for sands and sandy soils and sometimes in small degree for heavy clays. Little or no change occurs for the light to medium clays. Decreases in density may occur in silts. Decreases in optimum moisture content occur for clays. Increases occur for the silts, and little or no change takes place for sands and sandy soils.

2. Soil Amendments and Additives. *a.* **Soil and Aggregate Admixtures.** Soil amendments and additives have been used since the earliest projects (9)* to improve the reaction between the soil and the cement. Normally reacting soils have been used to amend soils which showed poor reaction by altering the soil grading or by diluting the poorly reacting soil. The favorably reacting materials, for example, limestone screenings and crushed limestone, have been used. Fine-grain soil has been added to clean sands and, contrariwise, to sand gravels, and sands, sand gravels, and pulverized bituminous surfaces have been mixed into clays to improve the reaction, and in many instances to reduce cement requirements.

3. Hydrated Lime or Quicklime. Hydrated lime has been used as an admixture to cement-treated soil mixtures to improve the cement reaction of some organic soils that exhibit retarded setting or are productive of abnormally low strengths when mixed with portland cement alone. An example of this type of application is a uniformly graded fine sand (70 per cent between sizes 0.1 and 0.2 mm and 3 per cent silt and clay sizes) containing detrimental organic matter that was not evident on visual inspection (but showed 0.3 per cent organic matter on analysis by the dichromate method) (96). Lime produced a beneficial effect in the form of early hardening of this type of sand-cement mixture that exhibited retarded setting time up to 7 days when mixed with normal portland cement alone. The addition of 2 per cent hydrated lime reduced retardation to about 2 days.

Lime has also been used as an admixture to highly plastic materials to facilitate pulverization and mixing and to increase compressive strength and resistance to loss in the wet-dry test, for the wet-dry test is often a significant criterion for determining cement requirements for plastic high-volume-change soils. Studies by the Corps of Engineers (188) have shown that 2 per cent hydrated lime was effective in reducing wet-dry losses on plastic base material.

4. Fly Ash. Two investigations have been made using fly ash as a soil amendment. One of these (165) involved a coastal silty soil (18 per cent sand, 56 per cent silt, 26 per cent clay, LL = 30, PI = 5) and mixtures of 94-3-3 and 94-4-2 parts of soil, cement, and fly ash, respectively. After 28 days curing, the 94-3-3 mix showed a 5.9

---

* Numbers in parentheses refer to corresponding items in the references at the end of this section.

per cent weight gain in the freeze-thaw test and a compressive strength of 33 psi after 12 cycles of test; and the 94-4-2 mix showed a weight gain of 0.70 per cent and a compressive strength of 220 psi after 12 cycles of freeze-thaw. The second investigation (184) included tests on a friable loess, a plastic loess (PI = 12), an alluvial clay (PI = 47), and a nonplastic dune sand. Cement contents ranged upward to 12 per cent, and fly-ash contents ranged from 9 to 21 per cent in terms of soil replacement, and varying percentage in terms of cement replacement. The conclusions (184) were that fly ash was not markedly beneficial as an admixture for the soils tested except as it reduced shrinkage cracking in the clay soil. It had no marked effect on cement-treated-soil strength and was detrimental to freeze-thaw resistance.

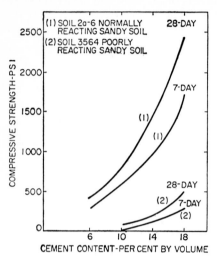

Fig. 11-1. Comparison of compressive strengths of a normally reacting sandy soil and a poorly reacting sandy soil (41).

5. Calcium Chloride. The influence of calcium chloride as a beneficial admixture to mixtures of cement and organic soils has been determined both in the United States (9, 37, 41, 147) and in Great Britain (96, 116, 136). The 7-day and 28-day compressive strengths of a normally reacting soil compared with similar data for a poorly reacting soil are shown in Fig. 11-1.

In the United States, tests performed on nine soils (41), four of which were poorly reacting and five normally reacting, showed that small percentages of calcium chloride had marked effect in improving the reaction of soils that showed poor reaction with cement alone. An example of data for one soil is shown in Fig. 11-2 for a soil without calcium treatment that required more than 26 per cent cement by volume for satisfactory hardening. With the addition of 0.4 to 1.0 per cent calcium chloride the soil was hardened satisfactorily with 14 per cent cement. It may be seen from Fig. 11-2 that compressive strength increases to an optimum and decreases with further increase in calcium chloride. Improvement in durability as indicated by wet-dry and freeze-thaw tests is of the order indicated by the compression-test data for the number of cycles tested. Generally, 0.6 per cent calcium chloride is an optimum value when both short-period and long-period effects are considered (41).

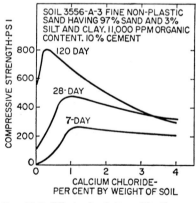

Fig. 11-2. Effect of calcium chloride content on the compressive strength of a poorly reacting organic sand (41).

D. Application of Types of Cement-treated Soil Mixtures to the Nature of the Facility

The various types of cement-treated soil mixtures were initially developed to serve under different conditions of use, depending upon the requirements for the facility.

The suggested use of each of the various types is based on the properties of each and the needs of the facility. The facilities listed here under each type of cement-treated soil mixture are in the main those found in the engineering literature. Information pertaining to preliminary testing, to criteria for mix design, and to items of geometric and structural design of soil-cement facilities is given later.

1. Soil Cement. This item includes soil-cement mixtures that satisfy accepted criteria for compressive strength and moisture gain, volume change, and brushing losses in the wet-dry and freeze-thaw tests when each is pertinent. Mixtures specified in Western states as Class A cement-treated base and some mixtures included under Class B cement-treated base satisfy criteria for soil cement. The use of soil-cement mixtures that satisfy those criteria is suggested for the following applications:

1. Subbases and base courses for rigid and flexible pavements for roads, streets, and airfields
2. Surfaced shoulders for highways and airfields (32, 45, 175)
3. Surfaced parking areas (157)
4. Surfaced storage areas for aggregates, miscellaneous materials, and equipment (75, 152)
5. Unsurfaced horizontal multiple-lift thick slope facings subject to periodic or continuous inundation and wave action (15, 110, 132)
6. Earth-dam cores (65)
7. Unsurfaced linings for reservoirs (68, 71, 132)
8. Foundations for some types of structures (110, 157)
9. Maintenance reconstruction and patching of granular bases (45, 48)
10. Modification of frost-susceptible soils (21, 161, 171)
11. Miscellaneous constructions including surface drains, culverts, surfaced sidewalks and paths (25), small arch bridges, masonry units (brick or block) for building construction (77, 114, 127), rammed monolithic construction of small buildings, etc. (77, 114, 127) where special conditions warrant (53, 83, 139)

2. Cement-modified Granular Soils (21, 38, 45, 59, 93, 118, 128, 140, 144, 147, 157, 178). Many sandy and gravelly soils are only slightly substandard as materials for bases, subbases, and subgrades, for flexible- or rigid-type pavements. They may contain excessive proportions of fine fraction material or excessively plastic fines, or both, and need improvement to bring them to a minimum acceptable quality. This may require only sufficient cement to modify the plastic properties of the soil, or it may require sufficient cement for substantial hardening to a quality only slightly less than that possessed by soil cement. Cement-modified granular soils are used in the following facilities:

1. Base and subbase courses for flexible-type surfaces for roads, streets, and airfields where type of material, traffic, and climatic conditions permit
2. Subbase and subgrade treatment under rigid-type pavement to prevent erosion by pumping action of the slabs
3. Patching and reconstruction of failed granular bases
4. Maintenance strengthening of subgrades and bases in patching operations
5. Modification of frost-susceptible granular soils

3. Cement-modified Silty and Clayey Soils (5, 6, 10, 11, 18, 60, 66, 84, 93, 110, 118, 140, 144, 152, 154, 185). Cement, in lesser proportion than for soil cement, is used to improve the performance of subgrade soils. The several purposes for which this type of soil-cement mixture is used are as follows:

1. As treatment to control shrinkage and expansion of high-volume-change subgrade soils
2. To improve the strength characteristics of subgrades
3. To reduce the effects of frost action on subgrades

4. Plastic Soil Cement (68, 94, 101, 158, 159). The difficulty of placing and compaction at optimum moisture content the usual type of soil-cement mixtures in other than installations permitting the use of flat surfaces led to the development of plastic soil cement. This type of mix is used in the following installations:

1. Linings for roadside drainage channels (ditches)

2. Linings for irrigation canals
3. Sacked riprap for erosion protection

III. BITUMINOUS-AGGREGATE-AND-SOIL STABILIZATION

A. General

In various regions of the United States, locally available materials are stabilized with bitumens in order to obtain economy in highway construction. The Great Plains area, the Basin and Range, and the Coastal Plains are especially suited for this type of stabilization. In large parts of these areas, coarse mineral aggregate is scarce and is not available within economical haul distances. These same regions are abundant with sands and silts that can be stabilized readily with bituminous materials. The weather, in addition, is generally suited for this type of construction. In these regions, then, bituminous stabilization is used now and probably will be used more extensively in the future as the coarse mineral aggregate becomes less plentiful.

Bituminous-stabilized materials are used primarily for base and subbase construction and may be composed of a variety of materials. A wide range of natural materials such as bank-run gravel, sand, sand clay, silty clay, limestone screenings, slag, and cinders are used. These materials are stabilized with a variety of bituminous materials, ranging from asphalt cements and asphalt cutbacks and emulsions to road oils and tars. The multiplicity of the number of naturally occurring materials and the bitumen types that can be used indicates the wide versatility of bituminous stabilization.

B. Types of Bituminous-stabilized Mixtures

When employed as a stabilizing agent, bitumens are used to produce various effects. These uses are divided into three principal groups:

1. To provide strength to cohesionless materials, such as clean sands, by acting as a binding or cementing agent. This material is usually termed sand bitumen.
2. To stabilize the moisture content of cohesive, fine-grain soils. This material is usually known as soil bitumen.
3. To provide cohesive strength and to waterproof materials that inherently possess frictional strength. When used with pit-run gravels, this type is usually known as sand-gravel-bituminous mixture.

Bitumens are also used on earth roads in order to make the surface dustproof, waterproof, and abrasive-resistant. This treatment is normally for plastic, clayey soils, where only a low-type road is desired.

1. Sand Bitumens. Theoretically, sands are stabilized with bituminous materials in order to provide the necessary cohesive strength. Although sand possesses high strength when confined, because of internal friction between the sand grains, it has no cohesive strength when dry and has a false type of cohesion when wet. If a thin film of bituminous material is supplied around each sand grain, the sand is bound or cemented together. This treatment provides cohesive strength without undue interference with the frictional resistance of the sand. An excess of asphalt must not be used, for it provides a thick film around the sand grains and the frictional resistance may be lost. An optimum amount of bitumen is needed for highest strength. Typical variations in stability of sand-bituminous mixtures with asphalt content are shown in Fig. 11-3.

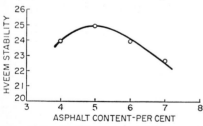

Fig. 11-3. Variation in stability with asphalt content of hot-mix sand asphalt.

Almost any type of sand is satisfactory for use in sand-bituminous mixtures provided it is graded well enough to have satisfactory strength for the use intended. Some sands, such as blow and dune sands, however, are disproportionately graded and may have to be adjusted by adding fine, inert mineral filler. In any case, the sand must be limited to practically cohesionless material. Plastic clay in large quantities cannot be tolerated.

Of the various types of bituminous-stabilized materials, the sand-bituminous mixture perhaps presents the widest latitude of use. When suitable bituminous binder and construction methods are employed, these mixtures can range in strength and durability from that needed for the highest type of construction down to that suitable for lightly traveled, inexpensive roads. Sand bitumen is widely used, because suitable sands are fairly well distributed throughout the world.

2. Soil Bitumens. Most cohesive soils have satisfactory bearing capacity for highway use at low-moisture contents. These soils, however, tend to lose stability at high moisture contents. Bituminous materials are used with them as waterproofing agents to maintain the low moisture content and to act as a stabilizing agent. Not only are the soil capillaries effectively plugged or blocked by the bituminous materials, but a partially protective bituminous film is also formed around the soil aggregation. This condition prevents an appreciable amount of moisture change within the soil mass and maintains the inherent cohesive strength of the soils. Care must be taken, for in some instances, the bitumen content required for waterproofing may be so much that it may be as detrimental to the strength as large amounts of moisture.

Practically, only those soils that can be pulverized can be stabilized with bitumens. Soil types then are limited to those with low liquid limits and low plasticity index numbers. Usually, satisfactory soils for bituminous stabilization have liquid limits less than 30 and plastic limits of about 18. Any soil that is of greater plasticity should be checked to determine if it can be satisfactorily pulverized before attempting to stabilize it with bituminous materials.

3. Sand-gravel-bituminous Mixtures. Inherently, naturally occurring sand gravels usually have fairly high strength. This characteristic results from aggregate interlock and frictional resistance due to the reasonably dense gradation of the material. The function of the bituminous binder, then, is to provide cohesive stability to the mass and to act as a waterproofer. Only a relatively small amount of bituminous material is needed, because thick films of bitumen on the aggregate, provided by large amounts of bituminous material, act as lubricants and lower the stability.

C. Factors Governing Application of Type

The type of bituminous-stabilized material to be used for a particular project depends upon such factors as characteristics of available materials, desired quality of bituminous-stabilized mixture, climate, local practice, etc. All factors should be evaluated when selecting the type to be used.

1. Soil Characteristics. Perhaps the most important factor in selecting the type of bituminous-stabilized material to be used is the availability of aggregate. For instance, if sands are prevalent, then sand-bituminous bases can be constructed. If, however, sands are scarce but suitable soils are available, then soil-bituminous stabilization can be used. Plastic clays should only be surface-oiled. If more than one type of soil is available and satisfactory for stabilization with bituminous materials, the more granular material should be used. Stability and density of bituminous-stabilized mixtures tend to increase as the maximum size of the particles increases.

Soils that exist in the roadbed are not the only materials that can be stabilized with bituminous materials. If the material existing on the roadbed does not meet requirements, then satisfactory materials can be carried to the site and added to the existing materials until the mixture meets requirements. Entirely new material may also be brought to the job site and used. In fact, in many instances, it is desirable to use entirely new material. Materials from deposits and borrow pits can be uniform in quality. Thus the entire project would have uniform material and the same bituminous content and controls would exist for the entire job.

2. Quality of the Bituminous-stabilized Mixtures. Bituminous-stabilized mixtures vary widely in quality. They range from the very-high-type sand-gravel-bituminous and sand-bituminous bases to nothing more than surface oiling. Quality is based upon the materials, construction methods, and controls which are used.

The highest type of soil-stabilized material is hot-mix sand asphalt and sand-gravel bitumen. High strength is obtained from the use of granular materials, hard asphalt-cement binder, and fairly rigid control of materials and construction. These types of bases can be satisfactorily used in the most heavily traveled roads if they are of high quality and are supported by adequate subbase and surfaced with asphalt concrete.

Usually sand-bituminous mixtures constructed with liquid bitumens are not quite as high in quality as the hot-mix sand asphalt. They are satisfactory, though, for use as bases on roads carrying medium amounts of traffic.

Although soil-asphalt mixtures are of medium quality, they can be used on roads carrying as many as 1,500 vehicles per day. Primarily, they are used on secondary roads where travel is light or medium.

Because they are usually used as they come from the pit and lack uniformity, sand-gravel-bituminous–stabilized materials may be practical only for use on more lightly traveled roads. Nevertheless, the quality of sand-gravel-bituminous–stabilized roads can be very high if rigid gradation requirements and close construction controls are required.

3. Other Factors. Different types of materials for bituminous stabilization and the processing of these materials vary widely in different areas because of such factors as climate and local practice.

In areas of heavy rainfall, soil-bituminous materials normally cannot be used. The moisture content of the soils is usually high, and adding additional liquid results in loss of strength. Aeration of the mixture is virtually impossible. Thus the volatile matter cannot be removed and the strength of the mix cannot be increased. On the other hand, sand-bituminous mixtures can be used in these areas, for they sometimes can have the water "rolled" out. Of the materials considered, hot-mix sand asphalt is the least affected by rainfall and can be used wherever asphalt concrete can be laid. Rainfall also influences the use of asphaltic material. Liquid asphalts are difficult to cure in areas of heavy rainfall. On the other hand, emulsions are especially suited to arid regions. The water in the emulsion helps to provide moisture necessary for optimum compaction.

Air temperature influences, to a great extent, the rate of evaporation of water and volatiles from the stabilized material. The higher the air temperature, the more quickly the materials will volatize. In fact, in very warm weather, rapid-curing cutbacks tend to cure too rapidly and should not be used. In any case, no attempt should be made to stabilize soils with liquid bituminous materials in cold weather. In warmer climates heavier grades of liquid bituminous materials can be used.

The type of mixing and construction equipment available also influences the quality of the bituminous-stabilized base material. If a hot-mix plant and finishing machine are not available, hot-mix sand asphalt cannot be laid down. If good mixing equipment, such as a traveling-plant mixer, is available, then heavier grades of asphalt can be used. Where only the minimum amount of equipment, such as a blade, is available for mixing, lighter grades and softer cutbacks must be used so that mixing can be thorough and volatilization does not take place too rapidly.

4. Limitations. In addition to the usual material and equipment limitations, probably the most serious limitation to the construction of bituminous-stabilized roads is the engineer's lack of knowledge of design methods and controls.

A number of different methods of designing bituminous-stabilized mixtures exist. In some instances there is a different method for each of the different types of bituminous materials used. These methods present different ideas of mix design. Usually, it is difficult for a materials engineer to differentiate between them and to decide which method is best suited to the type of stabilization he plans to use.

Even after an engineer has been able to evaluate the various available testing methods, he still has to decide on the exact criteria to be used for mix design. Insufficient correlation and research work have been done to establish the *exact* criteria needed for the various methods of test and for stabilizing the various bituminous mixes.

Another limitation to the development of good bituminous-stabilized materials is the lack of construction experience of the highway engineer himself. Most engineers are not familiar with soil-bituminous mixtures. They are also hesitant to experiment with new methods of construction. Familiarity will bring confidence and a willingness to use the new material.

D. Thickness and Wearing-course Requirement

Bituminous-stabilized materials are used primarily as bases and subbases for flexible pavements and must be designed as such. This flexible pavement must be covered with a surfacing, usually abrasion-resistant. Although the exact thickness of base or subbase and the type of surfacing should be determined by normal flexible-pavement design procedures, they are generally determined by current local practice.

1. Thickness of Various Types. The thicknesses of bituminous-stabilized bases and subbases depend upon the same design factors that apply to all bases, such as subgrade and subbase strength, intensity and type of wheel load, and moisture conditions. Various design methods are employed, such as the Corps of Engineers' CBR method, Hveem-Stabilometer method, and others.

Generally, the minimum thickness for the construction of bituminous-stabilized materials is 4 in. This depth is about the minimum that can be economically processed with existing equipment. Usually about 10 in. is the maximum thickness. The material may be placed in one or two lifts of 3 to 5 in. each or in as many as five lifts of approximately 2 in. each if high density is desired.

For the more lightly traveled roads carrying fewer than 400 vehicles per day, 4 to 6 in. of bituminous-stabilized base is usually satisfactory. When more traffic is carried, however, 8 to 10 in. of bituminous-stabilized base must be used. On heavily traveled routes, bituminous-stabilized materials should have the maximum thickness of 8 to 10 in. and should also have additional support provided by granular subbase and high-type bituminous-concrete surfacing.

More specifically, sand-bituminous mixtures, except hot-mix sand asphalt, are laid in layers 4 to 8 in. thick. The hot-mix sand asphalt is usually placed in 2-in. layers until the total thickness of 8 to 10 in. is reached. Soil-bituminous and sand-gravel-bituminous materials are usually 4 to 10 in. thick, laid down in one or two lifts.

2. Effectiveness as Wearing Course. Bituminous-stabilized materials of all types, except possibly hot-mix sand asphalt, cannot withstand the abrasive action of fast-moving pneumatic tires. Raveling soon occurs, and this leads to very rough surfaces and usually potholes. As a protection, bituminous-stabilized materials should be covered with a surfacing that is abrasive-resistant. Also, these surfaces provide additional protection, since they reduce deterioration of the bitumens in the base material.

The type of wearing course to be used on a bituminous-stabilized base depends upon the amount of traffic to be carried by the road. For roads having light to medium traffic, a single- or double-surface treatment provides satisfactory service. Whenever heavy traffic is to be carried, a thicker surface, such as bituminous concrete, must be used.

On these roads, where the surface is stabilized with bituminous materials other than asphalt cement and where water and/or volatiles must be cured out, abrasive-resistant surfaces should not be placed on the base for at least 2 weeks after compaction. This length of time is needed for curing of the base material in order to remove water and volatile materials. Hot-mix sand asphalt, however, has considerable cohesive strength and is sometimes used without surfacing, although the use of a seal coat is good practice. When used with a surfacing, no curing time is needed after laydown, because there are neither volatiles nor water to be cured from the base. Hot-mix bituminous-concrete surfacing can be placed immediately.

E. Sand-bituminous Mixtures

When sand, with or without a mineral-filler admixture, and a bituminous binder are mixed so that the bituminous binder provides cohesive strength, the material is known as sand-bituminous mixture. The primary function of the bituminous material is to provide cohesive strength and not waterproofing, as is the case in other types of soil stabilization.

1. Sand and Bituminous-material Requirements. The sand can be river, beach, dune, pit, or any other type of sand as long as it is substantially free from clay balls, vegetable matter, and other organic materials. Gradation of the sand is not too restrictive, and a wide range of gradation can be used. The amount of fine material, however, is usually critical. In most cases, not more than 12 per cent of the fine material can be allowed to pass the No. 200-mesh sieve. In some sands, such as wind-blown or dune sands, as much as 25 per cent of the material may pass the No. 200-mesh sieve. In no case, though, should the final composition contain much clay. The plastic limit of the material should be less than 10 per cent, and if possible less than 6 per cent. For those sands, such as wind-blown and dune sands, which cannot meet the requirements, the gradation can be adjusted by adding mineral filler or fine nonplastic soil such as loess.

Angular and rough-textured sand grains provide high stability. Many sands, however, are rounded and smooth-surfaced and have low strength. For such sands, the strength can usually be increased by adding fine material such as mineral filler.

Various types of bitumens have been used for stabilizing sand. The exact type depends upon the quality of the base desired and upon experience with and availability of the bituminous materials. These bituminous materials may be one of the following:

ASPHALT CEMENT: Penetration grades 85–100 and 120–150. The use of asphalt cement requires heated sands, hot mixing, and hot laydown. It provides the highest type of sand asphalt. If higher stability is needed, harder asphalt cements can be used within limits.

ASPHALT CUTBACK: Grades RC-1, 2, and 3. The grade used is that best suited for the climatic and mixing conditions. Sometimes a special RC-1 that has a harder residue (45–60 penetration), instead of the normal 80–120 penetration, is used (219).

ASPHALT EMULSION: Grade SS-1. Only this standard grade of emulsion is generally used, because the other grades may tend to break too quickly. However, some states such as Indiana and Iowa have specifications for MS types of emulsions which are suitable for this purpose (see Sec. 9).

ROAD TARS: Grades RT-6 to RT-10. The exact grade of tar depends upon the mixing and the climatic conditions.

2. Mixture Design. *a.* **Sand Stabilized with Asphalt Cement.** The oldest and most widely used test for designing hot-mix sand asphalt is the Hubbard-Field stability test. Specimens of the hot-mix sand-bituminous mixture, 2 in. in diameter and 1 in. in height, are molded under a static load. After they are cured, the specimens are immersed in a 140°F water bath before testing. They are then placed in the Hubbard-Field testing apparatus, which also contains water at 140°F, and the force required to extrude the samples through the hole is measured.

The Hubbard-Field design method uses only the strength value for determining the asphalt content and does not take additional criteria into account. Hot-mix sand asphalt must have a minimum Hubbard-Field stability of 800 lb at 140°F (219). The Hubbard-Field test must be used with care, for principally cohesive strength is measured. Mixtures that contain an excess of asphalt may have satisfactory cohesion but low frictional resistance.

Because of the deficiencies in the Hubbard-Field design method, hot-mix sand asphalt has been designed in recent years in a manner similar to that used for the design of hot-mix bituminous concrete. The same procedure is used for molding and for testing sand-asphalt and bituminous-concrete specimens. The only difference in the design procedures is that different criteria are used for selecting the optimum asphalt content. Usually no voids criteria are used with sand asphalt, and the optimum asphalt content is selected solely on the basis of strength. The Hveem gyratory method (220) has been used for sand-asphalt design. In the upper 2 in. of the sand-asphalt base the Hveem stability must be 20 or more; below the upper 2 in. a minimum Hveem stability of 17 is allowed (226). The Marshall method of design has also been used for designing hot-mix sand-asphalt mixtures with a minimum allowable Marshall stability of 500 lb.

Usually the asphalt content of hot-mix sand asphalt is between 5 and 10 per cent, based on the total weight.

3. Beneficial Additives. Sands that have low stability because of poor gradation or because of round, smooth grains can usually be increased in stability by the addition of filler material. This admixture may be finely ground crushed stone, rock dust, other fine sands, portland cement, loess, or any fine inert material. The material should definitely be nonplastic.

There is an optimum amount of mineral fines that can be used for a particular sand in order to produce maximum stability. Any quantity of fines less or more than this optimum amount does not produce as high stability. The optimum amount varies with the type of filler and with different sands. The exact amount of a particular fine material to be used with a specific sand in order to obtain maximum stability can be determined only by stability tests. Care must be taken that too much fine material is not used, for it causes brittleness in the mixture.

Hydrated lime has also been used as an admixture for sand-bituminous mixtures to aid the coating of sand grains. Usually, 1 to 2 per cent of the lime is added to wet sand, and then 4 to 10 per cent of the cutback material is added. Since excessive quantities of bitumens are not added, there is not too much difficulty in squeezing the excess water out by rolling. This method, of course, should be used only in those areas where rainfall is high and normal drying is difficult (230).

4. Recommendations for Construction and Field Control. Sand-asphalt bases are constructed in a manner similar to hot-mix asphalt concrete. The sand is heated in a dryer before use. All materials are accurately proportioned and combined at a central mixing plant. The hot sand-asphalt mixture is carried to the site by truck and is laid down with a finishing machine. Finally, like asphalt concrete, the sand-asphalt mixture is compacted with heavy rollers. Usually the sand-asphalt layers are not more than 2 in. in thickness and are compacted to at least 95 per cent of the density obtained in the laboratory. The techniques that are so important for the construction of bituminous concrete are also important for the construction of hot-mix sand-asphalt bases.

Different construction techniques must be used when liquid bituminous materials are employed to stabilize sands. The moisture content in the sand must be reduced to less than 5 per cent by aeration before the bituminous material is added. Unless this is done, there will be difficulty in mixing and the possibility of satisfactorily coating the sand grains will be greatly reduced.

Liquid bituminous materials should be mixed with sand by satisfactory equipment. Travel-plant mixers and blade graders are generally used. The type and efficiency of the mixing apparatus influence the grade and type of bituminous material used.

After the bitumen and sand are mixed, the mixture must be aerated until the volatiles and moisture contents are reduced. High strength is not obtained until these materials have been released from the mixtures. Usually, a maximum of 5 per cent should be allowed in these mixtures at the time of compaction. Of course, hot-mix sand asphalt requires no aeration prior to compaction. The sand-bituminous material has been aerated enough if the stability of the mixture is satisfactory when compacted. The materials should be compacted with the pneumatic-tired or smooth, steel-wheel rollers.

Most sand-bituminous mixtures do not have the abrasive resistance needed to be used as a surfacing. Accordingly, sand-bituminous mixtures should be surfaced by applying a seal coat or surface treatment. If a higher type of pavement is desired, it should be covered with a plant-mixed bituminous surfacing material.

F. Soil-bituminous Mixtures

When a fine cohesive soil, such as silty clay or sandy clay, is mixed with bituminous materials, the mixture is known as soil bitumen. Before mixing, however, the soil must be moist. Water is necessary since it is needed to help diffuse the bituminous material throughout the soil and to assist in breaking down the large soil aggregations.

1. Soil and Bituminous-material Requirements. Although a wide range of soil types can be effectively stabilized with bituminous materials, soils that meet the requirements have been known to give satisfactory service. If possible, the soils should be fairly well graded. Some plasticity is desired, but highly plastic soils cannot be used. In addition to these general requirements, the soil should contain no roots or other organic materials. Some acid organic materials are quite detrimental to the stability of soil-bituminous mixtures.

A few soils that do not meet the gradation requirements can be adequately stabilized with bitumens for limited use, provided they can be pulverized. Usually, soils that have a liquid limit less than 30 and a plastic index less than 12 can be pulverized and thus can be effectively covered with bitumen. When the soils have plastic indices greater than 12 to 15, the soil particles are not well coated and a type of mixing results that is termed phase mixing. When the plastic index of a soil is too high and pulverization cannot take place readily, sandy soil usually is added. This procedure can adequately reduce the plasticity of the soil and make it suitable for stabilization with bitumens.

Only liquid bituminous materials are suitable for stabilizing soils. Hard asphalt cement requires that the soils be heated. This heating is not feasible with fine-grain soils. The liquid bituminous materials for stabilizing soils commonly are:

ASPHALT CUTBACKS: Grades RC-1 to RC-4, Grades MC-1 to MC-4, and Grades SC-1 to SC-4. It is desirable to use the RC's for the more sandy soils because easier mixing permits their use and a harder-residue asphalt is deposited on the soil grains. With more plastic soils, though, it is desirable to use the MC's. For very plastic soils, the SC's are sometimes used. In actual practice, though, RC's present a fire hazard; and since the residues from the SC's are very soft, MC's are usually the most satisfactory asphalt cutbacks to use for stabilizing soils.

ASPHALT EMULSIONS: Grade SS-1.

ROAD TARS: Grades RT-3 to RT-6.

In all cases, the heaviest grade of a given type of material that can be worked into the soil according to the working conditions, mixing methods, and the climate at the time of mixing should be used.

Although the fact is not generally considered, similar bituminous materials may have different origin and thus different characteristics, just as soils do. For instance, cutbacks produced from cracked asphalt have properties different from that of vacuum-refined asphalt. Even though the bituminous materials may be of the same type and grade, cognizance should be taken that bitumens may vary considerably in their water-proofing and stabilizing properties.

2. Beneficial Additives. When soils contain a preponderance of carbonates and oxides, they are relatively hydrophobic and a good bond is developed between the soil particles and the bitumen. When, however, the material is siliceous and hydrophylic, the adhesion between the bitumens and the soil particles is generally poor. The adhesion of the bitumens for these poor soils can sometimes be increased by adding a small amount of materials, which are usually termed wetting agents or antistripping additives. These include such materials as fatty amines, amine salts, etc. Usually these additives are mixed with the soil just before the bitumen is added in order to change the charge on the surface of the soil particle and to increase the bond between the bitumen and the soil. In some instances, the additives are put in the bituminous material.

The amount of additive needed varies from very small amounts, perhaps 1 per cent, when used with the more granular materials, to very large amounts for heavy clays, in which a great deal of surface ionic exchange must take place. Small amounts of some of the antistripping additives improve the wet-strength and water-absorption resistance of the soil-bituminous mixtures a significant amount. These additives, however, have little effect upon the dry strength of the material. Care must be taken when using these antistripping agents. Some agents that improve the stabilization of the coarser-grain materials sometimes are detrimental when used with fine-grain soils.

Substances which tend to combine chemically with asphalts, such as epoxy resins and emulsifiers, significantly increase the compressive strength of stabilized soils when added in small amounts. Both the dry and wet strength of soil-asphalt specimens are usually helped by these types of additives.

3. Recommendations for Construction and Field Control. The soil is processed either at a central mixing plant or on the road by a blade or a travel-plant mixer. Because of the high degree of control, best processing is obtained when the bitumens are correctly added and mixed at the mixing plant. When bitumens are applied with the pressure distributor and mixed with a blade, poor control exists. Stabilization with the travel-plant mixer is more satisfactory, for there is better processing and higher-quality control than with the blade mixing. This process is also less expensive than the stationary plant mixing. Many good travel plants that provide excellent service are available on the market.

The soil must be in a moist condition before processing. This is necessary in order for the bituminous material to be thoroughly mixed in. The exact amount of water needed varies from soil to soil, but should be just enough to produce a light and fluffy condition. A measurement of this amount of water is determined by the fluff point (216). This point can easily be determined by stirring into the soil small amounts of water until the soil appears to be light and airy. The moisture content of the soil at this point is the fluff point. Actually, the fluff point varies over a wide range of moisture content but is approximately one-fourth of the liquid limit. Too much water should not be used to aid mixing, for there will possibly be a loss in the stability of the mixture and also drying of the mixture will be prolonged.

After the bitumen and soil are combined, the mixture contains detrimental volatile materials: water, from that used to aid mixing and from emulsions, and hydrocarbon volatiles from cutbacks and tars. When these are present in large amounts, the soil-bituminous mixtures have relatively low stability, as shown in Fig. 11-4. When the volatile matter is reduced by aeration, the stability of the mixture is increased. Soil-bituminous mixtures must be aerated until not more than 40 per cent of the total combined water and hydrocarbon volatiles are left in the mixture. The mixture has been dried enough when it compacts satisfactorily under the roller and does not rut or shove.

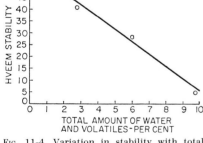

When aerated, soil-bituminous mixtures must not be dried too much. Dry mixtures have low unit weight when compacted. They also have large amounts of air voids and soak up water. The mixtures should be dried only just enough to permit satisfactory compaction.

Soil-bituminous mixtures can be compacted with either smooth-wheel steel rollers, pneumatic-tire rollers, or sheepsfoot rollers. Best compaction is obtained by blading the material in windrows to one side of the road, leaving about 2 in. of soil bitumen on the road surface. After the soil is compacted with the pneumatic-tire roller, the windrow is bladed to the other side of the road, again leaving a 2-in. layer. This layer is also compacted. By repeating the process until all the material is compacted, high density is obtained. If it is desired to compact the material in thicker layers, the sheepsfoot roller must be used. Final compaction of the top few inches must be done with a smooth-wheel or a pneumatic-tire roller. All soil-bituminous mixtures must be compacted to at least 95 per cent of standard proctor density. This density can easily be obtained provided the mixture has not been dried too much.

Fig. 11-4. Variation in stability with total amount of water and volatiles present at time of compaction for a soil-asphalt mixture.

Additional curing after compaction provides increased strength in the soil-bituminous mixture. Accordingly, it is desirable to let the compacted materials stand for at least 2 weeks before a seal coat or a bituminous-surface treatment is placed upon the surface.

G. Sand-gravel-bituminous Mixtures

In many areas a mixture of sand and gravel is obtained from local, river, or glacial deposits and is quite abundant. This material is usually known as bank-run gravel. It has high frictional strength, but usually little cohesive strength, for it normally has only a small amount of clay binder. A bituminous-stabilizing material provides cohesive binding strength and waterproofs the mixture. The waterproofing helps to prevent increases of moisture that result in loss of stability.

1. Materials. Economically, it is best that the sand-gravel mineral aggregate be used as it comes from the deposit. The material should be fairly well graded. Well-graded mixtures compact into a dense state and have high strength.

If the pit-run material does not meet requirements, adjustments can be made quite simply. When the necessary fines are not available in the material as it comes from the deposit, additional fine material must be added. Fines are very necessary for dense gradation and should be rigidly controlled. They should have a plastic index of less than 6. In some deposits scalping may need to be done because the top size of the aggregate should not be more than one-third the thickness of the compacted lift.

The liquid bituminous materials most commonly used to stabilize sand-gravel mineral aggregate are:

RAPID-CURING CUTBACK: Grades RC-1 to RC-3.

ASPHALT EMULSION: Grades SS-1 and certain MS types such as those specified by some states.

ROAD TARS: Grades RT-4 to RT-6.

Regardless of which bituminous material is used, it must be of such a grade and consistency that it can be uniformly dispersed throughout the gravel mixture with the available mixing equipment.

The amount of bituminous material to be used varies between 0.3 to 0.45 gal per sq yd per in. of compacted thickness of the base material (224). The right amount of bituminous material should be that amount just needed to stabilize the gravel and should not be as much as is needed to bind the material together as in asphalt concrete. After it is mixed, the material should have a lean appearance and be mealy in texture. Too much bitumen should not be used, for an excess acts as a lubricating agent for the round gravel and loss of strength results.

A check on the quantity of bituminous material needed may be obtained from the water-absorption test. Sand-gravel-bituminous specimens should not absorb more than 2 per cent moisture when they contain enough bitumen.

2. Field Control. Sand-gravel-bituminous mixtures are constructed in a manner similar to other types of bituminous stabilization. Some moisture in the aggregate aids the mixing and dispersion of the bitumens. Too much water must not be used, for it is detrimental to the strength and also makes it difficult to dry out the mixture before compaction.

The bituminous material is mixed in by blade or by a travel-plant mixer. Usually the bituminous material and the moisture needed for mixing provide enough liquid for satisfactory compaction to high density. If the mixture is too wet, drying of the mixture must take place before compaction.

Compaction is accomplished in 2- to 4-in. layers by steel-wheel or pneumatic-tire rollers. Since a large amount of water and volatiles is present in the material when compacted, the mixtures must be allowed to dry out after compaction. Usually 2 to 3 weeks is needed for satisfactory curing before a surface treatment or other surface can be placed on the base.

H. Road-oil Surface Application

When bituminous materials of low viscosity are applied to the exposed surface of a moist, densely compacted soil, it is usually known as a road-oil surface or sometimes as an oiled-earth surface. The construction is used primarily to waterproof the upper layers of a soil that cannot be stabilized effectively with bituminous materials by other methods. It is entirely different and should not be confused with the other types of bituminous-stabilized materials. Many engineers do not even consider this construction to be a bituminous-stabilized method.

Road-oiled surfaces can be made stable for certain periods of the year because the underlying soil is compacted at optimum moisture content to high density. It is successful only because the zone of heavy-oil penetration is relatively thin. This layer produces desirable waterproofing, which thus prevents the stable underlying soil from gaining moisture from rains and atmospheric condensations and losing strength. The top layer of oil concentration, however, contains too much oil to be stable. Since it is thin and has below it a stable base material, this lack of stability does not produce serious permanent surface distortions (221).

1. Materials and Construction. If the soils are granular enough and can be pulverized, they should be stabilized by previously discussed methods. Road-oil surface application can be made on any type of soil, even plastic soils, just as long as it absorbs the bituminous material.

The most commonly used bituminous materials are:

SLOW-CURING CUTBACKS: Grades SC-1 to SC-3. This is the most desirable bituminous material to be used for oiling earth roads. In many places it has been used so much that slow-curing cutbacks are called road oils. These materials are especially desirable, for they are composed of all intermediate fractions of the topped crude oil instead of asphalt cement and a lighter volatile material as are medium-curing and rapid-curing cutbacks. Slow-curing cutbacks are more desirable for oiling because the soil layers tend to fractionate by absorption the applied asphalt material, leaving the heavier material on the surface (221). Both cracked and uncracked slow-curing cutbacks have been successfully used for road-oil surface applications.

MEDIUM-CURING CUTBACKS: Grades MC-1 and MC-2. This type of asphalt material has been used in some instances. Its use has not been as successful as that of slow-curing cutbacks, because the material has a tendency not to penetrate completely into the soil. Excess bituminous material remaining on the surface is hazardous and must not be allowed to exist.

ROAD TARS: Grades 1, 2, and 3. The grades of bituminous materials to be used are determined on the job. Heavier grades can be used on warmer soils and on soils that are more granular.

The primary requirement for a good oiled-earth road is that the base soil be well compacted and drained. Unless the soil has high strength, potholes soon develop where weak soil exists. Adequate cross slope and good compaction of soil must be obtained before final oiling.

Loose dust and the upper crust of the soil should be removed by blading just prior to oiling. A clean moist surface is a must in order for oil to penetrate readily into the surface. If the surface is not moist, a *light* sprinkling of water may be desirable.

All of the oil is not applied in one application. About 0.5 gal per sq yd of bituminous material is applied in the first application. This is followed up by applying 0.25 gal per sq yd in each of two additional applications, making a total of approximately 1 gal per sq yd. Multiple applications are better than a single application of the total quantity, because less oil tends to run off and better penetration can be expected. Succeeding applications should not be applied until the free bitumen is thoroughly soaked into the surrounding surface and does not adhere to the wheels of the distributor. The time between applications varies with weather conditions from a couple of hours to one or more days.

After the final application is made, the surface should be allowed to cure for 24 hr. If after 2 days all of the bituminous material has not penetrated into the ground, clean sand should be used to blot the excess material.

2. Maintenance. Maintenance of road-oil surfaces is extremely important, perhaps even more important than the maintenance of other bituminous surfaces. Good drainage must be maintained at all times. Whenever raveled spots appear, they must be quickly repaired. Chuckholes should be filled with soil-bituminous mixtures and thoroughly compacted. Unless excellent maintenance is provided, the life of the road-oil surface treatment is greatly impaired.

After one season of use, the road-oil surface should be given an additional treatment of approximately ½ gal per sq yd. During the third season, it is advisable to apply 1 qt of bituminous material per sq yd. Such applications should be made on clean, freshly bladed surfaces after any necessary patching and compaction of the base.

IV. LIME AND LIME-POZZOLAN STABILIZATION

A. General

Soil treatments with lime or lime-pozzolan are among the oldest techniques for road construction, dating back to the Roman highways several hundred years B.C. For Roman roads the lack of compaction equipment necessitated placement of soil-lime or soil-lime-pozzolan as a mortar. In modern practice the lime and water contents are greatly reduced, and the moist mix is spread and compacted with rollers. It is then covered with a base course or a bituminous wearing surface. Lime- or lime-pozzolan-stabilized soil may be used as a subbase or base course for roads, airfields, parking lots, highway shoulders, and similar applications. In the United States, lime stabilization has been most widely used in Texas, but both lime and lime-pozzolan are gaining in use throughout the country and in various parts of the world.

B. Kinds of Lime

Lime, strictly defined, is calcium oxide, CaO, but as commonly used the term includes forms of quicklime and hydrated lime, which are oxides and hydroxides of calcium or calcium-magnesium. "Lime" is also loosely used for carbonate minerals such as in boiler scale or waste lime from sewage-treatment plants. Such carbonates are unsatisfactory for soil stabilization.

Commercial lime is manufactured by heating a crushed carbonate rock such as limestone, $CaCO_3$, to above 2000°F, causing release of carbon dioxide, CO_2, and leaving the high-calcium lime or CaO. A second type of carbonate rock termed *dolomite* is often used for lime manufacture. Dolomite consists of equal molar parts of $CaCO_3$ and $MgCO_3$. The resulting lime, called *dolomitic lime,* is a mixture of CaO and MgO.

Quicklime is another term for lime composed of calcium or calcium and magnesium oxides. Quicklime comes from rotary kilns in pebble form (¼ to 2½ in.), whereas vertical kilns produce quicklime in lumps (6 to 8 in.). Both types are then ground and sieved prior to slaking or use as a powder.

Quicklime reacts readily with water to produce *slaked lime,* or *hydrated lime,* in which all the CaO is converted to $Ca(OH)_2$. The magnesium oxide of dolomitic limes hydrates less readily, and most of the magnesium ordinarily remains as MgO. The resulting hydrated lime, $Ca(OH)_2 + MgO$, is termed *normal hydrated.*

Since about 1940 new processes for hydration have been developed to yield *highly hydrated* dolomitic lime, which is essentially $Ca(OH)_2 + Mg(OH)_2$. The normal and highly hydrated types of dolomitic lime are sometimes called "monohydrate" and "dihydrate," respectively. On the basis of physical tests most high-calcium and normal-hydrated dolomitic lime classifies as Type N, or "normal hydrated." The highly hydrated dolomitic type usually conforms to specifications for Type S, or "special hydrated." Following hydration, limes are usually passed through a hammer mill or ring-roll mill and graded with an air separator. Standard hydrated lime has 95 per cent passing the No. 325 sieve.

Ease of slaking of quicklime is greatly affected by both composition and the temperature of burning, and quicklimes may be classified as rapid-, medium-, or slow-slaking. In general, higher calcination temperatures result in slower slaking. Temperature becomes relatively more important for dolomitic quicklime that for the high-calcium variety since $MgCO_3$ has a lower dissociation temperature. In *"hard-burned"* dolomitic lime the MgO becomes practically nonreactive. *"Dead-burned dolomite"* is used as a refractory lime.

Hydraulic lime is obtained by calcining impure limestones containing 15 to 20 per cent clay. The resulting mixture will set under water, but retains some of the plastic properties of lime. It may be regarded as an intermediate between lime and portland cement. Hydraulic limes are widely used in many countries but relatively little used in the United States.

Lime-manufacturing plants are necessarily located close to workable deposits of limestone or dolomite. Limestone occurs throughout the geologic column and is abundant in the central and eastern United States. Oyster shells are used for lime production in the Gulf Coast area of the United States. Dolomite is common only in older geological deposits, particularly those of the early Paleozoic, which outcrop in the United States mainly in the Middle West and in New England. Dolomite is a result of mineral replacement, and not uncommonly both limestone and dolomite occur within the same bed or in the same quarry.

C. How Lime Stabilizes Soil

Three types of reactions occur when lime is mixed with moist soil. First, calcium ions cause a reduction in plasticity of cohesive soils so that they become more friable and more easily worked. The mechanism is either a cation exchange or a crowding of additional cations onto the clay, both processes acting to change the electric-charge density around the clay particles. Clay particles then become electrically attracted to one another, causing flocculation or aggregation. The clay, now occurring as aggregates, behaves like a silt, which has a low plasticity or cohesion. Aggregation takes place rather quickly and is caused by addition of only 1 or 2 per cent lime. Aggregation lowers the maximum density obtainable with a given compactive effort and reduces the optimum moisture content for compaction.

The efficiency of lime at reducing soil plasticity has been utilized to advantage in soil-cement and soil-bituminous mixtures. Preliminary addition of 1 to 2 per cent lime to fine-grained soils lowers plasticity and improves mixing with portland cement, road tar, or asphalt. Better mixing correlates with better strength and durability. U.S. Army, Corps of Engineers, data show that in one instance treatment with 2 per cent lime and 6 per cent cement was essentially equal to treatment with 10 per cent cement.

A second chemical reaction which can occur and does occur in lime plaster is *carbonation* of lime by carbon dioxide from air, resulting in $CaCO_3$ and $MgCO_3$. Calcium and magnesium carbonates are weak cements, and carbonation must be prevented for best strengths to develop in a road. Prevention involves care in storage of lime prior to its use, avoidance of undue delays during construction, and protection of the road base or subbase against infiltrating carbon dioxide–charged water, for example, rain water or from surface sprinkling used as a cure. Carbonation dangers are most pronounced in industrial areas, where CO_2 content in air may be twice that in rural areas, and the CO_2 content of rain water is sometimes increased several hundred per cent.

A third class of reactions, termed *pozzolanic reactions*, results in a slower, long-term cementation of compacted mixtures of lime and soil. The reactions are little understood, but apparently involve interaction between hydrated lime and siliceous and aluminous minerals in soils. The resulting gel cements the soil and may be similar to certain reaction products from the hydration of portland cement. A major difference is that under normal curing conditions considerably more time is required for pozzolanic-cementation reactions to contribute much strength. Pozzolanic reactions are greatly accelerated by heat or by the addition of certain chemicals.

D. Pozzolans

Some natural materials such as pumice or volcanic ash react much more readily with lime than do ordinary soils. These reactive materials are named *pozzolans** after a volcanic ash utilized by the Romans and obtained from near Pozzuoli, Italy. Pozzolans characteristically have a glassy or noncrystalline ionic structure. The most abundant natural pozzolans are volcanic materials. *Volcanic ash* is sand- and silt-size

---

* Also variously spelled pozzuolana, pozzolana, pozzuolane.

fragments of lava which have been blown into the air to later cool and settle to the ground. *Fine volcanic ash,* or *volcanic dust,* is predominately silty-size ash which blows farther and occurs in thinner layers. *Tuff,* or *trass,*† is consolidated volcanic ash, sometimes reworked by water. *Pumice* is not ash but a frothy volcanic lava which is sometimes so porous it will float on water. It is pulverized before use as a pozzolan.

The other major class of natural pozzolans depends for its activity on opal, or amorphous hydrous silica. *Diatomaceous earth* is composed of opaline shells of microscopic plants called diatoms. One of the most active pozzolans is opaline *chert,* a hard, siliceous rock which must be ground for use as a pozzolan.

Pozzolanic materials vary in activity, depending mainly on the degree to which they are crystalline, since crystallinity reduces activity. Volcanic glasses are noncrystalline, but weather easily to crystalline clay minerals and also devitrify or crystallize with time. Therefore volcanic pozzolans have not been found geologically older than the Tertiary. Opal also tends to crystallize with time.

With the exception of some volcanic dusts all natural pozzolans are ground before use to increase the area of reactive grain surfaces. A surface area of at least 3,000 sq in. per g (Blaine method) is sometimes recommended for pozzolans for use in concrete. (Portland cements are often in the range of 1,800 to 2,000 sq in. per g.)

1. Fly Ash. The most common artificial pozzolan and the pozzolan most widely used in soil stabilization is *fly ash,* a fine gray dust resulting from burning of pulverized coal. The major sources of fly ash are electric power plants, most of which are located near cities and industrial areas not serviced by hydroelectric power. The production of fly ash in the United States amounts to many thousands of tons per day, and disposal, particularly of the lower-quality ash, often constitutes a major problem. Except very close to a power plant, the major cost of fly ash is for hauling.

Fly ash is collected from smoke either mechanically by cyclone collectors or electrically by Cottrell precipitators. The electrical process is more efficient for collecting fines. Fly ash is composed predominantly of spherical particles of noncrystalline silica and alumina and rounded particles of magnetic iron oxide, Fe_3O_4. Calcium oxide from burning of limestone in the original coal occurs alone or in combination with other ingredients. The content of unburned porous bits of carbon varies, depending on efficiency of combustion, but usually runs less than 10 per cent in a good-quality ash. Carbon acts as a diluent for the pozzolanic grains and increases the water requirement and decreases the maximum compacted density of stabilized soils. Pozzolanic activity of fly ash depends partly on fineness. In good-quality fly ashes over 80 per cent passes the No. 325 sieve.

Occasionally fly-ash particles hold a latent electric charge which is difficult to dispel. The electric repulsion causes bulking of fly ash in trucks or railroad cars. It is often dissipated during transport, but a short haul can result in the charge carrying over into field operations, preventing complete compaction of soil–lime–fly ash. Electrical bulking can be a cause of rejection of a fly ash, or methods can be worked out to speed dissipation of the charge.

2. Other Artificial Pozzolans. High heat causes a loss in crystalline structure of the clay minerals in clays and shales, converting them into pozzolans. Use of burned clay or shale as a pozzolan dates back to early India and Egypt. In Rome, ground clay bricks and tiles were used as substitutes for volcanic ash. Today ground bricks or burned oil shales are used as pozzolans, and where economically feasible, pozzolans are produced artificially by calcining clays or shales. Glauconite and aluminum oxide minerals (bauxite) also become pozzolanically active on calcination. On the other hand, ash or opaline materials already highly active are not often benefited by heating. Temperatures required for heat-treatment are ordinarily in the range 600 to 900°C, the best temperature being established by trial and error or with the aid of differential thermal analysis.

Other artificial pozzolans include Si-Stoff, a siliceous waste from the manufacture of alum, used to some extent in Germany. Pulverized blast-furnace slag has been called a pozzolan but is more properly classed as a hydraulic cement.

† Trass is a German name for a siliceous trachytic tuff.

3. Pozzolanic Aggregates. Coarse materials with chemical structures similar to pozzolans are pozzolanically active and may be cemented by reactions with lime. In fact, these granular materials could be converted into pozzolans by grinding. Volcanic tuff (unground) is an effective lightweight concrete aggregate. *Volcanic cinders* are glassy, gravel-size ejected material similar in origin and composition to volcanic ash; they usually occur close to the volcano, sometimes as cinder cones. Lava flows quenched in water disintegrate into black, glassy sands which should be pozzolanic. *Scoria* is a glassy, vesicular lava which may be broken to form aggregate or ground to form a pozzolan. Scoria and other lavas vary in their activity, depending on crystallinity.

Water-cooled bottom slag from cyclone furnaces in electric power plants is a black, angular, glassy, granular material which is highly pozzolanic. Mixtures of lime, fly ash, and water-cooled slag are marketed in the United States under the trade name Poz-O-Pac for use in road or parking-lot base courses. Other artificial aggregates subject to pozzolanic cementation include clean cinders or clinker or furnace bottom ash, brick fragments, and expanded shale and various lightweight aggregates manufactured for use in concrete.

The most highly pozzolanically reactive aggregate is opal, and the resulting concentration of gel around large particles has proved deleterious in portland cement concrete. Chert is also often detrimental because of the content of opal. Certain volcanic materials, particularly obsidian (volcanic glass) or rhyolite or andesite lavas and tuffs, have also reacted unfavorably. These reactions would ordinarily be less serious in a granular stabilized soil than in concrete because of lower rigidity and the availability of void spaces for gel growth. Reactive aggregates may even prove to be desirable from the standpoint of improved cementation.

E. Lime and Lime-Pozzolan Soil-stabilization Mix Design

The choice of lime versus lime-pozzolan stabilization depends on economic factors, kind of soil, climate, and use. Selection of proper mix proportions for soil-lime-pozzolan is complicated by the fact that there are three materials. Fortunately, the mix proportions are seldom critical for a given soil, and many mixes could be used. The problem is then to select proportions which give the needed amount of cementation for the lowest cost. Trial batches are mixed in the laboratory, compacted, cured, and tested, so that the best mix may be chosen.

1. Trial Mixes. Several methods are available for selecting trial-batch mix proportions. One is to prepare mixes to represent all feasible combinations of soil, lime, and pozzolan. In Fig. 11-5, batches were prepared to give a gridlike plot with points at each intersection of the marked percentage lines. Strengths are then shown by contours. Thirty separate batches were required for preparation of this graph.

One way to short-cut the testing program is to eliminate all uneconomical trial mixes. First a maximum allowable lime percentage is selected which is economically competitive with other types of construction, such as soil cement, crushed stone, etc. As a hypothetical example, this is plotted as point *A* in Fig. 11-6. Next the cost of handling an additional material (i.e., pozzolan) is estimated and expressed as its equivalent in percentage lime. This is subtracted from *A* to give point *B*. Starting at point *B*, an *equal-cost line* is drawn with a negative slope equal to the cost of fly ash divided by the cost of lime, both on a dry, delivered, per-ton basis.* Trial mixes are then selected from the area below this line, since proportions above the line are uneconomical. A second limitation which can be imposed is to require a minimum of 3 per cent lime, since lower lime contents may lead to lean areas in the field construction from imperfect mixing. This limit is represented by line *CD*. A lower minimum content may be permissible if lime is applied in a slurry.

Selection of trial mixes from within triangle *BCD* is partly a matter of judgment. If maximum strength is desired, equal-cost points are selected at *A* and along line *BD*. For example, in Fig. 11-6 one might select 10 per cent lime, 90 per cent soil (point *A*),

---

* In this method, mix percentages must be expressed on a total-dry-mix basis.

then 7 per cent lime plus 8 per cent pozzolan, abbreviated as $85:7:8$ soil:lime:pozzolan (point E), then $79:5:16$ (point F), and $73:3:24$ (point D). Intermediate points can be filled in if desired. Ordinarily one of these mixes will give the highest strength and durability. If the resulting strength is excessive for the proposed use, costs can be cut by using less pozzolan or less lime or less of both. For example, if F and D are found to be about equally overdesigned (Fig. 11-6), the more economical ratio $82:3:15$ (point G) could be tried. Or if F is the best mix, trials could be made at points intermediate to F and H, thus maintaining the same lime:pozzolan ratio. If none of the trial mixes gives satisfactory strength and durability, the lime or pozzolan or soil type may be at fault, or chemical accelerators can be tried.

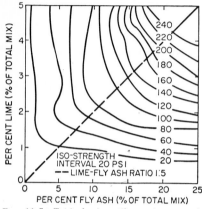

FIG. 11-5. Typical strength contour graph for soil-lime-fly ash. (The soil is a dune sand, stabilized with Detroit fly ash and monohydrate dolomitic lime.)

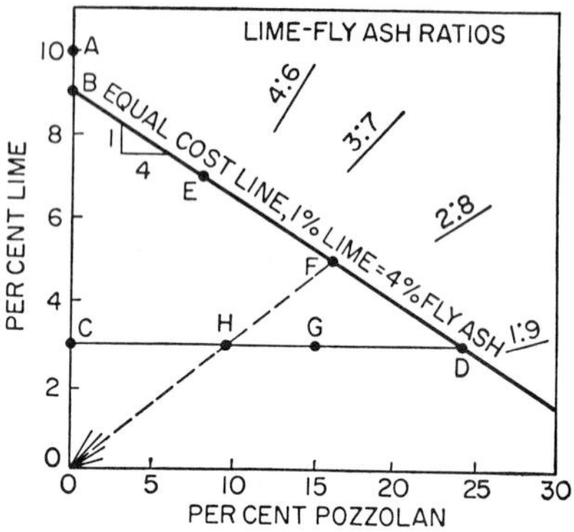

FIG. 11-6. Example of soil-lime-pozzolan mix design by the Iowa State equal-cost-line method.

Under some circumstances stabilization with lime alone will be most satisfactory and economical. For example, if economy dictates a lower equal-cost line and thus limits the cost triangle to low percentages of pozzolan (Fig. 11-6), lime stabilization should be given first consideration. Trial mixes may be prepared with different percentages of lime. Often strengths are increased relatively little by additions beyond a certain percentage lime, the optimum being in the range 2 to 10 per cent.

2. Chemical Accelerators. The slow rate of pozzolanic hardening has led to a search for trace chemicals to speed reactions so that the stabilized soil will better withstand early traffic and winter freeze-thaw. Two classes of chemicals have been found to be satisfactory. Calcium chloride may benefit the reaction by holding moisture and also by providing more calcium ions for incorporation into the cementitious gel, increasing long-term strengths but doing little for early strengths. Another class of chemicals includes alkalis, which act more to increase early strength. Of these sodium carbonate is most commonly used.* Sodium carbonate gives two reactions: First it reacts immediately with lime to produce sodium hydroxide and calcium carbonate, the latter acting as a cement. Best results are obtained when the sodium carbonate is used in powder form. Sodium hydroxide released by the reaction alters the surfaces of minerals and pozzolans to increase their pozzolanic reactivity. Sodium ions temporarily incorporated into the reaction product are replaced by calcium ions from the lime, releasing sodium hydroxide to act on other grain surfaces.

3. Kind of Soil. Accurate, representative soil samples are a prerequisite for any stabilization job. For mixed-in-place jobs separate samples are required to represent

* Patent applied for Feb. 7, 1958.

each soil type encountered, or if areas of specific types are small, preliminary cross blading and mixing must be specified prior to field construction. In this case an accurate composite sample is obtained by random sampling at a number of locations.

Preliminary tests required on any soil are grain-size analysis, plastic limit, liquid limit, plasticity index, and engineering classification. Knowledge of the clay minerals is often helpful for trial mix design. Knowledge of other factors, such as type and amount of organic matter, pH, sulfate content, Great Soil Group, etc., would probably aid in selection of a mix, but unfortunately the influences on strength and durability are not yet known. Supplemental use of such tests and identifications will eventually provide the needed clues. Effects of deleterious ingredients will of course show in strength- and durability-test results, whether the ingredients are identified or not.

4. Selected Trial Mixes. The very limited data available on proper mix proportions for different kinds of soils give the following guide for preparation of lime or lime-pozzolan trial mixes. Nearly all previous experience has been with fly ash as the pozzolan.

a. **Clays.** Proper treatment of *clays* depends mainly on the kind of clay minerals. Expanding clays containing montmorillonite react readily with lime, immediately losing plasticity and slowly gaining pozzolanic strength. However, one result of the change in plasticity is a depletion of calcium ions, which apparently hinders the pozzolanic reaction with fly ash, and fly ash is sometimes of little benefit with low percentages of lime. Montmorillonites containing sodium as the dominant exchangeable cation would presumably cause the greatest calcium depletion. Lime-stabilization trial mixes recommended by the National Lime Association for clays are with 5, 7, and 10 per cent lime. A major benefit from treatment with lime is the reduction in PI, usually to below 10.

Clays containing mainly illite, chlorite, vermiculite, or kaolinite are less effective robbers of lime and may be slightly pozzolanic. However, performance may be improved by addition of more active pozzolan, both from the standpoint of cementation and by improvement of the granular nature of the soil. Best ratios are usually in the range 1:9 to 4:6 lime to fly ash, and total amount of admixture being governed by economics and usage.

Treated clays are often susceptible to damage by freeze-thaw, and the total lime-pozzolan required may then be of the order of 25 to 30 per cent. With lime alone the percentages needed will be 8 to 12 per cent. At early ages a lime- (or lime-pozzolan) treated base course with its greatly reduced PI is less resistant to freeze-thaw than is the natural clay.

b. **Silty Soils.** Silty soils with less than 10 or 12 per cent clay may be somewhat pozzolanic, depending on mineral composition. In this case stability can be obtained with lime alone, or the best lime-pozzolan ratio will be low, of the order of 1:2. Recommended lime-stabilization trial mixes are the same as for clays. Silty soils containing sufficient montmorillonitic clay to coat the grains will not be as pozzolanically active and will benefit from larger additions of pozzolan. Before they gain strength, stabilized silty soils are highly susceptible to damage by freeze-thaw.

c. **Sandy Soils.** These materials are too coarse to react well with lime alone, and pozzolan is nearly always required. Best ratios are usually of the order of 1:5 lime to pozzolan. Strengths show a somewhat linear relationship to percentage total admixture. Sands are not as susceptible to freeze-thaw damage as are silts.

Sandy soils containing 10 to 30 per cent combined silt and clay have a better gradation for cementation bonding and may stabilize with lime alone, depending mainly on the pozzolanic nature of the clay mineral. Usually, however, lime-pozzolan will give best results. Strength and durability are often improved by the better gradation. Field attempts to artificially blend in clay have given varying success, depending mainly on intimacy and ease of mixing.

d. **Coarse Granular Soils.** Coarse-graded soils and mixtures with crushed stone possess inherent mechanical stability which may need only slight improvement to meet the requirements of base course. Well-graded mixes containing binder material such as clay or calcium carbonate (caliche) may therefore gain enough added strength with as little as 2 to 4 per cent lime.   If mixes are less well graded or if better cementation and flexural strength are needed, requirements can be met by use of a pozzolan. Unless the binder material is pozzolanic, the best lime-pozzolan ratio will be in the neighborhood of 1:5.

e. **Pozzolanic Granular Materials.** Pozzolanic materials such as scoria, cinders, chert, or water-cooled slag frequently develop very high strengths when stabilized with lime or lime-pozzolan.   Pozzolan is often required to give a satisfactory gradation for cementation at grain contacts.   In some cases flexural strengths have been obtained equal to those for portland cement concrete.   Typical mixes incorporate 4 to 7 per cent lime and 10 to 15 per cent fly ash.

5. Lime Specification. Pozzolanic reactions are extremely sensitive to chemical class of lime, and the type desired must be specified in some detail.   An exception is where only a change in soil plasticity is desired, as this is easily obtained with any class of lime.

Although high-calcium limes are often used in soil-lime and soil-lime-pozzolan stabilization, monohydrate dolomitic lime is now proved more effective for pozzolanic reactions and gives the first major indication of enough pozzolanic hardening to satisfactorily resist freeze-thaw.   (The dihydrate dolomitic lime is much less effective.) Pozzolanic reactions are ordinarily sluggish with $Ca(OH)_2$ unless they are accelerated by the presence of an equal amount of unhydrated MgO, by a hot climate, or by trace chemical accelerators such as sodium carbonate.   With low lime contents some accelerating influence is also found in stabilization of clays, but this disappears at high lime contents.   Even with low lime contents the strengths are seldom as high as with the same amounts of monohydrate dolomitic lime.   Sodium carbonate is the most effective accelerator, and addition of trace amounts to mixes with high-calcium lime will cause the soil-lime or soil-lime-pozzolan strengths to equal those with monohydrate dolomitic lime.

The choice of quicklime versus hydrated lime is largely a matter of convenience and economy.   Quicklime is cheaper on an active-ingredient basis since it contains no water.   However, quicklime cannot be spread dry because the dust is highly caustic and dangerous to workers.   Quicklime is therefore hydrated in a tank and sprayed as a slurry.   Hydrated lime can be purchased in bulk and spread with spreaders, purchased in bags and spread by hand, or applied in a slurry.   The amount of water in a slurry is adjusted to give the optimum moisture content for compaction, except that the slurry must be at least one-half to two-thirds free water if it is to be easily sprayed.   About 70 per cent water may be required for sprayable slurries of high-calcium quicklime.

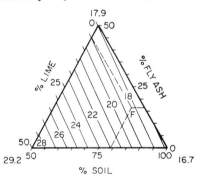

Fig. 11-7. Example of optimum-moisture-content interpolation by means of a triangular chart. Corner values are determined by trial. The optimum moisture content of composition *F* (79:5:16) reads 18.4 per cent.

Because both quick and hydrated lime tend to recarbonate readily by combination with CO_2 from air, storage must be in tight containers and under dry conditions. Quicklime ordinarily should not be stored more than 3 to 6 months; hydrated lime should not be stored longer than about 1 year.   The maximum carbon dioxide contents specified for quicklime and hydrated lime are 10 and 7 per cent, respectively.

6. Optimum Moisture Content of Soil-Lime and Soil-Lime-Pozzolan Mixtures. The optimum moisture content for compaction of soil-lime and soil-lime-pozzolan varies, depending on percentage lime and percentage pozzolan.   A moisture-density curve may be run for each trial mix, or a shorter procedure is to estimate optimum moisture contents by interpolation between two or three selected mix-test values.   For example, in Fig. 11-7, the corners of the triangle represent mixes of 50:50 soil-lime,

50:50 soil-fly ash, and 100 per cent soil. The optimum moisture contents determined in the laboratory are 29.2, 17.9, amd 16.7 per cent, respectively. The intermediate unit percentages, 18.0, 19.0, 20.0, etc., are sealed off on the appropriate sides of the triangle and connected by straight lines. The approximate optimum moisture content of any intermediate mix composition may then be read directly from the chart. Different plots are required for different kinds of soil, lime, or fly ash.

The mix finally selected for field use is put through the standard moisture-density test to more accurately appraise the moisture requirements and to measure the density expected in the field. Results from this test are also used for molding specimens for durability tests.

In the event of a rush program allowing little time for testing, optimum moisture contents of trial mixes can be estimated by texture or "feel." Accuracy depends on experience of the test personnel with the particular types of soils involved.

The above outline applies equally well for compaction to standard Proctor density or to a higher density employing greater compactive effort and a lower optimum moisture content. Higher compaction greatly improves pozzolanic cementation because of better grain contact.

F. Surfacing

Soil-lime or soil-lime-pozzolan may be used as a subbase underneath a flexible base or a portland-cement-concrete slab, or it can be used as a base course underneath a bituminous surfacing. To promote better curing the surfacing is often applied as soon as the base course is strong enough to support construction vehicles. Cohesive mixes or well-graded granular mixes can usually be surfaced immediately. Otherwise drying is prevented by occasional moderate sprinkling or by application of a seal coat. Purposes of the surfacing are to act as a seal against water and carbon dioxide, to withstand traffic abrasion, to distribute wheel loads to the base, and to provide some protection against freeze-thaw. Failure of the surfacing, by allowing entry of water, will often lead to failure of the base.

Thin bituminous surface treatments such as the direct or inverted (chip-coat) penetration macadam are satisfactory only for light traffic and where the base course is well drained, strong enough to resist shearing forces from wheel loads, and very resistant to freeze-thaw. These requirements are generally met only by A-1 or A-2 soils or with pozzolanic aggregates. Good drainage is necessary to prevent peeling of the surfacing and to reduce damage from freeze-thaw.

Well-drained stabilized fine-grained soils not subject to freeze-thaw may develop enough strength to allow use of thin surface treatments for light traffic. With poor drainage conditions a thicker, more competent surfacing is needed. Stabilized fine-grained soils are seldom as resistant to freeze-thaw as granular soils, and in freezing climates the additional protection from a thicker surfacing is needed. Here a major function of the surface course is to absorb the large number of minor or diurnal freeze-thaw cycles which penetrate only the upper 2 in. A minimum 2- to 3-in. surface-course thickness is recommended.

Even when base-course conditions are ideal, a thin surfacing will not stand up under prolonged heavy traffic. Ordinarily 2 to 3 in. of plant-mix asphaltic concrete is recommended.

V. CHLORIDES

Modern use of sodium and calcium chlorides in roads dates from the early 1900s. The earliest use was to allay dust, particularly in crushed stone or waterbound macadam roads. Treatment with chloride salts was found to keep roads moist in dry weather, since the salts retard evaporation and attract moisture from the air. An advantage of these treatments is the abundance of chlorides in rock salt, sea water, natural brines, and artificial by-product brines.

Chloride-treated roads remain dust-free until the chlorides are leached out. Presence of chlorides also aids mechanical stability in the road and can give a firm, hard-packed wearing surface with smoothness and riding qualities approaching those of more permanent, higher-cost roads. Occasional blading and surface reapplication of chlorides are necessary to prevent deterioration and pitting of the road surface. Where higher traffic volumes justify the additional expense, an alternative procedure is to use the chloride-treated layer as the base course for a bituminous wearing surface.

A. Grades and Specifications

1. Brines. Practically any commercial source of chlorides is satisfactory for road stabilization. Although chemical requirements are not strict, many differences exist, and chemical data are needed for intelligent use. Specifications for Solvay liquors sometimes limit the amount of magnesium and alkali (sodium and potassium) chlorides. The Iowa State Highway Commission standard specifications require at least 33 per cent calcium plus magnesium chlorides, at least 21 per cent calcium chloride, and not more than 4 per cent alkali (sodium and potassium) chlorides.

2. Sodium Chloride. Salt ordinarily is of rather high chemical purity, and most chemical specifications are designed around food requirements. Rock salt is not purified, but is merely crushed and screened prior to marketing. The purity often runs 95 per cent or higher. Size grades and their nomenclature are not standardized. Representative rock-salt grades are given in Table 11-1.

Table 11-1. Grades of Rock Salt, Retsoff Mine, N.Y.

Size	Passing sieve	Retained on sieve
No. 2	$\frac{3}{4}$ in.	No. 4
No. 1	$\frac{3}{8}$ in.	No. 8
Coarse C	$\frac{3}{8}$ in.	No. 12
Fine C	No. 8	

Salt from evaporation of brines is also available in different grades, not standardized. Typical grainer salt (crystallized in open pans, or grainers) may be graded as in Table 11-2. Vacuum salt, or granulated salt, crystallizes in vacuum pans and is graded as "industrial salt," passing the 20-mesh sieve, "table salt," passing the 30, and "flour salt," passing the 60.

Practically any fineness of sodium chloride is satisfactory for roads, although finer grades sometimes tend to cake when stockpiled along the road. Coarse C rock salt is commonly used. A tentative revision* of the ASTM specification (ASTM Designation D 632-43) is 20 per cent or less retained on the No. 3 sieve, 50 to 95 per cent retained on the No. 8, and at least 90 per cent retained on the No. 30. A minimum purity of 97 ± 0.5 per cent is also suggested.

3. Calcium Chloride. Marketed flake calcium chloride ordinarily meets ASTM standards for use in concrete (ASTM Designation D 98-48): $CaCl_2$, not less than 77.0

* Feb. 15, 1952.

Table 11-2. Grades of Grainer (Evaporated) Salt

Size	Passing sieve	Retained on sieve
GA ("ground alum")	$\frac{3}{8}$ in.	No. 30
Coarse	No. 4	No. 30
Medium	No. 20	No. 50
Fine	No. 30	No. 100

per cent; MgCl₂, not more than 0.5 per cent; total alkali chlorides, not more than 2.0 per cent; and other impurities, not more than 1.0 per cent. One hundred per cent passes the ⅜-in., and 90% is retained on the 20-mesh sieve. Calcium chloride is also sold as round pellets about ⅛ to ¼ in. in diameter, sometimes preferred for easier spreading. Anhydrous chloride, usually about 95 per cent pure, is also used, and the higher cost may be offset by saving in freight.

B. How Chlorides Stabilize Roads

Chlorides operate in several ways to increase stability of roads, but it must be remembered that they are not good cements and the road material must have mechanical stability in itself. Chlorides are therefore effective only with granular road surfaces and bases containing sand, crushed stone, gravel, or artificial aggregates.

A first function of chlorides in road construction is to allow easier moisture-content control by reducing evaporation. Prevention of drying also aids in reducing segregation of fines.

1. Lubrication. Dissolved chlorides improve compaction by lubricating between grains. The amount of improvement varies with the soil, but the increase in compacted density with the same compactive effort is usually of the order 1 to 7 per cent. Alternatively, the number of roller passes for the same density may be reduced by as much as two-thirds. Calcium chloride solutions are better lubricants than those with sodium chloride, but sodium chloride is believed to contribute indirectly by partial dispersion of clay binder material. Unfortunately, the dispersion also contributes to stickiness and more difficult mixing prior to compaction. Clay binder material should therefore be well distributed in the mix prior to the addition of salt.

2. Moisture Retention. One of the greatest virtues of chlorides in roads is their ability to retain moisture, minimizing dust in the finished roads. Dust-laying properties are a function of hygroscopicity and deliquescence, or ability to absorb moisture from the air and dissolve. These in turn are directly proportional to the decreases in vapor pressure of the solution compared with the vapor pressure of water at the same temperature.

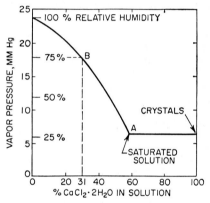

FIG. 11-8. Vapor pressure of calcium chloride solutions at 77°F. Ordinate (0 per cent CaCl₂·2H₂O) also shows the vapor pressure of water at various relative humidities.

For example, the vapor pressure of a saturated calcium chloride solution at 77°F (25°C) is 7.0 mm of mercury, whereas that of water is 23.8. The rate of evaporation from a free surface is reduced 23.8 ÷ 7.0 = 3.4 times. The deliquescence, or relative humidity at which CaCl₂·2H₂O will start absorbing water from air, is 7.0 ÷ 23.8 = 29.4%. The hygroscopicity, or ultimate amount of water absorbed, will be enough to reduce the vapor pressure of the solution to equal the vapor pressure of water in the atmosphere. In Fig. 11-8 a relative humidity of 75 per cent will result in water becoming absorbed until vapor pressures are equalized at point B. At this point the calcium chloride solution will be 100% − 31% = 69% water, or 1 lb of flake dihydrate will have absorbed 42 ÷ 58 = 0.725 lb of water. At humidities below this, absorption ceases, water evaporates, and crystals will re-form. At the other extreme, at 95 per cent RH a pound of chloride absorbs 8.4 lb of water; at 100 per cent RH the absorption becomes infinite—it is raining.

3. Flocculation of Clays. Excess sodium, calcium, or magnesium ions all cause clay to flocculate or form silt-size aggregates. Actually, most natural clays used in soil-aggregate mixtures are flocculated already. An important factor in the case of sodium chloride is that as sodium chloride becomes diluted or leaches out, clay passes through a stage of dispersion. During a prolonged rain, clay in the road surface theoretically disperses and swells and plugs the pores, preventing further leaching. Conversely, dispersed clay shrinks more on drying and would be more susceptible to blowing. Sodium chloride roads are usually maintained by occasional surface applications of calcium chloride.

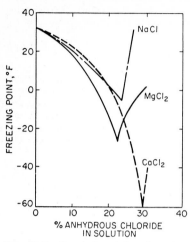

Fig. 11-9. Freezing points of various chloride solutions.

4. Solubility of Road Aggregate. Although the effectiveness of this factor has not been accurately evaluated, data show that the presence of sodium chloride in water greatly increases the solubility of some common road aggregates, particularly limestone and dolomite. About a 5 per cent solution of sodium chloride is optimum and will dissolve twice as much calcium carbonate as pure water. In a road, moisture content fluctuations related to atmospheric conditions can thus cause solution and recrystallization of tiny amounts of limestone. In a well-graded mix the precipitated carbonates could become a very good cement.

5. Freezing Point. A widely recognized property of chlorides is their ability to lower the freezing point of water, as evidenced by their use in antifreezes or to thaw icy streets.

Freezing-point depression by different chlorides is shown in Fig. 11-9. Particularly important is the occurrence of minimum freezing points, or eutectics, at certain solution concentrations. If the concentration of solution in the road is less than the eutectic, freezing will start at whatever temperature is indicated (Fig. 11-9). However, freezing will not be completed until the temperature is much lower because the material freezing will be pure water, which causes a gradual increase in the concentration of the remaining solution and a further lowering of its freezing point. As temperature further decreases, water continues to freeze out until the composition of the unfrozen solution reaches the eutectic composition. Then everything freezes. (Solution concentrations initially to the right of the eutectic will result in freezing out of chloride crystals until the eutectic composition is reached.) Thus any NaCl-treated road must reach $-6°F$ before all freezing is completed; any $MgCl_2$-treated road must reach $-27°F$, and any $CaCl_2$-treated road must reach $-59.8°F$. Therefore freezing of a chloride-treated road not only begins at a lower temperature, it is gradual and seldom completed. Destructive pressure build-ups by ice are minimized because the unfrozen chloride solutions leak free. True freeze-thaw cycles may never occur, because not everything freezes. Field tests have indicated that winter strength loss of chloride-treated roads is far less than in similar untreated roads.

Laboratory tests at Purdue University have shown that frost heave and moisture migration during freezing are minimized by inclusion of calcium chloride. The exact reason for this is not known.

6. Surface Tension. Solution of chlorides causes an increase in the surface tension of water, important in roads because the water-film surfaces are like elastic bridges binding adjacent rock or mineral particles. Chlorides thus increase the strength of these water-film bonds. As drying takes place and pore water decreases, the water films can cause a further tightening and consolidation of compacted material. Calcium chloride has been shown to cause up to 15 per cent increase in density of already compacted soil, increasing strength or stability several times. The slow rate of drying allowed by the presence of the calcium chloride produces uniform shrinkage and densification throughout the treated road layer. A 15 per cent increase in density would give a linear shrinkage of about 5 per cent, but the resulting fine cracks do not appreciably weaken the road. The initial drying after compaction is therefore very important as a period of cure. Figure 11-10 compares the effectiveness of different chlorides for increasing surface tension.

7. Crystallization. Relative humidity in a soil is difficult to evaluate, but undoubtedly runs high. Soil humidity is usually sufficient to keep any of the chlorides dissolved. In dry weather sodium chloride often crystallizes in the upper fraction of an inch of road surface, and the crystals cement the surface into a dense, hard crust which withstands traffic abrasion. Sodium chloride crystals also fill the voids and inhibit further drying and shrinkage. Shrinkage in an untreated road causes disruption of the clay bonds, loosening of coarse aggregate, and accompanying raveling, or "washboarding," and dust.

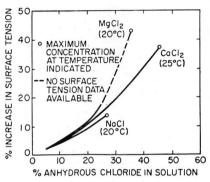

Fig. 11-10. Surface tensions of various chloride solutions.

C. Rain and Loss of Chlorides

A major question in the use of chlorides in roads is, how long will they last? Maintenance of a proper crown to the road becomes essential to prevent ponding of water,

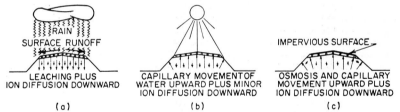

Fig. 11-11. Mechanisms of moisture movement and chloride movement in roads.

leaching, and development of pits and ruts. The recommended crown is the A type, with a high center and straight lateral slopes of ½ in. per ft.

There are three major directions of moisture movement in a granular roadway—up, down, and sideways (Fig. 11-11). Rain water tends to infiltrate and leach out chlorides, but if the road has a proper crown, most of the rain is deflected sideways into the ditches. During dry periods there is an opposite movement as capillary moisture moves upward to replace water evaporating at the road surface. The upward-moving water carries salts with it and may cause considerable chloride accumulation in the surface zone. While this action forms beneficial cementation, chlorides crystallized right at the surface are dissolved and washed away by the next sudden rain.

Another mechanism probably causing weak continual upward movement of water is osmosis, or water movement toward zones of chloride concentration. This could result in a gradual dilution of chlorides, except that water so brought up is usually quickly lost through evaporation.

A related kind of movement not of water but of chloride ions may account for a continual gradual loss. This is downward diffusion of ions in solution away from the concentration zones—in this case away from the treated surface or base and down into the subgrade. This process would go on continually, but is effectively opposed by drag from the upward movement of capillary water (Fig. 11-11).

To summarize, movement of water and chlorides is strongly influenced by climate, mix gradation, and capillarity and permeability. In open-graded mixes chlorides are leached rapidly downward by rain. In dense-graded mixes with improper road crown or in very humid climates the net movement is also downward. Usually, however, the net chloride movement in a dense-graded mix is upward, causing a surface crust or concentration zone. Use of an impervious surfacing prevents both rain infiltration and surface evaporation except on the embankment foreslopes, with the result that the major chloride loss is lateral by capillary movement or downward by ion diffusion.

D. Maintenance

Chloride roads perform very satisfactorily without bituminous surfacing but require occasional maintenance. Neglected chloride-treated roads will rapidly leach out and deteriorate. The cost of maintenance is usually less than the cost of maintaining similar nontreated roads.

Wearing surfaces of sodium or calcium chloride–treated roads are maintained by periodic application of calcium chloride either as flakes, pellets, or brine. Flakes or pellets should be applied only to wet roads to prevent loss by traffic or by blowing. The quantity is calculated to give $\frac{1}{2}$ lb of 78 per cent calcium chloride per sq yd of road surface. Sodium chloride roads are treated about 10 days after construction. Reapplication of calcium chloride is usually required two or three times a year, depletion being evidenced by the appearance of dust, which unless stopped leads to the loss of valuable fines.

Potholes or road roughness can be corrected by blading, which can only be done when the road is damp. It is particularly important to maintain adequate crown, which as previously mentioned is A type with lateral slopes of $\frac{1}{2}$ in. per ft of roadway. The crown should never be allowed to deteriorate below $\frac{3}{8}$ in. However, blading should never be done until road surface conditions make it necessary, since blading destroys the tightly knit surface and brings up loose stones.

An alternative for repairing potholes is to clean them out and fill and tamp with premixed patching material. This avoids unnecessary blading, or it can be done in dry weather when blading is impossible without prewetting. When a road surface needs both blading and addition of fines and more chlorides, one procedure is to rebuild the upper road by scarifying $\frac{1}{2}$ to 1 in. deep, mixing, and allowing traffic compaction.

VI. CONSTRUCTION

Soil-stabilization procedure involves preparation of the subgrade, addition of water and additives to the soil, mixing, spreading, compaction, and curing. Some operations are sometimes combined. Many types of stabilized-soil roads are covered with a bituminous surfacing.

A. Preparation of the Subgrade

Routine subgrade preparation is required for any road construction and involves removal of undesirable soil, rubble, and stumps from the upper 12 in., compaction of the soil, and shaping to grade and crown. The goal is a firm, well-drained subgrade not susceptible to volume change or damage by frost action.

A common method for stabilization is to use the subgrade soil in the mix, in which case the subgrade must be built up to near base or finish grade. This type of construction may require a more positive identification and more careful selection of subgrade soils.

B. Mixing

Farm equipment was used for mixing on most of the early soil-stabilization jobs of the 1920s and 1930s. However, the day of the plow, disk, and cultivator for road building is almost past. Special mixers have been developed which are much more efficient and much faster, giving a better job at lower cost. These mixers fit into two categories: traveling mixing machines, which literally "eat up the road," and stationary mixing plants.

1. Traveling Mixing Plants. The simplest type of traveling mixing plant is the small single-rotor mixer designed to move rapidly and mix an area in several passes. Some are self-propelled and incorporate pump, spray-bar, and metering devices for accurate addition of water or liquid-chemical stabilizers. In simpler types these features are left off and liquids are sprayed in a separate operation from a tank truck. The simplest rotary mixers are towed behind a tractor but usually contain a separate power unit to drive the mixing rotor.

Single-rotor mixers are satisfactory for all kinds of stabilization, and even when a more elaborate mixer is used, a single-rotor machine is often kept on hand for preliminary scarification of hard or compacted material and for occasional remixing or mixing in tight quarters. The depth of processing is adjustable to a maximum of from about 4 to 12 in., depending on the kind of machine. Mixing width is usually 6 to 7 ft. The rate of travel while mixing is 60 to 200 ft per min. A unique feature of these rotary mixers is a back hood which when lifted allows better aeration of the soil and promotes drying.

A more elaborate and more expensive traveling mixer is the slower-moving, multiple-rotor mixer designed to do all mixing in one pass. Spray-bar and metering attachments are standard. Mixing in one pass is sometimes preferred for faster processing of quick-setting materials such as soil cement and for ease of construction programming. Maximum processing depth with these units varies from about 7 to 9 in., depending on the model, and mixing width varies from 5 to 11 ft. The mixing speed may be varied from about 6 to 30 ft per min. Control of pulverization and mixing is by adjustment of the forward speed, slower speeds giving longer mixing.

A third class of traveling mixers is the traveling pugmill mixer, which picks up soil from a windrow, mixes it in an enclosed drum, and deposits it in a windrow behind the machine. Advantages of this method are thorough mixing plus positive control of soil quantities. (The amount of soil picked up by other types of traveling mixers is a function of cutter depth.) Disadvantages are the necessity for preliminary pulverization, usually with a single-rotor machine, and the need for an even windrow. However, if borrow soils are to be used, preliminary pulverization is necessary anyway and the soil can be dumped and spread into an even windrow. Machines are either tractor-drawn or self-propelled, and the different rates of forward travel available are from about 13 to 26 ft per min. Two or three windrows are required for the average two-lane road with a compacted thickness of 6 to 8 in.

2. Stationary Mixing Plants. Stationary pugmill mixing plants used for bituminous mixtures are readily adapted to soil stabilization. These fall into two classes—batch type and continuous-flow type. Advantages of stationary mixers are ease of ingredient proportioning and possibility for very thorough mixing. However, extra hauling is often required, and haul time of quick-setting mix materials such as soil cement must be limited to perhaps 30 min. In addition, mixed material must be covered during long hauls to prevent drying to below the optimum moisture content.

3. Addition of Stabilizing Agents. Soil stabilizers are added before or during the mixing operation, depending on whether they are liquid or powder and what type of mixing machine is employed.

For traveling mixers, powder-type stabilizers such as cement or lime are dispensed beforehand, either by spacing the bags and opening them by hand or by bulk distribution from mechanical spreaders. A minimum of 2.5 to 3 per cent of the nonsoluble powders is believed necessary to assure complete distribution, because of mixing inefficiencies and unavoidable occurrence of lean areas.

Water or liquid chemical additives are conveniently added during mixing, if the facilities are available, or they can be sprayed on from a tank truck. Water is sometimes sprayed on the road prior to mixing as an aid to pulverization, but usually liquids will be added after a preliminary pulverization and dry mixing to gain better mixing and avoid runoff. The goal of mixing is to obtain as uniform a mix as economically feasible, with a moisture content near the optimum for compaction.

4. Control. Inspection should provide for checks on proportioning, degree of pulverization, and uniformity of final mix. Proportioning for rotary traveling mixers depends in part on cutter depth, which should be measured directly or checked indirectly by measurement of thickness of loose or compacted thickness of processed material. Too small a cutter depth will result in a richer mix laid in a thinner layer.

Pulverization of granular noncohesive soils usually offers no problem. On the other hand, pulverization of clays is at best difficult, and complete pulverization may be impossible. The goal is usually to pulverize sufficiently for material to pass the $\frac{1}{4}$-in. sieve. The effect of clay lumps larger than $\frac{1}{4}$ in. is not so serious as might be supposed, so long as the lumps are enclosed in a matrix of stabilized soil. Part of the reason is that lumps increase the richness and often the strength of the remaining mix. In some kinds of stabilization, particularly those containing or evolving lime, clay lumps have been observed to gradually harden and become water-stable after placement in the road. However, clay lumps are a source of weakness, particularly at early age, and under no circumstances should the amount of clay lumps exceed 40 to 50 per cent of the total mix.

Mix uniformity is judged by digging a trench and noting the soil color. Unmixed streaks or layers are a sufficient reason for remixing. At the same time width and thickness of the stabilized layer should be measured to see if these meet specifications.

C. Compaction

Single- and multiple-rotor traveling mixers leave mixed soil in a layer ready for compaction, whereas soil from the traveling pugmill or stationary mixers must first be spread. Windrows from the traveling pugmill are spread with a blade grader; plant-mixed material is trucked to the site and either dumped and then spread by blade grader or spread into a uniform layer through mechanical spreader boxes.

Standard soil-compaction equipment is satisfactory for stabilized soil mixtures. Cohesive soils such as silts and clays may be compacted with pneumatic-tired rollers or sheepsfoot rollers but not with the newer vibratory types more suited to granular soils such as sand. Difficulty may be experienced in attempts to compact poorly graded sands with a sheepsfoot because of internal shearing and failure of the feet to "walk out" in consecutive passes. Well-graded granular soils may be compacted with any of the common types of equipment, including sheepsfoot, rubber-tired, steel-wheel, grid, segmented, or vibratory rollers. Crawler tractors and large-tired vehicles are used for supplemental compaction. Where time and ultimate density are not critical, low-cost granular roads are sometimes allowed to be compacted by traffic.

Compaction of some soils often results in formation of horizontal shear planes giving soil a thin, platy structure in the upper part of the compacted layer. In an unstabilized road the planes ordinarily do little harm and may be overlooked, but where stabilization depends on cementation, the planes must be removed.

Compaction planes are removed by light scarification with a weeder or spike-tooth harrow or a nail drag built for the purpose on the job. After scarification or dragging, addition of water is usually necessary to counteract drying. The loose material is then recompacted with a pneumatic-tired roller. Final shaping to grade is done by thin blading with a blade grader. A broom drag then may or may not be used to smooth out ridges, and the surface cut by the blade grader is final-rolled with a pneumatic roller or a light tandem steel-wheel roller, the latter being useful to remove ridges and punch down stones.

Compaction planes, which are essentially shear planes from overcompaction, may be hard to avoid in compaction of sandy, silty, or clayey soils. Gravelly mixes do not show this tendency, and scarifying may turn buried rock fragments and cause harm.

1. Control. An experienced inspector can judge when soil-moisture content is near optimum on the basis of feel. The test is whether the soil can be molded into a firm cast which is moist but not wet.

The prime control on compaction is the end result, or density. Density measurements are sometimes the most important part of inspection, since strength and hardening may critically depend on the density obtained. Ordinarily 95 or 100 per cent of standard AASHO density is specified. Measurements are by standard oil, sand-cone, or rubber-ballon techniques (AASHO Designation T 147-47).

Other than selection of improper compaction equipment for the soil types encountered, the most common obstacles thwarting compaction are weak subgrade and presence of excess water in the material being compacted. For this reason rainy weather usually halts construction, and material compacted on top of a saturated granular subgrade is liable to overwetting from below. When either overwetting or a weak subgrade prevent compaction, the remedy is bad news to the builder—tear out everything and start over. Continued efforts to compact will be futile. Stabilized mix material torn up and aerated will require addition of more stabilizer if (1) the chemical has a quick set, such as portland cement, (2) the chemical is destroyed, such as lime coming in contact with carbon dioxide in rain water or in air, (3) the chemical is leached out

D. Curing

Curing must be tailored to the stabilization method used. Since inorganic cementing reactions take place only in the presence of water, moist-curing is required for soil cement, soil-lime, and soil-lime–fly ash. Alternate methods for moist-curing are to spray on a bituminous coating or to cover with plastic, paper, or moist earth or straw. Another method is to keep the surface moist by periodic spraying from a water truck. Where rains are expected and would be harmful, as for soil-lime or soil-lime–fly ash, a sealed cure is recommended.

Bituminous materials commonly used for curing are RC or MC grade 2 or 3 cutbacks, RT-5 road tar, and asphalt emulsions, the exact type depending on climate and road conditions. Rate of application is 0.15 to 0.30 gal per sq yd. Excessive penetration of sprayed bituminous material into stabilized granular soil, resulting in weakening of the upper layer, is prevented by preliminary light sprinkling with water. After application the bituminous coat is sanded prior to use by traffic. The curing seal coat gives a worthwhile resistance to abrasion and forms part of the later surfacing.

Most stabilization methods not benefited by moist curing are benefited by air-drying, and air-drying is required for some to become effective. Rate of drying depends on soil and kind of additive, and in particular on weather. In some climates air-drying may be difficult to achieve, but ordinarily 7 days of favorable weather will give satisfactory drying. During this period traffic should be kept off or regulated according to the stabilizer used.

E. Surfacing

Stabilized soil can serve as a surface course, as in chloride- or lignin-treated soil-aggregate roads, or it can serve as a subbase under portland-cement-concrete pavement, or as a base course under a bituminous surfacing. Cheapest roads have no additional surfacing, but where traffic volumes increase maintenance, a bituminous surface is usually chosen. Furthermore, many types of soil stabilization require surfacing wherever the paving must withstand abrasion.

The common bituminous wearing course for light-traffic roads is a double surface treatment (double chip coat, double inverted penetration macadam) about $\frac{3}{4}$ in. thick. Higher traffic volumes warrant a thicker, higher type of surfacing such as $1\frac{1}{2}$ to 3 in. of plant-mix asphalt concrete. Where stabilized soil is susceptible to freeze-thaw damage, two important functions of the surface course are to keep out water and to absorb the high number of freeze-thaw cycles that penetrate the pavement only 1 to 2 in. These cycles are mostly diurnal in character and may total 40 or 50 in a single season. Stabilized silty and clayey soils are most apt to be damaged and usually need the thicker surfacing. Construction techniques will vary, depending on local experience and practice.

REFERENCES

Cement-treated Soil Mixtures

1. Mills, W. H., Jr.: "Road Base Stabilization with Portland Cement," *Engineering News-Record*, vol. 115, no. 22, pp. 751–753, Nov. 28, 1935.

2. Portland Cement Association: *Progress Report on Exploratory Laboratory Investigation of Soil-cement Mixtures, Development Department*, Chicago, 106 pp., May 1, 1936.

3. Mills, W. H., Jr.: "Stabilizing Soils with Portland Cement, Experiments by South Carolina Highway Department," *Proceedings, Highway Research Board*, vol. 16, pp. 322–347, 1936.

4. Allen, H.: "Soil Stabilization," *Proceedings, Highway Conference*, Boulder, Colo., January, 1937, pp. 46–58, University of Colorado.

5. Allen, H.: "Road Soil Base Bound with Cement," *Engineering News-Record*, vol. 119, no. 11, pp. 437–439, Sept. 9, 1937.

6. Texas State Highway Department: "The Effect of Various Admixtures on the Soil Constants of a Soil Binder," *Binders Investigational Project 8, Research Project 2*, Nov. 16, 1937.

7. Mills, W. H., Jr.: "Cement-soil Stabilization," *Proceedings, Highway Research Board*, vol. 17, part I, pp. 513–520, 1937.

8. Catton, M. D.: "Basic Principles of Soil-cement Mixtures and Exploratory Laboratory Results," *Proceedings, Highway Research Board*, vol. 17, part II, pp. 7–31, 1937.

9. Larson, G. H.: "Experimental Soil-cement Road in Wisconsin," *Proceedings, Highway Research Board*, vol. 17, part II, pp. 83–91, 1937.

10. Catton, M. D.: "Laboratory Investigation of Soil-cement Mixtures for Subgrade Treatment in Kansas," *Proceedings, Highway Research Board*, vol. 17, part II, pp. 92–105, 1937.

11. Sheets, F. T., and M. D. Catton: "Basic Principles of Soil-cement Mixtures," *Engineering News-Record*, vol. 120, no. 25, pp. 869–875, June 23, 1938.

12. Sheets, F. T., and M. D. Catton: "How to Process Soil-cement Roads," *Engineering News-Record*, vol. 121, no. 1, pp. 20–24, July 7, 1938.

13. Lancaster, C. M.: "Base Stabilization with Portland Cement," *Roads and Streets*, vol. 81, no. 12, pp. 17–24, December, 1938.

14. Catton, M. D.: "Soil-cement Mixtures for Roads," *Proceedings, Highway Research Board*, vol. 18, part II, pp. 314–321, 1938.

15. Finch, H. A.: "Earth-cement Mixture in Sacks Used for River-bank Revetment," *Engineering News-Record*, vol. 122, no. 19, p. 659, May 11, 1939.

16. Portland Cement Association: *Soil-cement Roads—Construction Handbook*, Chicago, 88 pp., 1939.

17. Mills, W. H., Jr., L. D. Hicks, F. W. Vaughan, R. R. Litehiser, H. E. Brooks, J. E. Wood, C. R. Reid, E. J. Sampson, and H. G. Henderson: "Progress in Soil-cement Construction," *Proceedings, Highway Research Board*, vol. 19, pp. 517–558, 1939.

18. Reid, C. R.: "Concrete Pavement Subgrade Design, Construction, Control," *Proceedings, Highway Research Board*, vol. 19, pp. 541–551, 1939.

19. Texas Highway Department: "Caliche-cement Stabilization," *Information Exchange*, 83, pp. 5–22, Jan. 1, 1940.

20. Johnson, A. W.: "Soil-cement Stabilization," *Proceedings, Highway Conference*, January, 1940, pp. 55–71, University of Colorado, Boulder, Colo.

21. Winn, H. F., and P. C. Rutledge: "Frost Action in Highway Bases and Subgrades." *Purdue University Engineering Bulletin*, vol. 24, no. 3, 100 pp., May, 1940.

22. Mills, W. H., Jr.: "Condition Survey of Soil-cement Roads," *Proceedings, Highway Research Board*, vol. 20, pp. 812–820, 1940.

23. Catton, M. D.: "Research on the Physical Relations of Soil and Soil-cement Mixtures," *Proceedings, Highway Research Board*, vol. 20, pp. 821–855, 1940.

24. "Test Road Compares Bases of Varying Thickness," *Engineering News-Record*, vol. 126, no. 3, pp. 115–116, Jan. 16, 1941.

25. "A Bicycle Pathway of Soil-cement Construction," *Public Works*, vol. 72, no. 2, p. 20, February, 1941.

26. Portland Cement Association: *Soil-cement Roads: Construction Handbook*, Chicago, pp. 94, 1941.

27. Carson, L. S., and C. R. Reid: "Chemical Determination of Cement Content of Soil-cement Mixtures from Cement-hardened Bases," *Proceedings, Highway Research Board*, vol. 21, pp. 471–483, 1941.

28. Mills, W. H., Jr.: "Condition Survey of Soil-cement Roads," *Proceedings, Highway Research Board*, vol. 21, pp. 484–492, 1941.

29. Watson, J. D.: "The Unconfined Compressive Strength of Soil-cement Mixtures," *Proceedings, Highway Research Board*, vol. 21, pp. 493–501, 1941.

30. "Soil-cement Pavement Utilizes Mill Tailings from Metal Mines," *Construction Methods*, vol. 24, no. 3, pp. 40–52, March, 1942.

31. "Cement Spreader and Road-mix Methods Speed Large Surfacing Jobs," *Engineering News-Record*, vol. 129, no. 11, pp. 366–368, Sept. 10, 1942.

32. Pirie, J. E., and J. R. Ward: "Soil-cement Stabilized Shoulders in Texas," *Roads and Streets*, vol. 85, no. 11, pp. 32–37, November, 1942.
33. Portland Cement Association: *Soil-cement Mixtures: Laboratory Handbook*, Chicago, 80 pp., 1942.
34. Hicks, L. D.: "Soil-cement Design in North Carolina," *Proceedings, Highway Research Board*, vol. 22, pp. 415–418, 1942.
35. Winterkorn, H. F., H. J. Gibbs, and R. G. Fehrman: "Surface Chemical Factors of Importance in the Hardening of Soils by Means of Portland Cement," *Proceedings, Highway Research Board*, vol. 22, pp. 385–414, 1942.
36. Highway Research Board: "Use of Soil-cement Mixtures for Base Courses," *Wartime Road Problems*, No. 7, Washington, D.C., 30 pp., October, 1943. (Second Edition, 1949, includes reprints of *ASTM Standard Test Procedures*.)
37. U.S. Corps of Engineers, Waterways Experiment Station: *Unconfined Compressive Strengths of Compacted Soil-cement (Fresh Water) and Soil-cement (Salt-water) Specimens*, Vicksburg, Miss., December, 1943.
38. Stanton, T. E., F. N. Hveem, and J. L. Beatty: "Progress Report on California Experience with Cement Treated Bases," *Proceedings, Highway Research Board*, vol. 23, pp. 279–295, 1943.
39. Catton, M. D., and E. J. Felt: "Weight-in-water Methods of Determining the Moisture Content of Soil-cement Mixtures in the Field," *Proceedings, Highway Research Board*, vol. 23, pp. 487–494, 1943.
40. Bauer, E. E.: "Discussion on Determination of Moisture in Soil-cement," *Proceedings, Highway Research Board*, vol. 23, pp. 494–496, 1943.
41. Catton, M. D., and E. J. Felt: "Effect of Soil and Calcium Chloride Admixtures on Soil-cement Mixtures," *Proceedings, Highway Research Board*, vol. 23, pp. 497–529, 1943.
42. Portland Cement Association: *Soil-cement Roads: Construction Handbook*, Chicago, 100 pp., 1943.
43. Worley, H. E.: "Triaxial Testing Methods Usable in Flexible Pavement Design," *Proceedings, Highway Research Board*, vol. 23, pp. 109–116, 1943.
44. Mizroch, J.: "Determination of Cement-content of Soil-cement Mixtures," *Public Roads*, vol. 23, no. 11, pp. 297–299, January-February-March, 1944.
45. "Soil-cement Shoulder and Patching Projects in Georgia," *Concrete*, vol. 52, no. 5, pp. 10–11, 23, May, 1944.
46. Friday, C. B.: "Soil-cement Paving Costs Reduced by Use of Clay Marl," *Engineering News-Record*, vol. 133, no. 22, pp. 681–684, Nov. 30, 1944.
47. Mainfort, R. C.: "The Development of a Mechanical Cement Spreader and Accessories for Use in Soil-cement Construction," *U.S. Civil Aeronautics Administration, Technical Development Report* 41, 32 pp., Washington, D.C., 1944.
48. Hicks, L. D.: "Use of Soil-cement for Patching Pavements in North Carolina," *Proceedings, Highway Research Board*, vol. 24, pp. 445–450, 1944.
49. Catton, M. D.: "Some Wartime Soil-cement Construction Experiences," *Proceedings, Highway Research Board*, vol. 24, pp. 450–466, 1944.
50. Touhey, B.: "Cement Treated Base Built in Arizona," *Engineering News-Record*, vol. 134, no. 2, pp. 108–110, Jan. 11, 1945.
51. "Building Sites Stabilized with Grout," *Engineering News-Record*, vol. 134, no. 2, pp. 140–141, Jan. 11, 1945.
52. Mihram, R. G., and Betty Brown: "Determination of Portland Cement in Drilling Muds and Soil-cement Mixtures," *Industrial and Engineering Chemistry, Analytical Edition*, vol. 17, no. 3, pp. 156–158, March, 1945.
53. Cooke-Yarborough, S. S.: "Experimental Soil-cement Culvert," *Indian Concrete Journal*, vol. 19, no. 5, pp. 64–65, May, 1945.
54. Whitton, R. M.: "Maintenance Methods for Preventing and Correcting the Pumping Action of Concrete Pavement Slabs," *Highway Research Board, Research Reports*, no. 1D, Special Papers on the Pumping Action of Concrete Pavements, pp. 3–12, 1945.
55. Frost, R. E.: "Correcting Pavement Pumping by Mud Jacking," *Highway Research Board, Research Reports*, no. 1D, Special Papers on the Pumping Action of Concrete Pavements, pp. 13–54, 1945.
56. Payne, G. W., T. A. Blair, L. H. Bond, H. F. Brown, H. H. Hopkins, Jr., M. C. Patton, J. W. Poulter, and C. S. Robinson: "Stabilization of Roadbed: Report on Methods Used," *Proceedings, American Railway Engineering Association*, vol. 47, no. 458, pp. 320–353, February, 1946. (Contains bibliography of 18 items.)
57. Weathers, H. C.: "Florida Experiments with Limerock," *Better Roads*, vol. 16, no. 4, pp. 25–26, 30, 32, April, 1946.
58. "Bulldozer Hoppers Spread Soil-cement Mixture for Landing Mat Base," *Construction Methods*, vol. 28, no. 7, pp. 95–97, 176–179, July, 1946.

59. "New Mexico Stabilizes Clay Subgrade with Two Percent Portland Cement Admixture," *Roads and Streets*, vol. 89, no. 8, pp. 91–93, August, 1946.
60. "California Builds Concrete Pavement with Soil-cement Base," *Roads and Streets*, vol. 89, no. 9, pp. 90–91, 93–98, September, 1946.
61. Portland Cement Association: "Application of Soil-cement to Low Cost House and Farm Building Construction," *Soil-cement Information*, No. ScB 13, Chicago, 4 pp., November, 1946.
62. Portland Cement Association: *Soil-cement Roads: Construction Handbook*, No. Sc3, 5th ed., Chicago, 86 pp., 1946.
63. Portland Cement Association: *Soil-cement Mixtures: Laboratory Handbook*, 2d ed., Chicago, 80 pp., 1946.
64. Smith, R., R. B. Peck, and T. H. Thornburn: "Second Progress Report of the Investigation of Methods of Roadbed Stabilization," *University of Illinois Bulletin*, vol. 44, no. 51, pp. 1–55; *Engineering Experiment Station, Reprint Series* 38, Apr. 22, 1947.
65. "Application of Soil-cement to Earth Dam Cores," *Pacific Builder and Engineer*, vol. 53, no. 8, pp. 76–77, August, 1947.
66. Willis, E. A.: "Experimental Soil-cement Base Course in South Carolina," *Public Roads*, vol. 25, no. 1, pp. 9–19, September, 1947.
67. Highway Research Board: "Maintenance Methods for Preventing and Correcting the Pumping Action of Concrete Pavement Slabs," *Current Road Problems*, No. 4-R, rev. ed., September, 1947.
68. Byrne, W. S., and W. G. Holtz: "Soil-cement Placed Mechanically," *Engineering News-Record*, vol. 139, no. 126, pp. 48–51, Dec. 25, 1947. (Also see *Civil Engineering*, vol. 17, no. 12, p. 751, December, 1947.)
69. State Highway Commission of Kansas: "Design of Flexible-pavement Using the Triaxial Compression Test," *Highway Research Board, Bulletin* 8, 63 pp., 1947.
70. "New Bulk-cement Unit Spreads and Measures," *Contractors and Engineers Monthly*, vol. 45, no. 1, p. 43, January, 1948.
71. "Soil-cement Ditch Lining—Economical Means of Preventing Erosion," *Roads and Streets*, vol. 91, no. 5, p. 67, May, 1948.
72. Proctor, C. S.: "Cap Grouting to Stabilize Foundation on Cavernous Limestone," *Proceedings, Second International Conference on Soil Mechanics and Foundation Engineering*, Rotterdam, Holland, June, 1948.
73. Loughborough, T. F.: "Performance of Soil-cement Roads in Virginia," *Virginia Highway Bulletin*, vol. 14, no. 11, pp. 3–6, September, 1948.
74. Reid, C. R.: "Report of Committee on Soil-cement Roads," *Highway Research Board. Bulletin* 14, pp. 12–17, October, 1948.
75. "Soil-cement Floor for Aggregate Stockpiles," *Roads and Streets*, vol. 91, no. 12, p. 58, December, 1948.
76. Hansen, R.: "Service Behavior Tests, Barksdale Field, Shreveport, La.," *Proceedings, American Society of Civil Engineers*, vol. 75, no. 1, pp. 45–55, January, 1949.
77. Burkhart, E. J.: "Investigation of Soils and Building Techniques for Rammed Earth Construction," *Texas Engineering Experiment Station, Research Report* 6, 23 pp. Texas A & M College System, College Station, Tex., May, 1949.
78. Highway Research Board: "Prevention of Moisture Loss in Soil-cement with Bituminous Material," *Research Report* 8-F, 34 pp., September, 1949.
79. Hicks, L. D.: "The Use of Agricultural Soil Maps in Making Soil Surveys," *Highway Research Board, Bulletin* 22, p. 108, October, 1949.
80. State Highway Commission of Kansas: *Lawrence Experimental Concrete Pavement: A Twelve Year Study*, Topeka, Kan., 102 pp., 1949.
81. Portland Cement Association: *Soil-cement Roads: Construction Handbook*, 6th ed., Chicago, 1949.
82. Highway Research Board: "Thickness of Flexible Pavements," *Current Road Problems* 8-R, Washington, D.C., 49 pp., 1949.
83. Jones, J. E.: "The Use of Stabilized Soil as a Structural Material," *Proceedings, International Association for Bridge and Structural Engineering*, pp. 263–270, 1949.
84. Spangler, M. G., and O. H. Patel: "Modification of a Gumbotil Soil by Lime and Portland Cement Admixtures," *Proceedings, Highway Research Board*, vol. 29, pp. 561–566, 1949.
85. Marshall, H.: "A Typical Ohio Soil-cement Project," *American Road Builders Association, Technical Bulletin* 172, 16 pp., 1950.
86. Mehra, S. R., and H. L. Uppal: "Use of Stabilized Soil in Engineering Construction: Section 2, Resistance of Cement-soil Mixture to Action of Water; and Section 3, Compressive Strength of Cement-soil Mixtures," *Journal of the Indian Roads Congress* (India), vol. 15, no. 1, pp. 184, 196, 197–204, August, 1950.

87. Mehra, S. R., and H. L. Uppal: "Use of Stabilized Soil in Engineering Construction: Section 4, Shrinkage of Compacted Soils," *Journal of the Indian Roads Congress* (India), vol. 15, no. 2, pp. 320–335, November, 1950.
88. Robeson, F. A., and W. E. Webb: "Cofferdam Grouting at Jim Woodruff Dam," *Engineering News-Record*, vol. 145, no. 1, pp. 35–37, July 6, 1950.
89. Portland Cement Association: *Soil-cement Mixtures: Laboratory Handbook*, 6th ed., Chicago, 1950.
90. Mehra, S. R., and H. L. Uppal: "Use of Stabilized Soil in Engineering Construction: Section 5, (A) Thermal Expansion of Compacted Soils, and (B) Thermal Conductivity of Compacted Soils," *Journal of the Indian Roads Congress* (India), vol. 15, no. 3, pp. 469–482, January, 1951.
91. Lovering, W. R.: "Uniformity of Class 'C' Cement Treated Base Subject of Tests," *California Highways and Public Works*, vol. 30, nos. 1–2, pp. 43–45, January–February, 1951.
92. Meyers, R. L.: "Successful Test-road Mixing Machine Proves Satisfactory in District 1," *California Highways and Public Works*, vol. 30, nos. 1–2, pp. 54–57, January–February, 1951.
93. Withycombe, E. A.: "Cement-treated-base Construction in California," *California Highways and Public Works*, vol. 30, nos. 7–8, pp. 40–43, July–August, 1951.
94. "Novel 'Sack' Rip-rap Used in Lieu of Culvert Headwalls," *Roads and Streets*, vol. 94, no. 8, p. 81, August, 1951.
95. Clare, K. E., and P. T. Sherwood: "The Determination of the Cement Content of Soil-cement Mixtures," *Journal of Applied Chemistry* (London), vol. 1, part 12, pp. 551–560, December, 1951.
96. Clare, K. E., and A. E. Pollard: "The Relationship between Compressive Strength and Age for Soils Stabilized with Four Types of Cement," *Magazine of Concrete Research* (London), vol. 3, no. 8, pp. 57–64, December, 1951.
97. Winterkorn, H. F., and E. C. Chandrasekharan: "Laterite Soils and Their Stabilization," *Highway Research Board, Bulletin* 44, pp. 10–29, 1951.
98. Worley, H. E.: "Triaxial Design, Correlated with Flexible Pavement Performance in Kansas," *ASTM, Special Technical Publication* 106, Triaxial Testing of Soils and Bituminous Mixtures, pp. 112–137, 1951.
99. Gilliland, J. L., and H. M. Hunter: "Rapid Method for Estimating Cement Content of Soil-cement and Blended Cements," *ASTM, Bulletin* 180, pp. 29–30, February, 1952.
100. Johnson, G. E.: "Silt Injection Checks Seepage Losses from Water Supply Canal in Nebraska," *Engineering News-Record*, vol. 148, no. 6, pp. 32–34, 37, Feb. 7, 1952.
101. Handa, C. L., C. L. Dhawan, and J. C. Bahri: "Engineering Properties of Cement Stabilization," *Indian Concrete Journal* (India), vol. 26, no. 3, pp. 65–70, Mar. 15, 1952.
102. Olinger, D. J.: "Soil-cement Stabilized Base," *University of Colorado, 25th Annual Highway Conference, Engineering Experiment Station Circular, Highway Series*, no. 25, pp. 43–46, April, 1952.
103. Rodes, V. H.: "Incorporation of Admixtures with Soil," *Proceedings of the Conference on Soil Stabilization*, pp. 169–173, Massachusetts Institute of Technology, Cambridge, Mass., June 18–20, 1952.
104. Robinson, P. J. M.: "British Studies on the Incorporation of Admixtures with Soil," *Proceedings of the Conference on Soil Stabilization*, pp. 175–182, Massachusetts Institute of Technology, Cambridge, Mass., June 18–20, 1952.
105. Jones, R. P.: "Commercial Stabilization Equipment: Development of the P & H Single Pass Processing Method," *Proceedings of the Conference on Soil Stabilization*, pp. 185–190, Massachusetts Institute of Technology, Cambridge, Mass., June 18–20, 1952.
106. Hurst, J.: "Hetherington-Berner Stabilization Equipment," *Proceedings of the Conference on Soil Stabilization*, pp. 191–192, Massachusetts Institute of Technology, Cambridge, Mass., June 18–20, 1952.
107. Seaman, H. J.: "Mechanical Stabilization and the Processing of Stabilized Soils—The Seaman Pulvi-Mixer," *Proceedings of the Conference on Soil Stabilization*, pp. 193–197, Massachusetts Institute of Technology, Cambridge, Mass., June 18–20, 1952.
108. Wood, C. W.: "Soil Stabilization with Wood Roadmixers and Allied Equipment," *Proceedings of the Conference on Soil Stabilization*, pp. 198–206, Massachusetts Institute of Technology, Cambridge, Mass., June 18–20, 1952.
109. Heacock, R. C.: "Barber-Greene Company's Experience and Observations on Soil Stabilization," *Proceedings of the Conference on Soil Stabilization*, pp. 207–213, Massachusetts Institute of Technology, Cambridge, Mass., June 18–20, 1952.

110. Catton, M. D.: *Soil-cement: A Construction Material, Proceedings, Conference on Soil Stabilization*, pp. 26–59, Massachusetts Institute of Technology, June 18–20, 1952.
111. Johnson, S. J.: "Soil Stabilization of Building Foundations," *Proceedings, Conference on Soil Stabilization*, pp. 241–254, Massachusetts Institute of Technology, June 18–20, 1952.
112. Powers, K. L., J. R. Benson, and V. S. Meissner: "Asphaltic Concrete and Soil-cement Tested as Rip-rap Substitutes at Bonny Reservoir," *Civil Engineering*, vol. 22, no. 6, pp. 34–36, June, 1952.
113. Maclean, D. J., P. J. M. Robinson, and S. B. Webb: "An Investigation of the Stabilization of a Heavy Clay Soil with Cement for Road Base Construction," *Roads and Road Construction* (London), vol. 30, no. 358, pp. 287–292, October, 1952.
114. Stone, R.: "Soil-cement for House Construction," *Civil Engineering*, vol. 22, no. 12, pp. 29–31, December, 1952.
115. Maner, A. W.: "Curing Soil-cement Bases," *Proceedings, Highway Research Board*, vol. 31, pp. 540–558, 1952.
116. Road Research Laboratory: *Soil Mechanics for Road Engineers*, p. 258, Department of Scientific and Industrial Research, H.M. Stationery Office (London), 1952.
117. Reynolds, J. H., Jr.: "Traffic Tests of Soil-cement Lanes," *American Road Builders Association, Technical Bulletin* 187, 1952.
118. Withycombe, E.: "Base Stabilization with Portland Cement," *Proceedings, Fifth California Street and Highway Conference*, pp. 34–38, University of California, The Institute of Transportation and Traffic Engineering, Berkeley, Calif., Feb 4–6, 1953.
119. Streed, E. R., and U. W. Stoll: "Flame Method for Estimating Cement Content of Soil-cement and Pozzolan-cement Mixtures," *ASTM Bulletin* 189, pp. 58–60, April, 1953.
120. Menon, T. M.: "Cement Stabilization of Soils with Particular Reference to the Effect of Curing Conditions on Their Compressive Strength," *Indian Concrete Journal* (India), vol. 27, no. 4, pp. 203–205, Apr. 15, 1953.
121. Clare, K. E., and A. E. Pollard: "The Effect of Curing Temperature on the Compressive Strength of Soil-cement Mixtures," *Road Research Note* RN/1980/KEC. AEP., Department of Scientific and Industrial Research, Harmondsworth, England, May, 1953.
122. Ver Brugge, M. A.: "Cement-stabilized Bases," *Proceedings, 14th Annual Highway Conference, University of Utah, Bulletin*, vol. 43, no. 6, pp. 72–76, June, 1953.
123. Maclean, D. J., and P. J. M. Robinson: "Methods of Soil Stabilization and Their Application to the Construction of Airfield Pavements," *Proceedings of the Institution of Civil Engineers* (London), part 2, vol. 2, pp. 447–502, June, 1953.
124. Leadabrand, J. A., and L. T. Norling: "Soil-cement Test Data Correlation in Determining Cement Factors for Sandy Soils," *Highway Research Board, Bulletin* 69, pp. 29–44, 1953.
125. Lacey, D. L.: "Flexible Pavement Design Correlated with Road Performance," *Highway Research Board, Research Report* 16-B, pp. 59–67, January, 1954.
126. Portland Cement Association: *Essentials of Soil-cement Construction*, Soil-Cement Bureau, SC-110, Chicago, January, 1954.
127. Portland Cement Association: *Application of Soil-cement to Low-cost House and Farm Building Construction*, SC-B-13, 10-53, Chicago.
128. Thompson, R. A. J.: "The Stabilization of Low Quality Gravels," Symposium on Soil Stabilization (Australia), pp. 64–71, Jan. 18–22, 1954.
129. Marshall, T. J.: "Some Properties of Soil Treated with Portland Cement," Symposium on Soil Stabilization (Australia), pp. 28–34, Jan. 18–22, 1954.
130. Foulkes, R. A.: "Soil Stabilization in Great Britain," Symposium on Soil Stabilization (Australia), pp. 114–137, Jan. 18–22, 1954.
131. Foulkes, R. A.: "German Soil Stabilization," Symposium on Soil Stabilization (Australia), pp. 14–23, Jan. 18–22, 1954.
132. Portland Cement Association: *Soil-cement for Paving Slopes and Lining Ditches*, SCB 14-3-54, Chicago.
133. Norling, L. T.: "Soil-cement—Present Design and Construction Practices," *Proceedings, 15th Annual Highway Engineering Conference, Bulletin* 65, pp. 126–133, University of Utah, Utah Engineering Experiment Station, July, 1954.
134. Massachusetts Institute of Technology: *Soil Stabilization for Highways*, Massachusetts Department of Public Works and Joint Highway Research Project, July, 1954.
135. California Division of Highways: *Standard Specifications*, August, 1954.
136. Clare, K. E., and P. T. Sherwood: "The Effect of Organic Matter on the Setting of Soil-Cement Mixtures," *Journal of Applied Chemistry* (London), vol. 4, part II, pp. 625–630, November, 1954.

137. Whitehurst, E. A.: "The Stabilization of Local Base Materials," *Tennessee Highway Research Program, Bulletin* 1, University of Tennessee, Knoxville, 52 pp., April, 1955.
138. *The ABC's of Soil-cement Stabilization*, Pettibone-Wood Manufacturing Company, Hollywood, Calif., 35 pp., April, 1955.
139. Mehra, S. R., and L. R. Chadda: "Construction of Soil-cement Drains," *The Indian Concrete Journal* (India), vol. 29, no. 5, pp. 161–162, May, 1955.
140. Alzueta, C.: "Review of Cement-treated Bases in California," *California Highways and Public Works*, vol. 34, nos. 5–6, pp. 40–44, May-June, 1955.
141. Handy, R. L., D. T. Davidson, and T. Y. Chu: "Effect of Petrographic Variations of Southwestern Iowa Loess on Stabilization with Portland Cement," *Highway Research Board, Bulletin* 98, pp. 1–20, 1955.
142. Baker, C. N.: "Strength of Soil-cement as a Function of Mixing," *Highway Research Board, Bulletin* 98, pp. 33–52, 1955.
143. Reinhold, F.: "Elastic Behavior of Soil-cement Mixtures," *Highway Research Board, Bulletin* 108, pp. 128–137, 1955.
144. Felt, E. J.: "Factors Influencing Physical Properties of Soil-cement Mixtures," *Highway Research Board, Bulletin* 108, pp. 138–163, 1955.
145. Whitehurst, E. A.: "Stabilization of Tennessee Gravel and Chert Bases," *Highway Research Board, Bulletin* 108, pp. 163–174, 1955.
146. California Division of Highways: "Method of Calculating the Design Thickness of Pavement Sections Based on Stabilometer and Expansion Pressure Measurements," *Materials Manual*, vol. 1, Test Method No. Calif. 301-B, part 7, rev. Sept. 1, 1957.
147. Harris, F. A.: "Selection and Design of Semi-flexible and Conventional Type Pavements," *Proceedings, Highway Research Board*, vol. 35, pp. 110–138, 1956.
148. Maclean, D. J.: "Considerations Affecting the Design and Construction of Stabilized-soil Road Bases," *Journal of the Institution of Highway Engineers* (London), vol. 3, no. 9, pp. 16–33, January, 1956.
149. California Division of Highways: "Method for the Determination of Compressive Strength of Cement Treated Bases, Classes 'A' and 'B' and Cement Treated Subgrade," *Materials Manual*, vol. 1, Test Method No. Calif. 312-B, Jan. 3, 1956.
150. California Division of Highways: "Method of Freezing and Thawing Test for Compacted Test Specimens of Cement Treated Bases, Classes 'A' and 'B' and Cement Treated Subgrade," *Materials Manual*, vol. 1, Test Method No. Calif. 313-B, Jan. 3, 1956.
151. California Division of Highways: "Method of Wetting and Drying Test for Compacted Test Specimens of Cement Treated Bases, Classes 'A' and 'B' and Cement Treated Subgrade," *Materials Manual*, vol. 1, Test Method No. Calif. 314-B, Jan. 3, 1956.
152. Chadda, L. R.: "Construction of Soil-cement Floors," *The Indian Concrete Journal* (India), vol. 30, no. 4, p. 105, Apr. 15, 1956.
153. Chadda, L. R.: "Effect of Moisture on the Compressive Strength of Soil-cement Mixtures," *Indian Concrete Journal* (India), vol. 30, no. 4, pp. 124, Apr. 15, 1956.
154. Leadabrand, J. A.: *Some Engineering Aspects of Soil-cement Mixtures*, American Society of Civil Engineers, Mid-South Section, Little Rock, Ark., Apr. 27, 1956.
155. Cruchley, A. E.: "Stabilization of a Heavy Clay—An Experiment at Hornchurch, Essex," *Roads and Road Construction* (London), vol. 34, no. 402, pp. 164–168, June, 1956.
156. Felt, E. J., and M. S. Abrams: "Strength and Elastic Properties of Compacted Soil-cement Mixtures," *ASTM, Special Technical Publication* 206, 1957.
157. U.S. Corps of Engineers: "Summary Reviews of Soil Stabilization Processes," *Report* 3, *Soil-cement, Miscellaneous Paper* 3-122, Waterways Experiment Station, Vicksburg, Miss., 38 pp., September, 1956.
158. Portland Cement Association: *Soil-cement Construction Handbook*, Chicago, 99 pp., 1956.
159. Portland Cement Association: *Soil-cement Laboratory Handbook*, Chicago, 63 pp., 1956.
160. Leadabrand, J. A., and L. T. Norling: "Simplified Methods of Testing Soil-Cement Mixtures," *Highway Research Board, Bulletin* 122, pp. 35–47, 1956.
161. Lambe, T. W.: "Modification of Frost Heaving of Soils with Additives," *Highway Research Board, Bulletin* 135, pp. 1–23, 1956.
162. Spencer, W. T., H. Allen, and P. C. Smith: "Report on Pavement Research Project in Indiana," *Highway Research Board, Bulletin* 116, pp. 1–56, 1956.
163. Portland Cement Association: *Summaries of Soil-cement Construction to January 1, 1957*, SC-104-1956 Supplement, Chicago. 7 pp., mimeo.
164. Leyder, J. P.: *Stabilization du sol au ciment, réport de recherche*, Centre de Recherches Routières, Brussels, Jan. 15, 1957.

165. Brennan, P. J.: "The Stabilization of a Fine Grained Soil Using Low Percentages of Portland Cement and Fly-ash," *University of Delaware, Report* S-2, Newark, N.J., Apr. 30, 1957.

166. Lambe, T. W., and Z. C. Moh: "Improvement of the Strength of Soil-cement with Additives," *Highway Research Board, Bulletin* 183, 1958.

167. Leadabrand, J. A., L. T. Norling, and A. C. Hurless: "Soil Series as a Basis for Determining Cement Requirements for Soil-cement Construction," *Highway Research Board, Bulletin* 148, 1957.

168. Fielding, R. V.: "Mudjacking with Fly Ash," *Virginia Highway Bulletin*, vol. 23, no. 8, pp. 11–12, August, 1957.

169. "The Work of the Road Research Laboratory," *Roads and Road Construction*, vol. 35, no. 418, p. 308, October, 1957.

170. American Association of State Highway Officials: *Standard Specifications for Highway Materials and Methods of Sampling and Testing*, part III, 1957.   Part III contains the recent revisions of the following standard methods of test:
    (1) Moisture-Density Relations of Soil-cement Mixtures, AASHO Designation T 134-57.
    (2) Wetting-and-drying Test of Compacted Soil-cement Mixtures, AASHO Designation T 135-57.
    (3) Freezing-and-thawing Test of Compacted Soil-cement Mixtures, AASHO Designation T 136-57.

171. American Association of State Highway Officials: *Standard Specifications for Highway Materials and Methods of Sampling and Testing*, part II, 1955:
    (1) Standard Method of Test for Cement Content of Soil-cement Mixtures, AASHO Designation T 144-49.
    (2) Standard Method of Test for Organic Impurities in Sands for Concrete, AASHO Designation T 21-42.

172. American Society for Testing Materials: *Procedures for Testing Soils*, 1958.   This publication includes the following test methods which are useful in testing for soil-cement.
    (1) Standard Methods of Test for Moisture-Density Relations of Soil-cement Mixtures, ASTM Designation D 558-57.
    (2) Standard Method of Wetting-and-drying Test of Compacted Soil-cement Mixtures, ASTM Designation D 559-57.
    (3) Standard Method of Freezing-and-thawing Test of Compacted Soil-cement Mixtures, ASTM Designation D 560–57.
    (4) Standard Method of Test for Cement Content of Soil-cement Mixtures, ASTM Designation D 806-47.
    (5) Suggested Method of Test for Making and Curing Soil-cement Compression and Flexure Test Specimens in the Laboratory.
    (6) Suggested Method of Test for Compressive Strength of Molded Soil-cement Cylinders.
    (7) Suggested Method of Test for Flexural Strength of Soil-cement Using Simple Beam with Third-point Loading.
    (8) Suggested Method of Test for Compressive Strength of Soil-cement Using Portions of Beams Broken in Flexure—Modified Cube Method.

173. American Society for Testing Materials: "Standard Method of Test for Organic Impurities in Sands for Concrete," *ASTM Standards*, part 3, ASTM Designation C 40-48.   (See also "Comparison of Standards for Colorimetric Test of Sand," *ASTM Proceedings*, vol. 34, and "A Critical Laboratory Review of Methods of Determining Organic Matter and Carbonates in Soils," *ASTM Technical Bulletin* 317, June, 1932.)

174. Texas Highway Department: *Strength Test of Soil-cement Mixtures*, THD 96, part I, *Test Procedure for Compressive Strength*, part 2, *Punching Shear Test*, November, 1953.

175. Shepard, C. H.: "Highway Shoulders Construction Practices," *Highway Research Board, Bulletin* 151, pp. 15–19, 1957.

176. Sherwood, P. T.: "The Effect of Sulphates on Cement-stabilized Clay," paper presented at the 37th Annual Meeting of the Highway Research Board, 1958.

177. Norling, L. T., and R. G. Packard: "Expanded Short-cut Test Method for Determining Cement Factors for Sandy Soils," paper presented at the 37th Annual Meeting of the Highway Research Board, 1958.

178. Aichhorn, W., and W. Steinbrenner: "Stabilization of Disintegrated Granite for Base Courses Exposed to Severe Frost Conditions," *Proceedings, Fourth International Conference on Soil Mechanics and Foundation Engineering* (London), vol. 2, pp. 89–93, 1957.

179. Grimer, F. J., and N. F. Ross: "The Effect of Pulverization on the Quality of Clay Cement," *Proceedings, Fourth International Conference on Soil Mechanics and Foundation Engineering* (London), vol. 2, pp. 109–113, 1957.

180. Sand-cement Road Stabilization in Holland," *Highways and Bridges and Engineering Works* (Great Britain), vol. 25, no. 1211, pp. 1, 3, 4, Oct. 9, 1957.

181. Rosenak, B.: "The Use of Soil-cement for Low Cost Housing in the Tropics," *Civil Engineering and Public Works Review*, vol. 52, no. 616, October, 1957.

182. Portland Cement Association: *Suggested Specifications for Soil-cement Construction*, SCB 12-6, Chicago.

183. Redus, J. F.: "Study of Soil-cement Base Courses on Military Airfields," paper presented at the 37th Annual Meeting of the Highway Research Board, 1958.

184. Davidson, D. T., R. K. Katti, and D. E. Welch: "Use of Fly Ash with Portland Cement for Stabilization of Soils," paper presented at the 37th Annual Meeting of the Highway Research Board, 1958.

185. Jones, C. W.: "Stabilization of Expansive Clay with Hydrated Lime and with Portland Cement," paper presented at the 37th Annual Meeting of the Highway Research Board, 1958.

186. Diamond, S., and E. B. Kinter: "A Rapid Method Utilizing Surface Area Measurements to Predict the Amount of Portland Cement Required for the Stabilization of Plastic Soils," paper presented at the 37th Annual Meeting of the Highway Research Board, 1958.

187. Handy, R. L.: "A Hypothesis of Cementation of Minerals in Soil-cement," paper presented at the 37th Annual Meeting of the Highway Research Board, 1958.

188. Fuller, M. G., and G. W. Dabney: "Stabilizing Weak and Defective Bases with Hydrated Lime," *Roads and Streets*, vol. 95, pp. 64–69, March, 1952.

189. "Symposium on Cement and Clay Grouting of Foundations," *Journal of the Soil Mechanics and Foundations Division, Proceedings of the American Society of Civil Engineers*, vol. 84, SM1, February, 1958, part 1.

190. Diamond, S., and E. B. Kinter: "Surface Areas of Clay Minerals as Derived from Measurements of Glycerol Retention," *Proceedings, Fifth National Clay Conference*, Urbana, Ill., 1956.

191. Haley, J. F., and C. W. Kaplar: "Cold Room Studies of Frost Action in Soil," *Highway Research Board, Special Report* 2, pp. 246–266, 1952.

192. Kersten, M. S.: "Thermal Properties of Soils," *University of Minnesota, Engineering Station Bulletin* 28, June 1, 1949.

193. Kawala, E. L.: "A Review of the Construction and Performance of Soil-cement Subbases for Concrete Pavement," paper presented at the 56th Annual Convention, American Roadbuilders Association, Washington, D.C., Jan. 20–23, 1958.

194. Portland Cement Association: *A Report of Cement-treated Subbases for Concrete Pavements*, Chicago, January, 1958.

195. Portland Cement Association: *Stabilizing Railroad Track by Pressure Grouting*, Chicago.

196. Associated General Contractors of America, Inc.: *Contractors' Equipment Ownership Expense*, 1227 Munsey Building, Washington, D.C.

197. Associated Equipment Distributors: *Compilation of Rental Rates for Construction Equipment*, 360 North Michigan Avenue, Chicago.

198. Johnson, A. W.: "Frost Action in Roads and Airfields," *Highway Research Board, Special Report* 1, *National Academy of Sciences, National Research Council, Publication* 211, Washington, D.C., 1952.

199. Sherwood, P. T.: "The Determination of the Cement Content of Soil-cement. III. An Investigation of Some of the Factors Involved," *Journal of Applied Chemistry*, vol. 7, pp. 596–604, November, 1957.

200. Mehra, S. R., L. R. Chadda, and R. N. Kapur: "Role of Detrimental Salts in Soil Stabilization with and without Cement. I. Effect of Sodium Sulphate," *The Indian Concrete Journal*, vol. 29, no. 10, pp. 336–337, October, 1955.

201. Chadda, L. R., and Raj Hem: "Role of Detrimental Salts in Soil Stabilization with and without Cement. II. Effect of Sodium Carbonate," *The Indian Concrete Journal*, vol. 29, no. 12, pp. 401–402, December, 1955.

202. Uppal, I. S., and B. P. Kapur: "Role of Detrimental Salts in Soil Stabilization with and without Cement. III. Effect of Magnesium Sulphate," *The Indian Concrete Journal*, vol. 31, no. 7, pp. 228–231, July, 1957.

203. *Journal of the Soil Mechanics and Foundations Division, American Society of Civil Engineers*, vol. 84, no. SM 1, February, 1958, part 1: A. W. Simonds, *Paper* 1544, "Present Status of Pressure Grouting Foundations"; S. J. Johnson, *Paper* 1545, "Grouting with Clay-cement Grouts"; G. A. Kravetz, *Paper* 1546, "The Use of Clay

in Pressure Grouting"; A. Klein and M. Polivka, *Paper* 1547, "The Use of Admixtures in Cement Grouts"; J. P. Elston, *Paper* 1548, "Suggested Specifications for Pressure Grouting"; F. H. Lippold, *Paper* 1549, "Pressure Grouting with Packers"; A. Mayer, *Paper* 1550, "French Grouting Practice"; E. P. Burwell, Jr., *Paper* 1551, "Practice of the Corps of Engineers"; G. K. Leonard and L. M. Grant, *Paper* 1552, "Experience of TVA with Clay-cement and Related Grouts."
204. "Soil Survey Manual," *U.S. Department of Agriculture, Handbook* 18, 1951.
205. Hicks, L. D.: "The Use of Agricultural Soil Maps in Making Soil Surveys," *Highway Research Board, Bulletin* 22, October, 1949.
206. Leadabrand, J. A., L. T. Norling, and A. C. Hurless: "Soil Series as a Basis for Determining Cement Requirements for Soil-cement Construction," *Highway Research Board, Bulletin* 148, 1957.
207. "Soils of the United States," *Atlas of American Agriculture*, part III, U.S. Department of Agriculture, 1935.
208. Kellog, Charles E.: "Development and Significance of the Great Soil Groups of the United States," *U.S. Department of Agriculture, Miscellaneous Publication* 229, 1936.
209. "Soils and Men," *Yearbook of Agriculture*, U.S. Department of Agriculture, 1938.
210. Jenny, H.: *Factors of Soil Formation*, McGraw-Hill Book Company, Inc., New York, 1941.
211. Riecken, F. F., and G. D. Smith: "Lower Categories of Soil Classification: Family, Series, Type and Phase," *Soil Science*, vol. 67, p. 107, January–June, 1949; J. Thorp, and G. D. Smith: "Higher Categories of Soil Classification: Order, Suborder and Great Soil Groups," *ibid.*, page 117.
212. *Highway Research Board Bulletin* 13, 1948; 22, 1949; 28, 1950; 46, 1951; 83, 1953; 22-R 1958.

Bituminous-aggregate-and-soil Stabilization

213. ASTM Committee D-18: "Tentative Method of Testing Soil-Bituminous Mixtures," ASTM Designation D 915-47T, pp. 334–345.
214. Baskin, C. M., and N. W. McLeod: "Waterproofed Mechanical Stabilization," *Proceedings, Association of Asphalt Paving Technologists*, vol. 12, pp. 276–316, 1940.
215. Benson, J. R.: "Bituminous Soil Stabilization," *Roads and Streets*, May, 1956, pp. 165–172.
216. Benson, J. R., and C. J. Becker: "Exploratory Research in Bituminous Soil Stabilization," *Proceedings, Association of Asphalt Paving Technologists*, vol. 13, pp. 120–181, 1942.
217. Dent, G. H.: "Stabilization of Soils with Asphalt," *American Road Builder's Association, Technical Bulletin* 200, pp. 3–25, 1953.
218. Endersby, V. A.: "Fundamental Research in Bituminous Soil Stabilization," *Proceedings, Highway Research Board*, vol. 22, pp. 442–456, 1942.
219. Florida State Road Department: *Standard Specifications for Road and Bridge Construction*, 1954.
220. Herrin, M., and J. R. Rogers: "Principles of Asphaltic Concrete Pavement Design, Control and Construction," *Oklahoma State University, Engineering Experimental Station, Publication* 96, May, 1955.
221. Highway Research Board: "Soil-Bituminous Roads," *Current Road Problems*, no. 12, September, 1946.
222. Holmes, A., J. C. Roediger, H. D. Wirsig, and R. C. Snyder: "Factors Involved in Stabilizing Soils with Asphaltic Materials," *Proceedings, Highway Research Board*, vol. 23, pp. 422–449, 1943.
223. Martin, G. E.: "Soil Stabilization with Tar," *Proceedings, Highway Research Board*, vol. 18, part 2, pp. 275–282, 1938.
224. McKenzie, Ora, Jr.: "Designing Mixtures of Bank-run Gravels and Bituminous Materials," *Highway Superintendent Digest*, Cornell University, Ithaca, N.Y., 1957.
225. McKesson, C. L.: "Suggested Method of Test for Bearing Value of Sand-asphaltic Mixtures," *Procedure for Testing Soils*, ASTM Committee D-18, pp. 369–371, July, 1950.
226. Oklahoma State Highway Commission: *Standard Specifications for Highway Construction*, 1954.
227. Reagel, F. V.: "Asphaltic Binder Stabilized Roads," *Proceedings, Highway Research Board*, vol. 18, part II, pp. 292–298, 1938.
228. Rhodes, E. O., and P. F. Phelan: "Development of a Method of Test for Soil Bituminous Mixtures," *Proceedings, ASTM*, vol. 46, pp. 1415–1428, 1946.

229. Roediger, J. C., and E. W. Klinger: "Soil Stabilization Using Asphalt Cutbacks as Binders," *Proceedings, Association of Asphalt Paving Technologists*, vol. 10, pp. 1–36, 1939.
230. *Soil Mechanics for Road Engineers*, H.M. Stationery Office, London, 1954.
231. Thurston, R. R., and B. Weetman: "The Stabilization of Soils with Emulsified Asphalt," *Proceedings, Association of Asphalt Paving Technologists*, vol. 10, 1939.
232. Winterkorn, H. F.: "Granulometric and Volumetric Factors in Bituminous Soil Stabilization," *Proceedings, Highway Research Board*, vol. 36, pp. 773–782, 1957.
233. Winterkorn, H. F., and G. W. Eckert: "Physico-chemical Factors of Importance in Bituminous Soil Stabilization," *Proceedings, Association of Asphalt Paving Technologists*, vol. 2, pp. 204–257, 1940.
234. Wooltorton, E. L. D.: *The Scientific Basis of Road Design*, Edward Arnold & Co., London, 1954.

Lime and Lime-Pozzolan Stabilization

235. Davidson, D. T., R. K. Katti, and R. L. Handy: *1956–57 Field Trials of Soil-Lime-Fly Ash*, Final Report, The Detroit Edison Company, Detroit.
236. Goecker, W. L., et al.: "Stabilization of Fine and Coarse-grained Soils with Lime-Fly Ash Admixtures," *Highway Research Board Bulletin* 129, pp. 63–82, 1956.
237. G. & W. H. Corson, Inc., *Poz-O-Pac* Manual, Corson, Plymouth Meeting, Pa.
238. "Lime," *Iowa Engineering Experiment Station, Screenings from the Soil Research Laboratory*, vol. 1, no. 5, September–October, 1957.
239. "The Story of Fly Ash, or, Smoke Got in Their Eyes," *Iowa Engineering Experiment Station, Screenings from the Soil Research Laboratory*, vol. 1, no. 6, November–December, 1957.
240. Kirk, Raymond E., and Donald F. Othmer: *Encyclopedia of Chemical Technology*, vol. 8, "Lime and Limestone," pp. 346–382, The Interscience Encyclopedia, Inc., New York, 1952.
241. Laguros, J. G., et al.: "Evaluation of Lime for Stabilization of Loess," *Proceedings, ASTM*, vol. 56, pp. 1301–1319, 1956.
242. Lu, L. W., D. T. Davidson, R. L. Handy, and J. G. Laguros: "The Calcium:Magnesium Ratio in Soil-Lime Stabilization," *Proceedings, Highway Research Board*, vol. 36, pp. 794–805, 1957.
243. Minnick, L. John, and Ralph Williams: "Field Evaluation of Lime-Fly Ash-Soil Compositions," *Highway Research Board Bulletin* 129, pp. 83–99, 1956.
244. National Lime Association: "Lime Stabilization of Roads," *Bulletin* 323, The Association, Washington 5, D.C.
245. Woods, K. B., and E. J. Yoder: "Stabilization with Soil, Lime, or Calcium Chloride as an Admixture," *Proceedings of the Conference on Soil Stabilization*, pp. 3–25, Massachusetts Institute of Technology, June 18–20, 1952.

Chlorides

246. Allen, Harold, et al.: "Use of Sodium Chloride in Road Stabilization," *Proceedings, Highway Research Board*, vol. 18, part II, pp. 257–274, 1938.
247. Burggraf, Fred, et al.: "Use of Calcium Chloride in Road Stabilization," *Proceedings, Highway Research Board*, vol. 18, part II, pp. 209–256, 1938.
248. Calcium Chloride Institute: *Calcium Chloride for Stabilization of Bases and Wearing Courses*, Manual SM-1, The Institute, Washington 6, D.C.
249. Kirk, Raymond E., and Donald F. Othmer: *Encyclopedia of Chemical Technology*, vol. 12, "Salt," pp. 67–82, The Interscience Encyclopedia, Inc., New York, 1952.

Lignin Derivatives

250. Harmon, John P.: "Use of Lignin Sulfonate for Dust Control on Haulage Roads in Arid Regions," *Bureau of Mines Information Circular* 7806, U.S. Department of the Interior, October, 1957.
251. Sinha, S. P., D. T. Davidson, and J. M. Hoover: "Lignins as Stabilizing Agents for Northwestern Iowa Loess," *Proceedings Iowa Academy of Science*, vol. 64, pp. 314–348, 1957.
252. The Lake States Roadbinder Association: *Sulphite Roadbinder*, The Association, Appleton, Wis.

Other Chemicals

253. American Chemical Society: "Altering Soil Properties with Chemicals," Symposium, *Industrial & Engineering Chemistry*, vol. 47, pp. 2230–2281, November, 1955.
254. Kardoush, F. B., J. M. Hoover, and D. T. Davidson: "Stabilization of Loess with a Promising Quaternary Ammonium Chloride," *Proceedings, Highway Research Board*, vol. 36, pp. 736–754, 1957.
255. Lyons, J. W.: "Stabilizing a Problem Soil—Cheaply," *Engineering News-Record*, Aug. 15, 1957.
256. Mainfort, R. C.: "Soil Stabilization with Resins and Chemicals," *Highway Research Board Bulletin* 108, pp. 112–120, 1955.
257. Massachusetts Institute of Technology: *Proceedings of the Conference on Soil Stabilization*, Cambridge, June 18–20, 1952.
258. Nicholls, R. L., and D. T. Davidson: "Soil Stabilization with Large Organic Cations and Polyacids," *Proceedings Iowa Academy of Science*, vol. 64, pp. 349–381, 1957.
259. Sheeler, J. B., J. C. Ogilvie, and D. T. Davidson: "Stabilization of Loess with Aniline-Furfural," *Proceedings, Highway Research Board*, vol. 36, pp. 755–772, 1957.

Construction

260. Hewes, Lawrence I., and Clarkson H. Oglesby: *Highway Engineering*, John Wiley & Sons, Inc., New York, 1954.
261. Massachusetts Institute of Technology: *Proceedings of the Conference on Soil Stabilization*, Cambridge, Mass., June 18–20, 1952.
262. Portland Cement Association: *Soil-cement Construction Handbook*, The Association, 33 West Grand Avenue, Chicago 10, 1956.
263. Department of Scientific and Industrial Research, Road Research Laboratory: *Soil Mechanics for Road Engineers*, H.M. Stationery Office, London, 1952; reprinted 1954.

Section 12

HIGHWAY MAINTENANCE

REX M. WHITTON, *Chief Engineer, Missouri State Highway Commission, Jefferson City, Mo.* (Highway Maintenance).

I. INTRODUCTION

A. Scope

Highway maintenance is the preserving and keeping of each roadway, structure, and facility as nearly as possible in its original condition as constructed or as subsequently improved, and such additional work as is necessary to keep traffic moving safely. This involves patching, filling ruts, removing surface corrugations, pouring cracks, blading surfaces, cleaning ditches and culverts, repairing bridges, and fighting floods. It also includes operation of the highway as a safe, efficient, pleasing link in the nation's highly important transportation system. This adds such duties as erecting signs and traffic controls, painting traffic stripes, mowing grass and weeds, clearing away trash, removing snow, spreading cinders, and other such services which the public has come to expect. The end result of adequate highway maintenance is smooth, safe, and efficient roadways, clear waterways, and clean and attractive rights-of-way.

B. Influence of Location and Design on Maintenance

All procedures from the original location reconnaissance to the final construction operation have a bearing on maintenance costs and conditions.

In the original route-location study, much thought should be given to drainage problems, proper stream crossings, the avoidance of possible landslide conditions and unsuitable soils, directness of route—because a shorter distance generally means less maintenance cost, horizontal and vertical alignment, and other similar problems.

Design features of the highway have an important bearing on the future maintenance of the highway, for ease of maintenance as well as cost. The predicted volume and character of traffic play an important part in the selection of the surface type and geometric design of the highway. The location of the final grade line with respect to the natural ground, particularly in regions of snowfall, is most important in that if the elevation of the roadbed is somewhat higher than the adjacent ground areas, the wind will generally keep the roadway clear of snow. The grade percentage through cuts should be designed to result in the least amount of erosion possible and at the same time have enough fall to provide good drainage. The cut-and-fill slopes should be flat enough to encourage growth of vegetation and permit power mowing if at all practical. The waterways and structures should be large enough to handle the runoff drainage.

Good and proper construction methods in accordance with approved specifications make a definite and recognized contribution to the ease and cost of future highway maintenance.

12-1

II. SNOW REMOVAL AND ICE CONTROL

Snow removal and ice control should start in the planning stage of the highway. The snow and ice problem varies greatly from one locality to another because of latitude, prevailing temperatures, and precipitation.

A. Planning

Designs to alleviate snow problems should provide for high grades, wide rights-of-way, flat slopes, careful designation of borrow-pit locations, and proper selection and placing of roadside plantings.

B. Grades

Grades should be kept above the general elevation of the adjacent land. If all grades are constructed from 1 to 2 ft higher than the land on either side of the highway, practically no snow will drift onto the traveled portion of the road; even wet packing snow will tend to clear off the surface of such grades. It is not necessary to have deep dangerous ditches adjacent to the roadway in order to accomplish this snow-free type of highway. Actually, less drifting will occur if the ditches are shallow and with flat slopes than if they are deep and sharp.

C. Wide Rights-of-way

A narrow right-of-way is a snow hazard. Vegetation, buildings, fences, and other structures built entirely on private property cause serious snow drifting on the roadway unless the right-of-way is wide enough so that these drifts run out before they reach the traveled portion of the highway. Wide rights-of-way also provide ample borrow material for building the grades to proper heights.

D. Flat Slopes

Flat backslopes and inslopes aid materially in the prevention of drifting. Surface winds tend to follow up and down flat slopes and keep the snow moving. In the case of steep backslopes, the wind tends to cut horizontally across the highway and the snow is dropped in the cut. Cuts with steep backslopes are much more apt to become blocked with snow than cuts of exactly the same depth but with flat back- and foreslopes.

E. Borrow Pits

Borrow pits are wonderful for snow storage. If successive storms occur with little melting of the snow between such storms, snow fences will fill up and road ditches will become clogged with snow blown in from the fields or plowed from the highway surface. When such a condition occurs, a good wide borrow pit is most useful. The snow from many more storms can be handled in such a borrow pit before conditions become critical. To be of most benefit such borrow pits should be constructed on the windward (snow-fence) side of the highway.

F. Roadside Plantings

Care must be exercised in the selection of roadside plantings and in locating them on the rights-of-way. No planting should be made closer than 50 ft to the center line of the highway. Even beyond 50 ft, generally speaking, only trees well spaced or very-low-growing shrubs should be planted. Clumps of shrubs or trees close enough to retard the wind materially should not be planted closer than 60 to 75 ft to the center line of the highway, except where such plantings are adjacent to high fills where the snow can pile up to the top of the plantings and still leave the highway free from snow (see also Sec. 13).

G. Drift Control

Wood-slat snow fence is most commonly used for snow-drift control. This fence is usually 4 ft high, and the vertical lath spaced approximately 2 in. center to center, although a wider spacing would probably be as effective and cheaper. Snow fence is placed on the prevailing-wind side of the highway. The theory of snow fence is not that it prevents the drifting of snow, but that it facilitates drifting and confines it to the space between the fence and the highway. Until such time as the snow fence is filled, that is, the drift is built up to the top of the fence, the fence will slow the air currents and cause them to drop their load of snow between the fence and the highway. Such air currents will be comparatively free from snow when they pass over the highway. Even when the snow fence is full, it will still tend to cause the air currents to lift from the ground surface and carry at least a part of their snow load across the highway. Snow fence should ordinarily be placed 100 to 150 ft from the center line of the road. The exact location will be determined by cut-and-try methods and will be influenced by the lay of the land and the prevailing direction and intensity of the wind.

After the snow fence is filled with snow, additional fence may be erected and the procedure repeated. This additional fence may be placed as an extension above the old fence or may be constructed as a new fence 75 to 100 ft back from the first fence.

H. Advance Preparation

It is important that both men and equipment are made ready in advance of snow emergencies. In the spring, when it is quite certain that there will be no more snow, all snow and ice equipment should be inspected. Necessary repairs and replacements should be provided. The deadline for having all damaged equipment either repaired or replaced should be at least 2 months prior to the first possible snow of the next winter.

Careful planning for the use of men and equipment during snow and ice storms is necessary for good and successful results. The experience, ability, and dependability of all men, type and adaptability of all available equipment, and finally the type and relative importance of the roads to be cleared or treated during the storm must be considered. The plan should assign adequate equipment, well manned, to the most important specific sections of road down to and including the lightly traveled rural feeder road. This plan in brief and the assignments of men and equipment for all shifts around the clock should be distributed to all involved and concerned. During the storm, as the work progresses, each crew should report by radio or telephone at predetermined intervals to a control center. Under this system, in case of a shift in storm conditions or equipment breakdowns, personnel requirements can quickly be determined and steps taken to correct any deficiencies.

During the snow-fighting season the men should be alerted at all times. Regardless of what day of the week or what time of day or night the snow occurs, the men should be ready. Frequent reports (at least daily) from the U.S. Weather Bureau office will be a great help in preparing the men for coming storms.

I. Types of Storms

Snow may occur in various intensities and conditions, and different procedures are required to combat the different kinds. A good experienced foreman will usually be able to determine the proper course of action soon after the storm starts; however, too much must not be expected from the foreman along this line—snow storms sometimes refuse to follow a pattern and fool even the most experienced weather observers.

J. Light, Dry Snows

Light, dry snows are of little concern. Even with a light wind and moving traffic, the pavements (if properly constructed) will blow clear. Usually no special work by the maintenance division is required, but maintenance men must be alert at all times for any changes in the snowstorm.

K. Wet Snows

Wet snows with heavy winds are a problem, as will be pointed out later, but wet snows even without wind also result in dangerous conditions if not properly combated. Wet snows under traffic pack on the pavements and, particularly if followed by a drop in temperature, create very dangerous conditions. The best method of preventing the packing of wet snow is to have a large fleet of light, high-speed, straight-blade snowplows available. These plows should be started with the storm and kept going until the storm ends and the pavements are clear—or until the snow gets so deep that snowplows cannot handle the situation.

L. Drifting Snow

Drifting of either wet or dry snow will cause blockades under the following conditions:
1. If a road is not properly designed and constructed
2. If the snowplows are not adequate
3. When the amount of snow is in excess of the capacity of the plows available
4. In extreme blizzard conditions

When snowdrifts form in sufficient depths to seriously impede traffic, the straight-blade plows become ineffective. If the drifting is not too severe, the V plows on 2-ton trucks will handle them. If the drifting is continuous, the opened sections will soon drift full again and the new drift will be deeper than the original, since the snow will build up to the height of the windrow formed by the first plowing. Successive plowing and drifting will build up drifts so deep that only the heaviest equipment can cope with them.

M. Blocked and Slippery Roads

Most blocked roads can be opened by heavy-duty V plows on heavy, four-wheel-drive trucks or on heavy-duty, tandem-drive motor graders. Drifts which cannot be penetrated by these large Vs can always be opened by rotary equipment. A good procedure is to work a large V and a rotary together. The V opens as far as it can go, the rotary following up and widening out the heavy cuts. When the V can go no farther, the operator backs his plow out and the rotary proceeds to open through the excessive drift. The rotary then pulls to the side of the road, and the V again takes over. Many more miles of blocked road can be opened in this manner than by either the V or the rotary working independently.

N. Winging Back

When the road has been plowed open, great mounds of snow will be left piled on both shoulders of the road. It is imperative that these ridges of snow be removed or pushed beyond the shoulder before the next snowstorm or severe blow or thaw occurs. Long wings on the side of the snowplow trucks or motor graders may be used to blow such ridges clear of the right-of-way.

Probably the most important requirement in snow and ice work is for the foreman to keep a clear head. The public is apt to become panicky in extreme blizzards and blockades. It behooves the foreman and road workers to remain "cool" and attack the problem determinedly and systematically.

O. Packed Snow and Ice

Packed snow forms when the plows cannot keep ahead of the storm. Ice is usually caused by a freezing rain, drizzle, or sleet. Any of these conditions results in a slippery surface that is very dangerous. Curves, hills, and sections adjacent to railroad crossings and highway intersections should be covered with a chemically treated abrasive. If traffic is sufficiently heavy to warrant it, the entire road should be treated with abrasives. This type of work must be continuous. Frequently freezing rains and drizzles continue over many hours or days, thus erasing all good done by the abrasives, so that treatments must be frequently repeated.

P. Treatment of Icy Pavements

Application of abrasive material either alone or in combination with calcium or sodium chloride is the most common method of treating icy pavements. In some cases sodium or calcium chloride is applied without abrasives. The pavement should be treated whenever slippery conditions exist and for as long a period as necessary.

1. Equipment. A truck equipped with a special cinder or aggregate bed having a conveyor in the bottom feeding the abrasive material to a spinner-type spreader at the rear, all powered by the power take-off of the truck, has proved to be the fastest and most economical spreader yet developed. Its use is not confined to ice control, since it can have a snowplow mounted for snow removal, and during summer seasons it makes an excellent chat or sand spreader on bituminous work and gravel spreader on gravel roads.

When the conveyor-type truck bed is not available, then any 2- to 5-ton truck with a dump bed can be used, and with these there are two general types of spreaders. One is attached to the truck, the other is on a trailer. Power may be obtained from a belt driven by the rear wheels of the truck, from a separate motor, or from a wheel attached to the spreader and driven by contact with the road surface. Speed of discharge and uniformity in distribution are the important functions of a satisfactory spreader.

2. Materials. Calcium chloride should conform to the requirements of ASTM D 98. Sodium chloride should conform to the requirements of ASTM D 632.

3. Abrasives. Local availability and unit cost determine the type of abrasive to be used. The cost of labor and equipment for storing and applying abrasives is several times that of the material itself. Thus a small initial saving in material cost might be more than offset by the necessity for larger quantities of abrasives per square yard or more frequent applications. The following abrasives are the most commonly used.

a. **Sand.** Sand containing an excess of fines does not provide good traction, and coarse material is likely to whip off and is therefore not effective in combating the skid hazard. Sand for ice control should be clean, hard, sharp, and free from loam, clay, or frozen lumps, with 100 per cent passing a $\frac{3}{8}$-in. sieve, not less than 45 per cent passing a No. 16 sieve, and not more than 30 per cent passing a No. 50 sieve.

b. **Cinders.** Cinders are sharper than sand and cling to tires and cut into ice better than sand particles. The melting or imbedding action of cinders with their greater porosity is better than that of sand because of the larger quantity of moisture and chloride held at or near the surface of the cinder particles. Because of their dark color, cinders absorb more heat than sand when the sun is shining, which results in greater imbedment. Also, cinders have less tendency to clog drainage structures.

c. **Washed Stone Screenings.** Stone screenings from which the fines have been removed by washing produce a good abrasive; 100 per cent should pass the $\frac{3}{8}$-in. sieve, with not less than 45 per cent passing a No. 16 sieve and not more than 30 per cent passing a No. 50 sieve.

The percentage of wear of crushed rock from which screenings for abrasives are prepared should not be more than 45 per cent, as determined by AASHO Method T 96 or ASTM Method C 131.

4. Treatment of Abrasives. Abrasives should be treated to prevent freezing before application and as an aid to anchoring the material in the ice or packed snow so that it will not be blown off the road. Heating the abrasives does not work well in practice. Treatment with sodium or calcium chloride is the best method of preparing abrasives for treating icy pavements.

Experience has shown that on heavily traveled highways untreated abrasives whip off the road and that even moderate winds will sweep the material from the surface. Only on light-traffic roads or streets free from wind is it possible to keep abrasives on the road without the addition of chemicals.

a. **Treatment with Dry Calcium or Sodium Chloride.** When abrasives are placed in storage they should be treated with chloride, the quantity depending on the moisture content of the abrasive and the temperature and as indicated in Table 12-1.

The chloride should be well mixed with the abrasive. Each particle of abrasive

Table 12-1. Treatment for Sand or Stone Screenings When Placed in Storage

H$_2$O in abrasives before treating, %*	Lb 77–80 % flake calcium chloride per cu yd, predicted min. temp.				Lb sodium chloride (rock salt) per cu yd, predicted min. temp.		
	+10°F	0°F	−10°F	−20°F	+10°F	0°F	−5°F†
4	15	20	30	35	20	30	35
6	25	30	40	50	30	40	50
8	30	40	55	65	40	55	65
10	40	55	70	80	50	70	80
			Treatment for cinders				
15	20	30	40	50	30	40	45
20	30	45	55	70	40	60	65
25	40	55	70	85	55	75	85
30	50	70	90	100	70	95	105

* Per cent H$_2$O calculated on wet basis.
† No protection below − 6°F.

should be coated with a film of chemical to prevent freezing while stocked and to assure quick anchorage when applied to the road.

b. **Brine-spray Method.** A convenient and efficient method of applying calcium chloride to the abrasive to obtain quick and uniform dispersal of the chemical is by spraying with brine. The brine is usually made up of 5 lb of flake calcium chloride to 1 gal of water.

A practical way to prepare such a solution is to fill a 50-gal drum about two-thirds full of water (never add the calcium chloride first), then put in 200 lb of dry flake calcium chloride, stir until it is dissolved, and fill with water to the top.

The amount of chloride, applied in brine, which can be held by the abrasive without loss from drainage is controlled by the moisture content of the abrasive as received.

c. **Dry or Damp Sand or Stone Screenings.** With these materials it is generally possible to use about 8 gal of brine per cu yd of abrasive with a negligible loss of chloride. The operator, after a few initial tests and trials, learns to soak the abrasive without oversaturation. The brine is usually sprayed on the abrasive as it is being loaded into bins or into covered stockpiles. Eight gallons of brine introduces about 32 lb of chloride into 1 cu yd of abrasive. This amount is sufficient for pretreatment to prevent freezing in the bins and for use on the road at temperatures down to about 20°F. For lower temperatures, it becomes necessary to add additional flake chloride as given in Table 12-2.

Table 12-2. Treatment for Brine-sprayed Dry or Damp Sand or Stone Screenings Stored and Used at Temperatures Below 20°F (Based on Total Absorption of 18 Gal per Cu Yd)

Brine, gal per cu yd	CaCl$_2$, lb per cu yd	Add. CaCl$_2$ for +10°F, lb per cu yd	Add. CaCl$_2$ for −5°F, lb per cu yd	Add. CaCl$_2$ for −20°F, lb per cu yd
6	24	16	26	36
8	32	8	18	28
10	40	0	10	20
12	48	0	2	12

d. **Dry or Damp Cinders.** With dry or damp cinders, it is necessary on account of absorption to use 15 to 20 gal of brine per cu yd of abrasive.

e. **Wet Sand, Stone Screenings, or Cinders.** When the sand, stone screenings, or cinders are more than damp, the brine-spray method is not recommended. However, if its use is necessary with wet sand or screenings, add only enough brine to assure saturation and then supplement this with dry flake calcium chloride.

f. **Sodium Chloride.** Sodium chloride may be used in the form of a brine spray as described for calcium chloride, except that it is not recommended for use in climates or during seasons when temperatures below zero Fahrenheit might be encountered.

5. Storage of Abrasives. The best method of storing abrasives is in bins so located that they will serve more than one road. A bin, to be used efficiently, should be accessible at two elevations, one at an upper level so that trucks can discharge into the bin, the other on a lower level so that trucks can be loaded directly by gravity. Generally the capacity should be 50 to 200 cu yd, depending on the needs of the roads to be served.

When storage is required at a location which does not have the necessary difference in elevation for a bin, a shed gives the next-best protection for treated abrasives.

Where it is not possible to have either a bin or shed, the abrasives should be placed in a stockpile. Such a pile should have a triangular cross section and be protected by a waterproof covering such as waterproof paper. The degree of protection will depend upon the watertightness of the cover. Since abrasives practically always contain moisture, they will freeze if exposed to the weather. Treatment with chloride will prevent freezing for a period, but the chemical will leach out in time if not in watertight storage, and the outer crust, if not the whole pile, will freeze. This effect can be minimized by spreading dry chloride over the top of the stockpile.

6. Application of Abrasives. Before treating a road surface with abrasives or chemicals to prevent skidding on icy surfaces, the snow or loose ice should be removed as completely as possible.

An effective method for preparing the surface is to score the ice or packed snow with a saw-tooth blade grader to roughen the surface. The loosened ice or snow should be removed as soon as possible with conventional plowing blades.

7. Quantities. An application of ½ to 1 lb of treated sand or cinders per sq yd will give a good cover and should make the road nonskid if it is uniformly distributed. When traffic or wind has whipped off the material, it may be necessary to re-cover some sections, especially on grades, curves, and intersections.

8. Application of Calcium and Sodium Chloride without Abrasives. The rising demand for modern highway and urban expressway design and construction to adequately provide facilities for the free and uninterrupted movement of dense traffic has presented a maintenance problem of ice prevention.

On expressways in urban areas the combination of abrasives with calcium creates considerable dirt and causes drainage systems to become inoperable in many instances, and in all cases the abrasives eventually have to be picked up after storms.

Snow quickly packs to icy conditions under dense traffic and has the same effect as ice on the movement of traffic. Obviously, the pavement surfaces must be treated immediately as the snow- or icestorm starts. Timing of the treatment is of utmost importance, since an hour's delay can completely stop traffic.

Some states are successfully handling the problem of ice control by the application of sodium or calcium chloride directly to the road surfaces. Application at the rate of 200 to 500 lb per mile of two-lane roadway has been found to be very effective. The amount applied is governed by the prevailing temperature and amount of snow or ice on the road surface.

Mechanical spreaders specially adapted to the distribution of chlorides are available and are generally used attached to the tail gates of dump-bed trucks and light, fast vehicles such as pickup trucks.

The movement of traffic over such treated surfaces helps to keep the snow or ice cut up and in a slushy condition, making it possible for snowplows to remove the unpacked snow and ice as traffic continues to move. It has been found that vehicles will carry the chloride treatment for a considerable distance beyond the area actually treated and spread its effectiveness.

Calcium and sodium chloride should meet designation M 144 and M 143, respectively, of AASHO Specification for Highway Materials.

9. Treatment of Portland Cement Concrete. Calcium- or sodium chloride-treated abrasives or these chlorides used alone should be applied as sparingly as possible to portland-cement-concrete pavement. This is because repeated freezing and thawing of concrete in contact with these salts may be conducive to surface pitting or scaling.

Unscaled pavements over 4 years old are less susceptible to damage than new pavements. Those that have been given protective surface treatments or those made from air-entrained concrete are generally not susceptible to damage from ice-control chemicals.

III. ROADWAY DRAINAGE MAINTENANCE

Water in liquid or frozen form is one of the contributing factors, and often the sole contributing factor, to many maintenance difficulties in all types of roadway surfaces, shoulders, side ditches, slopes, and roadsides. Water helps to disintegrate pavements and surfaces, softens and reduces supporting value of subgrades and shoulders, erodes aggregate surfaces, subgrades, shoulders, slopes, and ditches, and deposits sediment in waterways and drainage structures. Uncontrolled floodwater does damage to structures and roadbeds. Frozen underground water may result in frost heaves and landslides.

A. Importance of Adequate Drainage Design

Improper highway design has contributed to pavement distress due to inadequate drainage. Many miles of lip-type pavements have been constructed with shoulders sloping toward the pavement and, in some instances, with no side ditches. This design causes the pavement to be a waterway and in many cases results in failure of the pavement by disintegration or pumping.

B. Drainage of Pavement Bases and Subbases

Another design feature which contributes to maintenance difficulties is the construction of pavement and bases and subbases in a trench. It is imperative that subgrade drainage be provided either by constructing shoulders of porous material or by placing pipe drains through the shoulders.

C. Surface and Subsurface Drainage

Water may reach the highway by direct rainfall or snowfall which melts afterward, by surface runoff from adjacent areas, by underground flow, or by capillary action from the water table below the pavement. The control and disposal of surface water is called surface drainage, and the removal of underground water is called subsurface drainage.

1. Surface Drainage Good surface drainage is accomplished by having the pavement surface as waterproof as possible and by having such crown and surface smoothness that the water will flow freely to the pavement edge and onto the shoulder. In turn, the shoulder should be kept flush with the pavement and smooth enough to provide for the free flow of water away from the pavement edge and across the shoulder to the side ditch in cut sections or the downslope of fill sections.

The side ditch is constructed longitudinally with and outside the shoulder. It should be below the grade of the subgrade in order that water from the subgrade can be drained to the side ditch. The difference in the elevation of the side ditch and the top of the subgrade can vary somewhat, but it should not be less than 12 in., and 18 in. will provide better results in most soils. The elevation of the side ditch will have some control over the elevation of the water table of the subgrade. The side ditch should be sufficiently large to carry the runoff of the area draining to it. If too small, excessive erosion will result. The side ditch should be kept clean and free from water pockets at all times. Water pocketed in side ditches will tend to seep to the subgrade, which causes loss in supporting value.

Side ditches may be cleaned with motor graders, or if the ditch is too wet or deep, it may be cleaned with a pull grader and wheel or track-type tractor.

The gradient of the side ditch may be great enough to cause erosion. As covered in detail in Sec. 4, it is highly desirable to have the inslope and backslope of the side ditch covered with turf in order to control erosion. In some highly erosive types of soil where sod will not control the erosion, it is desirable to construct checks in the side ditch, line the ditch with loose rock, or, in extreme cases, pave the ditch with concrete or rock masonry. Ditch checks can be built with timber, woven wire, stone, concrete, blue-grass sod, etc.

2. Interception Ditches. Interception ditches are desirable back of the cut slope when the runoff from the adjacent area is sufficient to erode the cut slope. The interception ditch is carried to an outlet beyond the cut. Paved or pipe waterways are often necessary to carry the water down the cut-or-fill slopes.

3. Subsurface Drainage. Underground water from springs, seepage, or other sources will cause many types of damage to the road surface and should be removed from the roadbed by underdrains. An underdrain usually consists of clay tile, concrete tile, or perforated metal pipe laid in a trench and backfilled with porous crushed aggregate, coarse gravel, or sand. The outlets of underdrains are usually in the side ditches or culverts.

Free water under the pavement is a contributing factor to slab pumping of concrete pavement, and under bituminous pavements, water softens the subgrade, which leads to bituminous-pavement failures. Water under the pavement can be removed by drain tile or pipe through the shoulder to the inslope or downslope.

IV. MAINTENANCE OF BITUMINOUS SURFACES, PAVEMENTS, AND BASES

Bituminous surfaces and the maintenance work necessary to keep them in good condition vary widely; yet the average highway user does not notice much difference between a surface of oiled earth and a bituminous-concrete pavement on an adequate base so long as they are both smooth and safe for travel.

Maintenance of bituminous pavements consists primarily of patching, base repairs, surface treatments, and resurfacing. These processes are common to all types of bituminous-surface highways.

Regular and frequent inspections are essential to proper maintenance and make possible the repair of failures or impairments in the early stages, thereby providing lower costs and better service. Rainy-weather inspections are valuable in that surface defects can be detected readily and marked for future correction.

A. Bituminous Patching

Bituminous patching involves the use of mineral aggregate and bitumen in smoothing a rough or uneven riding surface. There are two types: patches made in proper season and patches made out of season. The latter type is required under emergency conditions in the winter or spring and is often temporary because of adverse weather.

Corrugations are often created by excessive bitumen in the mix, incorrect gradation of aggregate, excessive moisture in the bituminous mix, or traffic overloads. Any type of unevenness which might create a hazard or discomfort to the occupants of a fast-moving vehicle should be repaired immediately. For this reason, it is imperative to have a supply of bituminous mix available for patching at all times.

The areas of bituminous patches vary in size from small potholes to several hundred square yards covering the full width of the roadway. The premixed material must be readily available for use in order to prevent delay in repairing an impaired surface. Early repair is important in order to stop the enlargement of the surface failure. If slow- or medium-curing liquid bituminous material is used, the patching mixture can be prepared, stored inside or outside, and used as needed. If stored outside for any length of time, the mixed material should be stockpiled, with the top well rounded and sprayed with a light coat of bitumen to retard drying and reduce moisture absorption from rain or snow.

B. Patching Materials

For patching during periods when the aggregate is wet, an antistripping agent should be used as an additive. The mixture, after drying, can be placed with good results even on wet surfaces. There are several types of antistripping additives that can be obtained commercially. Detailed directions for the use of each is supplied by the manufacturer.

A satisfactory grading for mineral aggregate is as follows:

Sieve size	Percentage passing
½ in.	100
No. 4	40–65
No. 10	30–55
No. 40	8–25
No. 200	0–5 or 5–10

Other gradations have been satisfactorily used, and local conditions will govern the gradation to a great extent. The aggregate should comply with the AASHO Standard Specification M 63 (see Sec. 7). The bitumen used in the patching mix can be slow-curing (SC) liquid asphalt, medium-curing (MC), or rapid-curing (RC) cutback asphalt, emulsified asphalt, or tar. MC-3 or MC-4 grade asphalts have proved satisfactory and provide a mixture that can be stored for several months before being used (see Sec. 9). In some localities, the patching mixture can be obtained commercially, and the determining factors in its use are the cost and the time saved in preparing the mix.

C. Road-mix Method

In preparing a patching mixture by the road-mix method, a road surface of ¼ to ½ mile in length, depending on the quantity of material to be mixed, is generally used as a "mixing board." Areas alongside or adjacent to the highway are sometimes acquired for use as mixing boards so that the mixed material may be prepared and left in open storage without interference to the road user. If the patch is of sufficient length, the aggregate may be placed along the impaired area, mixed with bitumen, and laid without additional handling. From 8 to 12 cu yd of aggregate per 100 lineal ft is a satisfactory rate. The spreading of aggregate must be uniform in order that the bitumen can be applied at a uniform rate. This can be accomplished by use of a windrow evener—a horizontal V-shaped drag made of wood or metal having a vertical height of 18 to 30 in., with an adjustable opening at the bottom of the V and a cross-section area that will permit the predetermined rate of aggregate to pass through the opening. Any wood surface coming in contact with the aggregate should be covered with metal to prevent wear.

The aggregate should be surface-dry, which can be accomplished by blading back and forth across the road. The aggregate need not be dry if asphalt emulsion is used or if an antistripping additive is used.

The bitumen is ordinarily transported in insulated tank cars by rail from the refinery to the shipping point nearest the job, heated with a tank-car heater or a circulating booster heater, unloaded and hauled to the job in insulated relay tanks. The bituminous distributor is often used on short hauls instead of relay tanks. If asphalt refineries are located within economical truck-hauling distance of the job, the bitumen may be transported very advantageously by insulated tank trucks. The bitumen is transferred from relay tanks or tank trucks to bituminous distributors and applied at predetermined rate to the windrow of aggregate at a temperature that varies with different types and grades of bitumen. The recommended range of temperature for application of MC-3 cutback asphalt is 150 to 200°F, and for RT-5 tar, the range is 80 to 150°F. For asphalt emulsion, no heating is necessary under normal temperatures. The application temperature must be high enough to provide uniform distribution with the available equipment. The percentage of bitumen to be used varies slightly with both the type and grade of aggregate. The final determination of the quantity should be decided from past experience and field performance of the materials used or by recognized design methods (see Sec. 9). With MC-3 cutback asphalt the rate is 16 to 20 gal to the cubic yard of aggregate. The bitumen should be applied in two or more applications in order to avoid possible loss of bitumen and to ensure even distribution throughout the aggregate. Multiple application of bitumen reduces mixing operations, in that additional coating of aggregate is accomplished with each application.

In the road-mix method, the actual mixing can be accomplished with motor graders by blading the aggregate and bitumen back and forth across the road surface. Motor graders with 75- to 100-hp engines do a better job of mixing than the lower-powered machines. A soil tiller or aggregate mixer is sometimes used to supplement the mixing-and-drying operation. This machine mixes or aerifies with curved prongs or blades revolving around a horizontal axle, and it has been found that as much as 4 gal of bituminous material per cu yd of aggregate can be saved by the use of this type of mixing machine. The aggregate tiller-type mixer is pulled by a wheel tractor. It will replace two motor graders in the mixing operation.

A road-mix crew consists of a foreman, one or more motor-grader operators, distributor operator, tank-car-heater operator, and truck drivers. The equipment needed is a tank-car heater or booster heater, a bituminous distributor, one or more motor graders, rotary broom, trucks, a tiller-type aggregate mixer, and rollers.

D. Plant-mix Method

The plant-mix method of mixing patching material provides more positive control of the proportioning of the aggregate and bitumen and less segregation of the aggregates in sizes than the road-mix method. A common portland-cement-concrete mixer of 1- or 2-sack capacity or a small bituminous mixer which has been designed and built for this type of operation can be used for mixing aggregate and bitumen. The proportions of bitumen and aggregate can be measured either by volume or by weight. For design of satisfactory mixtures see Sec. 9.

The mixed material that is not used immediately for patching should be stockpiled in a convenient off-the-road location for future use.

E. Patching Operation

Holes in the bituminous-road surface should be cleaned of all loose aggregate and dust. The bottom and sides of the hole should be squared and lightly primed. If the bitumen used for mixing is too heavy for priming, it can be thinned with kerosene. The patching mixture should be placed as soon as possible after application of the prime coat. If the hole is more than 3 in. deep, the patch should be made in two or more layers and each layer tamped or rolled. Small holes can be hand-tamped, and larger patches should be rolled. Either pull-type or power rollers do a satisfactory job. Care should be taken to secure a satisfactory riding surface by proper placement and compaction of the patching material.

Depressions in bituminous surfaces, which are usually the result of subgrade settlement or lateral displacement caused by a combination of heavily loaded traffic and weak subgrade, can be corrected by patching if the base or subgrade is not too weak. The surface of the depression should be thoroughly cleaned and swept free of all foreign particles and primed with about a 0.1 gal per sq yd of bituminous material to secure a good bond. Too much prime will result in the excess bitumen being absorbed by the patch mixture, with possible shoving as the final result. The mixed bituminous material should be placed immediately after the application of the prime, not to exceed 2-in. layers. Each layer should be leveled and rolled. Depending on the size of the patch, the leveling can be done either by hand raking or by blading with a motor grader or a pull-type blade. Around the edges of the patch, all aggregate of a size greater than the thickness of the featheredge should be removed so that the edges of the patch can be raked and rolled to a smooth juncture with the old surface. Before rolling is completed, the surface of the patch should be checked with a 10-ft straightedge or a string, and areas showing more than ¼-in. variation from a true surface should be corrected by adding more mixed material to the low places and removing material from the high places. The rolling and surface checking should continue until the patch is satisfactorily completed.

Corrugations in a bituminous surface can be removed by cutting off the high ridges with a heavy motor grader with blade set at a very acute angle to the direction of operation so that there will be a shearing action of the cutting edge. A disk attachment is now available for the motor grader which does good work in removing corrugations or ridges. The depth of cut should be very limited with each pass of the blade in order to prevent the surface from being torn out in pieces of considerable depth. Best results can be obtained when the temperature is high and the surface is in its most pliable condition. The removal of the corrugations usually leaves the bituminous surface in a weakened condition, and it should be strengthened by placing a patch over the entire area. The same procedure should be followed as in patching a depressed area. If the pavement was originally laid using liquid bituminous material, a more permanent method for correcting a corrugated surface is to remove the pavement to the affected depth by scarifying and blading. The loosened material should be aerated and often can be relaid with or without the addition of more bitumen or aggregate or both.

Bituminous surfaces sometimes oxidize or become dry, and the aggregate ravels out. Early stages of this oxidation and raveling can be controlled by an application of bitumen alone or bitumen and chips, but if the oxidation and raveling has progressed until patching is necessary to restore a satisfactory surface, the same patching procedures should be followed as previously described.

A bituminous-patching crew consists of a foreman, two or three men for spreading, raking, and tamping, operator for motor grader, if used, flagman, and truck drivers. Trucks, and possibly a motor grader and a pull-type or power roller, are necessary equipment, along with small tools.

F. Base Repairs

Bases under bituminous surfaces vary from the natural soil to soil aggregate, compacted aggregate, bituminous-soil aggregate, soil cement, bituminous concrete, brick, portland cement concrete, and other types. Investigation and a thorough study of the cause of the base failure and its correction should precede the repair work.

Shoving of the base and resulting depressions are usually due to inadequate thickness of base, inadequate subgrade support, excess moisture in the base or subgrade, or combinations of these causes. If the cause is from seepage or from improper removal of surface water, then tile, pipe, or aggregate drains should be used for underdrainage. Surface water can be controlled by surface ditches and construction of proper shoulder slopes. Inadequate base of subgrade support in small areas is economically corrected by increasing the thickness of the base. If the base failure is slight, this inadequacy may be corrected by adding additional surface thickness.

If it is determined that the base thickness must be increased, the surface limits of the failed area should be determined and marked. The bituminous surface and base should be removed, and materials that can be used in the repair should be salvaged. The subgrade should be removed to the depth necessary to give the needed additional thickness of new base. Often the subgrade is composed of unsuitable material that must be removed. Air hammers with spade bits and moil-point bits are very efficient for removing bituminous-concrete surface and base. The cold-mix surface can be removed by hand tools, or scarifiers if the affected area is sufficiently large. The sides of the area should be as nearly vertical as possible, and the subgrade should be thoroughly compacted. Densely graded local aggregate is generally the most economical and most satisfactory material to be used for repairing base failures. Maximum size of the aggregate should be from 1½ to 2 in. Satisfactory material salvaged from the old base can be used in the bottom of the new repair. The new base should be placed in layers not more than 3 in. thick, and each layer should be thoroughly compacted. The surface of the base repair should be covered with a bituminous surface treatment or a bituminous mat. A surface treatment may consist of one or more applications of liquid bituminous material and aggregate chips. The repaired base area may be temporarily covered only with a prime coat of bituminous material for the purpose of testing the adequacy of the base repairs before placing the wearing course.

Even though the failed base under a bituminous surface may be soil cement, bituminous concrete, or portland cement concrete, it can be successfully repaired by using densely graded aggregate as described. Aggregate costs may be such, however, as to make it more economical to use a bituminous mix, in view of the fact, as has been generally conceded, that less thickness of bituminous concrete is needed than plain densely graded and compacted aggregate to give the same traffic-carrying results. Bituminous concrete can be used successfully for making base repairs. The aggregate should meet AASHO Specifications for Materials, Designation M 76. The subgrade should be thoroughly compacted and primed. The bituminous mixture should be placed in layers not exceeding 2 in., and each layer compacted.

Soil-cement mixtures can be used for repairing bases (see Sec. 11).

Portland cement concrete as a base under a bituminous pavement is subject to "slab pumping" and "blowups" in some geographical areas and can be maintained by the same methods described for maintenance of portland-cement-concrete pavement.

Failing areas of brick or block pavement being used as a base for bituminous pavement can be repaired using densely graded aggregate, bituminous concrete, portland cement concrete or by re-laying the brick or block on an adequate subbase. The method used should be determined by economy and convenience.

G. Surface Treatments

Bituminous surfaces must have the correct proportion of bitumen and aggregate to give satisfactory results. If the surface contains too much bitumen, it will "bleed" and become slippery when wet and corrugations or rutting and shoving may develop. If the surface contains too little bitumen, aggregate may be lost from raveling under traffic.

Bleeding occurs ordinarily soon after construction of the pavement. Bleeding should be corrected as soon as possible after occurrence by an application of aggregate chips or coarse sand, reasonably free from dust. The aggregate should all pass the $\frac{1}{2}$- or $\frac{5}{8}$-in. sieve, and 100 per cent should be retained on the $\frac{1}{8}$-in. sieve. Rolling the cover aggregate into the surface is recommended. The prompt application of such blotting material will eliminate the tracking of the bitumen and the development of slick spots and corrugations.

A slight excess bitumen in the top portion of bituminous-concrete pavement may result in a slick surface and shoving and rutting in extreme stages. The treatment consists of an application of aggregate chips with or without additional bitumen and rolling. When additional aggregate cannot be made to adhere, removal of the top portion of the pavement to a slight depth sufficient to eliminate excess bitumen or to correct the corrugations or ruts is indicated. This work should be done with a motor grader when the surface of the pavement is hot.

When the surface oxidizes it may crack and permit water infiltration, thus causing further deterioration. Under these conditions a bituminous surface treatment or seal coat should be applied consisting of $\frac{1}{4}$ to $\frac{1}{3}$ gal of bitumen with 20 to 30 lb of cover aggregate per sq yd. Cutback asphalt grade SC, MC, or RC, emulsified asphalt, or asphalt cement is commonly used.

If the dry surface condition is not remedied in the early stages and the raveling has progressed until the surface of the pavement has become materially affected, it is often advisable to place a heavy surface treatment, sometimes referred to as a double surface treatment. Prior to the treatment, all holes and depressions should be properly patched as previously described and a prime coat added; the lighter grades of cutback asphalt, asphalt emulsion, or tar should be used. RC-1 or RC-2 cutback asphalts have given excellent results. The rate of application will ordinarily vary from 0.15 to 0.25 gal per sq yd. The prime coat should be allowed to cure thoroughly before the next application of bitumen is placed. The second, or binder, coat should be sufficiently heavy or ductile to hold the aggregate which is to be placed and imbedded by rolling and traffic.   It should be applied at a rate of 0.25 to 0.3 gal per sq yd and within the temperature application range indicated by specifications for the particular type and grade of bitumen used.   Immediately following the application of the bitumen, the first course of cover aggregate should be evenly spread and rolled.   It is important at this point to remove from the surface any loose cover aggregate before applying the next application of bitumen.   Another application of bitumen and cover aggregate is made in the same manner to obtain the double surface treatment.

It is not easy to establish a rule for determination of the best type and grade of bitumen to be used for surface-treatment work, and as a result, it is necessary to make this decision on the basis of local conditions (see Sec. 9).   The cover aggregate can be either ¾- or ½-in. size.   If the ¾-in. aggregate is used, it should comply with the gradation requirements specified for size No. 67 of AASHO Specification M 43 (see Sec. 9).   If the ½-in. maximum-size aggregate is used, it should comply with the gradation requirements specified for AASHO size No. 7.   The second surface-treatment course should use aggregate with ½-in. maximum size.   The first course of cover aggregate should be lightly rolled to imbed the aggregate into the bitumen.   The second cover should be more extensively rolled with the flat-wheel roller, followed by drag brooming and then rolled with a pneumatic-tire roller until the maximum quantity of cover aggregate is imbedded into the bitumen. When practical, better over-all results can be obtained by barring traffic from the surface treatment for 24 hr after completion.   It is very important that the quantities of bitumen and aggregate in double seal treatments be under rigid and accurate control.   If excess bitumen is used, excessive bleeding may occur and the surface may become slick, or the surface may remain soft and pick up under heavy traffic.

A bituminous-surface-treatment crew is made up of a foreman, heater operator, distributor operator, two to four laborers, roller operator, truck driver, two flagmen, and laborers.   The equipment consists of a bituminous heater, distributor, aggregate spreader, roller, and trucks.

H. Resurfacing

Bituminous pavements, though remaining relatively sound structurally, often become distorted, and the riding qualities of the surface impaired.   Under these conditions bituminous resurfacing, retreading, or upper-decking may be indicated.   An additional bituminous mat on top of the present surface may be more economical than tearing up the surface and strengthening the base.   If the surface has deteriorated badly, then the surface should be reshaped and additional thickness added to the base.

Bituminous mats are frequently constructed by road-mix methods.   The quantity of aggregate that can be mixed on the road is about 14 or 15 cu yd to 100 ft of roadway. A commonly used quantity per 100 ft is 10 cu yd, which yields a compacted thickness of 1¼ in. of mat 20 ft wide; also 12 cu yd, which produces a thickness of 1½ in.

Prior to placing the aggregate on the road surface for mixing, the depressions and holes in the surface should be patched.   The actual mixing operations, when motor graders or motor graders and aggregate mixers are used, are the same as the mixing operations for bituminous patching previously described.   If a traveling bituminous mixing plant is used, the aggregate is windrowed and evened on the road surface by methods previously described.

After applying a light tack coat of about 0.1 gal per sq yd of the same bitumen as used in the mix, the mixed material is then laid with motor graders.   Two motor graders have been found to work very efficiently and economically in laying the mixed material.   The laying operation should be done with as few passes with the motor grader as possible to prevent segregation and at the same time obtain the desired results in smoothness of the surface.   Longitudinal rutting is a common defect when the laying is done with motor graders under traffic, but skilled equipment operators can hold this difficulty to a minimum.   Rolling with pneumatic-tire rollers during the final blading operation is believed to produce better compaction.   Final rolling should be with a smooth-wheel roller of 5 to 8 tons.   Smooth-wheel rolling should start at the edge and finish at the center of the pavement.

A bituminous-surfacing crew for road mixing and laying with motor graders should be made up of a foreman, five or six motor-grader operators, tank-car-heater operator, relay-tank truck drivers, a driver and operator of the bituminous distributor, roller operators, two or three laborers, and one or two flagmen.   The necessary equipment consists of five or six motor graders, one circulating booster heater, relay-tank trucks, pneumatic-tire roller, 5-ton flat-wheel tandem roller, and rotary broom.   An aggregate mixer, sometimes referred to as a tiller or pulvimixer, has been found very efficient in mixing operations and will replace two motor graders.

The use of hot-mix asphalt concrete should be limited to those roadways where the base is known to be adequate.   Base movement will cause a hot-mix surface to break up more quickly than a cold-mix surface because the bitumen is generally of lower penetration.   Cold-mix bituminous concrete is also generally considered more salvageable than hot-mix.   Unit costs of hot-mix bituminous concrete are usually greater than cold-mix because more labor and a greater investment in equipment and materials are required for the completed surface.   Bituminous-concrete pavers are commonly required for laying the hot mix, and this usually provides a better riding surface than is obtained with cold mix laid with motor graders.   The determining factors between hot-mix and cold-mix resurfacing are usually the condition of base and costs.

If the bituminous surface or pavement condition has deteriorated because of base weakness to a point where it is of little value as a surface, it becomes more economical to scarify the surface and break it up completely.   The base should then be strengthened by the most economical method, which is usually determined by the availability and cost of local aggregate.

V. MAINTENANCE OF PORTLAND-CEMENT-CONCRETE PAVEMENT

The length of pavement life and the maintenance costs of a portland-cement-concrete pavement, plain or reinforced, are determined primarily by the volume and weight of traffic, by its subgrade support, by weather, and by its inherent stability resulting from proper design and construction.   Lack of subgrade support results in settlement and possible breakage of the pavement, and inadequate design and construction results in pavement cracking and disintegration or scaling or both.   Excess water under the pavement causes a reduction in subgrade support in many types of subgrades, and excess water in cracks in the pavement contributes to pavement scaling and pavement disintegration.   Traffic contributes to pavement deterioration in direct proportion to its weight and volume.

A. Cracks and Joints

Keeping the amount of water that reaches the subgrade under the pavement and that which collects in cracks and joints to the lowest possible amount will contribute greatly to the life, serviceability, and low maintenance cost.   For that reason it is desirable to seal all cracks and joints as completely as possible even though it is known that surface water reaches the subgrade in much greater volumes along the edge of the pavement than through cracks and joints.   Also, it is more difficult to prevent water from reaching the subgrade along the edge of the pavement.   Methods and materials have been developed by which cracks and expansion and contraction joints can be filled and sealed more or less successfully.

Many types of filler have been and are being used in the cracks and joints; however, all types of cracks need to be refilled and resealed at varying intervals as necessary to keep the crack or joint sealed against the entrance of water.   Even though there is a present-day trend toward a reduction in the number of joints deliberately placed in pavements, much thought has been and is continuing to be given to developing better methods and materials for filling and sealing.

The material used to fill and seal joints and cracks should adhere to the sides of the crack, should be pliable at all temperatures but should not become fluid under high temperatures nor brittle under low temperatures, and should have low-expansion qualities to hold extrusion to a minimum.   Most materials used to the present time have been bitumen or compounds whose basic ingredient is bitumen. Compounds having rubber as one of the ingredients are being used. The bituminous materials vary considerably in composition, often because of the personal experience of the user, the climatic conditions under which they are to be used, and the temperature range under which they are to serve.

In Northern states, asphalt having a penetration of 85–100 is common, while in Southern states, asphalt having a penetration of 40–50 has been used. RC-3 cutback asphalt is also being used satisfactorily. The asphalts should meet the requirements of AASHO Specification M 20 or equivalent, and RC-3 cutback asphalt should meet AASHO Specification M 81 or equivalent. Tar products can be used satisfactorily as fillers and sealers of joints and cracks. The AASHO Specifications for Highway Materials, Designation M 52, suggest RT-10, RT-11, and RT-12 for crack filler.

Maintenance engineers in various sections have developed many mixtures, with bitumen as the principal ingredient that has been used successfully for filling joints and cracks. A mixture of sawdust and cutback asphalt has given good results in some geographical areas where sawdust is readily available. Proprietary materials, some of which contain rubber, have been and are being used successfully.

Cracks and joints should be cleaned before being filled. If the old joint filler has extruded above the surface of the pavement to any great extent, it can be efficiently and economically removed by blading it off with a motor grader. However, crack cleaning with a portable kerosene burner and pressure tank mounted on a dolly has gained in popularity in recent years. The bituminous material along the crack is heated sufficiently to be scraped along the crack, filling the open crack and removing the surplus. The scraping tool is ordinarily a hoe straightened to serve as a scraper. This method is economical in that it fills the crack and improves the ridability of the pavement by removal of the surplus. Routine cleaning can be done with such small tools as mattocks, spades, hoes, and hand brooms. Compressed air can be used to blow the loose materials from the cracks and joints. Where rubber compounds are to be used, extreme care must be taken to clean the sides of the crack or joint to promote adherence of the compound. A machine has been designed and manufactured that does a very satisfactory job of cleaning the sides of the crack or joint, as well as removing the old filler to a desired depth in one operation. This crack-cleaning machine consists of a power unit which operates a small wheel some 8 in. in diameter, with cutting knobs extending about 1 in. from the rim. The cutting wheel is made of very hard material and revolves at high speed. The complete unit is mounted on small rubber-tired wheels and is operated by one man. The machine can be obtained commercially. Concrete power saws can also be used. An air compressor is conveniently and efficiently used in the machine-cleaning method for blowing the loose particles from the crack or joint.

Asphalt cement or cutback asphalt must be heated to be properly poured. Asphalt cement for crack pouring is usually obtained in solid form in drums and is heated in asphalt kettles and poured at a temperature between 225 and 275°F. Care should be taken to avoid overheating, since this changes the character of the bitumen so that it becomes brittle when cold and will not adhere to the concrete. RC-3 cutback asphalt is obtained in railroad tank cars (insulated if available) and usually stored in bituminous storage tanks to be used as needed. The cutback can be heated by steam asphalt heaters or circulating booster heaters, the latter being more efficient and economical. Trailer bituminous distributors of from 400 to 600 gal capacity are very efficient for transporting the cutback from storage to job site and for holding the material at the desired temperature through the pouring operations. The asphalt is drawn out of the distributor into conical-shaped pots with $\frac{1}{4}$-in. openings in the bottom and poured by hand into the cracks and joints. A hand-pouring nozzle with cutoff valve connected by a flexible metal hose direct to the distributor is also used, but it is difficult to do neat work with it. Care should be taken to fill the crack or joint to the proper level and to avoid the use of excess material. Repouring after the poured material in the joint has cooled is sometimes necessary to bring the poured material to the desired level in the joint. On windy days, special precautions must be taken to control the pouring, and if the wind is too strong, pouring should be postponed. Hot-poured crack filler should be covered immediately with sand, rock dust, sawdust, or similar material to prevent removal of the material by traffic.

A crew for pouring hot cutback asphalt consists of two men for cleaning the cracks and joints before the pouring, a truck driver to operate the truck pulling the bituminous unit, a truck driver to operate the cover-material truck, two or three men with pouring pots, one or two men for sanding, one or two flagmen, depending on the volume of traffic, and a foreman. As a safety precaution, it is advisable to pour the cracks and joints in one lane of the pavement at a time, with the crew working between the bituminous unit and the cover-material truck moving with the direction of traffic.

Kettles for heating rubber-compound crack and joint fillers under controlled pouring temperature, which must be in a range of 400 to 440°F, are designed and constructed on the double-boiler principle, having one kettle placed inside another with oil between the two. The direct-flame heat is applied to the outside surface of the outer kettle. A hand pot for pouring the hot rubber compound has been made to give accurate control of the flow from the spout. The spout is rubber-covered or -cushioned, and the pouring is done by drawing the spout along the surface of the pavement over the crack or joint. This method of pouring eliminates wastage of material.

B. Undersealing

Asphalt undersealing (9)* consists of pumping highly viscous asphalt heated to a temperature of 400 to 450°F through a drilled hole in the concrete pavement. The purpose is to fill the void under the pavement and to cover the affected area of the subgrade with a sheet of asphalt, thus supporting and undersealing the pavement by filling the cracks and joints from below.

The practice of undersealing concrete pavements with asphalt has been developed primarily to control slab "pumping." Slab pumping is defined as the ejection of water, carrying soil particles, through cracks or joints in the pavement or along the edge of the pavement, caused by vertical movement of the pavement under heavy axle loads (2).

Pavement "blowing" has developed during the past few years and is the result of some of the same causes as slab pumping, but occurs where soil-aggregate bases with an excess of soil fines have been used under the pavement. Blowing is defined as the movement of water under and along the edge of the pavement, which may or may not result in the displacement of base material. Voids may occur under the pavement from other causes, such as fill settlement, slides, or the erosion of the subgrade by movement of water under the pavement on heavy grades (3).

Raising pavement settlements by undersealing should not be attempted since there is no control possible. It is good practice to underseal with asphalt during the earliest detectable stages of slab pumping and thus keep pavement damage from this cause to a minimum. The earliest stages can be detected by slight spalling of the pavement along the longitudinal center joint and adjacent to a transverse crack or joint and by seepage of clear water through the cracks and joints from the subgrade to the surface of the pavement.

C. Mudjacking

Mudjacking is a method of filling voids under concrete pavement and of raising sections of concrete pavement to a desired elevation. It is accomplished by mixing and pumping materials, such as soil, cement, and water, through drilled holes in the pavement with a mudjack. A mudjack is a machine consisting of a power unit, continuous-type pugmill in which the soil, cement, and water are mixed, and piston pumps which force the mixed materials through a flexible hose and a tapered nozzle to the point of discharge. The complete unit is ordinarily mounted on a two-axle, pneumatic-tire trailer.

---

* Numbers in parentheses refer to corresponding items in the references at the end of this section.

For void-filling purposes, the mudjack mix is usually a slurry mixture of the consistency of thick cream composed of top loam soil (preferable), cement, and water. Frequently used proportions of this mix are 4 sacks of cement to the cubic yard of soil to which is added about half as much water by volume. Many other mixtures have been used, some containing only cement and water, some a soil-cement-water mixture with various types of asphalts (4, 5, 7).

1. Soils. Reference (6) should be consulted for detailed information.

2. Slurries. The term "slurry" refers to the soil-water-admixture combination used to fill the void between the slab and subgrade. It usually has the consistency of thick cream, which is the proper fluidity to fill the majority of the voids under the pavements. The most widely used mixtures are as follows:

Volume exclusive of water, %

1. Soil... 60–84
   Cement.. 16–40

2. Soil... 77
   Cement.. 16
   Cutback asphalt (SC-2, MC-1, RC-3).......... 7

The final mixture of soil and cement must have low shrinkage but should not be hard when dry. The mix should be fluid so as to spread rapidly under the pavement and flow into small cavities. The exact fluidity required can be determined only by experience. Good practice requires a fairly stiff mixture, to fill the large voids under the slab or to raise it, and a thinner mixture, to fill the smaller voids.

3. Procedure. The soil and cement in proper proportions are transported to the job in trucks and there hand-shoveled into the mudjack and followed with the proper amount of water. The mudjack mixes the soil, cement, and water to the proper consistency and then pumps the slurry under the pavement through a heavy hose and a tapered metal insert nozzle.

Experience has proved that slurry-mudjacked pavements will again develop voids under traffic with heavy axle loads and will require continued mudjack slurry treatment. Instead of repeating the mudjack treatment, it has been found that asphalt underseal of the pavement some 2 or 3 weeks after the mudjack slurry treatment gives better and more lasting results.

Settlements in concrete pavements often occur at bridge ends, over culverts, or at other places to such an extent as to cause an unsafe condition to traffic. If the slab has not broken too badly, it can be raised back to the desired grade line by the use of a mudjack. The procedure to be followed in raising the pavement is the same as outlined for filling voids under the pavement, with the exception that the consistency of the slurry is a little thicker or stiffer.

The crew for mudjack operation consists of one foreman, one nozzle operator, one mudjack operator, one man to shovel soil from truck to mudjack hopper, one man to plug holes and clean pavement, two truck drivers, one man to load trucks at soil pit, one man to operate a jackhammer for drilling holes in the pavement, and one or two flagmen, depending on traffic density. The equipment for mudjack work consists of an air compressor with truck to pull it, mudjack, three dump trucks for hauling soil and cement, and one truck to haul water.

The amount of soil-cement slurry pumped under concrete pavement either for filling voids or lifting the slab varies according to the condition of the pavement and the subgrade, but has averaged about 30 cu yd per mile.

D. Patching with Bituminous Mixtures

Concrete pavements develop failures, depressions, and irregularities that can be successfully and economically patched and repaired with bituminous mixtures.

E. Blowups (8)

Blowups occur at transverse cracks or joints in concrete pavement when the crack or joint is unevenly filled with a noncompressible material and when high temperature and moisture conditions cause the concrete to expand so that sufficient expansion results to cause the pavement to be raised off the subgrade. This raise has been as much as 24 in. and has occurred in a length of pavement as short as 10 ft. The pavement will often remain in raised position until broken down by maintenance operations. In other cases the damage from blowups is a crushing or crumbling of the pavement 1 or 2 ft each side of the crack or joint.

Blowups are a hazard to traffic and must receive immediate attention. Normally, the width of the transverse section of pavement to be removed need not be greater than 6 in. It can be controlled by marking and cutting the concrete with a chisel bit in a pavement breaker along a straight transverse line the determined distance away from, parallel to, and on either side of the center break caused by the blowup. Often the pavement is so badly broken and shattered that the affected pieces of concrete cannot be stabilized sufficiently to be satisfactory for traffic. The small pieces of concrete can be broken to smaller pieces and used as base for a bituminous surface patch or the crushed concrete may be removed and replaced with a full-depth bituminous patch as a temporary or more or less permanent repair, depending on local conditions. If necessary, very temporary repairs can be made with roadside soil. This has the advantage of allowing the pavement to settle back to its original position under traffic before the more permanent repairs are made with bituminous materials or with new portland-cement-concrete patches.

F. Patching

A hole, depression, or sharp break in the surface of an otherwise smooth concrete pavement is a traffic hazard and should be repaired without delay. It is not always easy to define the limits of a bituminous patch, and extreme care should be used in outlining the edges of the proposed patch by chalk or paint. The area to be patched should be dry and free of dust, dirt, or other foreign materials, including former bituminous patches if they are not dry and solid. Procedures are similar to those used in patching bituminous surfaces.

G. Bituminous Surface Treatments

Concrete pavements are subject to scaling, disintegration, and wear, which in the beginning appear to be of minor importance but which, if allowed to progress, will cause serious impairment and eventual failure of the pavement.

Scaling is a shelling off of thin pieces from the surface of the pavement. While the causes of scaling are often obscure, it may result from use of too much salt or salt compounds for snow and ice removal and control and for concrete curing by surface application. Scaling may also be caused by impurities in the original concrete mix, too much water in the mix, too much finishing, and unsuitable aggregate. Scaling is accelerated by traffic in direct proportion to the volume and weight of the traffic. It can be retarded and the life of the pavement extended by bituminous patches and surface treatments of the affected areas.

Disintegration of concrete pavement is the crumbling or breaking of the pavement into small pieces. Some of the contributing causes may be unsound aggregate, freezing and thawing, and bad drainage. The first sign of pavement disintegration is surface cracking, and it usually occurs immediately adjacent to the intersection of the longitudinal center crack with a transverse crack or joint. The disintegration extends in all directions until the full width is affected, and then extends longitudinally. This surface cracking is sometimes referred to as map cracking, or D cracking. If allowed to progress with no control efforts, disintegration will reduce a concrete pavement to a course of completely unbound pieces of aggregate which will ravel and rut under traffic.

Disintegration can be retarded and the life of the pavement prolonged by correcting poor subdrainage, where it is one of the contributing causes.

Wear due to abrasion of traffic on concrete pavement causes very little maintenance except in minimum amounts at isolated spots such as stop areas where brakes are sharply applied, sharp curves where cars tend to skid, pavement areas adjacent to commercial entrances where surfacing aggregate from the entrance tends to be whipped onto the pavement and serves as an abrasive agent between tires and pavement, and pavement intersections subject to heavy cross traffic from all directions. Abrasion of pavement is a minor difficulty, but it can be retarded by bituminous surface treatment of the affected area.

Bituminous surface treatments consist in application of a liquid bituminous material followed immediately by an application of small and carefully sized aggregate. Procedures are similar to those described for bituminous surfaces.

H. Patching with Concrete

When concrete pavements become broken into such small pieces that they cannot be stabilized by undersealing or mudjacking to permit bituminous patching, or when disintegration or scaling has progressed both in severity and area until bituminous patching is deemed inadequate, or when the impaired surface is very limited in area and it is desired to use concrete patches in order to retain a uniform surface appearance, it becomes advisable to patch with concrete. The decision between concrete and bituminous patching methods merits the attention and study of the best and most experienced maintenance engineers. Concrete patches should be used to replace cuts in the pavement for the purpose of installing or maintaining public utilities. The type of pavement break or failure determines the shape of the patch. Slab-pumping failures result in the need for two types of patches. These are triangular, diamond, rectangular, or square patches to replace center breaks and full-width or half-width patches to replace full- or half-width breaks in the pavement. Blowup failures usually require full-width patches. Disintegration failures in the early stages usually require diamond-shaped concrete patches, and in the later stages of failure, full-width patches are required. Failures from a combination of loading and inadequate subgrade support result in the need for patches of irregular shapes and sizes.

Experience indicates that irregularly shaped patches whose limits are determined by the outside breakage lines, diamond-shaped patches with the center of the diamond along the center of the pavement, triangular patches with the base of the triangle along the center or edge of the pavement, and conventional inside- or outside-edge rectangular patches and inside plug patches will all give satisfactory service. Good practice indicates that the dimension of any side of a patch should not be less than 4 ft, that a side of a patch parallel to the flow of traffic should not be placed under a wheel track, and that the inside angle of a diamond- or triangular-shaped patch should not be less than 60° with the points flattened. Irregular, diamond-, and triangular-shaped patches are economical because they save money both in the removal of the old concrete and the placing of the new concrete. Experience also indicates that expansion and contraction joints in concrete patches are unnecessary even though there was a joint in the old pavement at the location of the new patch, the only exception being in patches placed at bridge ends.

Along with the locating and marking of proposed concrete patches, a record must be kept of the surface area of each patch in order to estimate the amounts of materials needed for the patching. And at the same time, or before any replacement work is started, a study should be made to determine the cause of the failure and the replacement methods that will prevent a recurring failure from the same cause.

Asphalt undersealing of old pavement immediately adjacent to the proposed patch will stabilize the old pavement when the failure was due to slab pumping, and experience indicates that the patches will last longer. It is good practice to make the patch thicker than the old pavement by 1 in. or slightly more. The thickness should be determined from the subgrade conditions and the volume of heavily loaded traffic. Good results have been obtained both with and without underpinning the old pavement with the new patch. If the original failure was due to slab pumping, or if the subgrade is found unstable, the use of 3 or 4 in. of aggregate under the proposed patch is worthwhile. Use of reinforcing in the patches does not seem to be justified. For traffic safety, concrete-patching operations should be confined to one lane of pavement at a time.

VI. MAINTENANCE OF SHOULDERS

A highway shoulder is the transitional area between the paved surface of the roadway and the natural ground surface of the roadside, or it could be defined as the transitional area between the vehicular travelway and the lateral waterway. A shoulder should provide a safe place of refuge for the vehicle operator during a traffic emergency. A shoulder is an area over which melted snow and the rain water falling on the pavement pass to the side ditches. It gives lateral support to some types of pavement.

A. General Maintenance

To the road user, the importance of maintaining a smooth and stable shoulder alongside the pavement is second in importance only to the surface maintenance of the pavement itself. Deep ruts, abrupt holes, or slippery surfaces on the shoulder present a terrific hazard to fast-moving traffic. Depressions in the shoulder along the edge of the pavement serve as reservoirs which feed water to the subgrade, causing many types of pavement difficulties. Methods must be devised by which shoulders can be maintained in smooth, stable, and nonslippery condition at all times. Shoulders should slope away from the edge of the pavement with sufficient fall so that rain water will flow freely from the surface of the pavement to the side ditches, but the shoulder slope should not be steep enough to be hazardous for traffic. The proper degree of slope will vary slightly with the type of shoulder surface and with the gradient of the roadbed. Shoulders on steep gradients should have enough slope to cause the water to flow into the side ditch and not parallel with the road surface. The amount of work necessary to maintain the shoulder varies directly with the type of shoulder construction, volume of traffic, particularly dual-tired truck traffic, and indirectly with the width of the pavement. Adequate maintenance methods and costs vary in accordance with the types of materials used in building shoulders.

A large part of the mileage of shoulders on the present highways was constructed with available natural soil, and it has been the responsibility of the maintenance organization to maintain the shoulder in its natural condition or to improve it as the volume of traffic demanded and the available funds permitted.

Possibly the first step in shoulder improvement in the early maintenance was the development of turf. As traffic increased and turf was destroyed along the edge of the pavement, the practice of placing crushed aggregate on the surface of the shoulder for a width of 1 or 2 ft adjacent to the pavement was followed. This method resulted in an aggregate-surfaced shoulder immediately adjacent to the pavement, with the remaining width covered with turf. The next stage in shoulder improvement was probably the stabilization of the aggregate next to the pavement with soil, bitumen, or soil cement, with the outer area of the shoulder covered with turf. Another type of shoulder improvement has been soil-aggregate stabilization over the entire width of the shoulder and the growth of turf on the stabilized layer. Possibly the final stage in shoulder improvement is to pave a portion or all of the width with bituminous or portland cement concrete.

Shoulder maintenance has usually included improvement of the shoulder from a natural-earth shoulder to the type found most effective for the volume of traffic, using the highway in so far as is possible with the available funds. Maintenance methods vary with the type of shoulder construction. The color of the shoulder surface should contrast with the color of the adjacent roadway surface so that motor-vehicle operators can easily distinguish between the roadway and the shoulder. This is especially important for night traffic. The shoulder types may be listed as natural soil, turf-covered soil, turf-covered soil aggregate, and bituminous-treated partially paved and paved shoulders.

B. Natural-soil Shoulders

The natural-soil shoulder is the most difficult to maintain and should be improved to some other type as soon as possible. It is slippery during wet weather, dusty during dry weather, and subject to erosion during periods of heavy rainfall. It is subject to rutting by traffic during wet periods. Soil shoulders must be bladed or dragged at frequent intervals to eliminate ruts. Blading can be done with motor graders or with pull-type blades.

C. Turf Shoulders

Turf or grass-covered shoulders are far more desirable and economical to maintain than natural-soil shoulders. Turf shoulders are considered adequate if traffic is not too heavy. In most geographical areas, it is possible to find some type of grass that will grow in the topsoil of the area. If the shoulder is built entirely of subsoil, it may be necessary to remove and replace a portion with topsoil. Investigation of the farm pastures, farmyards, school- and churchyards, and cemeteries in the area will reveal the types of grasses that are being grown, and from this information, a seed mixture can be designed. The most desired type of turf is usually a combination of grasses and legumes which will develop a dense growth with a minimum of height in the shortest period of time and which will be self-perpetuating from either roots or seed or both. Root perpetuation of grasses is preferred, because the plants whose roots do not die every year tend to hold the soil, thereby reducing erosion and increasing load-carrying capacity of the shoulder, while the plants whose roots die each year tend to loosen the soil, thereby increasing erosion and decreasing load-carrying capacity. See Sec. 13 on "Landscaping" for additional information on grasses.

Fertilizing the shoulder soil at the time of or immediately prior to the seeding operation has been found to be highly desirable in helping to establish quickly a heavy growth of grass. Commercial fertilizers have been proved most adaptable to shoulder-seeding operations, because of ease in handling. An analysis of the soil as well as a careful visual study of the site to be seeded will indicate the amount and kind of fertilizer needed. Grasses can be established on a very poor soil by the use of a complete fertilizer containing nitrogen, phosphate, and potash, but in many cases it is only necessary to add nitrogen and phosphate to the soil. This can best be done by applying nitrogen in the form of ammonium nitrate (33 to 35 per cent) and phosphate in the form of triple phosphate (45 per cent). These materials are highly concentrated, and there is less bulk to handle to obtain the desired application of nutrients to the soil. Fertilizers can be hand-broadcast with the seed or, to ensure uniform application, can be applied with seed drills having fertilizer attachments.

Seeding can be done with hand seeders, but much better seeding is done with a seed drill. Harrowing with a spike-tooth harrow or some similar implement should follow the seeding. Covering the seeded area with a straw or hay mulch is desirable. Incidentally, a hay mulch containing seed is a satisfactory and economical seeding method. The mulch is effective in controlling erosion until the grasses are established. Mulch aids in the germination of the seed and the growth of the young grass by conserving the soil moisture. Mulch should be tied down at intervals with soil, to prevent loss from wind and water. A pulvimixer or disk can be used to work the mulch partially into the soil.

Well-timed mowing operations on the shoulders and inslopes will control the growth of weeds and keep the grass at an attractive height. If shoulders are mowed at frequent intervals, the clippings may be allowed to remain on the ground, where they will serve as a mulch and have some fertilizing value.

Holes and ruts in the turfed shoulder should be immediately repaired by filling with selected soil. Reseeding or sodding repaired areas can be done if advisable. Turf shoulders will build up in elevation until the shoulder becomes higher than the pavement. This build-up can be reduced to some extent by rolling the shoulder following wet periods, but eventually and at irregular intervals the build-up of the shoulder must be cut off with a blade and the proper elevation and slope of the shoulder reestablished. It is considered good practice to remove the build-up in the spring months because the turf will usually reestablish itself from the roots without the expense of reseeding.

Turf shoulders have the advantage of good color contrast between pavement and shoulder and provide inexpensive maintenance when traffic does not prevent or offer too much resistance to the growth of turf and maintenance of the shoulder. Turf shoulders are slippery when wet and become soft during prolonged periods of rain or during the thawing-out period following a freeze. However, turf shoulders are less slippery, more stable, less subject to erosion, and less expensive to maintain than bare and untreated natural-soil shoulders. When the pavement is so narrow that there is considerable wheel traffic on the shoulder, it becomes practically impossible to maintain the turf adjacent to the pavement, and then, of course, this type of shoulder is unsatisfactory, at least next to the pavement.

Turf shoulders have been found to be detrimental where adjacent to most types of bituminous surfaces, especially those with inadequate bases. The growth of turf on the shoulders retards the lateral flow of surface water from the pavement to the side ditches and tends to increase the saturation of the topsoil, during periods of thaw and rainfall, beacuse of the loosening effect of the root system of the turf. This shoulder saturation naturally spreads to the subgrade under the bituminous surface, beginning, of course, at the pavement edge, but eventually extending to the full-surfaced width, resulting in settlement, displacement, and cracking of the bituminous surface under traffic loads. Evaporation of excess moisture in the shoulder and subgrade is retarded by a completely turfed shoulder since the root system tends to retain the excess moisture in the soil, and the turf growth aboveground tends to insulate against the evaporating effect of sun and wind.

Soil or soil-aggregate shoulders, practically void of turf, have proved to be satisfactory immediately adjacent to bituminous surface, since they are readily shaped and bladed, resulting in shoulders that are flush with the pavement edge, well drained, quick-drying after periods of rainfall and generally presenting a neat and comparatively safe condition.

D. Strip-shoulder Stabilization

On highways with pavement not exceeding 18 or 20 ft in width and carrying large volumes of traffic, especially dual-tired truck traffic, the problem of maintaining the shoulder flush and smooth adjacent to the pavement becomes acute, and it becomes necessary for the shoulder to have something more than a turf covering to obtain satisfactory results. In fact, 9- and 10-ft traffic lanes are too narrow for large volumes of fast traffic, and the solution of the shoulder-maintenance problem lies in the proper widening of the traffic lane. In view of the fact that many highway organizations are not sufficiently well financed to adequately convert large mileages of 9- and 10-ft traffic lanes into 12-ft paved lanes, it falls to the lot of the maintenance group in the highway organization to bridge the gap by widening the narrow traffic lanes on the shoulder side with a low-type surfacing or stabilization process or both. The pavement widening or shoulder surfacing usually ranges in width from 1 to 3 ft on each shoulder.

Probably the most common type of shoulder surfacing consists of placing a crusher-run rock or a pit-run gravel on a low or rutted shoulder to bring it back to grade and to smooth out ruts and inequalities. Subsequent blading of the aggregate-surfaced shoulder tends to mix the soil and aggregate and results in a soil-aggregate-stabilized shoulder surface. Grass will grow on a soil-aggregate surface if there is approximately 10 per cent or more soil in the mixture and the seeding practices described in the paragraphs on turf shoulders are followed. A grass-covered, soil-aggregate surface has greater superior supporting value because of the aggregate and has superior resistance to water erosion and traffic wear because of the turf. See Sec. 11 for information on granular stabilization.

The turf on a soil-aggregate surface will withstand so much traffic wear, and then it dies out. The problem becomes one of adding additional binder in the surface. This can be achieved by the use of bituminous material. The bitumen may be applied to the surface as a penetrating binder and as a seal for holding a subsequent application of aggregate chips and by using it as a binder for soil stabilization or for aggregate or both.

E. Bituminous Shoulders

Road oil, grade SC-1 or SC-2, has been successfully used to treat soil or soil-aggregate shoulders for widths of 2 or 3 ft or the full width of the shoulder. The SC road oil is applied by a pressure bituminous distributor at a rate varying from 0.5 to 1 gal per sq yd to the shoulder which has been made smooth and flush with the pavement and sloping away at the desired rate of fall. The oil should be placed in two or three applications of from 0.2 to 0.4 gal each, and each application allowed to penetrate the shoulder and cure before the subsequent application of aggregate. The shoulder should be barricaded to traffic until the road oil is soaked in. Oiled shoulders are good for 2 or 3 years with a minimum of maintenance and then should be reshaped and re-oiled. Holes or ruts in the oiled shoulder should be patched with a mixture of road oil and soil as presented in the material on oiled-earth-road maintenance in this section.

F. Soil-cement Shoulders

Shoulders may be stabilized by mixing portland cement with the soil. The type of work is commonly known as soil-cement stabilization. Details relative to moisture content and quantities of cement per cubic yard of soil can be found in Sec. 11 on "Soil Stabilization."

G. Portland-cement-concrete Shoulders

Narrow traffic lanes can be widened and shoulders paved with portland cement concrete with satisfactory results.

VII. GRAVEL- AND EARTH-ROAD MAINTENANCE

A large mileage of all roads in the United States is still classified as gravel, crushed stone, or earth and as such are low-type roads. Such roads do not carry a large amount of traffic compared with higher-type surfaces, but to the man who lives on one of them they are just as important as any road anywhere. Those in responsible charge of such roads must maintain them in a manner to give as good service as can reasonably be expected from such a type of road.

A. Earth Roads

Earth roads may be classified as graded and ungraded. Ungraded earth roads are the lowest class, and they serve very few people. In many cases they are mere trails. The maintenance of such roads consists largely in smoothing the surface after rain. However, they may frequently be greatly improved by blading with a blade grader or a motor grader. Such work can justifiably be classed as maintenance.

B. Graded Earth Roads

It is assumed that roads in the graded-earth classification have been built to a satisfactory grade and cross section, that the gradients have been improved, and adequate right-of-way has been secured. Such roads should be in condition for further improvement by surfacing. In fact, they are often considered stage construction and are maintained as earth roads until funds are available for further improvement. The first maintenance obligation on a graded earth road is to keep the surface smooth. Usually a good blading after each rain, after the road is partially dried out, will suffice.   A motor grader is the best machine for this purpose.   The motor grader should be of sturdy construction capable of handling a 14- or 16-ft blade.

The second most important maintenance need on this type of road is the control of erosion.   Usually such roads will not justify expensive erosion-control work such as sodding of ditches and seeding of slopes.

In the snow belt, snow removal, or at least the opening of drifts, is justified on the heavier-traffic roads of this type.   Here again the tandem-type motor grader with V plow and wing is very useful.

C. Gravel Roads

Gravel roads include roads constructed with crushed stone, shale, slag, or other aggregate as well as those built with gravel.   The aggregate used should be densely graded or have sufficient amount of binder (soil) to cause the gravel to pack or set up under traffic.   Many untreated gravel roads may be considered as stage construction.   Such roads will have some sort of dust-alleviation work done on them when funds are available.   However, many miles of untreated gravel roads will have to be maintained as such for many years.

D. Untreated Gravel Roads

Gravel roads which are maintained without any surface treatment are, of course, low-class roads usually called upon to serve light traffic.   Their maintenance is, in general, the same as prescribed for low-class earth roads, consisting in general of keeping drainage lines open and the surface smooth.   The ditching should be done with motor graders or blade graders.

More crown must be maintained in a gravel road than in a higher-type road, particularly on flat grades.   The amount of crown will vary somewhat with different types of aggregate and under varying climatic conditions, but in general the slope from the center of the road to the shoulder should be not less than $\frac{1}{2}$ in. per ft, and probably not greater than $\frac{3}{4}$ in. per ft, $\frac{5}{8}$ in. per ft usually sufficing.   This would give a $7\frac{1}{2}$-in. crown on a road 24 ft shoulder to shoulder.   The crown should be parabolic for best drainage and safe driving.   If adequate crown is not maintained, water will tend to pool on the surface of the road and be splashed out by passing vehicles, thus accentuating the corrugations.

E. Calcium Chloride–treated Gravel Roads

Gravel roads which are called upon to carry fairly heavy traffic should be maintained as dustless surfaces by the application of a dust palliative.   Calcium chloride may be used for this purpose (see also Sec. 11).

If the road was originally built with calcium chloride incorporated in the structure, it can be maintained by light surface applications of calcium chloride.   The chloride can be placed with an ordinary lime spreader and should be applied at the rate of from $\frac{1}{4}$ to $\frac{1}{2}$ lb of chloride per sq yd, depending upon the original mixture and type of aggregate.

Gravel roads originally built without incorporation of calcium chloride may be maintained with surface application of calcium chloride if the traffic is sufficient to cause a serious dust nuisance or if the aggregate is expensive enough to warrant a special conservation measure.   Such roads should be bladed smooth and to proper crown and stabilized cross section before the chloride is applied.   More chloride is required on a first application than in the case of roads in which calcium chloride is incorporated.   Ordinarily about 1 lb per sq yd is sufficient for the original application, which should be made in the late spring or early summer.   If possible the chloride should be applied to a damp road, either after a rain or during prolonged dry spells after the application of water with a sprinkling tank.   After the original application has been made, whenever the road begins to show a tendency to dust, additional chloride should be applied at the rate of $\frac{1}{4}$ to $\frac{1}{2}$ lb per sq yd.   This also should be applied when the road is damp.

If holes develop in a calcium chloride–treated gravel road during dry weather, no attempt should be made to fill the holes by blading.   However, a mixture of gravel with clay binder and calcium chloride can be tamped into the holes in a moist condition, and a smooth surface thus obtained.   Calcium chloride should meet designation M 144 of AASHO Specification for Highway Materials.

F. Bituminous-treated Gravel Roads

Gravel roads carrying heavy traffic may be maintained as dustless surfaces, and the surfacing material may be conserved by the application of thin inverted penetration coats of liquid bituminous material.*   This type of work is done as follows.   First, the road should be well ditched to ensure good drainage and brought to a smooth and proper cross section.   Additional surfacing material should be added to places, indicating base weakness to ensure uniform stability, and all the surfacing material should be stabilized or tightly bound down.   Secondly, just prior to applying the first, or primer, coat of bituminous material, the surface should be tight-bladed with a motor grader to give a final smoothing up to the surface and remove any loose surfacing material to the edge of the surface.

Following the tight blading, the primer coat should be applied.   Ordinarily this consists of an application of 0.40 to 0.60 gal per sq yd of RT-1 tar, SC-1 road oil, or MC-1 cutback asphalt applied hot with a bituminous pressure distributor.   The rate of application will vary with the various soil types encountered, light loam or sandy types requiring more and tight-clay types requiring less.

After the prime coat has cured, it is then followed by the surface, or final, treatment. Again the application is made with a pressure distributor.   Various grades and amounts of bituminous material have been used on this application; commonly the material used is tar RT-5, RT-6, or RT-7 or MC-3, MC-4, or MC-5 cutback asphalt. The rate of application varies from 0.20 to 0.35 gal per sq yd.   Tar should meet designation M 52, SC road oil should meet designation M 141, and MC cutback asphalt should meet designation M 82 of AASHO Specifications for Highway Material.

This application of bitumen should be covered immediately with stone chips or pea gravel.   Any excess surfacing material which may have been windrowed at the edge of the surface by tight blading may be lightly bladed back on the surface as cover material.   The entire surface, however, should be covered, and additional material will be required.   A total of 10 lb per each 0.10 gal of bituminous material is sufficient. The additional material required can best be applied with an aggregate box spreader or spinner-type spreader.   The surface should then be rolled with a pull-type or powered 5-ton roller and then opened to traffic.

Traffic can use the road after the prime coat has cured and prior to the final surface treatment; however, the prime coat should not be left too long under traffic before making the final application or the prime coat will tend to ravel or pothole.

Bleeding spots in the final treatment should be covered lightly with cover material until all bleeding has stopped.

Such surfaces should not be classed as a bituminous pavement, but should be considered as a dust-alleviated gravel road commonly called "bituminous-aggregate surface treatment."

REFERENCES

1. "Recommended Practice for Snow Removal and Treatment of Icy Pavements," *Highway Research Board, Current Road Problems* 9-3R, January, 1954.
2. Allen, Harold: "Final Report of Committee on Maintenance of Concrete Pavements as Related to the Pumping Action of Slabs," *Proceedings, Highway Research Board*, vol. 28, pp. 281–310, 1948.

* See also portions of Sec. 11 dealing with bituminous-soil stabilization.

3. Vogelgesang, Carl E.: Effectiveness of Granular Bases for Preventing Pumping of Rigid Pavements, "Performance of Concrete Pavement on Granular Subbase," *Highway Research Board, Bulletin* 52, 1952.
4. Goetz, W. H.: "Development of Cement-slurry Mixtures for Use in Correcting Pumping Pavements," *Proceedings, Highway Research Board*, vol. 27, pp. 232–244, 1947.
5. "Maintenance Methods for Preventing and Correcting the Pumping Action of Concrete Pavement Slabs," *Highway Research Board, Current Road Problems*, 4-R, 1947.
6. Wintermeyer, A. M.: "Laboratory Tests Assist in Selection of Materials Suitable for Use in Mudjack Operations," *Public Roads*, vol. 14, No. 10, December, 1933.
7. McKain, A. G.: "A Study of Mudjacking Operations," Texas Highway Department, *Roads and Streets*, July, 1938.
8. Woods, K. B., H. S. Sweet, and T. E. Shelburne: "Pavement Blowups Correlated with Source of Coarse Aggregate," *Proceedings, Highway Research Board*, vol. 25, pp. 147–168, January, 1946.
9. Linzell, Samuel O.: "Subsealing Concrete Pavements," *Proceedings, Association of Asphalt Paving Technologists*, vol. 16, pp. 3–30, 1947.

Section 13

LANDSCAPING

NELSON M. WELLS, *Director, Landscape Bureau, New York State Department of Public Works.*

I. INTRODUCTION

A. The Landscape Architect in Highway Planning

Recognition of the value of the collaboration of landscape architects has grown with the growth of motor highways. Planting on state highways was approved in 1912 in Massachusetts, and in 1927 a special group was assigned to landscape work in the state of Connecticut. About this time also in Michigan, tree planting along roadsides had its start. These early assignments were mostly of a maintenance nature, such as the care of roadside trees. In some cases, there was considerable emphasis on the mowing of grass and repairing of erosion.

Every state highway department today has one or more men who are responsible for some phases of roadside improvement. The maintenance of roadsides will always be a part of their work, and there is a growing responsibility for more complete roadside improvement as an engineered part of highway location and cross-section design, including measures of conservation, design of plantings for functional reasons, and the design of appropriate roadside rest areas to favor safety and for the pleasure of travelers.

Landscaping a highway means designing and building a highway as an integral part of the general landscape and developing the roadsides for the best use and enjoyment of the traveler. The attainment of these objectives for safety, economy, efficiency, and beauty is best assured through the cooperation of professionally trained landscape architects with engineers in the various special fields of highway design.

The growth of landscape architectural personnel in highway organizations is largely due to an increasing appreciation by the general public of the economic and aesthetic values of roadside improvements since the stipulation in 1933 by the U.S. Public Roads Administration that each state highway department include in its program of construction on the Federal-aid highway system a definite number of projects to provide for the appropriate landscaping of parkways or roadsides as a part of the construction work. The Federal Highway Act of 1940 allowed up to 3 per cent of Federal aid without requiring the states to furnish matching funds to acquire land for the preservation of natural beauty and, with matching state funds, to provide specific roadside and landscape developments.

The 1944 Report on Inter-regional Highways recognized that highway design in its broadest sense rests upon a balanced agreement of landscape principles with the more commonly recognized engineering principles. It also stated that these can be combined in complete consistency with utilitarian functions. More recently, the standards of the Committee on Planning and Design Policies of the American Association of

State Highway Officials (7)* have emphasized many of the basic principles of roadside improvement growing out of the designs of landscape architects employed in designing parkways and the corresponding recommendations of the Highway Research Board Committee on Roadside Development.

The Federal-aid Highway Act of 1956 set in motion an expanded, long-range program of vision requiring close collaboration of all available talent in every state and the need for imagination and foresight in initial highway planning and design. Improved design standards for the National System of Interstate and Defense Highways, as required by the Act, were adopted by the AASHO on July 12, 1956, and approved by the Commissioner of Public Roads for use on Federal-aid interstate projects on July 17, 1956. The objectives of these standards can be realized by conscious attention in design to their attainment, especially in the design of divided highways as two separate one-way roads to take advantage of terrain and other conditions for safe and relaxed driving, economy, and pleasing appearance. The standards require that all known features of safety and utility be incorporated in each design to result in a National System of Interstate and Defense Highways that will be a credit to the nation.

B. Beauty as a Factor in Highway Design

Good appearance is also intimately integrated in almost every element of highway design, for whatever contributes to safety and efficiency becomes inherently more attractive and adds to the pleasure of all who see the work and who use the highway.

Beauty has become a recognized element of the safety and value of a highway facility as well as having reached a recognized value in public life. For many years the law of aesthetics has ruled that neither police power nor the power of eminent domain could be based solely on aesthetic considerations without relation to such factors as health, safety, convenience, comfort, or welfare. A unanimous decision of the United States Supreme Court in 1954, however, has recognized aesthetics in its own right, as quoted in part:

"The concept of the public welfare is broad and inclusive. The values it represents are spiritual as well as physical, aesthetic as well as monetary. It is within the power of the legislature to determine that the community should be beautiful as well as healthy, spacious as well as clean, well-balanced as well as carefully patrolled."

II. ESTABLISHING GRASS ON ROADSIDES

A. Seeding, Sprigging, and Sodding

In all parts of the country where a turf of grasses can be maintained, there is a growing practice of establishing a turf on disturbed areas as promptly as the new grades are finished (14). Turf has great value in controlling erosion, preventing siltation, and improving the appearance of a highway. Many contractors whose main business may be grading, laying pavements, or building bridges have recognized these values and are becoming equipped and acquainted with the techniques of establishing turf as a part of their regular operations.

Grasses which can be established by seeding methods are suitable for roadsides in most sections of the country. In other sections, the best kinds of grasses are better established by vegetative methods such as transplanting individual plants or distributing pieces of sod containing the roots of the living plants. The former is called sprigging. The latter, which is not commonly used, except with such grasses as Bermuda grass in the south, is called topsoil planting.

Where immediate protection against erosion is required, such as in drainageways or

---

* Numbers in parentheses refer to corresponding items in the references at the end of this section.

on slopes on highly erodible soils which might be washed out before grass plants could become well rooted, it is common practice to transplant pieces of turf and immediately establish a continuous cover of growing plants. This is called sodding or solid sodding. Strips of sod are often used around drain-inlet gratings and next to paved gutters.

UPDATE

Considerable success has been achieved with spraying pregerminated, wet grass seed onto presoaked embankments. The embankments are then covered loosely with straw to reduce erosion while rooting takes place—usually within a few days. Local housing contractors often have considerable experience with particular grass types and preparation periods that work best with local soils.

Seed can be forced to germinate by keeping it soaked in barrels for approximately a week before spraying; 55-gal drums have worked well for small batches.

State highway departments have also received excellent public acceptance of embankments and medians planted with native wildflowers—plants that would in some cases be considered weeds if they were to be seen in a garden.

B. Soils and Amendments

In at least the moderately humid regions of the country where sufficient moisture is present, a satisfactory turf can be established on nearly all soils provided there is an adequate proportion of the fine particles known as silt or clay size. No organic matter is required. Fertilizing and mulching or restricted seasons of seeding are requisites for success, however.

Where the layer of surface soil is of such a quality that it cannot be permitted in the embankment, it may be stripped in advance of grading and replaced to a depth up to 6 in. for the growing of turf.

In general, the quality of soil required for growing grass is one having more than a 15 per cent silt-clay content. If it has less than 5 per cent, it is doubtful whether grass can be established or satisfactorily maintained on it. The probability of success increases to a 20 per cent content. Over this there is apparently no top limit on the proportion of fines in relation to good turf growth.

Frequently, a desirable soil gradation can be obtained in the upper several inches of a finished grade by an admixture of a clayey or silty soil. This may be a considerable economy over importing topsoils. It also avoids the undesirable practice of stripping topsoil from agricultural lands.

The flattening and rounding of slopes favor a reduction in erosion; also in the establishment of turf. In rock fills where the surface water leaches rapidly and no capillarity exists, a surface cover of embankment soil may be necessary to support plant growth. Where the fills are very high and inaccessible, this soil may be advantageously placed at the outer edges of successive fill layers (15). In similar situations, cut brush or straw has been placed at the outer edges of the fill layers, or placed in shallow trenches, to help hold soil which is subsequently spread over the fill.

Where a layer of soil is placed on the surface for growing grass the subsoil may need to be scarified to form a bond to prevent slippage and to favor root penetration into the subsoil. A newly placed topsoil layer thicker than 4 in. is subject to slipping.

C. Tillage

Although a seedbed is good provision for germinating grass seed and the seed may advantageously be covered lightly with soil, an increasing amount of successful seeding is being done with the seed placed on the surface of the soil and covered by a mulch. Without a mulch it is at least necessary to cultivate the soil sufficiently to break up any surface crust which would permit the seed or fertilizer to wash away. The condition of the soil immediately following grading is ordinarily ideal for seeding either without or with a mulch.

D. Fertilizers and Lime

The application of fertilizers at the time of seeding is desirable on topsoils as well as subsoils. Commercial fertilizers for turf usually contain the proportion of 1 part of nitrogen, 2 parts of phosphorus, and 1 part of potassium, such as 5-10-5 or 6-12-6, although variations are generally satisfactory (16). There should be from 60 to 100 lb of nitrogen per acre applied during the first growing season. The customary rate is about 80 lb of nitrogen at the time of seeding; this would be 1,600 lb of a 5-10-5 formulation or 800 lb of a 10-6-4 commercial fertilizer. The proportion of the other two elements is not so significant as long as they are represented in fair amounts.

Each kind of grass and legume has its own range of tolerance to soil acidity and alkalinity, and it also has its zone of preference. Most of the common turf grasses for roadsides prefer a slightly acid soil. Soils which exceed this acidity can be modified by the addition of lime applied either in the form of hydrated lime or as ground limestone, preferably the latter.

Lime releases nitrogen from the fertilizers, and the practice for many years has been to apply it somewhat in advance of fertilizing and seeding operations. This loss is not significant, however, and except when using the more rapidly acting hydrated lime, it is satisfactory to apply lime, fertilizer, and seed at the same time and in the same operation if desired.

E. Grasses and Legumes

The kinds of grasses desirable for roadsides vary with the geography and climate of a region and must be determined for local conditions. In the northeast semihumid region, it has been found best to use a large percentage of a permanent grass like red fescue or Kentucky bluegrass. Because these require a year or more to become well established, a small amount of a faster-germinating and shorter-lived grass such as ryegrass or redtop is usually added as a mixture. The latter should not exceed about 12 or 15 per cent of the seed mixture (*proportion* of mixture should control, not *rate* of seeding pounds per acre) so as not to compete with the permanent grasses.

Variations occur in the selection of kinds of seed to meet local conditions, such as the use of Canada bluegrass, bents, and different fescues on certain sandy soils. Also, different legumes such as ladino clover and wild white clover in rates up to about 5 or 6 lb per acre are sometimes included in the mixtures. Clovers are usually transient in a turf mixture. They may be prominent the first year and practically disappear at other times.

Rates of seeding as low as 20 lb per acre have been satisfactory, although a rate of 40 or 60 is more customary. (*Note:* 50 lb is *high* rate for highway seeding.) There appears to be little justification for the formerly popular rates of from 100 to 300 lb per acre.

The kinds and rates of grasses for various regions of the country are best ascertained from local state agricultural colleges, and in general they will be more simple than for lawns or golf-course requirements.

The system followed by the Federal government where the quality of the seed is defined according to pure live seed, or is purchased by the weight of pure live seed, is adequate for roadside use and avoids the complexity of detailed specifications. It is generally desirable to require a minimum percentage of germination, however, and to have a test made to verify that all minimum requirements have been furnished.

F. Mulching

A mulch of inorganic or organic matter is of value in the culture of grass (17). Various materials such as stone chips, cinders, and coarse sand have been successfully used, but hay and straw have several advantages. When used at a rate of from 1 to 3 tons per acre, or enough to essentially hide the surface of the earth, hay or straw dissipates raindrops and aids materially in checking erosion from surface water. It shades the ground and, because of its rough surface, cuts down air motion and thus further conserves moisture. The inorganic materials are chiefly beneficial where fire is a hazard.

The use of a mulch of hay or straw is generally accepted as an erosion-preventive measure and is applied as promptly as possible following grading. Fertilizer and seed can be applied before or after the mulch is placed. The use of mulch will permit seeding at almost any season, although late June to mid-August is usually too dry in most parts of the country. Mid-August to early October is generally most favorable.

Various methods of anchoring hay and straw have been used. Cereal grains sprout promptly and are effective even if added as a small amount to the seed mixture. Small piles of earth on the mulch, disking or pulvimixing it into the soil, spreading branches, winding twine around pegs, and numerous other methods have been used. A growing use is being made of asphalt emulsions as a sticker applied by a spreader after the mulch is placed, or as droplets blown onto the mulch as it is being placed by air pressure.

G. Methods

Mechanical methods for distributing fertilizers, seeds, and mulch are today in a stage of progressive improvement in keeping with the newly accepted methods of omitting seedbed preparation and placing materials on the surface of the soil without incorporation. Formerly seed and fertilizer were distributed by drills or spreaders or broadcast by hand. The method of mixing seed, fertilizer, and whatever lime is required in a tank of water and pumping the mixture on to the surfaces to be seeded has gained popularity. Larger equipment on the same principles has more recently been used with tank capacities of from 1,000 to 3,000 gal. Progress has also been made in using granular fertilizers and pelleted seeds blown by an air compressor. These methods are economical on large-scale operations and for placing the material up to 70 ft from the machines even on steep slopes. Flexible pipes of lightweight materials have been used to reach higher slopes.

Hay and straw are commonly blown out by an ensilage or hay blower to form a mulch. One of the difficulties with this equipment is the delivery of the material without breaking it into small pieces, which are less beneficial in controlling erosion than the longer straws.

H. Maintenance

The current methods of seeding, fertilizing, and mulching are generally successful in creating a stand of grass where grass is a natural ground cover. The character of the cover from that time forward is largely a matter of maintenance.

Normally, grass should be mowed two or three times each year to cause it to form a dense turf. Its thrift and appearance will determine the need of maintenance fertilizing. The phosphorous and potassium used at the time of seeding normally last several years, but the nitrogen may have to be replaced as often as once a year. Broadleaved weeds can be eliminated by using selective, hormone-type herbicides, and some grasses like red fescue in the North, when well established, are satisfactory as to height and appearance without any mowing.

With even a sparse cover to start with, the desired quality of turf can usually be attained through maintenance procedures.

I. Mowing Limits

Once grass has been established, there is need of determining how it had best be maintained to serve its purpose. Many turf areas will have to be mowed. These include medium-width and narrow medians, areas where sight distance is required on the inside of curves, near built-up communities and certainly in all urban areas. In other sections, such as where the highway is passing through lands not in cultivation, there may be little excuse for mowing, and so long as the turf checks erosion, it may be left undisturbed to grow up to woody plants from natural sources.

The application of this principle can be a considerable economy. In hilly country, it may eliminate mowing many of the cut-and-fill slopes. When applied to a wide right-of-way, it often permits an attractive combination of mowed areas adjoining cultivated fields, in contrast to unmowed roadsides in back of the ditch line, so that the scenery of swamplands, abandoned farms, or woods comes close to the pavement. This broad-scale relationship to the scenery of the countryside is in keeping with the scale of high-speed highways.

III. PLANTING

A. Highway Planting Design

The term "landscaping" in connection with highways means planting to most people. Certainly plants are an important part of the scenery, and different plants serve many purposes. Grasses are the kinds of plants most frequently used. Trees, shrubs, and vines require design considerations regarding their kinds and locations and more special provisions for their establishment and care (18, 19).

The three broad categories of landscape work that contribute to the pleasing appearance of a highway, parkway, or urban arterial are (1) suitable and pleasing ground forms, (2) a continuous and well-kept cover of turf, and (3) attractive trees and shrubs. The proper functioning of the areas and the basic ground forms and turf background are essential. Taller-growing plants cannot cover over their omission. Trees and shrubs are particularly significant in built-up areas because they contribute a natural and pleasing appearance to an otherwise man-made and often tiresome kind of scenery.

In the over-all design of new highways, or the maintenance of existing highways, it is desirable to first take into account the purposes which vegetation will serve. Frequently the trees and shrubs which are existing on the site can be protected or managed to meet these requirements. Otherwise it may be necessary to introduce new plantings.

Plantings along highways should be planned objectively, and on a broad scale, before consideration is given to the actual kinds of plants to be used. Some specific purposes are to help stop erosion, to guide traffic, or to screen a view. Planning at an early stage can be limited to a determination of where trees or shrubs are needed and their sizes and shapes when fully grown.

Another objective might be called a negative determination wherein no plant would be saved or planted if it were a hazard to safety or if it could become an undue maintenance burden. Such an analysis often leads to the elimination of many existing trees and shrubs which serve no justifiable purpose but if saved would add to the maintenance burden. Trees too close to the pavement or those which restrict sight distance or are an existing or potential hazard and could fall on the pavement constitute the greatest percentage of trees which should be removed.

B. Planting in Urban Areas

Having trees and shrubs in evidence along highways in an urban or suburban situation is probably more significant than having them along highways in any other environment (18). Trees occur naturally in most country situations, and even though they may be at some distance from the highway, they are conspicuous and pleasing in the landscape.

As one approaches a built-up area, the shade trees on home grounds are usually quite prominent, and the addition of trees close to the route will serve chiefly as an addition to these existing trees in subordinating the prominence of buildings in the scenery. In principle, if not in volume, they serve the valuable function, of unifying and confining the views of the motorists to the highway itself for reasons of safety, and insulating the traveled way from the neighborhood by screening out sounds, dust, and headlights. They improve the appearance of the neighborhood and, to an appreciable degree, the land values.

Where the roadsides consist of more intensively built-up areas, the amount of natural green growth is reduced and sometimes none is to be seen. It is in these backgrounds probably more than anywhere else along highways that plantings become the greatest benefit to the people of the community and to the users of the highway. With the increasing urban and suburban populations and the accompanying increase of through highway routes interconnecting vast concentrations of people, their homes, and the places where they commute to work or travel to seek recreation, the value of roadside improvement including appropriate planting is fast approaching its greatest significance. In this field perhaps more than any other the landscape architect will play the role of a co-planner in designing the highway work of the future.

C. Planting for Erosion Control

Considerable planting has been done for the prevention of erosion (18, 20). Vines like Hall's honeysuckle have been planted generously on slopes in the Northeastern states, kudsu vine in the South, and ice plant in California. Woody vines like bittersweet, low-spreading shrubs like dwarf blueberry, and even black locust trees have also been used in some cases. The latter as an example of a deep-rooted plant has limited value in stabilizing unstable subsoils. Each of these has undoubtedly been successful on particular areas, whereas in other situations, either the plants have outgrown their usefulness or have been crowded out by other plants or they have constituted too great a burden of keeping cleared of debris.

It has been determined in some regions that a cover of turf at a cost of about a tenth of even a cheap ground-cover planting of vines was adequate for erosion control, and should the cost of mowing prove excessively high, the grass could be left unmowed to grow up to woody plants from natural seeding.

There are, however, situations where vines or low suckering types of shrubs have been more desirable than turf because of the steepness or roughness of the slope and the difficulty of mowing; where the deeper roots of the woody plants might better survive drought conditions; or where the foliage effect was superior in appearance (22). These latter reasons are somewhat aside from erosion control, however.

D. Planting for Traffic Guidance (18)

Trees have been used along the outside of a curve on many roads with low volumes of traffic, where under varying lighting and weather conditions they have helped to inform the driver regarding the changing alignment of the road. When planted along the sides of a road, they have also served through their diminishing size, because of perspective, to inform the driver that he is approaching a descending grade in the highway. These two principles of using planting to guide traffic are less significant on broad, high-speed highways having long-sight-distance provisions and where it is necessary to keep trees at a considerable distance from the pavement.

Plantings in the angle between diverging lanes have sometimes been helpful in emphasizing the traffic pattern. The difficulties with shrub plantings in such situations have been the extra work in mowing around them, picking up the wind-blown litter which they collect, and keeping the growth from crowding the traveled way or hindering sight distances. An increasing tendency is to provide signs and pavement markings rather than adding plantings in these situations.

The growth of trees and shrubs on fill slopes may form a desirable background and accentuate a guide rail. If these plants are at least as tall as the level of the road shoulder, they will contribute a sense of safety, which enables the driver to stay in the outside lane without fear of being too close to the edge of the embankment.

E. Planting in Relation to Pavements

The setback of trees from the edge of the pavement, whether it applies to existing trees or to the location for new planting, should be based on the speed and volume of traffic. The determination of such relationships is of course arbitrary. A tree 2 ft from the pavement may have caused no accident, whereas a tree 30 ft away may have been hit several times. An adequate setback ranges from 10 to 12 ft on two-lane, light-traffic highways to 20 ft on divided highways of four or more lanes. Sight-distance requirements are more specifically determined along arterial highways, and trees should cause no reduction in the accepted minimums.

The branches of trees should be a minimum height above the pavement and shoulders according to local laws to provide clearance for high vehicles and to allow for the bending of branches in windstorms and under the load of snow. The clearance of vision under the branches should be carefully checked at depressed vertical curves.

F. Planting at Traffic Intersections and Circles

The effect of trees and shrubs is probably more significant near a traffic interchange than along any portion of intervening highways (18). A bridge structure is prominent in the landscape, and its appearance will normally be benefited by trees, or trees and shrubs, to screen and thus subordinate the most massive elements of the construction, to project above the railings and soften the silhouette as seen against the sky, or to otherwise accentuate or add to the best elements of the design.

In some cases, the bridge may be relatively unobtrusive, whereas the approaching ramps will be the conspicuous man-made elements of the scenery. Planting can greatly soften these effects, and it takes only a few trees to relieve the apparent extent of the slopes or harshness of their silhouettes.

Plantings may also occur at the outside of a turn or beyond a junction of lanes to emphasize a change of direction. If such plantings could be hit by a vehicle out of control, it is preferable to omit trees and to use only shrubs with flexible stems which can receive an impact without causing too great injury.

Shrubs may be used to screen oncoming headlights or to screen headlights from sweeping across a lane from one side as if a car were about to enter the flow of through traffic.

A grove of trees has many advantages at an interchange which covers several acres of land. The trees will essentially obscure all the interweaving traffic and at the same time focus attention on the one or more lanes of travel immediately ahead of the driver. They constitute the best long-range and low-cost cover for all the property, except for those areas which are maintained in turf for reasons of adequate sight distance.

Smaller intersections usually benefit from having sufficient planting around their perimeter to screen any distractions in the neighborhood. Elsewhere, high-headed trees may serve to accentuate a difference in the traffic pattern from the approaching open stretches of highway. Vines and ground covers may be planted on slopes too steep to be easily mowed, and the balance of the area maintained as turf.

Traffic circles require different planting designs according to their size and surrounding conditions. It is usually desirable to have trees and shrubs so disposed, or so restricted, that all traffic movements can be readily comprehended. Sometimes, however, a mass of dense planting is placed near the center of the circle to obscure headlights from across the circle. At other times, low masses of evergreen shrubs are used along the edges of the pavements to emphasize the pattern. In general, it is desirable to plant traffic circles in some distinctive manner different from the approaching highways so that attention is directed toward the abrupt change of speed and direction.

G. Screening with Plants

1. Screening Views. The extent of planting required to screen a view, or even a particular feature from a highway, is surprisingly extensive. Where the surrounding views are undesirable, such as where a parkway passes through a built-up area, the screening usually becomes a continuous belt along the property lines. Scattered groups of plants often provide an adequate foil without being a complete screen. In many cases, a screen planting is added at a later date, or a scattered planting is thickened up as nearby properties become developed in a character different from expectations. This is continually happening along parkways in the outlying areas of the larger cities.

2. Screening Headlight Glare. Screening the glare from headlights with planting in a median strip is essentially a corrective measure which is not necessary if the median is 40 ft or more wide or if the opposing lanes of traffic are on different grades (18). The plants for the screen should be evergreen if possible or densely twiggy or form a considerable mass in depth. A clipped hedge may be used on very narrow medians, and naturalistic groups of trees and shrubs may be appropriate on wide medians. The plants need to be quite tall to cut out oncoming headlights across depressed vertical curves. Care must be taken not to interfere with necessary sight distance, such as where other roads are crossing at grade. It is a moot question whether the screen should occur only at curves, only on tangents, or throughout the extent of the highways and whether intermittent screening is more objectionable than no screening whatever.

Several objections may arise in connection with such screen planting. Shrubs are expensive to install and so costly to maintain that their safety value as a screen needs to be carefully appraised for the particular situation involved. The plantings normally complicate the mowing pattern. They may cause snow drifting and use up space for snow storage. They sometimes require much work in cleaning out papers and other debris, and if the median is narrow, one lane of traffic may have to be closed off to protect maintenance workmen. Any continuous belt of uniform planting is conspicuous and likely to detract from the best landscape appearances.

3. Screening Sound. Research is continuing to determine ways and means of reducing the sounds of traffic, particularly in built-up areas where the noise may affect living conditions (23). Trees and shrubs, and particularly evergreens for their year-round effectiveness, have been found to reduce the noise of traffic appreciably. When planted along the sides of the route, the amount of sound reduction is of course related to the height and to the density of the planting screen.

4. Snowdrift Control. Living snow fences have been effectively installed in certain areas to cause the snow to be deposited off from the traveled way (18, 24). The location of such barriers should be determined on the site because of the variations in which snow is deposited by wind currents and the prevailing winds during and following snowstorms. Other local conditions may dictate a preference for either tree- or shrub-group patterns and their necessary heights to affect drifting.

Many patterns of barriers and kinds and combinations of evergreen and deciduous trees and shrubs have been tried, and each varies in its effect on clogging snow or in causing snow to deposit on the windward or leeward sides. Masses of plant growth like a woods cause the snow to be trapped before it can blow onto a highway. One-, two-, and three-row barriers with the plants spaced at about the interval of their individual heights have also proved to be effective.

The fact that the snowdrift should be formed between the barrier and the pavement and that the width of the drift is normally 15 to 20 times the height of the barrier means that the barrier should ordinarily be outside the right-of-way. This complication, together with its costs and uncertainty of effectiveness, usually resolves in the use of removable snow fencing instead of plantings.

H. Plants and Planting Methods

1. Plants. Plants suitable for roadside planting can be either native or exotic, but their habit, which includes their ultimate height and spread, their branching character, and leaf and flower quality should conform to the general design requirements (25). Growing conditions along highways are usually more difficult than in parks or on lawns, so the tolerance of the plants is more critical. They should be naturally free of pests and diseases, hardy for the latitude and exposure, and tolerant of the poor soils and droughty conditions so prevalent in highway construction. The quantity of usable kinds of plants, ranging from creeping vines to trees of forest stature, is so great across the continent and their individual characteristics so different, that advice from an experienced landscape architect or similar authority should be secured to aid in their selection.

The effective life of many shrubs is only a few years. They soon lack vigor and become open in habit, and the old wood must be pruned away. The lack of maintenance facilities to rejuvenate such shrubs or replace them has emphasized the desirability of using mostly large- and small-growing trees. Few shrubs are being used today on modern parkway and arterial-route plantings.

Although it is possible to transplant any plant at any time it is most practical in public-works programs to transplant plants in relatively small sizes and at the season of the year that has proved best in local experience. Trees in a 3- to 4-in.-caliber size, for example, will usually overtake a transplanted 6-in.-caliber tree in a few years' time because of the greater vigor of the younger size.

2. Planting Contracts. The quality of the plants, which includes height in proportion to either the stem size or the width of the crown, the diameter of the ball if moved with a ball of earth, and similar dimensions are quite adequately defined in a book called *American Standard for Nursery Stock*, which is a code of standards sponsored by the American Association of Nurserymen (26).

The quality of the plants and the entire planting operations can be readily defined as a written specification and carried out as a planting contract (18). The requirements specified for soil preparation should be checked as the work is being done, and each stage of the planting operations should be carefully and intelligently inspected.

Many aspects of planting work are impossible to control, such as climatic conditions or field conditions in the nursery, adequate root pruning, storage conditions, and protection of the plants from sunlight and drying winds before their arrival on the job. These conditions, plus the fact that planting operations over extensive areas are often difficult to inspect and local forces are either untrained or unavailable for maintenance at the most critical times, have led to the desirability of paying the contractor for maintenance of the work for the first year after planting and being responsible for replacing all plant losses.

3. Soil Preparation. Good agricultural soil is normally adequate for a planting soil. It is customary to provide a layer of planting soil called a plant bed where vines, ground covers, or shallow-rooted shrubs are to be planted at comparatively close spacings, say, up to 3 ft on centers. This bed should be a minimum of 6 in. deep for vines like Hall's honeysuckle and a minimum of 16 in. deep for shrubs like sweet fern and low blueberries. It can usually be placed most economically as a part of the grading operation.

The diameter of plant pits for individual plants with a root spread less than 24 in. should be twice the spread of the roots. Plants with a root spread of 2 to 4 ft should have a pit equal to the spread of the roots plus 2 ft. For plants with a greater spread of roots the pit diameter should be $1\frac{1}{2}$ times the root spread. The depth of the pit should be great enough to accommodate a 6-in. layer of planting soil below the roots. In very poor growing conditions such as in a porous stone fill or on some city streets, the volume of soil should be considerably increased above the given minimums because the roots may have little to draw upon beyond the limits of the prepared soil.

It is customary to add fertilizers to the planting soil at the time of planting. Stable manure in a ratio up to one-seventh of the soil mass, or bonemeal at the rate of about 10 lb per cu yd, has been customarily used, whereas today a greater use is made of a commercial fertilizer such as a 5-10-5 or 10-10-10 at the rate of about 5 lb per cu yd. Where it is not efficient to premix the fertilizer with the soil, it can be satisfactorily applied on the surface over the roots after planting at a rate of, say, 3 lb per sq yd.

I. Drainage

Where the soil is impervious or excess water will be retained, each plant pit or plant bed should be thoroughly drained.

J. Planting

The care of plants between the time they are dug until they are planted is extremely important to prevent loss of moisture. Precautions include the shortest possible time for the plant to be out of the ground, shading the roots from sunlight, protecting them from wind, keeping the roots covered with moist materials like wet moss or sticky clay, or transplanting a ball of soil so that the roots are undisturbed. Other precautions include spraying the trunks, stems, and leaves with plasticlike coatings to reduce transpiration.

The planting operation requires a good firm placement of planting soil against the roots or earth ball without leaving air pockets, accompanied and followed by generous but not excessive watering.

K. Staking

Most trees and some shrubs need to be tied to stakes or guyed to pegs or deadmen to hold their tops from swaying in the wind, preventing proper root establishment. These supports are usually kept in place for a year after planting.

L. Mulching

It is desirable to place a 3- or 4-in. layer of peat, muck, strawy stable manure, leaves, peat moss, or similar material over the planting soil to conserve moisture in the soil and to moderate temperature variations. If the mulch is over 7 in. thick, it will usually preclude the growth of weeds.

M. Pruning

Except when plants are grown in containers of some sort, the process of transplanting means the loss of roots, usually the ends of roots, which function in picking up water and plant foods from the soil. It is desirable to cut away a portion of the top of the plant to provide a balance with the remaining root system. Depending on the kind of plant, the probable amount of root loss, and numerous other factors, the amount of necessary pruning will vary up to one-third of the whole top of the plant. Pruning is accomplished variously for different conditions, from taking out complete branches as a thinning operation to cutting back portions of each branch, as well as favoring the branching structure or the shape of the plant (27).

N. Maintenance

Watering and keeping weeds pulled out from around the plants are the essential maintenance operations. Other work includes pruning in relation to any drying out of the branches and keeping mulch, wrappings, and guys in proper condition.

REFERENCES

1. Johnson, A. W.: "Erosion Control along American Highways," Golden Anniversary Meeting of the American Society of Agricultural Engineers, June 24, 1957. A symposium on soil erosion.
2. "Erosion Control-Trends and Techniques," *Highway Research Board Circular* 156, February, 1952. Methods of erosion control and bibliography.
3. "Expressway or Parkway: Desirable Design Factors Underlying Highway Construction," Highway Research Board, reprint, 1955. Parkway principles in expressway design.
4. "Photogrammetry and Aerial Surveys: A Symposium," *Highway Research Board, Bulletin* 157, 1957. Highway location and design from air photographs. Electronic computing methods.
5. "Stabilized Turf Shoulders," *Highway Research Board, Special Report* 19, 1954. Summary of tests, committee reports, and bibliography.
6. "The Design of Roadside Drainage Channels," *Proceedings, 23d Annual Meeting, Highway Research Board*, 1943, pp. 264–266. Formulas for runoff computations and cross-section designs.
7. *A Policy on Geometric Design of Rural Highways*, American Association of Highway Officials, 1955. Drainage channels, pp. 181, 209–211; earth slopes for design of cuts

and fills, pp. 211–213; easement curves and profiles, pp. 143–147; grading design, pp. 406, 407, 572, 573; coordinating horizontal and vertical alignment, pp. 79–181.

8. "Transitional Grading of Highway Slopes," *Proceedings, 28th Meeting, Highway Research Board*, 1948. Slope grades and transitional grading.

9. "Our Streets Can be Beautiful and Useful," *Proceedings, First Street Tree and Utility Conference, Cleveland, Ohio*, March, 1955. Forum discussions on street trees and utilities.

10. *Proposed Guide Specifications for Roadside Improvement*, U.S. Bureau of Public Roads, 1956, Commerce Department, District of Columbia. 11049. Suggested specifications with summary of practices in highway departments.

11. "Mechanization of Roadside Operations," *Highway Research Board Special Report 16*, 1953. Review of equipment in current use for roadside development and maintenance.

12. "Reducing Damage to Trees from Construction Work," *U.S. Department of Agriculture, Farmers' Bulletin 1967.* Tree wells, tree walls, barricades, etc.

13. "Tree Care," *U.S. Department of the Interior, National Park Service, Tree Preservation Bulletins:* no. 3, Tree bracing; no. 4, shade-tree pruning; no. 6, general spraying practices; no. 7, ropes, knots, and climbing; no. 8, safety for tree workers; no. 9, transplanting trees and other woody plants.

14. *Roadside Vegetative Cover Research Project: Final Report*, New York State Department of Public Works, Landscape Bureau, June 30, 1955. Report of roadside-turf production and maintenance. Use of herbicides. Seeding, fertilizing, and mowing equipment.

15. *Erosion Control on California State Highways. California Highways and Public Works.* Department of Public Works, Division of Highways, State of California, 1950. Use of brush and wattles, straw mulch, and check dams.

16. "Farm Chemicals: Dictionary of Plant Foods," Allied Chemical and Dye Corporation, Nitrogen Division, 1955. A guide to fertilizer materials and terms.

17. "Mulching on Roadsides," *Highway Research Board, Circular 189*, January, 1953. Materials and methods for mulching.

18. "Planning and Management of Roadside Vegetation. An Analysis of Principles," *Highway Research Board, Special Report 23*, Nelson M. Wells, Chairman, Special Task Committee: Vegetation in relation to the value of adjacent lands, pp. 5, 6; in relation to the highway cross section and interchanges, pp. 14–19; vegetation for traffic guidance, pp. 6–8; legal control of vegetation, pp. 11–41; administrative control, pp. 29–31; vegetation in relation to erosion control, medians, snow drifting, views, structures, waysides and memorials, urban areas, pp. 19–29.

19. "Smaller Street Trees Needed," *Arnoldia* Arnold Arboretum, Harvard University, Cambridge, Mass., vol. 2, no. 6, June, 1951. Trees appropriate for narrow streets and for combining with utility lines.

20. "Ground Cover Plants for Erosion Control," *Highway Research Board, Circular 166*, June, 1952. A list of plants used for erosion control in eight climatic zones of the United States.

21. Committee on Roadside Development: "Selective Cutting and Vista Clearing," *Highway Research Board, Publication 496*, 1957.

22. "The Selection and Use of Ground Covers on Highway Areas," *Proceedings, 19th Annual Meeting, Highway Research Board*, December, 1939.

23. "Abatement of Highway Noise and Fumes," *Highway Research Board, Bulletin 110*, 1955. Methods of reducing noise and fumes and bibliography.

24. "Snow Control by Tree Planting," *Proceedings, 16th Annual Meeting, Highway Research Board*, November, 1936. Kinds of plants and their arrangement for controlling snow drifting.

25. "The Selection and Use of Trees on Highway Areas," Report of Subcommittee on Plant Ecology, *Proceedings, 20th Annual Meeting, Highway Research Board*, December, 1940.

26. *American Standard for Nursery Stock*, A Code of Standards Sponsored by the American Association of Nurserymen, Inc., Washington, D.C., and approved by American Standards Association, as Z60.1, 1949.

27. "Pruning Highway Trees," *Highway Research Board, Circular 227*, January, 1954.

INDEX

D

S